Nonlinear Dynamics in Equilibrium Models

John Stachurski • Alain Venditti
Makoto Yano
Editors

Nonlinear Dynamics in Equilibrium Models

Chaos, Cycles and Indeterminacy

Selected Papers of Kazuo Nishimura

Springer

Editors
Prof. John Stachurski
Australian National University
Research School of Economics
Canberra, ACT
Australia

Prof. Alain Venditti
National Center for Scientific Research
CNRS-GREQAM and EDHEC
Marseille
France

Prof. Makoto Yano
Kyoto University
Faculty of Economics
Kyoto
Japan

ISBN 978-3-642-22396-9 e-ISBN 978-3-642-22397-6
DOI 10.1007/978-3-642-22397-6
Springer Heidelberg Dordrecht London New York

Library of Congress Control Number: 2011945087

Printed on acid-free paper

Springer is part of Springer Science+Business Media (www.springer.com)

Preface

This volume contains a collection of research papers on international trade, general equilibrium, and economic growth by Kazuo Nishimura, one of the most outstanding economic theorists of his generation. Like others who have read his work, we have all been inspired by Kazuo's insight and his vision. Unlike most of these economists, we have also had the great privilege to work as his colleagues and co-authors, and to share in the warmth and generosity of his friendship.

Compilation of these collected works was greatly assisted by Aiko Tanaka, who contributed many hours to the task of producing a publishable document, and the first class editorial team at Springer. We also benefited from the excellent work environment provided at Kyoto Institute of Economic Research, Groupement de Recherche en Economie Quantitative d'Aix-Marseille, and the Research School of Economics at Australian National University.

Finally, we would like to thank all the co-authors of Kazuo, Elsevier, the Econometric Society, Springer, Wiley and World Scientific for permission to reprint.

Makoto Yano
John Stachurski
Alain Venditti

Acknowledgments

It is a great honor for me to have selected papers of my work published. I would like to thank John Stachurski, Alain Venditti and Makoto Yano, who as colleagues provide such constant academic stimulation, and who took the initiative to edit this volume. Needless to say, my thanks also go to my supervisor, the late Professor Lionel McKenzie, who played such an important role in the very development of equilibrium dynamics; I began my research on macro dynamics under his guidance in the graduate school at the University of Rochester. I have been also influenced by several distinguished economists who have made pioneering contributions to general equilibrium theory, capital accumulation theory and international trade theory respectively, namely Professors Takashi Negishi, Ronald W. Jones and Hirofumi Uzawa.

Tragically, one of my co-authors, Koji Shimomura, is no longer with us. He passed away on February 24, 2007, at the age of 54. Koji was one of the most prolific Japanese economists of our generation and published many excellent and insightful papers, mainly on trade and dynamics. The paper included in Chapter 10 of this book is the first paper we wrote on indeterminacy in the dynamic two country model. Koji and his work will be sorely missed.

I would like to close by mentioning that I am fully aware that I owe a great deal to my co-authors. Without their selfless support, I would not have been able to continue my work on the frontier of economic dynamics. I would like to convey my heartfelt appreciation to those who have given me the opportunities that allowed me to write the joint papers in this volume.

December 1, 2011 *Kazuo Nishimura*
Kyoto

Contents

Chapter 1
Introduction

John Stachurski, Alain Venditti, and Makoto Yano

Kazuo Nishimura has made outstanding contributions to three main fields of economic theory: international trade, general equilibrium, and the theory of economic growth. With his characteristic mix of originality, technical clarity, elegance, and rigor, Kazuo's insights into nonlinear dynamics have transformed our understanding of economic growth, business cycles, and the relationship between them. For young scholars beginning work in these areas, Kazuo's papers are essential and inspiring.

Kazuo Nishimura obtained his Master's degree from Tokyo University in 1972. In 1973, he entered the graduate program in economics at the University of Rochester. His supervisor was Lionel McKenzie—one of the giants of twentieth-century neoclassical economics. After spending three years at Rochester, Kazuo joined the Department of Economics at Dalhousie, Canada, in 1976. He obtained his PhD in 1977. Subsequently, he returned to Japan with an appointment at the Tokyo Metropolitan University. He spent 10 years there, during which time he also taught at the State University of New York at Buffalo, and at the University of Southern California. In 1987, he joined the Institute of Economic Research at Kyoto University. He served as its director in 2006–2010.

Kazuo has had a highly distinguished career. He was elected fellow of the *Econometric Society* in 1992. He served as the vice president of the *Japanese Economic Association* in 1999–2000, and as its president in 2000–2001. He was co-editor of

J. Stachurski
Australian National University, Canberra, ACT, Australia
e-mail: john.stachurski@anu.edu.au

A. Venditti
National Center for Scientific Research, CNRS-GREQAM and EDHEC, Marseille, France
e-mail: alain.venditti@univmed.fr

M. Yano
Institute of Economic Research, Kyoto University, Yoshida-Honmachi, Japan
e-mail: yano@kier.kyoto-u.ac.jp

J. Stachurski et al. (eds.), *Nonlinear Dynamics in Equilibrium Models*,
DOI 10.1007/978-3-642-22397-6_1, © Springer-Verlag Berlin Heidelberg 2012

the *Japanese Economic Review* during 1995–1997, and its editor-in-chief from 1997 to 2000. In 2004, he established the *International Journal of Economic Theory*, a new journal for the profession that is already highly regarded within its field. He has served as a managing editor since its inception. He has also served on the editorial board of numerous leading journals in economics and applied mathematics, and he has edited a number of special issues for prestigious journals. He has organized successful conferences around the world.

Recently Kazuo has become an active and leading participant in the public debate in Japan about the science curriculum taught in high schools and universities. Through the popular press and scholarly journals on science and mathematics education, Kazuo has strongly advocated the need for improved training in science and mathematics for high school students and humanities students at universities. To further promote scientific and mathematical education, he founded the International Society for Education. He currently serves as president of the society.

To say that Kazuo is an extremely busy person is still an understatement. However, as his numerous co-authors, friends and colleagues all over the world will attest, he is always available to discuss research, provide insight, encouragement, and the benefit of his remarkable energy. A perfect day for Kazuo is to work through till evening with some co-authors on a difficult research problem, then to break for dinner and fine sake together in a Kyoto restaurant, all the while still discussing new ideas. If, late into night, a break for sleep appears to be necessary, there is no doubt that Kazuo will be back the next day with some hand-written notes providing the solution to the problem.

We, Kazuo's friends, colleagues and co-authors, have been privileged to benefit over many years from his loyalty, warmth and generosity. It is with great pleasure that we present this collection of selected papers, containing his most important and enduring contributions.

1.1 Overview of the Papers

In this section we provide an overview of the papers selected for this volume. While the optimal growth model provides a common framework linking these papers, they have been classified according to the particular qualitative property of the equilibrium that is analysed:

1. Optimal growth and endogenous cycles
2. Optimal growth and chaotic dynamics
3. Global dynamics in optimal growth models
4. Dynamic models with non-concave technologies
5. Stochastic optimal growth models
6. Indeterminacy in exogenous growth models
7. Indeterminacy in endogenous growth models

The complete list of papers can be found in Sect. 1.2.

1.1.1 Introductory Survey

As the title of the book suggests, all the selected papers involve nonlinear dynamics, a major field of research within economic theory. In collections such as this, it is usually desirable to provide the reader with a review of basic material and concepts common to the field. This task has already been accomplished by Kazuo, in a paper titled "Nonlinear dynamics in the infinite time horizon model," published with Gerhard Sorger in the *Journal of Economic Surveys* in 1999. Starting from the standard aggregate formulation of the optimal growth model, in which the unique optimal equilibrium converges toward a long-run balanced growth path, the paper then goes on to introduce the main features of nonlinear dynamics, including periodic cycles, chaotic dynamics, stochastic shocks, non-concave technologies, externalities, local indeterminacy, sunspot fluctuations and analysis of multisector models. It provides the perfect lead article of the present collection. It remains only for us to provide the reader with an overview of the selected papers, in order that he or she might quickly locate those papers most relevant to his or her field.

1.1.2 Optimal Growth and Endogenous Cycles

This section contains three important contributions on the existence of endogenous fluctuations in the perfect competitive framework of the multisector optimal growth model. The first and second belong to the set of papers written during a long and fruitful collaboration between Kazuo and Jess Benhabib (New York University). They deal with closed economies. The third paper is the first of many collaborations with Makoto Yano of Keio University. It studies a model of trade between two countries. These two collaborations have produced highly influential contributions on the analysis of non-linear dynamics under perfect competition.

The first paper, "The Hopf bifurcation and the existence and the stability of closed orbits in multi-sector models of economic growth," was published with Jess Benhabib in the *Journal of Economic Theory* in 1979. It shows that, under very general circumstances, the standard optimal growth model with two or more capital goods can give rise to optimal trajectories that are limit cycles. This was a striking result, as research on optimal growth in the 1960s and 1970s had typically found these models to be characterized by a strong form of stability, particularly when the discount rate is small (see, for instance, Brock and Scheinkman, 1976; Cass and Shell, 1976; McKenzie, 1976; and Scheinkman, 1976). In contrast, this paper, by treating the discount rate as a parameter, establishes via bifurcation theory that optimal paths can become closed orbits as the steady state loses stability.

Bifurcations for differential equation systems arise when, for some value of a parameter, the Jacobian of the function describing motion acquires eigenvalues with zero real parts at a stationary point. If a real eigenvalue becomes zero, this results in multiplicity of steady states. In this paper, the focus is on a bifurcation that results from pure imaginary roots: the Hopf bifurcation. In contrast to bifurcation from a real eigenvalue, no additional stationary points arise from a Hopf bifurcation. Instead, closed orbits emerge around the stationary point. Nonjoint production and a single, possibly composite, consumption good are assumed to assure the uniqueness of the steady state. Under some additional assumptions, it is shown that the Jacobian of the functions describing the motion of the system cannot vanish at a steady state, and, as a result, that only the Hopf bifurcation is possible. An example is given of a totally unstable steady state, giving rise to optimal paths that are closed orbits. The stability of orbits that arise from the optimal growth problem is discussed.

The second paper, "Competitive equilibrium cycles," was published in 1985 with Jess Benhabib in the *Journal of Economic Theory*, and provides a discrete-time extension of the previous contribution. Considering a two-sector optimal growth model with one pure consumption good and one investment good, it gives sufficient conditions for generating robust period-two cycles in outputs, stocks, and prices. These conditions are expressed in terms of the discount rate and differences in relative factor-intensities between the two industries.[1] Finally, some intuitive explanation for the existence of cyclical equilibria that arise in neoclassical technologies are also given. The paper is notable for the sophisticated techniques from differential topology used to prove the global existence of period-two cycles, corresponding locally to the existence of a flip bifurcation.

In the last paper of this section, "Interlinkage in the endogenous real business cycles of international economies," published in 1993 with Makoto Yano in *Economic Theory*, the transmission of business cycles between trading countries is analyzed. In a perfect foresight model with many consumers, it is well-known that competitive equilibrium paths behave like optimal growth paths.[2] Therefore, a perfect foresight equilibrium path may exhibit business cycles, even in a many-consumer model such as a large-country trade model. In this paper, a two-country optimal growth model in which the countries have independent technologies is considered. First, the determinants of each country's global accumulation pattern in autarkic equilibrium are analyzed using the results of Benhabib and Nishimura contained in the second paper of this section, providing conditions on the sectoral technologies that lead to existence of business cycles (in particular, period-two cycles). Global accumulation under free-trade is then addressed. Conditions for the international transmission of business cycles are given in terms of the parameters of each country's technologies.

[1] In particular, the discount factor must be sufficiently large, and the consumption good sector must be more capital intensive than the investment good sector.

[2] See, for instance, Becker (1980), Bewley (1982), Epstein (1987), and Yano (1983, 1984).

1.1.3 Optimal Growth and Chaotic Dynamics

The two papers by Jess Benhabib and Kazuo Nishimura contained in the previous section had a strong impact on the group of researchers studying macroeconomic dynamics and business cycles, initiating a common program of research that aimed to provide detailed analysis of the qualitative properties of capital accumulation paths in optimal growth models. For example, the landmark paper of Boldrin and Montrucchio (1986), which provides an anything-goes result for optimal paths in standard multisector optimal growth models, was initiated at the beginning of the 1980s after Luigi Montrucchio attended a presentation of the "Competitive equilibrium cycles" paper.

The findings of Boldrin and Montrucchio indicated that equilibrium time paths in deterministic equilibrium economies may exhibit complex—even chaotic—dynamics. Building on the results of McKenzie (1976) and Scheinkman (1976), they showed specifically that almost any capital accumulation path can be realized as a solution of an optimal growth model, provided that the discount rate is large enough. However, this leaves open the question as to whether or not chaotic optimal paths can appear under more reasonable rates of discounting. The two papers with Makoto Yano in this section, titled "Nonlinear dynamics and chaos in optimal growth: an example," and "Chaotic solutions in dynamic linear programming," published in *Econometrica* in 1995 and in *Chaos, Solitons and Fractals* in 1996 respectively, showed that they can.

The first paper demonstrates the possibility of ergodically chaotic optimal accumulation in a two-sector model with linear preferences and Leontief production. A condition is provided under which the optimal transition function is unimodal and expansive. It is then shown that the set of parameter values satisfying this condition is nonempty, no matter how weakly future utilities are discounted. The second paper is a complement to the first, as it provides a simpler and more intuitive proof of the same result.

While these two papers provide strong conclusions on the compatibility between chaotic optimal paths and low discounting of future utilities, their main limitation concerns the fact that preferences and technologies are specified. This extra structure inhibits the formation of a general picture. In consequence, the paper "On the least upper bound of discount factors that are compatible with period-three cycles," published in the *Journal of Economic Theory* in 1996 with Makoto Yano, studies a class of general, reduced form optimal growth models. In these models time is discrete, and an indirect utility function summarizes both preferences and the social production function. The authors derive the least upper bound of the set of discount factors such that a cyclical optimal period-three path emerges in a standard model of optimal growth.[3] For this purpose, they characterize a cyclical optimal period-three

[3]The existence of a cyclical path of period three is a fundamental criterion for the emergence of complex nonlinear dynamics, in particular of cyclical paths of any other periodicity (see Sarkovskii (1964) and Li and Yorke (1975)).

path in terms of the dual path of prices. They then demonstrate that a condition for the existence of a cyclical optimal period-three path can be expressed by the existence of a solution to a system of linear equations. By solving this system, they finally demonstrate that the least upper bound mentioned above is equal to $(3-\sqrt{5})/2$.

1.1.4 Global Dynamics in Optimal Growth Models

Although they focus on the particular case of chaotic optimal paths, the contributions contained in the previous section are concerned more generally with the following question: "How is long-run optimal behavior affected by changes in the rate at which the future is discounted?" The paper contained in the present section, "Intertemporal complementarity and optimality: a study of a two-dimensional dynamical system," published in *International Economic Review* in 2005 with Tapan Mitra, provides additional insights on this point. It studies the underlying structure of the two-dimensional dynamical system generated by a class of optimal growth models that allow for intertemporal complementarity between adjacent periods, but preserves the time-additively separable framework of standard Ramsey models. The purpose of the article is to complete the program sketched in the contribution of Samuelson (1971), providing both a complete local and global analysis of the model under which the standard results of the Ramsey model continue to hold, even with dependence of tastes between periods. Global convergence results are established and related to the local analysis using the mathematical theory of two-dimensional dynamical systems. The local stability property of the stationary optimal stock is also related to the differentiability of the optimal policy function near the stationary optimal stock, using the stable manifold theorem.

1.1.5 Dynamic Models with Non-Concave Technologies

Traditionally, optimal growth theory is concerned with the derivation of an optimal consumption plan of a representative agent when the production function in the economy is concave and preferences of the consumer exhibit decreasing marginal utility. In these settings, there exists a unique long-run steady state capital stock that is globally stable, in the sense that the optimal capital stock converges to the steady state level independent of the initial condition. The paper contained in this section, "A complete characterization of optimal growth paths in an aggregated model with a non-concave production function," published in the *Journal of Economic Theory* in 1983 with W. Davis Deckert, provides some extensions of this framework by considering a non-concave production function and thus the local existence of increasing returns. It appeared as a path-breaking study of optimal one-sector growth under non-convex technology. When the technology is convex, the Euler equation is sufficient to prove the monotonicity property of the equilibrium path. This is no longer the case when non-convexities are introduced. Deckert and Nishimura show that the Euler equation in conjunction with the Principle of Optimality is sufficient to prove monotonicity for non-concave production functions. For certain interest rates, they prove that the optimal path converges to a steady state only if the initial capital stock is above a critical level, otherwise it converges to zero. They also demonstrate that the set of points for which the value function is differentiable is precisely the set of initial capital stocks from which there is a unique optimal path.

1.1.6 Stochastic Optimal Growth Models

In the first paper in this section, "Stochastic optimal growth with nonconvexities," published with Ryszard Rudnicki and John Stachurski in the *Journal of Mathematical Economics* in 2006, the problem of optimal growth with non-concave production (i.e., the problem studied in the paper with W. Davis Deckert discussed in the last section) is revisited, but now with random shocks to productivity. A number of key results from the deterministic case are replicated, such as almost everywhere differentiability of the value function, and monotonicity of optimal policies. In addition, the authors establish a basic dichotomy in growth dynamics when sufficient mixing is present: Either global stability is observed, or the economy is globally asymptotically collapsing to the origin.

While the dichotomy result of this paper is useful in its own right, it does not provide a set of conditions on the primitives under which global stability is shown to hold. For the concave case, this problem was essentially solved as early as the 1970s (see, e.g., Brock and Mirman (1972)), but the non-concave case proved far less tractable. The third paper in this section, "Stability of stochastic optimal growth models: a new approach," published with John Stachurski in the *Journal of Economic Theory* in 2005, provided a breakthrough. The paper gave a set of stability conditions on the model primitives that could be viewed as natural analogues of the classical deterministic conditions. The breakthrough resulted from a new method developed by the authors for studying stability of optimal dynamic models, based on viewing marginal utility of consumption as a Lyapunov function.

1.1.7 Indeterminacy in Exogenous Growth Models

A wide-spread perception among economists is that macroeconomic business-cycle fluctuations can be driven by changes in expectations of fundamentals. A major strand of the literature focusing on fluctuations derived from agents' beliefs is based on the concept of sunspot equilibria, dating back to the early works of Shell (1977), Azariadis (1981), and Cass and Shell (1983). As shown by Woodford (1986), the existence of sunspot equilibria is closely related to the indeterminacy of perfect foresight equilibrium (i.e., the existence of a continuum of equilibrium paths converging toward one steady state from the same initial value of the state variable).

Following the seminal contribution of Benhabib and Farmer (1994), infinite horizon Ramsey-type models augmented to include external effects in production have been shown to exhibit multiple equilibria and local indeterminacy. Much of the research in this area has been concerned with the empirical plausibility of indeterminacy in markets with external effects, which exhibit some degree of increasing returns. While the early results on indeterminacy relied on relatively large increasing returns, Benhabib and Farmer (1996) showed that indeterminacy can also occur in two-sector models with small sector-specific external effects and mild increasing returns. Nevertheless, a number of empirical researchers on disaggregated US data (see for instance Basu and Fernald (1997)) have concluded that returns to scale seem to be roughly constant.

The three papers contained in this section show how indeterminacy can occur in standard growth models with sector-specific externalities, constant social returns, decreasing private returns and small or negligible external effects. The first one, "Indeterminacy and sunspots with constant returns," published with Jess Benhabib in the *Journal of Economic Theory* in 1998, is an influential contribution that has initiated many extensions and developments. Considering a two-sector economy with Cobb-Douglas technology and a utility function that is linear in consumption, Benhabib and Nishimura show that local indeterminacy arises if and only if the sector-specific externalities generate a capital intensity reversal between the private and the social levels. Specifically, indeterminacy requires that the consumption good sector is capital intensive at the private level but labor intensive at the social level. This conclusion is fundamentally based on the breakdown of the duality between the Rybczynski and Stolper-Samuelson effects, caused by introduction of market imperfections.

The second paper, "Trade and indeterminacy in a dynamic general equilibrium model," published with Koji Shimomura in the *Journal of Economic Theory* in 2002, introduces sector-specific externalities in a Heckscher-Ohlin, two-country, dynamic general equilibrium model. In their model, factors are internationally immobile, and the countries both use identical Cobb-Douglas technologies. Under the same capital intensity reversal condition as in the previous paper, Nishimura and Shimomura show that there are multiple equilibrium paths from the same initial distribution of capital in the world market, and that the distribution of capital in the limit differs among equilibrium paths. One equilibrium path converges to a long-run equilibrium in which the international ranking of factor endowment ratios differs from the initial ranking, while another equilibrium path maintains the initial ranking and converges to a different long-run equilibrium. Since the path realized is indeterminate, so is the long-run trade pattern. Therefore, the long-run Heckscher–Ohlin prediction, that each country exports those goods such that the abundantly endowed factor of production is intensively used for producing it, is vulnerable to the introduction of externality.

In most papers focusing on the existence of indeterminacy within two-sector models with constant social returns, the utility function is assumed to be linear, which implies an infinite elasticity of intertemporal substitution in consumption. However, as shown recently by Mulligan (2002) and Vissing-Jorgensen (2002) for instance, this elasticity is estimated around unity or even below. The last paper of this section, "Indeterminacy in discrete-time infinite-horizon models with nonlinear utility and endogenous labor," published with Alain Venditti in the *Journal of Mathematical Economics* in 2007, considers a two-sector economy in which preferences are given by a additively separable CES utility function, defined over consumption and leisure. Three particular cases are considered:

1. Labor supply is infinitely elastic,
2. The elasticity of intertemporal substitution in consumption is infinite, and
3. Labor supply is inelastic.

In the first case, the steady state is a saddle-point for any value of elasticity of intertemporal substitution in consumption, and any level of external effects. In the second case, local stability of the steady state is independent from elasticity of labor supply, and local indeterminacy relies only on the loss of duality between the Rybczynski and Stolper-Samuelson effects. In the third case, a geometrical method of Grandmont et al. (1998) is used to show that the steady state is locally indeterminate if and only if the elasticity of intertemporal substitution in consumption is large enough. On this basis, the main result of the paper is established: the steady state is locally indeterminate if and only if the elasticity of intertemporal substitution in consumption is sufficiently large, and that of labor supply is sufficiently low. Moreover, period-two cycles always occur as the elasticity of labor supply is increased, and the steady state becomes saddle-point stable.

1.1.8 Indeterminacy in Endogenous Growth Models

The fundamental contributions of Romer (1986) and Lucas (1988) showed that if preferences are homothetic and external effects in production generate constant returns to scale for reproducible factors, then endogenous growth can occur in infinite-horizon models. This means that the growth process is endogenously determined, perpetual, and can be characterized by the dynamics of growth rates for output and consumption. This fact, combined with some of the previous results provided in this collection, suggests the possible existence of growth rate fluctuations. The three contributions contained in this section are concerned precisely with this point.

The first, "Indeterminacy under constant returns to scale in multisector economies," published with Jess Benhabib and Qinglai Meng in *Econometrica* in 2000, is an extension to the endogenous growth framework of the first paper discussed in Sect. 1.1.7. The authors consider a multisector economy with one consumable capital good and n pure capital goods produced without fixed factors from Cobb-Douglas technologies. The economy is characterized by sector-specific externalities and constant social returns. Instantaneous utility is homogeneous with a constant intertemporal elasticity of substitution. Under some conditions on the matrices of private and social input coefficients, local indeterminacy exists. In the particular case of a two-sector endogenous growth model, these conditions imply that the consumable capital good is intensive in the pure capital good from the private perspective, but it is intensive in itself from the social perspective. Contrary to the case of exogenous growth, this result holds for any value of the elasticity of intertemporal substitution in consumption.

Most of the papers in this collection considering multisector economies with productive external effects assume sector-specific externalities, and show that a capital intensive consumption good sector at the private level is a necessary condition for the occurrence of local indeterminacy. However, such a formulation is quite different from the one initially considered by Romer (1986) and Lucas (1988), where global externalities are assumed. In this case, some additional inter-relationships between the sectors are introduced, and these new mechanisms may provide different channels through which local indeterminacy can arise. The second paper of this section, "Global externalities, endogenous growth and sunspot fluctuations," published with Harutaka Takahashi and Alain Venditti in *Advanced Studies in Pure Mathematics* in 2009, considers a two-sector economy with Cobb-Douglas technologies, labor-augmenting global external effects, and increasing social returns. The paper focuses on the local indeterminacy of endogenous growth paths. After proving the existence of a balanced growth path, existence of sunspot fluctuations is shown to be compatible with a labor intensive consumption good sector at the private level, provided that elasticity of intertemporal substitution in consumption admits intermediary values. In particular, the existence of period-two cycles requires precisely such a capital intensity configuration.

The last paper of this collection, "A homoclinic bifurcation and global indeterminacy of equilibrium in a two-sector endogenous growth model," published with Paolo Mattana and Tadashi Shigoka in the *International Journal of Economic Theory* in 2009, also focuses on the consequences of global externalities, discussed in the context of an endogenous growth model similar to that of Lucas (1988). In the Lucas model, an external effect appears in the physical-goods sector. In this paper, it appears in the educational sector. Contrary to the Lucas formulation, this external effect yields multiple balanced growth paths. Moreover, the model undergoes a global homoclinic bifurcation, and exhibits global indeterminacy of equilibrium.

1.1.9 Summary

This selection of Kazuo Nishimura's papers, since the contribution published in the *Journal of Economic Theory* in 1979, provides a precise description of the usefulness of nonlinear dynamics for the analysis of infinite-horizon intertemporal optimization models. As co-authors and friends, it is our pleasure to show how Kazuo's works contributed to the development of nonlinear dynamics in economic theory. We look forward to seeing even further progress along the lines of these selected papers in the coming years.

1.2 List of Papers

The selected papers of Kazuo Nishimura chosen for this volume are as follows:

Introductory Survey:

1. "Nonlinear Dynamics in the Infinite Time Horizon Model," (with Gerhard Sorger), *Journal of Economic Surveys,* Vol. 13, 619–652, 1999

Optimal Growth and Endogenous Cycles:

2. "The Hopf Bifurcation and the Existence and the Stability of Closed Orbits in Multi-Sector Models of Economic Growth," (with Jess Benhabib), *Journal of Economic Theory,* Vol. 21, pp. 421–444, 1979
3. "Competitive Equilibrium Cycles," (with Jess Benhabib), *Journal of Economic Theory,* Vol. 35, pp. 284–306, 1985
4. "Interlinkage in the Endogenous Real Business Cycles of International Economies," (with Makoto Yano), *Economic Theory,* Vol. 3, pp. 151–168, 1993

Optimal Growth and Chaotic Dynamics:

5. "Nonlinear Dynamics and Chaos in Optimal Growth: An Example," (with Makoto Yano) *Econometrica,* 63, 981–1001, 1995
6. "Chaotic Solutions in Dynamic Linear Programming," (with M. Yano), *Chaos, Solitons and Fractals,* 7, 1191–1953, 1996
7. "On the Least Upper Bound of Discount Factors that are Compatible with Optimal Period-Three Cycles," (with Makoto Yano), *Journal of Economic Theory,* 66, 306–333, 1996

Global Dynamics in Optimal Growth Models:

8. "Intertemporal Complementarity and Optimality: A Study of a Two-Dimensional Dynamical System," (with Tapan Mitra), *International Economic Review,* Vol. 46, No. 1, pp. 93–131, 2005

Dynamic Models with Non-concave Technologies:

9. "A Complete Characterization of Optimal Growth Paths in an Aggregated Model with a Non-concave Production Function," (with W. Davis Dechert), *Journal of Economic Theory,* Vol. 31, pp. 332–354, 1983

Stochastic Optimal Growth Models:

10. "Stochastic Optimal Growth with Nonconvexities," (with Ryszard Rudnicki and John Stachurski), *Journal of Mathematical Economics,* Vol. 42(1), pp. 74–96, 2006
11. "Stability of Stochastic Optimal Growth Models: a New Approach," (with John Stachurski), *Journal of Economic Theory,* Vol.22, pp. 100–118, 2005

Indeterminacy and Externalities in Exogenous Growth Models:

12. "Indeterminacy and Sunspots with Constant Returns," (with Jess Benhabib), *Journal of Economic Theory,* Vol. 81, pp. 58–96, 1998
13. "Trade and Indeterminacy in a Dynamic General Equilibrium Model," (with Koji Shimomura), *Journal of Economic Theory,* Vol. 105, pp. 244–259, 2002
14. "Indeterminacy in discrete-time infinite- horizon models with nonlinear utility and endogenous labor," (with Alain Venditti), *Journal of Mathematical Economics,* Vol. 43(3–4), pp. 446–476, 2007

Indeterminacy and Externalities in Endogenous Growth Models:

15. "Indeterminacy Under Constant Returns to Scale in Multisector Economies," (with Jess Benhabib and Qinglai Meng), *Econometrica,* Vol. 68, pp. 1541–48, 2000
16. "Global Externalities, Endogenous Growth and Sunspot Fluctuations," (with Harutaka Takahashi and Alain Venditti), *Advanced Studies in Pure Mathematics,* Vol. 53, pp. 227–238, 2009
17. "A Homoclinic Bifurcation and Global Indeterminacy of Equilibrium in a Two-Sector Endogenous Growth Model," (with Paolo Mattana and Tadashi Shigoka), *International Journal of Economic Theory* Vol. 5, pp. 25–60, 2009

Chapter 2
Non-linear Dynamics in the Infinite Time Horizon Model*

Kazuo Nishimura and Gerhard Sorger**

2.1 Introduction

There are two important reasons why economic growth theory became (again) a popular field during the last 15 years. The first one is that recently developed growth theoretic models capture more aspects of economic growth than the models used in the fifties and sixties. Nowadays, models of economic growth try to explain technological progress and innovations, whereas the purpose of traditional models was mainly to describe the process of capital accumulation. The second reason for the resurrection of growth theory is that, even in its restricted sense as a theory of optimal capital accumulation, it has been found to be able to explain a much wider range of phenomena than it has previously been believed Using results from the theory of non-linear dynamical systems, it has been shown that optimal growth theory can provide new explanations for business cycles and for international differences in growth and development. The present survey concentrates on this aspect of optimal growth theory, that is on the possibility of fluctuations and non-uniqueness in models of optimal capital accumulation. It is a selective survey on non-linear dynamics in infinite time horizon models of optimal growth, which extends and updates our earlier paper Nishimura and Sorger (1996).

*Journal of Economic Surveys 13, 619–634, 1999

**We thank Donald George, Michael McAleer and Les Oxley for many helpful comments.

K. Nishimura (✉)
Institute of Economic Research, Kyoto University, Japan
e-mail: nishimura@kier.kyoto-u.ac.jp

G. Sorger
Department of Economics, University of Vienna, Austria
e-mail: gerhard.sorger@univie.ac.at

J. Stachurski et al. (eds.), *Nonlinear Dynamics in Equilibrium Models*,
DOI 10.1007/978-3-642-22397-6_2, © Blackwell Publishing 1999. Published by
Springer-Verlag Berlin Heidelberg 2012. All Rights Reserved

The general framework for all the models discussed in this paper is the same: one seeks a Pareto-optimal allocation in an intertemporal equilibrium model with production. The social welfare function is the discounted sum over an infinite time horizon of utility derived from consumption. The constraints of the optimization problem are given by the production technology. We discuss models with a single production sector (the aggregative model), models with separate sectors for the production of consumption goods and investment goods (the two-sector model), and general n-sector models. For the most part we do not allow for external effects in production, such that the models can be stated as dynamic optimization problems. In that case, our main concern is with the occurrence of optimal cycles and optimal chaos. In Sects. 2.4 and 2.5, however, we introduce production externalities. This requires that the optimization models are augmented by an equilibrium condition. In the analysis of these models one is primarily interested in the possibility of non-uniqueness and indeterminacy of equilibria.

Except for Sect. 2.5 we concentrate on models that are stated in a discrete-time framework. On the one hand, some of the results we present for discrete-time models have exact analogues in continuous-time models. Discussing both versions would lead to unnecessary repetitions. On other hand, there are important differences between modelling time as a discrete variable or as a continuous variable. To obtain periodic solutions in a continuous-time model, it is necessary that the state space has at least dimension two, and to obtain chaotic dynamics the minimum dimension of the state space is three. In contrast, cycles of arbitrary length and chaos are already possible for one-dimensional difference equations. Therefore, most of the literature on complicated dynamics has used the simpler discrete-time approach and we follow this tradition.

We start in Sect. 2.2 by introducing the simplest of all optimal growth models, namely the aggregative model. Under the standard assumption of a neoclassical production function, all optimal paths converge monotonically to a unique non-trivial steady state. If one allows for increasing returns to scale at small capital-labour ratios, then there can be two steady state solutions and it depends on the initial capital-labour ratio to which steady state the economy converges. The remainder of Sect. 2.2 deals with two-sector models of optimal growth. After introducing the general framework of two-sector models, we discuss the possibility of optimal cycles and optimal chaos. The former is shown to occur in an economy in which both sectors have Cobb-Douglas production functions whereas the latter is demonstrated in an economy with technologies that are (approximately) described by Leontief production functions.

In Sect. 2.3 we consider optimal growth models in reduced form. Because these models do not specify the details of the production technologies, they have a more compact and, hence, simpler mathematical structure. We present results that characterize to a large extent the set of all possible optimal policy functions for this general class of models. Section 2.3.3 deals with the role of the time-preference for the occurrence of complicated dynamic phenomena. We present various results that show precisely in what sense a high time-preference rate is responsible for the optimality of complicated dynamics. A survey of some prominent examples of complicated dynamics in reduced form optimal growth models concludes Sect. 2.3.

Section 2.4 introduces external effects and shows how the resulting equilibrium problem can have a continuum of different solutions. Section 2.5 surveys a few results on optimal growth models that are formulated in continuous time. Since continuous-time dynamical systems can exhibit complicated dynamics only for a sufficiently high dimension of the state space, the results on cycles and chaos are somewhat harder to derive than in the discrete-time case. We state results on the occurrence of Hopf bifurcations and closed orbits in multi-sector optimal growth models without external effects, and on indeterminacy of equilibria in models with externalities. For both cases we use a framework with Cobb-Douglas technologies.

2.2 Basic Models of Economic Growth

This section introduces one-sector and two-sector models of optimal growth. These are the most commonly used models in growth theory. We show in Sect. 2.2.1 that all equilibrium paths converge monotonically to a unique steady state if the production function satisfies the usual neoclassical assumptions. Section 2.2.2 show that monotonicity still holds if the production technology exhibits increasing returns to scale for small capita-labour ratios. However, in this case there may be two steady states and the initial conditions determine to which one the economy converges. Section 2.2.3 deals with models with separate production sectors for consumption goods and investment goods. It is shown that in this class of models optimal paths may be non-monotonic and even chaotic. This result holds true even if the production functions satisfy the usual neoclassical properties.

2.2.1 An Aggregative Model with a Neo-Classical Production Function

The basic premises of the aggregative model can be described as follows: in each period t, a single homogeneous output, Y_t, is produced from the two homogeneous input factors labour, L_t, and capital, K_t. The technically efficient possibilities for production are summarized by an aggregate production function $F(K_t, L_t)$ which exhibits constant returns to scale, positive marginal utility, and decreasing marginal rate of substitution. Because of constant returns to scale, the output-labour rate $y_t = Y_t/L_t$ is given by

$$y_t = f(k_t), \tag{2.1}$$

where $k_t = K_t/L_t$ is the capital-labour ratio and $f(k) = F(k, 1)$. The following assumptions on the function f will be used.

A1 $f : [0, +\infty) \mapsto R$ *is continuous, and twice continuously differentiable on* $(0, +\infty)$.

A2 $f(0) = 0$, $f(k) > 0$, $f'(k) > 0$, *and* $f''(k) < 0$ *for all* $k > 0$. *Moreover,* $\lim_{k\to 0} f'(k) = +\infty$ *and* $\lim_{k\to+\infty} f'(k) = 0$.

The labour force is assumed to be constant and the capital stock is depreciating at the positive rate δ. Per capita output may be allocated between consumption and gross investment. Denoting per capita consumption by c_t this implies

$$y_t = c_t + k_{t+1} - (1-\delta)k_t. \tag{2.2}$$

The initial per capita capital stock k_0 is historically given. Social welfare over the infinite planning period is presumed to be represented by the functional

$$\sum_{t=0}^{+\infty} \rho^t u(c_t), \tag{2.3}$$

where $\rho \in (0,1)$ is the discount factor. Thus, social welfare is the discounted sum of period-wise utility of per capita consumption. The utility function is assumed to be increasing and to have decreasing marginal utility.

A3 $u\colon [0,+\infty) \mapsto R$ *is continuous, increasing, and twice continuously differentiable on* $(0,+\infty)$.

A4 $u'(c) > 0$ *and* $u''(c) < 0$ *for all* $c > 0$. *Moreover,* $\lim_{c\to 0} u'(c) = +\infty$ *and* $\lim_{c\to+\infty} u'(c) = 0$.

A sequence of stocks $(k_t)_{t=0}^{+\infty}$ is called a feasible path from k_0 if it satisfies the condition $0 \leqslant k_{t+1} \leqslant f(k_t) + (1-\delta)k_t$ for all $k \geqslant 1$. For each feasible path there is a corresponding sequence of consumption rates $(c_t)_{t=0}^{+\infty}$ determined by (2.1) and (2.2). A feasible path is called an interior path if $c_t > 0$ holds for all $t \geqslant 0$, and it is called stationary if $k_t = k$ for all $t \geqslant 0$ and some constant $k \geqslant 0$. An optimal path from k_0 is a feasible path from k_0 that maximizes the objective functional in (2.3).

An interior path will be called an Euler path if it satisfies the discrete-time Euler equation

$$u'(c_{t-1}) = \rho u'(c_t)[f'(k_t) + 1 - \delta]. \tag{2.4}$$

If an interior path is optimal then it must be an Euler path. We shall outline a proof of this statement in a more general framework in Sect. 2.3 below. Under the assumptions stated above, any Euler path which satisfies the transversality condition

$$\lim_{t\to\infty} \rho^t u'(c_t)[f'(k_t) + 1 - \delta]k_t = 0$$

is an optimal path.[1] The economic interpretation of the transversality condition is based on the observation that, in a competitive economy, $f'(k_t) + 1 - \delta$ is equal

[1]It has to be noted that this sufficient optimality condition depends crucially on the concavity assumptions stated in A2 and A4. In the following subsection we shall modify A2. In that case

to the interest rate and $f(k_t) - k_t f'(k_t)$ is equal to the wage rate. Using this fact it can be shown that the transversality condition says that the present value of lifetime consumption is exactly equal to the present value of lifetime labour income plus the value of the initial capital stock.

Because we assume that the marginal utility of consumption becomes unbounded as the level of consumption goes to 0, every optimal path from a positive initial stock $k_0 > 0$ is an interior path and, consequently, it must be an Euler path. A stationary optimal path $(k, k, k, \ldots)$ with $k > 0$ must satisfy the Euler equation (2.4) and, hence,

$$\rho^{-1} = f'(k) + 1 - \delta. \tag{2.5}$$

We call a solution of (2.5) a steady state. The following theorem is the discrete-time version of a result proved by Cass (1965) and Koopmans (1965). The basic structure of the aggregative model was first laid out by Ramsey (1928).

Theorem 1. *Consider the model defined by (2.1)–(2.3). Under assumptions A1–A4 there exists a unique steady state k^*. Moreover, an optimal path from $k_0 > 0$ is a monotonic sequence converging to k^*.*

2.2.2 An Aggregative Model with Convex–Concave Production Function

For this section we assume $\delta = 1$ so that the capital stock fully depreciates in one period. The production function is assumed to exhibit increasing returns to scale for sufficiently small capital stocks. Thus, assumption A2 is replaced by the following one.

A2′ $f(0) = 0$, $f(k) > 0$, *and* $f'(k) > 0$ *for all* $k > 0$. *There exists* $k_I > 0$ *such that* $f''(k) > 0$ *for all* $k \in (0, k_I)$, *and* $f''(k) < 0$ *for all* $k \in (k_I, +\infty)$. *Moreover,* $\lim_{k\to 0} f'(k) \geq 1$ *and* $\lim_{k\to +\infty} f'(k) = 0$.

Recall that a steady state of the economy is a solution of $\rho f'(k) = 1$. If the condition

$$\lim_{k\to 0} f'(k) < \rho^{-1} < \max\{f(k)/k | k > 0\} \tag{2.6}$$

holds, as then there exist two steady states k_* and k^*. Without loss of generality we assume $k_* < k^*$. The aggregative optimal growth model with a production function that shows increasing returns for small capital-labour ratios and decreasing returns for large capital-labour ratios was studied by Majumdar and Nermuth (1982), Dechert and Nishimura (1983), and Mitra and Ray (1984). The following result is due to Dechert and Nishimura (1983).

---

the Euler equation with the transversality condition will no longer provide a sufficient optimality condition.

Theorem 2. *Consider the model defined by (2.1)–(2.3). Under assumptions A1, A2′, and A3–A4 the following is true.*

(a) The optimal path from $k_0 > 0$ is a monotonic sequence.
(b) If (2.6) holds, then there exists $k_c \in (0, k^*)$ with the following property: an optimal path from $k_0 < k_c$ converges to 0 and an optimal path from $k_0 > k_c$ converges to k^*.

2.2.3 Two-Sector Models

2.2.3.1 The Basic Framework

The model we consider in this section is a discrete-time version of the two-sector optimal growth model from Uzawa (1964). There are two goods: the pure consumption good, C, and the pure capital good, K. Each sector uses both capital and labour as inputs. Capital input must be made one period prior to the production of output. Labour input is made in the same period as output is produced. Denote by $F_C(K_C, L_C)$ and $F_K(K_K, L_K)$ the production functions of sectors C and K, respectively, where K_i and L_i denote the factor inputs in sector $i \in \{C, K\}$. The production functions are assumed to have all the standard neoclassical properties.

The labour endowment of the economy is constant and time-independent. Without loss of generality we normalize it to 1. Denote by c_t and y_t the (per capita) outputs of sectors C and K, respectively, in period t. Moreover, denote by $K_{C,t-1}$ and $L_{C,t}$ the factor inputs used in sector C for the production of c_t, and by $K_{K,t-1}$ and $L_{K,t}$ those used in sector K to produce y_t, i.e.,

$$c_t = F_C(K_{C,t-1}, L_{C,t}), \tag{2.7}$$

$$y_t = F_K(K_{K,t-1}, L_{K,t}). \tag{2.8}$$

Moreover, denote by k_{t-1} the aggregate capital input, i.e.,

$$K_{C,t-1} + K_{K,t-1} = k_{t-1}. \tag{2.9}$$

The output of the capital good sector, y_t, represents the gross accumulation of capital, that is

$$y_t = k_t - (1-\delta)k_{t-1}, \tag{2.10}$$

where $\delta \in (0, 1)$ is the rate of depreciation. Since the total labour force in the economy has been normalized to 1, we have

$$L_{C,t} + L_{K,t} = 1. \tag{2.11}$$

Denote by $u(c)$ the representative consumer's instantaneous utility when he consumes c units of the consumption good. With these notations, the two-sector optimal growth model is described by the maximization problem

$$\max \sum_{t=1}^{+\infty} \rho^t u(c_t) \tag{2.12}$$

$$\text{subject to } k_0 = k \text{ and constraints (2.7)–(2.11),}$$

where $\rho \in (0, 1)$ is the discount factor.

In order to analyze the dynamics of the above model it is convenient to express, for each given amount of capital input k, the trade-off between the two outputs by $c = T(y, k)$, i.e.,

$$T(y,k) = \max F_C(K_C, L_C) \tag{2.13}$$

$$\text{subject to } \begin{cases} F_K(K_K, L_K) = y, \\ L_C + L_K = 1, \\ K_C + K_K = k. \end{cases}$$

Note that for a given amount of capital input, k_{t-1}, the relation $c_t = T(y_t, k_{t-1})$ describes the production possibility frontier in the output plane. The domain of the function T is defined as $\Omega = \{(y,k)|k \geqslant 0, 0 \leqslant y \leqslant F_K(k,1)\}$. With this definition, the optimal growth model (2.12) can be transformed into the form

$$\max \sum_{t=1}^{+\infty} \rho^t U(k_{t-1}, k_t)$$

$$\text{subject to } k_0 = k \text{ and } k_t \leqslant F_K(k_{t-1}, 1) + (1-\delta)k_{t-1},$$

where $U(x,z) = u(T(z-(1-\delta)x, x))$. This is the reduced form of the two-sector optimal growth model. We shall discuss models in reduced form in detail in Sect. 2.3 below.

2.2.3.2 Optimal Cycles

In the following we assume that the instantaneous utility function is linear, i.e., $u(c) = c$ and that capital fully depreciates within one period, i.e., $\delta = 1$. Then the reduced form utility function is identical to the social production function, i.e., $U(k,y) = T(y,k)$.

The Euler equation in the two-sector optimal growth model is

$$U_2(k_{t-1}, k_t) + \rho U_1(k_t, k_{t+1}) = 0, \tag{2.14}$$

where $U_1(k, y) = \partial U(k, y)/\partial k$ and $U_2(k, y) = \partial U(k, y)/\partial y$. The steady state k^* corresponds to a stationary solution $(k^*, k^*, k^*, \ldots)$ of (2.14). The local behaviour around k^* is determined by the roots of the characteristic equation evaluated at the steady state. This equation is

$$\rho U_{12}(k^*, k^*)\lambda^2 + [\rho U_{11}(k^*, k^*) + U_{22}(k^*, k^*)]\lambda + U_{21}(k^*, k^*) = 0. \quad (2.15)$$

We consider the case in which the production functions in both sectors are of the Cobb-Douglas form. Specifically, we assume

$$F_C(K_C, L_C) = K_C^{\alpha} L_C^{1-\alpha}, \quad (2.16)$$

$$F_K(K_K, L_K) = K_K^{\beta} L_K^{1-\beta}, \quad (2.17)$$

where α and β are positive numbers smaller than 1. From the first order conditions for problem (2.13) we can derive

$$(K_K/L_K)/(K_C/L_C) = [\beta/(1-\beta)]/[\alpha/(1-\alpha)].$$

Hence

$$(K_K/L_K) - (K_C/L_C) = \begin{cases} > 0 & \text{if } \beta > \alpha, \\ < 0 & \text{if } \beta < \alpha. \end{cases} \quad (2.18)$$

As long as the powers in both Cobb-Douglas production functions differ from each other, no factor intensity reversal takes place. If $\beta > \alpha$, the production of consumption goods is more labour intensive than the production of capital goods, and if, $\beta < \alpha$, the converse is true. For this economy the steady state value is

$$k^* = \frac{\alpha(1-\beta)(\rho\beta)^{1/(1-\beta)}}{\beta[1-\alpha+\rho(\alpha-\beta)]} \quad (2.19)$$

and the roots of Eq. (2.15) are $\lambda_1 = (\beta - \alpha)/(1-\alpha)$ and $\lambda_2 = (1-\alpha)/[\rho(\beta-\alpha)]$. If $\alpha > \beta$, then both roots are negative, and if $\rho < (1-\alpha)/(\alpha-\beta) < 1$, then the steady state is totally unstable. In this case, there exists an optimal cycle of period 2 as shown in the following theorem from Nishimura and Yano (1995b).

Theorem 3. *Consider the economy described by the production functions (2.16) and (2.17), linear utility function $u(c) = c$, and full depreciation $\delta = 1$. If*

$$\rho < \frac{1-\alpha}{1-\beta} < 1$$

then the steady state k^* given in (2.19) is totally unstable and there exists an optimal path which is periodic of period 2.

In two sector model with Cobb-Douglas production functions and linear utility, the sign of the cross partial derivative $U_{12}(x, y)$ is determined by the factor intensity difference of the consumption good sector and the capital good sector (see Benhabib and Nishimura 1985). This fact together with the relation in (2.18) implies that

$$U_{12}(x, y) \begin{cases} > 0 & \text{if } \beta > \alpha, \\ < 0 & \text{if } \beta < \alpha. \end{cases} \tag{2.20}$$

Optimal paths in this model can be described by a difference equation of the form $k_{t+1} = h(k_t)$. The function h is called the optimal policy function and will be discussed in detail in Sect. 2.3 below. Equation (2.20) implies (see Theorem 14 below) that, in the case $\beta > \alpha$, the graph of the optimal policy function h is strictly increasing whenever it lies in the interior of Ω. Analogously, if $\alpha > \beta$, then the graph of h is strictly decreasing in every point in the interior of Ω. In the case of $\alpha > \beta$ the optimal policy function h becomes an unimodal map because the graph of h increase along the boundary of Ω and decreases in the interior of Ω. The global dynamics of the two sector model with Cobb-Douglas production functions is studied by Nishimura and Yano (1995b). The analysis has been extended to the case of partial depreciation by Baierl et al. (1998), from where the following result is taken.

Theorem 4. *Consider the economy described by the production functions (2.16) and (2.17), linear utility function $u(c) = c$, and depreciation $\delta \in [0, 1]$. If $\alpha > \beta$ and*

$$\frac{(1-\alpha)-(\alpha-\beta)}{1-\beta} < \rho(1-\delta) < \frac{1-\alpha-\rho(\alpha-\beta)}{1-\beta}$$

then the steady state is totally unstable and there exists an optimal path which is periodic of period 2.

2.2.3.3 Optimal Chaos for Arbitrarily Weak Discounting

Before we proceed, we have to discuss three possible definitions of complicated dynamics in systems of the form $x_{t+1} = h(x_t)$. Here, x_t is the state of the economy at time t (for example the capital-labour ratio) and $h: [0, 1] \mapsto [0, 1]$ is a continuous function which describes the evolution of the economy.[2]

We say that the dynamical system $x_{t+1} = h(x_t)$ exhibits ergodic chaos if there exists an absolutely continuous probability measure μ on the interval $[0, 1]$ which is invariant and ergodic under h. Here, a measure μ on the set $[0, 1]$ is said to be

[2]For simplicity we assume that $x_t \in [0, 1]$ holds for all t, although the definitions hold more generally.

absolutely continuous if it has a Radon-Nikodym derivative with respect to the Lebesgue measure, and it is said to be invariant under h if $\mu(\{x \in [0,1] | h(x) \in B\}) = \mu(B)$ for all measurable sets $B \subseteq [0,1]$. The invariant measure μ is said to be ergodic if for every measurable set $B \subseteq [0,1]$ which satisfies $\{x \in [0,1] | h(x) \in B\} = B$ we have either $\mu(B) = 0$ or $\mu(B) = 1$. The most important property of a dynamical system which exhibits ergodic chaos is that the (asymptotic) statistical properties of the deterministic trajectories can be approximated by the invariant and ergodic measure μ. In other words, time averages along a trajectory $(x_0, x_1, x_2, \ldots)$ can be replaced by space averages in the following form:

$$\lim_{\tau \to +\infty} \frac{1}{\tau} \sum_{t=0}^{\tau-1} \phi(h^{(t)}(x)) = \int_{[0,1]} \phi(z) \mu(dz),$$

where $h^{(t)}(x)$ is the t-th iterate of h evaluated at x, that is , $h^{(0)}(x) = x$ and $h^{(t+1)}(x) = h^{(t)}(h(x))$ for all $t \geqslant 1$ and all $x \in [0,1]$. This formula, which is known as the ergodic theorem, is valid for μ-almost all initial points $x \in [0,1]$ and for all continuous functions ϕ: $[0,1] \mapsto \mathbb{R}$.

We say that the dynamical system $x_{t+1} = h(x_t)$ exhibits geometric sensitivity if there is a real constant $\gamma > 0$ such that the following is true: for any $\tau = 0, 1, 2 \ldots$ there exists $\varepsilon > 0$ such that for all $x, y \in [0,1]$ with $|x - y| < \varepsilon$ and for all $t \in \{0, 1, \ldots, \tau\}$ it holds that

$$|h^{(t)}(x) - h^{(t)}(y)| \geqslant (1 + \gamma)^t |x - y|.$$

Geometric sensitivity implies that small perturbations of the initial conditions are magnified at a geometric rate over arbitrary but finite time periods. Of course, the geometric magnification cannot last indefinitely because the state space $[0, 1]$ of the dynamical system is bounded. Note also that geometric sensitivity implies that there is no stable periodic path of the dynamical system.

Finally, we say that the dynamical system $x_{t+1} = h(x_t)$ exhibits topological chaos if there exists a p-periodic solution for all sufficiently large integers p and there exists an uncountable invariant set $S \subseteq [0,1]$ containing no periodic points such that

$$\liminf_{t \to +\infty} \left| h^{(t)}(x) - h^{(t)}(y) \right| = 0 < \limsup_{t \to +\infty} \left| h^{(t)}(x) - h^{(t)}(z) \right|$$

holds, whenever $x \in S$, $y \in S$, and either $x \neq z \in S$ or z is a periodic point. The set S is called a scrambled set. The condition displayed above says that any two trajectories starting in the scrambled set move arbitrarily close to each other but do not converge to each other or to any periodic orbit. Thus, trajectories starting in the scrambled set are highly unpredictable over the long-run. Although the scrambled set is required to be uncountable, it can be a set of measure 0.

Let us now return to two-sector optimal growth models. Suppose that both sectors have Leontief production functions

$$F_C(K_C, L_C) = \min\{K_C, L_C\}, \tag{2.21}$$

$$F_K(K_K, L_K) = \mu \min\{K_K, L_K/b\}, \tag{2.22}$$

where $\mu > \rho^{-1}$ and $b > 1$. Note that $b > 1$ implies that the capital good sector is more labour intensive than the consumption good sector. We still assume that the utility function is linear and that capital is fully depreciated after one period. In this case the maximization problem (2.12) can have multiple solutions such that, in general, the solution cannot be described by an optimal policy function. Nishimura and Yano (1995a, 1996b), however, prove that optimal paths are described by an optimal policy function, if the parameter values are suitably chosen. Furthermore, they show that, under certain parameter restrictions, the optimal policy function is expansive and unimodal.[3] We are now going to describe this result. Let $\gamma = b - 1$ and define

$$h(k) = \begin{cases} \mu k & \text{if } \ 0 \leqslant k \leqslant 1/(\gamma+1), \\ -(\mu/\gamma)(k-1) & \text{if } \ 1/(\gamma+1) \leqslant k \leqslant 1. \end{cases} \tag{2.23}$$

Under the assumption $\mu/(1+\gamma) \leqslant 1$ the function h maps the closed interval $I = [0, \mu/(1+\gamma)]$ onto itself. For all practical purposes, we may therefore restrict our attention to the interval I and treat h as a function from I onto itself. Nishimura and Yano (1996b) prove the following result.

Theorem 5. *Let h_I be the function defined in (2.23) restricted to the interval $I = [0, \mu/(1+\gamma)]$. Suppose that the parameters μ, ρ, and γ satisfy*

$$0 < \rho < 1, \quad \gamma > 0, \quad \rho\mu > 1 \quad \text{and} \quad \gamma + 1 > \mu. \tag{2.24}$$

Then it follows that optimal paths of the two-sector model with linear utility function and Leontief production functions (2.21) and (2.22) satisfy the equation $k_{t+1} = h_I(k_t)$ provided that one of the following two conditions is satisfied:

A: $\mu \leqslant \gamma$,
B: $\gamma < \mu \leqslant \min\left\{\left(\gamma + \sqrt{\gamma^2 + 4\gamma}\right)/2, \left(-1 + \sqrt{1+4\gamma}\right)/(2\rho)\right\}$.

Under condition A the decreasing portion of h_I has slope larger than or equal to -1. More specifically, if condition A is satisfied with strict inequality, then the positive fixed point of the difference equation $k_{t+1} = h_I(k_t)$ is globally asymptotically

[3] A function $h: [A, B] \mapsto [A, B]$ is called unimodal if there exists $\overline{x} \in [A, B]$ such that h is strictly increasing on $[A, \overline{x}]$ and strictly decreasing on $[\overline{x}, B]$. Moreover, h is called expansive if it is piecewise differentiable with $|h'(x)| > 1$ for all x at which h is differentiable.

stable. If, instead, condition A is satisfied with equality, then every optimal solution from $k > 0$ converges to a period-two cycle, except for the unique path that corresponds to the positive fixed point.

Under condition B, the decreasing portion of h_I has slope smaller than -1. In this case, h_I is expansive and unimodal. It has been shown that these two properties imply that the dynamical system $k_{t+1} = h_I(k_t)$ exhibits ergodic chaos and geometric sensitivity. Nishimura and Yano (1996b) show that the set of parameter values (ρ, μ, γ) satisfying condition B and (2.24) is non-empty if $0 < \rho < 1/2$.

Condition B and (2.24) are sufficient conditions for h_I to be an optimal policy function that generates chaotic dynamics. There may be other sufficient conditions. In fact, Nishimura and Yano (1993, 1995a) provide an alternative and constructive method to fine parameter values (μ, γ) for which h_I describes optimal paths and is ergodically chaotic and geometrically sensitive. This method works for any given discount factor ρ, even if it is arbitrarily close to 1. In those papers it is shown that part of the graph of the optimal policy function lies on a von Neumann facet containing the stationary state and that any optimal path is confined in a small neighbourhood of the facet. In this respect, the result is closely related to the neighbourhood turnpike theorem of McKenzie (1983), which implies that any optimal path converges to a neighbourhood of the von Neumann facet. Nishimura et al. (1994) extend the results of Nishimura and Yano (1993, 1995a) to the case in which the von Neumann facet is trivial.

The above result is based on a model with Leontief production functions and a linear utility function, in which part of the graph of the optimal policy function lies on the boundary of the technology set Ω. Consequently, the consumption levels along an optimal path are at the minimum level (which is equal to 0) during the periods in which the activities lie on the boundary. In the following we demonstrate the existence of a topologically chaotic optimal policy function whose graph lies in the interior of the technology set, even if future utility is discounted arbitrarily weakly. In this case, the consumption levels are always strictly positive. For this purpose, we adopt a model with CES production functions and a utility function with a constant elasticity of intertemporal substitution. This model, which has been analyzed in Nishimura et al. (1998), contains that from Nishimura and Yano (1993, 1995a) as a limiting case. It is a parameterized model of optimal growth involving parameters $\pi = (\mu, a, b, \rho)$ that satisfy

$$\pi \in \Pi := \{(\mu, a, b, \rho) | b > a > 0, 1/a > \mu > 1/\rho, \mu > (b - a)/a\}.$$

In addition to those parameters we introduce a 'shift parameter' $\lambda \in [0, 1)$. We define the instantaneous utility function

$$u_\lambda(c) = \frac{1}{I - \lambda} c^{1-\lambda}, \tag{2.25}$$

and the production functions

$$F_{C,\lambda,\pi}(K_C, L_C) = \left[\frac{1}{2}K_C^{-1/\lambda} + \frac{1}{2}(L_C/a)^{-1/\lambda}\right]^{-\lambda}, \tag{2.26}$$

$$F_{K,\lambda,\pi}(K_K, L_K) = \mu\lambda\left[\frac{1}{2}K_K^{-1/\lambda} + \frac{1}{2}(L_K/b)^{-1/\lambda}\right]^{-\lambda}, \tag{2.27}$$

where $\mu_\lambda = (1/2)^\lambda[1 + \mu^{1/\lambda}]^\lambda$. The economy defined by these functions converges to the CES economy discussed before as the parameter λ approaches 0. The following result is proved in Nishimura et al. (1998).

Theorem 6. *For any* $\rho' \in (0, 1)$ *there exist* $\pi \in \Pi$, $\overline{\lambda} \in (0, 1)$, *and* $\rho \in (\rho', 1)$ *such that the two-sector economy defined by* (2.25)–(2.27), *depreciation rate* $\delta = 1$, *and discount factor* ρ *has the following properties for all* $\lambda \in (0, \overline{\lambda})$:

(a) An optimal path from $k_0 > 0$ *is an interior path.*
(b) Optimal paths are described by an optimal policy function which exhibits topological chaos.

2.3 Reduced Form Models in Discrete Time

This section deals with optimal growth models in reduced form. This means that the utility function and the sectoral production functions are not explicitly specified but that they are only implicitly given in the form of a reduced utility function and a transition possibility set. We continue to use a discrete-time framework and to exclude any external effects. The latter assumption implies that the models can be formulated as standard dynamic optimization problems. Section 2.3.1 introduces the model, discusses the assumptions, and states a few optimality conditions. Section 2.3.2 presents general results for the characterization of the optima (equilibria) of the model, Sect. 2.3.3 studies the interrelation between the size of the discount factor (i.e., a measure of the time-preference of the decision maker) and the dynamic complexity of the equilibrium paths, and Sect. 2.3.4 discusses various example from the literature.

2.3.1 *Definitions, Assumptions, and Optimality Conditions*

Reduced form optimal growth models have been used by many authors because of their simple mathematical structure and their wide applicability in economics. For a comprehensive survey of methods for and applications of such models we refer to McKenzie (1986) and Stokey and Lucas (1989). In this subsection we briefly summarize the definitions and results which are needed in the remainder of this section.

At each time t the state of the economic system is described by a vector $x_t \in X$ where the state space $X \subseteq \mathbb{R}^n$ is a compact and convex set with non-empty interior. A reduced form optimal growth model consists in finding

$$V(x) = \sup \sum_{t=0}^{+\infty} \rho^t U(x_t, x_{t+1}), \tag{2.28}$$

where the supremum is taken over the set of all sequences $(x_t)_{t=0}^{+\infty}$ satisfying the constraints

$$(x_t, x_{t+1}) \in \Omega \quad t \in \{0, 1, 2 \ldots\} \tag{2.29}$$

$$x_0 = x. \tag{2.30}$$

Here, ρ is the discount factor, U is the reduced utility function, Ω is the transition possibility set, $x \in X$ is the initial state, and V is the optimal value function.

It has been illustrated in Sect. 2.2.3.1 how an optimal growth model in primitive form can be converted into its reduced form. We now introduce a number of assumptions that will be used in this section.

A5 *$\Omega \subseteq X \times X$ is a closed and convex set such that the x-section $\Omega_x = \{y \in X | (x, y) \in \Omega\}$ is non-empty for all $x \in X$. The set $\cup_{x \in X} \Omega_x$ has non-empty interior.*[4]

A6 *$U : \Omega \mapsto R$ is a continuous and concave function.*

A7 $\rho \in (0, 1)$.

We shall refer to the dynamic optimization problem (2.28)–(2.30) as problem (Ω, U, ρ). Assumptions A5–A7 are standard assumptions in the relevant literature. They imply that (Ω, U, ρ) has an optimal path from any given initial state $x \in X$ and that the Bellman equation

$$V(x) = \max\{U(x, y) + \rho V(y) | y \in \Omega_x\}$$

holds for all $x \in X$. Moreover, a path $(x_t)_{t=0}^{+\infty}$ satisfying (2.29) and (2.30) is optimal if and only if $V(x_t) = U(x_t, x_{t+1}) + \rho V(x_{t+1})$ holds for all $t \in \{0, 1, 2 \ldots\}$. However, optimal paths for (2.28)–(2.30) need not be unique. To ensure that optimal paths are unique one has to add a strict concavity assumption. A very weak version of this assumption is the following.

A8 *The optimal value function V is strictly concave.*

If A5–A8 hold, then there exists for every $x \in X$ exactly one $y \in \Omega_x$ such that $V(x) = U(x, y) + \rho V(y)$. In other words, there exists a unique maximizer on the

---

[4]The assumption that $\cup_{x \in X} \Omega_x$ has non-empty interior in X is satisfied whenever Ω has non-empty interior in $X \times X$. The converse is not true as can be seen by simple examples.

right hand side of the Bellman equation. Let $h(x)$ denote this maximizer, that is,

$$h(x) = \arg\max\{U(x, y) + \rho V(y) | y \in \Omega_x\}.$$

The function $h: X \mapsto X$ defined in that way is called the optimal policy function of the optimization problem (Ω, U, ρ). It maps each state $x \in X$ to its optimal successor state $h(x)$. Optimal paths are uniquely determined as the trajectories of the difference equation $x_{t+1} = h(x_t)$ with (2.30) as the initial condition. Under assumptions A5–A8 both h and V are continuous functions.

Assumption A8 is sometimes awkward, because it involves the optimal value function V. A sufficient (but not necessary) condition for A8 is that the function U is strictly concave. Some of the results presented below involve an assumption that is based on the concepts of α-concavity and a-convexity. If α is any real number then V is α-concave, if $x \mapsto V(x) + (\alpha/2)\,\|x\|^2$ is a concave function. Analogously, we say that V is α-convex, if $x \mapsto V(x) - (\alpha/2)\,\|x\|^2$ is convex.[5]

A9 *There exist positive real numbers α and β such that V is α-concave and $(-\beta)$-convex.*

In many cases the optimization problem (Ω, U, ρ) is also assumed to satisfy the following monotonicity and smoothness assumptions.

A10 *If $x \leqslant \overline{x}$ then $\Omega_x \subseteq \Omega_{\overline{x}}$. The function $x \mapsto U(x, y)$ is non-decreasing and the function $y \mapsto U(x, y)$ is non-increasing.*

A11 *The partial derivatives $U_1(x, y)$ and $U_2(x, y)$ exist and are continuous for all (x, y) in the interior of Ω.*

An optimization problem (Ω, U, ρ) which satisfies A5–A8 and A10 has a non-decreasing optimal value function V. A feasible path $(x_t)_{t=0}^{+\infty}$ for problem (Ω, U, ρ) is called an interior path if x_{t+1} is in the interior of Ω_{x_t} for all $t \geqslant 0$. It is known that under A11 and interior optimal path must satisfy the Euler equation

$$U_2(x_t, x_{t+1}) + \rho U_1(x_{t+1}, x_{t+2}) = 0$$

for all $t \geqslant 0$. This can be seen by writing the objective function as

$$U(x_0, x_1) + \rho U(x_1, x_2) + \cdots + \rho^t U(x_t, x_{t+1}) + \rho^{t+1} U(x_{t+1}, x_{t+2}) + \cdots$$

Note that the variable x_{t+1} does not appear in any term other than those displayed above. Since $(x_t)_{t=0}^{+\infty}$ is interior and optimal, maximization with respect to x_{t+1} implies that the Euler equation holds. Conversely, if A5–A7 and A11 hold, and if a feasible path $(x_t)_{t=0}^{+\infty}$ satisfies the Euler equation as well as the limiting transversality condition

[5] For a discussion of these concepts we refer to Vial (1983). Some authors (e.g. Montrucchio 1994) use the term concavity-α instead of $(-\alpha)$-convexity.

$$\lim_{t \to +\infty} \rho^t U_1(x_t, x_{t+1}) x_t = 0,$$

then it is an optimal path.

2.3.2 *Characterization of Optimal Policy Functions*

It has been mentioned in the previous subsection that every optimal growth model (Ω, U, ρ) that satisfies assumptions A5–A8 has a continuous optimal policy function. Thus, continuity is a necessary condition for a function to be an optimal policy function. We state this result as part (a) of the following theorem. Part (b) was proved by Mitra and Sorger (1999b) as a corollary to Theorem 10 below. It shows that by strengthening continuity to Lipschitz continuity one obtains a sufficient condition for rationalizability.[6]

Theorem 7. *(a) Let (Ω, U, ρ) be an optimal growth model satisfying assumptions A5–A8 and let h be its optimal policy function. Then h is continuous.*
(b) Let $h: X \mapsto X$ be a Lipschitz continuous function with Lipschitz constant L. For every $\rho \leqslant 1/L^2$ there exists an optimization problem (Ω, U, ρ) satisfying A5–A10 such that h is the optimal policy function of (Ω, U, ρ).

Note that Lipschitz continuity of h not only guarantees the existence of a model satisfying A5–A8 with h as its optimal policy function, but that one can even require the model to satisfy the strong concavity assumption A9 as well as the monotonicity assumption A10. Weaker versions of part (b) of the above theorem (in which Lipschitz continuity is replaced by a much stronger smoothness condition) have been derived by Boldrin and Montrucchio (1986) and Neumann et al. (1988). It has to be noted that these results as well as Theorem 7(b) have constructive proofs. In principle, it is therefore possible to find an optimal growth model for any given (sufficiently smooth) optimal policy function.

Theorem 7 shows that the set of all possible optimal policy functions contains the set of Lipschitz continuous functions but is contained in the set of all continuous functions. This obviously raises the question of how tight this characterization is. Are there continuous functions that cannot occur as optimal policy functions of optimal growth models satisfying A5–A8? Are there optimal policy functions that are not Lipschitz continuous? The answers to both of these questions are affirmative. Hewage and Neumann (1990) were the first to demonstrate that not every continuous function can be an optimal policy function of a model satisfying A5–A8. This result has been replicated in various versions by Sorger (1995), Mitra (1996b), and Mitra and Sorger (1999a). They all involve continuous functions which are defined on a

[6]The function $h : X \mapsto X$ is called Lipschitz continuous if there exists a constant $L > 0$ such that $\|h(x) - h(y)\| \leqslant L \|x - y\|$ for all $x, y \in X$. The number L is called a Lipschitz constant for h. It is obvious that Lipschitz continuity implies continuity but not vice versa.

one-dimensional state space and have the slope $+\infty$ at an interior fixed point. Here we present the version taken from Mitra and Sorger (1999a).

Theorem 8. *Let $h\colon X \mapsto X$ be a continuous function, where X is a compact interval on the real line. Assume that h has the fixed point $x = h(x) \in X$ and that there exists z in the interior of X such that $x = h(z)$. If*

$$\lim_{y \mapsto x} \sup\, [h(y) - h(x)]/(y - x) = +\infty$$

then h cannot be the optimal policy function of an optimal growth model (Ω, U, ρ) satisfying A5–A8.

This result rules out any function as an optimal policy function which has slope $+\infty$ at a fixed point x which is either in the interior of the state space (this is the case if $z = x$ in the theorem) or can be reached along a trajectory of h emanating from the interior of X. For example, the function $h(x) = x^{1/3}$ defined on $X = [-1, 1]$ cannot be an optimal policy function of any problem (Ω, U, ρ) which satisfies A5–A8.

Now let us consider the second question posed above, namely whether every optimal policy function of a model satisfying A5–A8 is necessarily Lipschitz continuous. That this is not the case has been demonstrated by several examples. These examples also show that Theorem 8 is quite tight in the following sense: neither can any of the essential assumptions be weakened nor is the result true if one replaces the slope $+\infty$ by $-\infty$. First, it is easy to construct optimal growth models for which the optimal policy function has slope $+\infty$ at a fixed point on the boundary of X.[7] Second, an example of an optimal policy function that has slope $+\infty$ at an interior point that is not a fixed point is given in Mitra and Sorger (1999a) and, third, an example of an optimal policy function that has slope $-\infty$ at an interior fixed point is given in Mitra and Sorger (1999b).[8] Consequently, Lipschitz continuity is not a necessary condition for a function to be an optimal policy function under assumptions A5–A8.

Lipschitz continuity of h can be proved, if one adds a stronger concavity assumption. This was established under various assumptions by Montrucchio (1987, 1994) and Sorger (1994b, 1995). A version that is based on assumption A9 is presented in part (b) of Theorem 9 below (see, e.g., Sorger 1995). If only half of the strong concavity assumption A9 is assumed, namely that V is α-concave for some positive number α (but V is not necessarily $(-\beta)$-convex), then one obtains only Hölder continuity of the optimal policy function in the interior of X. This result is stated as part (a) of the theorem (see Montrucchio 1994; Sorger 1994b).

---

[7]Choose, for example, $X = [0, 1]$, $\Omega = \{(x, y)|x \in X,\ 0 \leq y \leq \sqrt{x}\}$ and $U(x, y) = ax - y$, where $a > 0$ is a sufficiently large real number. With the exception of A9 all assumptions from Sect. 2.3.1 hold. Moreover, the optimal policy function is $h(x) = \sqrt{x}$ which has the properties claimed in the text.

[8]The existence of these examples was proved by applying Theorem 10(b) below.

Theorem 9. *Let (Ω, U, ρ) be an optimal growth model satisfying assumptions A5–A8 and let h be its optimal policy function.*

(a) If there exists a positive number α such that V is α-concave, then the following is true: for every x in the interior of X there exist constants $\varepsilon > 0$ and $K \in \mathbb{R}$ such that the inequality $\|h(x) - h(y)\| \leqslant K \|x - y\|^{1/2}$ holds for all y such that $\|x - y\| \leqslant \varepsilon$.

(b) If A9 is satisfied then h is Lipschitz continuous.

So far we have discussed the problem of characterizing the set of optimal policy functions. Now we consider briefly the characterization of the set of all pairs (h, V) consisting of an optimal policy function and the corresponding optimal value function of models (Ω, U, ρ) satisfying A5–A8. It turns out that this set can be characterized in a much more precise way. Let $W: X \mapsto \mathbb{R}$ be a concave function. The subdifferential of W at z is defined as $\partial W(z) = \{p \in \mathbb{R}^n | W(y) \leqslant W(z) + p(y - z) \text{ for all } y \in X\}$. Elements of the subdifferential are called subgradients. The notion of a subgradient is one of several possible generalizations of the notion of a gradient a function which is not necessarily differentiable. If W is differentiable at z then $\partial W(z)$ contains a single element, namely the gradient. If W is not differentiable at z then $\partial W(z)$ may be empty or it may contain more than one subgradient. If z is in the interior of the domain X, then $\partial W(z)$ is necessarily non-empty. The subdifferential of a concave function may be empty at boundary points of its domain.

Let $g: X \mapsto X$ and $W: X \mapsto \mathbb{R}$ be any pair of function such that W is concave and consider the following condition $R(\rho, X; g, W)$.

$R(\rho, X; g, W)$: For every $x \in X$ such that the subdifferential $\partial W(x)$ is non-empty there exist subgradients $p_x \in \partial W(x)$ and $q_x \in \partial W(g(x))$ such that the inequality

$$W(g(x)) - W(g(y)) + q_x[g(y) - g(x)] \leqslant (1/\rho)[W(x) - W(y) + p_x(y - x)]$$

holds for all $y \in X$.

The following result is due to Mitra and Sorger (1999b).

Theorem 10. *Let $h: X \mapsto X$ and $V: X \mapsto \mathbb{R}$ be two given functions.*

(a) *If there exists a dynamic optimization problem (Ω, U, ρ) on X such that assumptions A5–A8 hold and such that h is the optimal policy function and V the optimal value function, then the following is true: h and V are continuous, V is strictly concave, and condition $R(\rho, X; h, V)$ holds.*

(b) *Assume that h and V are continuous, that V is strictly concave, and that there exists an open and convex set X^* containing X and a concave function $V^*: X^* \mapsto R$ which coincides with V on X such that $R(\rho, X; h, V^*)$ holds. Then there exists a dynamic optimization problem (Ω, U, ρ) on X such that assumptions A5–A8 hold and such that h is the optimal policy function and V the optimal value function. If, in addition, V is non-decreasing then (Ω, U, ρ) can be chosen such that A10 holds.*

Part (a) of this proposition states a condition that is necessary for the pair (h, V) to be the optimal solution of a dynamic optimization problem, whereas part (b) provides a sufficient condition. The only difference between these two conditions is that the sufficient condition requires that V can be extended as a concave function to some open set containing the state space X. Theorem 10 therefore provides an extremely tight characterization of the set of all pairs (h, V) which can arise in models satisfying A5–A8. The remaining gap between the necessary and the sufficient condition can be closed completely if one restricts the class of optimal growth models to those satisfying certain monotonicity and free disposal assumptions; see Mitra and Sorger (1999b) for details. As has already been pointed out before, Theorem 10 has several interesting consequences. The result stated in Theorem 7(b) is one of them, and others will be presented in Sect. 2.3.4 below.

2.3.3 The Influence of Time-Preference

This subsection deals with the influence of the size of the discount factor ρ on the dynamics of optimal growth paths. There are two classes of results: turnpike theorems and minimum impatience theorems. Turnpike theory was one of the most active research areas in economic growth during the sixties and seventies and has been surveyed, for example, by McKenzie (1986). We shall therefore only state one result from this literature, just to explain the idea of a turnpike theorem. Minimum impatience theorems, on the other hand, have been derived only recently.[9] We shall discuss two such theorems, namely those which we consider as the most powerful derived so far.

Essentially, a turnpike theorem proves that, under certain conditions, optimal growth paths stay close or even converge to a unique stationary optimal growth path, which is called the turnpike. An essential assumption of most turnpike theorems is that the discount factor ρ is sufficiently close to 1.[10] Thus, turnpike theorems prove that optimal growth paths are stable and highly predictable in the long-run if the decision maker is sufficiently patient. The following result is a version of the turnpike theorem by Scheinkman (1976).

Theorem 11. *Let Ω and U be given such that A5 and A6 are satisfied and assume that U is twice continuously differentiable on the interior of Ω. Assume that there exists $\overline{x} \in X$ such that $(\overline{x}, \overline{x})$ is in the interior of Ω, $\overline{x}$ maximizes the function $x \mapsto U(x, x)$ subject to the constraint $(x, x) \in \Omega$, and from any initial state in X there exists a feasible path that reaches $\overline{x}$ in finite time. Furthermore assume that the Hessian matrix*

[9] The first minimum impatience theorems were presented in Sorger (1992a,b).

[10] There are also a few turnpike theorems that hold independently of the size of the discount factor, e.g. Araujo and Scheinkman (1977). These results, however, depend on quite strong structural assumptions and will not be discussed in this paper.

$$\begin{pmatrix} U_{11}(\overline{x},\overline{x}) & U_{12}(\overline{x},\overline{x}) \\ U_{21}(\overline{x},\overline{x}) & U_{22}(\overline{x},\overline{x}) \end{pmatrix}$$

is negative definite and that the matrix $U_{12}(\overline{x},\overline{x})$ is non-singular. Then there exists $\overline{\rho} \in (0,1)$ such that the following is true for all $\rho \in (\overline{\rho},1)$.

(a) There exists a constant optimal growth path $(\overline{x}^{\rho},\overline{x}^{\rho},\overline{x}^{\rho},\ldots)$ for the model (Ω,U,ρ).
(b) If $(x_t)_{t=0}^{+\infty}$ is an optimal growth path for problem (Ω,U,ρ) then $\lim_{t\to+\infty} x_t = \overline{x}^{\rho}$.

It is worth emphasizing that the critical value $\overline{\rho}$ in Theorem 11 depends in general on Ω and U. Thus, the claim that a low time-preference rate of the decision maker implies simple (i.e., asymptotically stable) behaviour of optimal paths does not hold uniformly but only for given preferences and technology.

Let us now turn to minimum impatience theorems. They are similar to turnpike theorems because they state that, in order for complicated dynamics to occur in an optimal growth model, the decision maker has to have a high time-preference rate. Minimum impatience theorems, however, are different from turnpike theorems because they do not make this statement for given fundamentals Ω and U but for a given optimal policy function, or a given class of optimal policy functions. More precisely, they state that, if a function $h\colon X \mapsto X$ has certain characteristics (usually related to the complexity of the dynamics it generates), then h can be an optimal policy function of a model (Ω,U,ρ) satisfying A5–A8 only if ρ is sufficiently close to 0. We shall now discuss two such results, one dealing with Li-Yorke chaos on one-dimensional state spaces and the other one dealing with dynamics with positive topological entropy on general n-dimensional state spaces. Other examples of minimum impatience theorems for specific functions h will be presented in Sect. 2.3.4 below.

Consider a continuous function $h\colon X \mapsto X$, where $X \subseteq \mathbb{R}$ is an interval. The dynamical system $x_{t+1} = h(x_t)$ is said to exhibit Li-Yorke chaos if there exists a periodic point of period 3. It is known that a one-dimensional dynamical system having this property exhibits topological chaos (see Li and Yorke 1975). More precisely there exists a periodic orbit for any period $p \geqslant 1$ as well as an uncountable scrambled set $S \subseteq X$. The importance of the existence of a point of period 3 is further demonstrated by Sarkovskii's theorem which shows that a one-dimensional dynamical system having a periodic point of period 3 has periodic points of all periods (see Sarkovskii 1964 or Block and Coppel 1992). The following result was discovered independently by Mitra (1996a) and Nishimura and Yano (1996a).

Theorem 12. *Let $h\colon X \mapsto X$ be the optimal policy function of an optimal growth model (Ω,U,ρ) satisfying A5–A8 and assume that the state space X is an interval on the real line. If h exhibits Li-Yorke chaos then $\rho < \left(\sqrt{5}-1\right)^2/4 \approx 0.382$. Conversely, if $\rho < \left(\sqrt{5}-1\right)^2/4$ then one can find an optimal growth model*

(Ω, U, ρ) satisfying A5–A8 and A10 such that the optimal policy function of this model exhibits Li-Yorke chaos.

Note that a discount factor equal to 0.382 implies a discount rate approximately equal to 160%. A different way to understand the message of Theorem 12 is as follows: if the time-preference rate is 5% per year, then a discount factor of 0.382 implies that the period length of the model is roughly 20 years.

We now turn to a minimum impatience theorem that deals with the notion of topological entropy of a dynamical system. Let $h\colon X \mapsto X$ be a continuous map and $A \subseteq X$ a given compact and invariant subset of the state space.[11] The topological entropy $\kappa(h, A)$ of h on A measures the tare at which different trajectories of h become distinguishable by observations with finite precision as the number of observations increases. Positive topological entropy is often taken as a definition of complicated dynamics. To formally define $\kappa(h, A)$ we need a few preliminary definitions. Let T be a positive integer and $\varepsilon > 0$. We say that a subset $B \subseteq X$ is (T, ε)-separated if for any two different points x and y in B there exists $t \in \{0, 1, 2, \ldots, T-1\}$ such that $\left\|h^{(t)}(x) - h^{(t)}(y)\right\| > \varepsilon$.[12] Now assume that A is a compact and invariant subset of X. In that case the number

$$s_{T,\varepsilon}(h, A) = \max\{\#B | B \subseteq A \text{ and } B \text{ is } (T, \varepsilon) - \text{separated}\}$$

is well defined and finite. Here, $\#B$ denotes the cardinality of B. We call

$$c_{+}(A) = \lim_{\varepsilon \to 0} \sup \frac{\ln\ s_{1,\varepsilon}(h, A)}{-\ln\ \varepsilon}$$

the upper capacity of the set A (see also Brock and Dechert 1991). It measures the growth rate of the number of ε-balls which are required to cover A as ε approaches 0. It is clear that $c_{+}(A) \leqslant n$ must hold for every compact set $A \subseteq \mathbb{R}^n$. The topological entropy of h on A is defined as

$$\kappa(h, A) = \lim_{\varepsilon \to 0} \lim_{T \to +\infty} \sup \frac{\ln\ s_{T,\varepsilon}(h, A)}{T}.$$

The positivity of topological entropy is intimately related to the sensitive dependence on initial conditions and, consequently, to the poor accuracy of long term predictions if measurements are not exact. High entropy corresponds to very sensitive dependence on initial conditions and, therefore, to the impossibility of making good long term predictions if initial states are not exactly known. For dynamical systems defined on a one-dimensional state space X there is a close

[11] A subset A of the state space X is called invariant (under h) if for every $x \in A$ it holds that $h(z) \in A$. The reader should not confuse this concept with the invariance of a measure μ under h, discussed in Sect. 2.2.3.3 above.

[12] As before $h^{(t)}(x)$ denotes the t-th iterate of h evaluated at $x \in X$.

relation between the topological entropy and the presence of certain periodic points. To be precise, if $\kappa(h,X)=0$ then only periodic points of period $p=2^i$ can occur whereas if $\kappa(h,X)>0$ then h must have a periodic point with a period different from a power of 2 (see, e.g., Alseda et al. 1993). This shows in particular that dynamical systems exhibiting Li-Yorke chaos must have positive topological entropy. The following result was proved by Montrucchio and Sorger (1996) (see Montrucchio 1994 for an earlier version).

Theorem 13. *Let (Ω, U, ρ) be an optimal growth model satisfying A5–A8 and let $h: X \mapsto X$ be its optimal policy function. Assume that A is a compact subset of X which is contained in the interior of the state space X and which is invariant under h. Then it holds that $\kappa(h,A) \leqslant -(\ln \rho)c_+(A)$.*

This result shows that high topological entropy of the optimal policy function is only possible if ρ is close to 0. Because the upper capacity satisfies $c_+(A) \leqslant n$, it follows from Theorem 13 that $\kappa(h,A) \leqslant -n(\ln \rho)$. We would like to emphasize that Theorem 13 does not rule out that complicated dynamics (that is, dynamics with a positive topological entropy) can occur in optimal growth models with arbitrarily low time-preference rates. As a matter of fact, if one considers a sequence of functions $h_k: X \mapsto X$ such that $\kappa(h_k, X) > 0$ for all k and $\lim_{k\to+\infty} \kappa(h_k, X) = 0$, then it may be possible to construct a corresponding sequent of optimal growth models (Ω_k, U_k, ρ_k) satisfying A5–A8 such that h_k is the optimal policy function of (Ω_k, U_k, ρ_k) and such that $\lim_{k\to+\infty} \rho_k = 1$. Such a construction has been performed in Nishimura et al. (1994) for reduced form models; see also Sect. 2.2.3.3 for a discussion of the construction in two-sector optimal growth models.

2.3.4 Examples

This subsection presents several examples of optimal growth models with cyclical or chaotic solutions. Since one can use the constructive proof of Theorem 7(b) (or its weaker version presented by Boldrin and Montrucchio (1986)) to find optimal growth models with these features, we include only examples that are interesting from a historical point of view or examples that are extremely well studied.

One of the earliest example of optimal cycles in a reduced form optimal growth model is due to Sutherland (1970). In this example the state space is $X = [0,1]$, the transition possibility set is $\Omega = X \times X$, and the reduced utility function is

$$U(x,y) = -9x^2 - 11xy - 4y^2 + 43x.$$

There exists a unique optimal steady state $\overline{x}^\rho$ for each discount factor $\rho \in (0,1)$. For $\rho = 1/3$ it has been shown by Sutherland (1970) that $\overline{x}^\rho = 1/2$, that $\overline{x}^\rho$ is dynamically unstable, and that the periodic path $(0,1,0,1,0,\ldots)$ is optimal.

Subsequently, Samuelson (1973) reported an example due to Weitzman. It was later generalized by McKenzie (1983) and Benhabib and Nishimura (1985). The state space and the transition possibility set are again given by $X = [0, 1]$ and $\Omega = X \times X$ but the utility function is of the Cobb-Douglas type

$$U(x, y) = x^{\alpha}(1 - y)^{\beta},$$

where $\alpha > 0$, $\beta > 0$, and $0 < \alpha + \beta \leqslant 1$. Samuelson (1973) considers the case $\alpha = \beta = 1/2$, McKenzie (1983) studies the linearly homogeneous case $\alpha + \beta = 1$, and Benhabib and Nishimura (1985) analyses the general case. The optimal steady state is $\overline{x}^{\rho} = \rho\alpha/(\rho\alpha + \beta)$. If $\alpha \in (1/2, 1)$ and $\beta \in (0, 1/2)$ then the steady state is stable for $\rho \in (\rho_0, 1)$ and unstable for $\rho \in (0, \rho_0)$ where $\rho_0 = \beta(2\alpha - 1)/[\alpha(1 - 2\beta)]$. Under the same restrictions on α and β, Benhabib and Nishimura (1985) also prove the existence of optimal period-two cycles in this model.

In the above examples the optimal policy functions have negative slopes at the steady states. This may be checked by solving the characteristic equations of the linearized Euler equations at the steady state. A result characterizing the global monotonicity properties of the optimal policy function h in models with a one-dimensional state space is the following theorem due to Benhabib and Nishimura (1985).

Theorem 14. *Let (Ω, U, ρ) be an optimal growth model satisfying A5–A8. Assume that the state space is an interval on the real line and that U is twice continuously differentiable. Let $h: X \mapsto X$ be the optimal policy function of (Ω, U, ρ).*

(a) Assume $U_{12}(x, y) > 0$ for all (x, y) in the interior of Ω. If $(z, h(z))$ is in the interior of Ω then $h(z)$ is strictly increasing at $x = z$.
(b) Assume $U_{12}(x, y) < 0$ for all (x, y) in the interior of Ω. If $(z, h(z))$ is in the interior of Ω then $h(x)$ is strictly decreasing at $x = z$.

The cross-partial derivatives of U are negative in the above examples. Thus, the optimal policy functions must be strictly decreasing by Theorem 14(b), and either no cycles at all or period-two cycles arise in these examples.

If one modifies the Weitzman example by considering the utility function

$$U(x, y) = (x - ay)^{\alpha}(1 - y)^{\beta},$$

then more complicated dynamics may arise. This was suggested by Scheinkman (1984) and proved by Boldrin and Deneckere (1990). This leads us to examples of optimal growth models with chaotic solutions. The first to prove the possibility of chaotic optimal growth paths were Deneckere and Pelikan (1986) and Boldrin and Montrucchio (1986). In both of these papers a model with polynomial utility function was presented for which the optimal policy function is the so-called logistic map $h: [0, 1] \mapsto [0, 1]$ defined by $h(x) = 4x(1 - x)$. This is perhaps the best-known example of a chaotic dynamical system. It exhibits Li-Yorke chaos and its topological entropy is $\ln 2 > 0$. Deneckere and Pelikan (1986) used the utility function

$$U(x, y) = xy - x^2y - (1/3)y - (3/40)y^2 + (100/3)x - 7x^2 + 4x^3 - 2x^4$$

and the discount factor $\rho = 1/100$. Boldrin and Montrucchio (1986) found another example with a discount factor of about the same size. Note that $\rho = 1/100$ corresponds to 9,900% discount rate. The following result shows that the logistic map cannot be an optimal policy function in any model satisfying A5–A10 with a discount factor $\rho > 1/16$. Moreover, it is shown that this is a tight discount factor restriction.

Theorem 15. *Let $h: X \mapsto X$ be defined by $h(x) = 4x(1-x)$, where $X = [0, 1]$. Moreover, let $\rho \in (0, 1)$. The following two conditions are equivalent:*

(a) There exists a transition possibility set Ω and a utility function U such that the model (Ω, U, ρ) satisfies A5–A10 and such that h is the optimal policy function of (Ω, U, ρ).
(b) The discount factor satisfies $\rho \in (0, 1/16]$.

Note that the logistic map is Lipschitz continuous with Lipschitz constant $L = 1/4$. The implication (b) ⇒ (a) is therefore an immediate corollary of Theorem 7(b). The converse implication was shown by Montrucchio (1994) and can also be derived as a non-trivial implication of Theorem 10 (see Mitra and Sorger 1999a).

Another prominent example of chaotic dynamics is the tent map $h(x) = 1 - |2x - 1|$ defined on $X = [0, 1]$. This function has properties very similar to the logistic map: it exhibits Li-Yorke chaos, its topological entropy is $\ln 2 > 0$, and it is topologically equivalent to the logistic map.[13] The tent map, however, can arise as an optimal policy function in models with larger discount factors as shown by the following result (taken from Mitra and Sorger 1999a).

Theorem 16. *Let $h: X \mapsto X$ be defined by $h(x) = 1 - |2x - 1|$, where $X = [0, 1]$. Moreover, let $\rho \in (0, 1)$. The following two conditions are equivalent:*

(a) There exists a transition possibility set Ω and a utility function U such that the model (Ω, U, ρ) satisfies A5–A8 and such that h is the optimal policy function of (Ω, U, ρ).
(b) The discount factor satisfies $\rho \in (0, 1/4]$.

2.4 External Effects and Indeterminacy

So far we have considered models without any externalities. The presence of external effects may have a substantial influence on the dynamic properties of equilibrium paths. Most notably, it may render equilibrium paths indeterminate.

[13] Two maps $h_1 : X_1 \mapsto X_1$ and $h_2 : X_2 \mapsto X_2$ are called topologically equivalent if there exists a homeomorphism $f : X_1 \mapsto X_2$ such that $f(h_1(x)) = h_2(f(x))$ for all $x \in X_1$.

By indeterminacy we understand the existence of a continuum of equilibrium paths sharing a common initial condition. Thus, indeterminacy is a particularly strong form of non-uniqueness. This section first introduces external effects and indeterminacy by a simple example of a one-sector economy and then discusses a number of results from the literature concerning indeterminacy.

Consider the aggregative optimal growth model discussed in Sect. 2.2.1 above but assume that there is a production externality implying that output of each firm depends on its own capital stock and on the average capital stock of all firms. Formally, we suppose that period-t output of the representative firm is given by $f(x_t, \overline{x}_t)$, where x_t denotes its own capital stock and $\overline{x}_t$ is the average capital stock. Given any sequence $(\overline{x}_t)_{t=0}^{+\infty}$ an equilibrium relative to $(\overline{x}_t)_{t=0}^{+\infty}$ is a solution to the following dynamic optimization problem:

$$\max \sum_{t=0}^{+\infty} \rho^t u(c_t) \tag{2.31}$$

subject to

$$c_t + x_{t+1} \leqslant f(x_t, \overline{x}_t), \qquad c_t \geqslant 0, \qquad x_0 = x. \tag{2.32}$$

Maximization is carried out with respect to sequences of consumption $(c_t)_{t=0}^{+\infty}$ and capital $(x_t)_{t=0}^{+\infty}$ taking $(\overline{x}_t)_{t=0}^{+\infty}$ as given. Since x_t is the capital stock of the representative firm and $\overline{x}_t$ is the average capital stock, in equilibrium it must hold that $x_t = \overline{x}_t$ for all t. Defining the reduced utility function $U(x, y; \overline{x}) = u(f(x, \overline{x}) - y)$ and the transition possibility set $\Omega(\overline{x}) = \{(x, y) | x \geqslant 0, 0 \leqslant y \leqslant f(x, \overline{x})\}$ one can rewrite the optimization problem (2.31)–(2.32) in reduced form as

$$\max \sum_{t=0}^{+\infty} \rho^t U(x_t, x_{t+1}; \overline{x}_t)$$

$$\text{subject to } (x_t, x_{t+1}) \in \Omega(\overline{x}_t), \qquad x_0 = x.$$

Writing down the optimality conditions for an interior solution and using the externality condition $x_t = \overline{x}_t$ one obtains

$$U_2(x_t, x_{t+1}; x_t) + \rho U_1(x_{t+1}, x_{t+2}; x_{t+1}) = 0, \tag{2.33}$$

$$\lim_{t \to +\infty} \rho^t U_1(x_t, x_{t+1}; x_t) x_t = 0. \tag{2.34}$$

Let us start by assuming that the Euler equation (2.33) has a stationary solution x^* and that the initial state $x = x_0$ coincides with x^*. Since (2.33) is a second-order equation, different choices of x_1 lead to different trajectories of the Euler equation. If x^* is a locally asymptotically stable solution of (2.33), that is, if the linearization of (2.33) around x^* has two roots inside the unit circle, then it follows that every value x_1 sufficiently close to x^* generates a trajectory that converges to x^*. It is obvious that the transversality condition (2.34) is satisfied along all of these trajectories and one can conclude that there exists a continuum of equilibrium paths. The characteristic equation of the linearization of (2.33) around x^* is

$$\lambda^2 + \lambda\left(\frac{U_{22}^*}{\rho U_{12}^*} + \frac{U_{11}^* + U_{13}^*}{U_{12}^*}\right) + \frac{1}{\rho} + \frac{U_{23}^*}{\rho U_{12}^*} = 0,$$

where U_{ij}^* denotes $U_{ij}(x^*, x^*; x^*)$. A necessary and sufficient condition for both roots of this equation to be inside the unit circle is that the following three inequalities hold:

$$\begin{aligned} \frac{1}{\rho} + \frac{U_{23}^*}{\rho U_{12}^*} &< 1, \\ \frac{1+\rho}{\rho} + \frac{U_{23}^* + U_{22}^*}{\rho U_{12}^*} + \frac{U_{11}^* + U_{13}^*}{U_{12}^*} &> 0, \\ \frac{1+\rho}{\rho} + \frac{U_{23}^* - U_{22}^*}{\rho U_{12}^*} - \frac{U_{11}^* + U_{13}^*}{U_{12}^*} &> 0. \end{aligned} \tag{2.35}$$

If there is no external effect, then $U_{13}^* = U_{23}^* = 0$ and it follows that the first inequality cannot be satisfied because $\rho \in (0, 1)$. Thus, externalities are necessary to generate the kind of indeterminacy we have described above. If one allows for externalities, then the conditions in (2.35) can easily be satisfied. For example, if $\rho = 3/4$, $U_1^* = 8/3$, $U_2^* = -2$, $U_{11}^* = -34/9$, $U_{12}^* = 4/3$, $U_{13}^* = 16/3$, $U_{22}^* = -1$, and $U_{23}^* = -1$ then the Euler equation (2.33) as well as the stability conditions in (2.35) hold. Primitives u and f which generate a reduced utility function with these properties are, for example, all functions satisfying $u'(1) = 2$, $u''(1) = -1$, $f(1, 1) = 2$, $f_1(1, 1) = 4/3$, $f_2(1, 1) = -1$, $f_{11}(1, 1) = -1$, and $f_{12}(1, 1) = 2$.[14] In this example, which is due to Kehoe (1991), the external effect is negative because $f_2(1, 1) < 0$. Boldrin and Rustichini (1994) show that in the case of a positive externality (that is, if $f_2(x^*, x^*) \geqslant 0$), any stationary solution x^* is locally unique such that indeterminacy cannot occur.

The production externality in the above example is modelled by assuming that the average capital stock influences output. This is the most commonly assumed form of an external effect in models of capital accumulation (especially in the literature on endogenous growth). Examples of indeterminacy in the case of slightly different production externalities have been constructed by Spear (1991) and Kehoe et al. (1991). The former paper assumes that aggregate savings of all agents affect the production of output. This implies that tomorrow's aggregate capital stock appears as an argument of the production function for today's output. Kehoe et al. (1991), on the other hand, assumed that average consumption influences output.

[14] In the case, $x^* = 1$ is the steady state value.

Models with positive external effects of the kind discussed above have been used extensively in the endogenous growth literature, because they can help to explain positive long-run growth of per-capita output. Equilibria along which per-capita output, consumption, and capital are growing at constant rates are termed balanced growth paths. Boldrin and Rustichini (1994) have shown that balanced growth paths are determinate in a one-sector model if certain reasonable assumptions on the production function and the utility function hold. If negative externalities are allowed, then again indeterminacy can arise.

Although the discussion so far has focussed on one-sector models, the issue of indeterminacy is even more relevant for multi-sector models of capital accumulation. Boldrin and Rustichini (1994) prove that indeterminate equilibria may exist in two-sector models under standard assumptions on preferences and technology even in the case of positive externalities. They deal with indeterminacy of stationary solutions and balanced growth paths. The main reason why indeterminacy occurs in their model is that the social production function exhibits increasing returns to scale. Benhabib et al. (1999) and Nishimura and Venditti (1999) show that, in a multi-sector optimal growth model with positive externalities, indeterminacy may even arise under constant or decreasing returns to scale. The result on constant returns to scale technologies is a discrete-time analogue of a corresponding result for continuous-time models by Benhabib and Nishimura (1998a) (which we will discuss in Sect. 2.5 below).

Externalities may also be cause of chaotic dynamics. Boldrin et al. (2001) derived a set of conditions under which equilibrium paths in an endogenous growth model are indeterminate and behave chaotically.

Indeterminate equilibria have been found in a variety of other models, too. Models with overlapping generations (Kehoe and Levine 1985), models with finance constraints (Woodford 1986 or Sorger 1994a), and money-in-the-utility-function models (Matsuyama 1990) are just a few examples. We shall further discuss indeterminacy in the following section using a continuous-time framework.

2.5 Continuous-Time Models

So far we have concentrated on models which are formulated in discrete time. Some of the results that we have presented do only hold in this framework whereas others have continuous time counterparts. In this section we briefly discuss some of the work that has been done on infinite horizon models in continuous time. In one-sector growth models optimal paths converge to an unique steady state (see Cass 1965; Koopmans 1965). This result holds in the continuous-time model as well as the discrete-time model. In the two-sector growth model with a unique steady state, optimal paths converge to the steady state in the continuous time framework, while optimal paths may be chaotic in the discrete-time framework. This global convergence in the continuous-time frame work was proved by Shrinivasan (1964) and Uzawa (1964).

The main obstacle in constructing continuous-time economic models which generate complicated dynamics is the necessity of a high-dimensional state space. Whereas cycles of arbitrary length and chaos are possible for one-dimensional difference equations, they are not possible in one-dimensional differential equations. As a matter of fact, a differential equation defined on a one-dimensional state space can only have monotonic solutions (at least if solutions are unique). For cycles to be possible a dimension of at least two is required, and for chaotic dynamics one needs at least a three-dimensional system.

Another difficulty is the lack of simple sufficient conditions for the existence of chaotic dynamics. Existence theorems for periodic solutions, on the other hand, are easier to verify. In particular one can apply the Hopf bifurcation theorem, which basically only requires that a pair of complex conjugate eigenvalues crosses the imaginary axis as one of the model parameters is varied. The first application of this theorem in the context of an infinite horizon optimal growth model was presented in Benhabib and Nishimura (1979). In the framework of multi-sector capital accumulation models they studied general conditions for a Hopf bifurcation to occur. We shall present the multi-sector model with Cobb-Douglas technologies to illustrate the source of complex dynamics.

2.5.1 Basic Framework

The material presented in this subsection is taken from Benhabib and Nishimura (1979). We consider an economy with one consumption good sector and n capital good sectors. The production inputs are the n capital goods as well as labour. The total supply of labour in the economy at time t, $x_0(t)$, is constant and normalized to 1, that is $x_0(t) = 1$ for all t. A representative agent optimizes an additively separable utility function with discount rate $\rho > 0$. The objective functional is therefore

$$\int_0^{+\infty} e^{-\rho t} u(y_0(t))dt, \tag{2.36}$$

where u is a twice continuously differentiable, concave, and increasing instantaneous utility function, and $y_0(t)$ denotes the consumption rate at time t. The constraints of the problem are

$$y_j(t) = e_j \prod_{i=0}^{n} x_{ij}(t)^{\beta_{ij}} \quad j = 0, 1, \ldots, n, \tag{2.37}$$

$$\frac{dx_i(t)}{dt} = y_i(t) - \delta x_i(t) \quad i = 1, 2, \ldots, n, \tag{2.38}$$

$$\sum_{j=0}^{n} x_{ij}(t) = x_i(t) \quad i = 0, 1, \ldots, 2n. \tag{2.39}$$

Here, $x_i(t)$ is the stock of the i-th capital good if $i \in \{1, 2, \ldots, n\}$, and $x_0(t) = 1$ is the labour supply. Moreover, $x_{ij}(t)$ is the allocation of input factor i to the production of the j-th good for all $j = 0, 1, \ldots, n$ and all $i = 0, 1, \ldots, n$. The constant $\delta > 0$ is the depreciation rate. To begin with, e_j is assumed to be constant for all j. In a later subsection we shall introduce a production externality by assuming that e_j depends on the variables $x_{ij}(t)$. The parameters β_{ij} are assumed to be non-negative with $\sum_{i=0}^{n} \beta_{ij} = 1$ such that all production functions have constant returns to scale.

Equations 2.38 describe the accumulation of the n capital goods $x_i(t)$. The initial values of the stocks at time 0, $x_i(0)$, are given. The optimization is with respect to the inputs $x_{ij}(t)$ for all $i, j \in \{0, 1, \ldots, n\}$ and all $t \geqslant 0$. The Hamiltonian function associated with problem (2.36) is

$$H = u\left(e_0 \prod_{i=0}^{n} x_{i0}^{\beta_{i0}}\right) + \sum_{j=1}^{n} p_j \left(e_j \prod_{i=0}^{n} x_{ij}^{\beta_{ij}} - \delta x_j\right) + \sum_{i=0}^{n} w_i \left(x_i - \sum_{j=0}^{n} x_{ij}\right).$$

Here p_j and w_i are co-state variables and Lagrange multipliers, representing the utility prices of the capital goods and their rentals, respectively. The first order conditions for the maximization of the Hamiltonian function are

$$\begin{aligned} w_s(t) &= p_j(t)\beta_{sj} e_j x_{sj}(t)^{-1} \prod_{i=0}^{n} x_{ij}(t)^{\beta_{ij}} \\ &= u'\left(y_0(t)\right) \beta_{s0} e_0 x_{s0}(t)^{-1} \prod_{i=0}^{n} x_{i0}(t)^{\beta_{i0}} \end{aligned} \tag{2.40}$$

for $j = 1, 2, \ldots, n$, all $s = 0, 1, \ldots, n$, and all $t > 0$.

From now on we shall assume that the utility function is linear, i.e., $u(y_0) = y_0$. It can be shown that under this assumption and under a constant returns Cobb-Douglas technology, the static efficiency conditions given by (2.40) imply that factor rentals and outputs are uniquely determined by output prices, and that outputs can be expressed as a function of aggregate stocks and prices. Therefore, taking the consumption good as the numeraire, we can express factor rentals as $w_i(p)$ and outputs as $y_i(x, p)$, where $x = (x_1, x_2, \ldots, x_n)$ and $p = (p_1, p_2, \ldots, p_n)$. The necessary conditions for the optimal solution of problem (2.36) are given by the following equations of motion:

$$\frac{dx_i(t)}{dt} = \partial H / \partial p_i(t) = y_i(x(t), p(t)) - \delta x_i(t) \quad i = 1, 2, \ldots, n, \tag{2.41}$$

$$\frac{dp_i(t)}{dt} = \rho p_i(t) - \partial H / \partial x_i(t) = (\rho + \delta) p_i(t) - w_i(p(t)) \quad i = 1, 2, \ldots, n. \tag{2.42}$$

It is straightforward to show that under our assumptions the system (2.41)–(2.42) has a unique steady state (x^*, p^*). Linearizing around the steady state we obtain

$$\begin{bmatrix} dx(t)/dt \\ dp(t)/dt \end{bmatrix} = \begin{bmatrix} \left[\frac{\partial y(x^*,p^*)}{\partial x}\right] - \delta I & \left[\frac{\partial y(x^*,p^*)}{\partial p}\right] \\ [0] & -\left[\frac{\partial w(p^*)}{\partial p}\right] + (\rho+\delta)I \end{bmatrix} \begin{bmatrix} x(t) - x^* \\ p(t) - p^* \end{bmatrix}$$
$$= J \begin{bmatrix} x(t) - x^* \\ p(t) - p^* \end{bmatrix}. \tag{2.43}$$

We note that the matrix J is quasi-triangular, so that its roots are the roots of

$$\left[\frac{\partial y(x^*, p^*)}{\partial x}\right] - \delta I \quad \text{and} \quad -\left[\frac{\partial w(p^*)}{\partial p}\right] + (\rho + \delta)I.$$

Let B be the $n \times n$ matrix defined by

$$B = \left[\beta_{ij} - \beta_{0j}\beta_{i0}/\beta_{00}\right]$$

where the constants β_{ij} are the powers from the Cobb-Douglas production functions. Moreover, let W denote the $n \times n$ diagonal matrix with diagonal elements $w_i(p^*)$, $i = 1, 2, \ldots, n$. Similarly let P denote the $n \times n$ diagonal matrix with diagonal elements p_i^*, $i = 1, 2, \ldots, n$. One can show that

$$\left[\frac{\partial y(x^*, p^*)}{\partial x}\right] = P^{-1}B^{-1}W \quad \text{and} \quad \left[\frac{\partial w(p^*)}{\partial p}\right] = W(B')^{-1}P^{-1}.$$

Furthermore, one can show that the roots of the matrices B and

$$\left[\frac{\partial y(x^*, p^*)}{\partial x}\right]$$

have the same sign structure. It follows from this property that the roots of J come in pairs of the form $(\mu_i - \delta, -\mu_i + \rho + \delta)$.

2.5.2 *The Hopf Bifurcation*

In a three sector model with Cobb-Douglas production functions the steady state can be totally unstable and limit cycles may occur. The intuition for this result is as follows. From the form of the eigenvalues of J mentioned at the end of the previous subsection it follows that the steady state of system (2.43) is necessarily saddle point stable if $\rho = 0$. If ρ increases, then the steady state may lose the saddle point stability and become totally unstable. In Benhabib and Nishimura (1979) the three sector model with depreciation rate $\delta = 0.1$ and the following parameters are studied:

$$\begin{aligned} e_0 &= 10.2425 \;\; \beta_{00} = 0.9524 \;\; \beta_{01} = 0.9723 \;\; \beta_{02} = 0.3941 \\ e_1 &= 1.5084 \;\; \beta_{10} = 0.0017 \;\; \beta_{11} = 0.0265 \;\; \beta_{12} = 0.5635 \\ e_2 &= 3.0266 \;\; \beta_{20} = 0.0459 \;\; \beta_{21} = 0.0012 \;\; \beta_{22} = 0.0423 \end{aligned}$$

The Jacobian at the steady state of this system has four complex roots when $\rho = 0.148$. The real parts of two of them change sign from negative to positive and the real parts of the other two roots remain positive when ρ crosses the value 0.148 from below. Thus, a Hopf bifurcation takes place at $\rho = 0.148$ and periodic optimal paths exist for discount rates close to 0.148.

2.5.3 A Multi-Sector Model with Externalities

Recently there has been a renewed interest in the possibility of indeterminacy and sunspots or, put differently, in the existence of a continuum of equilibria in dynamic economies with some market imperfections. Much of the research in this area has been concerned with the empirical plausibility of the conditions that lead to indeterminacy in economies with external effects or monopolistic competition, in which the sectoral production functions exhibit some degree of increasing returns. While the early results on indeterminacy relied on relatively large increasing returns and high markups, more recently (Benhabib and Farmer, 1996) showed that indeterminacy can also occur in two-sector models with small sector-specific external effects and very mild increasing returns. Benhabib and Nishimura (1998a) demonstrated that indeterminacy can occur in a standard growth model with constant social returns, decreasing private returns, and small or negligible external effects. Furthermore Benhabib and Nishimura (1998b,c) provide a systematic method of treating multi-sector models with externalities.

In this subsection we consider the case that the coefficients e_j of the production functions are given by

$$e_j(t) = \prod_{i=0}^{n} x_{ij}(t)^{b_{ij}}.$$

Therefore, the social production functions are

$$y_j(t) = \prod_{i=0}^{n} x_{ij}(t)^{\beta_{ij}+b_{ij}} \qquad j = 0, 1, \ldots, n.$$

The representative agent, however, is not aware of the dependence of $e_j(t)$ on his choice variables. Instead, he treats $e_j(t)$ as an exogenous function of time. Thus, there is an external effect. The static first order conditions for the representative agent's problem are therefore given by (compare with (2.40))

$$w_s(t) = p_j(t)\beta_{sj}x_{sj}(t)^{-1}\prod_{i=0}^{n} x_{ij}(t)^{\beta_{ij}+b_{ij}}$$

$$= u'(y_0(t))\,\beta_{s0}x_{s0}(t)^{-1}\prod_{i=0}^{n} x_{i0}(t)^{\beta_{i0}+b_{i0}}$$

for $j = 1, 2, \ldots, n$, all $s = 0, 1, \ldots, n$, and all $t \geqslant 0$. We assume again that the utility function is given by $u(y_0) = y_0$. Following exactly the same procedure as in the case without externalities, one can derive the equations of motion from the necessary optimality conditions and linearize them around the steady state. This yields again

$$\begin{bmatrix} dx(t)/dt \\ dp(t)/dt \end{bmatrix} = \begin{bmatrix} \left[\frac{\partial y(x^*,p^*)}{\partial x}\right] - \delta I & \left[\frac{\partial y(x^*,p^*)}{\partial p}\right] \\ [0] & -\left[\frac{\partial w(p^*)}{\partial p}\right] + (\rho+\delta)I \end{bmatrix} \begin{bmatrix} x(t) - x^* \\ p(t) - p^* \end{bmatrix}$$

$$= J \begin{bmatrix} x(t) - x^* \\ p(t) - p^* \end{bmatrix}.$$

In the case of externalities, however, the submatrices of J are

$$\left[\frac{\partial y(x^*, p^*)}{\partial x}\right] = [a_{ij} - a_{0j}a_{i0}/a_{00}]^{-1} \quad \text{and} \quad \left[\frac{\partial w(p^*)}{\partial p}\right] = [\widehat{a}_{ij} - \widehat{a}_{0j}\widehat{a}_{i0}/\widehat{a}_{00}]^{-1},$$

where $a_{ij} = x^*_{ij}/y^*_j$ is the i-th input coefficient of the j-th output and $\widehat{a}_{ij} = a_{ij}(\beta_{ij} + b_{ij})/\beta_{ij}$ is an adjusted value of a_{ij} to reflect the externality. The roots of J are related to the input coefficient matrices. Moreover, one can show that

$$[a_{ij} - a_{0j}a_{i0}/a_{00}] = W^{-1}BP$$

and

$$[\widehat{a}_{ij} - \widehat{a}_{0j}\widehat{a}_{i0}/\widehat{a}_{00}] = P\widehat{B}'W^{-1},$$

where B and $\widehat{B}$ are $n \times n$ matrices given by

$$B = [\beta_{ij} - \beta_{0j}\beta_{i0}/\beta_{00}]$$

and

$$\widehat{B} = \left[(\beta_{ij} + b_{ij}) - (\beta_{0j} + b_{0j})(\beta_{i0} + b_{i0})/(\beta_{00} + b_{00})\right].$$

In the two sector case the matrices $\widehat{B}$ and B reduce to scalars. They reflect the factor intensities defined by the Cobb-Douglas exponents with and without the external effects, respectively. We may therefore say that the capital good is labour intensive from the private perspective if $\beta_{11}\beta_{00} - \beta_{10}\beta_{01} < 0$, but that it is capital intensive from the social perspective if $(\beta_{11} + b_{11})(\beta_{00} + b_{00}) - (\beta_{10} + b_{10})(\beta_{01} + b_{01}) > 0$. The expressions above allow us to state the following result taken from Benhabib and Nishimura (1998a).

Theorem 17. *In the two-sector model, if the capital good is labour intensive from the private perspective, but capital intensive from the social perspective, then the steady state is indeterminate.*

References

Alseda, L., Llibre, J. and M. Misiurewicz (1993), Combinatorial Dynamics and Entropy in Dimension One, World Scientific, Singapore.

Araujo, A. and J. A. Scheinkman (1977), "Smoothness, Comparative Dynamics, and the Turnpike Property," *Econometrica*, 45, 601–620.

Baierl, G., Nishimura, K. and M. Yano (1998), "The Role of Capital Depreciation in Multisectoral Models," *Journal of Economic Behavior and Organization*, 33, 467–479.

Benhabib, J. and R. E. Farmer (1996), "Indeterminacy and Sector Specific Externalities," *Journal of Monetary Economics*, 37, 397–419.

Benhabib, J. and K. Nishimura (1979), "The Hopf Bifurcation and the Existence and Stability of Closed Orbits in Multisector Models of Optimal Economic Growth," *Journal of Economic Theory*, 21, 421–444.

Benhabib, J. and K. Nishimura (1985), "Competitive Equilibrium Cycles," *Journal of Economic Theory*, 35, 284–306.

Benhabib, J. and K. Nishimura (1998a), "Indeterminacy an Sunspots with Constant Returns," *Journal of Economic Theory*, 81, 58–91.

Benhabib, J. and K. Nishimura (1998b), Indeterminacy under Constant Returns to Scale in Multisector Economies, mimeo.

Benhabib, J. and K. Nishimura (1998c), Indeterminacy Arising in Multisector Economies, mimeo.

Benhabib, J., Nishimura, K. and A. Venditti (1999) Indeterminacy and Cycles in Two Sector Discrete-Time Models, mimeo.

Block, L. and W. A. Coppel (1992), Dynamics in One Dimension, Springer-Verlag, Berlin.

Boldrin, M. and R. Deneckere (1990), "Sources of Complex Dynamics in Two-Sector Growth Models," *Journal of Economic Dynamics and Control*, 14, 627–653.

Boldrin, M. and L. Montrucchio (1986), "On the Indeterminacy of Capital Accumulation Paths," *Journal of Economic Theory*, 40, 26–39.

Boldrin, M., Nishimura, K., Shigoka, T. and M. Yano (2001), "Chaotic Equilibrium Dynamics in Endogenous Growth Models," *Journal of Economic Theory*, 96, 97–132.

Boldrin, M. and A. Rustichini (1994), "Growth and Indeterminacy in Dynamic Models with Externalities," *Econometrica*, 62, 323–342.

Brock, W. A. and W. D. Dechert (1991), Non-linear Dynamical Systems: Instability and Chaos in Economics, in *Handbook of Mathematical Economics: Vol. IV*, ed. by Hildenbrand, W. and H. Sonnenschein, North-Holland, Amsterdam, 2209–2235.

Cass, D. (1965), "Optimum Growth in an Aggregative Model of Capital Accumulation," *Review of Economic Studies*, 32, 233–240.

Dechert, W. D. and K. Nishimura (1983), "A Complete Characterization of Optimal Growth Paths in an Aggregated Model with a Non-concave Production Function," *Journal of Economic Theory*, 31, 332–354.

Deneckere, R. and Pelikan, S. (1986), "Competitive Chaos," *Journal of Economic Theory*, 40, 13–25.

Hewage, T. U. and D. A. Neumann (1990), Functions not Realizable as Policy Functions in an Optimal Growth Model, discussion paper, Bowling Green State University.

Kehoe, T. J. (1991), Computation and Multiplicity of Equilibria, in *Handbook of Mathematical Economics: Vol. IV*, ed. by Hildenbrand, W. and H. Sonnenschein, North-Holland, Amsterdam, 2049–2144.

Kehoe, T. J. and D. Levine (1985), "Comparative Statics and Perfect Foresight in Infinite Horizon Economies," *Econometrica*, 53, 433–454.

Kehoe, T. J., Levine, D. and P. Romer (1991), "On Characterizing Equilibria of Economies with Externalities and Taxes as Solutions to Optimization Problems," *Economic Theory*, 2, 43–68.

Koopmans, T. C. (1965), On the Concept of Optimal Growth, in *The Econometric Approach to Development Planning*, North-Holland, Amsterdam.

Li, T. and J. A. Yorke (1975), "Period Three Implies Chaos," *American Mathematical Monthly*, 82, 985–992.

Majumdar, M. and M. Nermuth (1982), "Dynamic Optimization in Non-convex Models with Irreversible Investment: Monotonicity and Turnpike Results," *Zeitschrift für Nationalökonomie*, 42, 339–362.

Matsuyama, K. (1990), "Sunspot Equilibria (Rational Bubbles) in a Model of Money-in-the-Ulitility-Function," *Journal of Monetary Economics*, 25, 137–144.

McKenzie, L. (1983), "Turnpike Theory, Discounted Utility, and the von Neumann Facet," *Journal of Economic Theory*, 30, 330–352.

McKenzie, L. (1986), Optimal Economic Growth, Turnpike Theorems and Comparative Dynamics, in *Handbook of Mathematical Economics: Vol. III*, ed. by Arrow, K. and M. Intriligator, North-Holland, Amsterdam, 1281–1355.

Mitra, T. (1996a), "An Exact Discount Factor Restriction for Period-Three Cycles in Dynamic Optimization Models," *Journal of Economic Theory*, 69, 281–305.

Mitra, T. (1996b), On the Nature of Policy Functions of Dynamic Optimization Models, working paper, Cornel University.

Mitra, T. and D. Ray (1984), "Dynamic Optimization on a Non-convex Feasible Set: Some General Results for Non-smooth Technologies," *Zeitschrift für Nationalökonomie*, 44, 151–175.

Mitra, T. and G. Sorger (1999a), "On the Existence of Chaotic Policy Functions in Dynamic Optimization," *Japanese Economic Review*, 50, 470–484.

Mitra, T. and G. Sorger (1999b), "Rationalizing Policy Functions by Dynamic Optimization," *Econometrica*, 67, 375–392.

Montrucchio, L. (1987), "Lipschitz Continuous Policy Functions for Strongly Concave Optimization Problems," *Journal of Mathematical Economics*, 16, 259–273.

Montrucchio, L. (1994), "Dynamic Complexity of Optimal Paths and Discount Factors for Strongly Concave Problems," *Journal of Optimization Theory and Applications*, 80, 385–406.

Montrucchio, L. and G. Sorger (1996), "Topological Entropy of Policy Functions in Concave Dynamic Optimization Models," *Journal of Mathematical Economics*, 25, 181–194.

Neumann, D., O'Brien, T., Hoag, T. and K. Kim (1988), "Policy Functions for Capital Accumulation Paths," *Journal of Economic Theory*, 46, 205–214.

Nishimura, K., Shigoka, T. and M. Yano (1998), "Interior Optimal Chaos with Arbitrarily Low Discount Rates," *Japanese Economic Review*, 49, 223–233.

Nishimura, K. and G. Sorger (1996), "Optimal Cycles and Chaos: A Survey," *Studies in Nonlinear Dynamics and Econometrics*, 1, 11–28.

Nishimura, K., Sorger, G. and M. Yano (1994), "Ergodic Chaos in Optimal Growth Models with Low Discount Rates," *Economic Theory*, 4, 705–717.

Nishimura, K. and A. Venditti (1999), Intersectoral Externalities and Indeterminacy, mimeo.

Nishimura, K. and M. Yano (1993), Optimal Chaos when Future Utilities are Discounted Arbitrarily Weakly, in *Nonlinear Analysis and Mathematical Economics*, Lecture Note Series, Institute of Mathematical Analysis, Kyoto University.

Nishimura, K. and M. Yano (1995a), "Nonlinear Dynamics and Optimal Growth: An Example," *Econometrica*, 63, 981–1001.

Nishimura, K. and M. Yano (1995b), Non-linearity and Business Cycles in a Two-Sector Equilibrium Model: An Example with Cobb-Douglas Production Functions, in *Nonlinear and Convex Analysis in Economic Theory*, ed. by Maruyama, T., and W. Takahashi, Springer-Verlag, Berlin, 231–245.

Nishimura, K. and M. Yano (1996a), "On the Least Upper Bound of Discount Factors that are Compatible with Optimal Period-Three Cycles," *Journal of Economic Theory*, 69, 206–333.

Nishimura, K. and M. Yano (1996b), "Chaotic Solutions in Dynamic Linear Programming," *Chaos, Solutions and Fractals*, 7, 1941–1953.

Ramsey, F. (1928), "A Mathematical Theory of Saving," *Economic Journal*, 38, 543–559.

Samuelson, P. A. (1973), "Optimality of Profit, Including Prices under Ideal Planning," *Proceedings of the National Academy of Sciences of the United States of America*, 70, 2109–2111.

Sarkovskii, A. N. (1964), "Coexistence of Cycles of a Continuous Map of a Line into Itself," *Ukrainian Mathematical Journal*, 16, 61–71.

Scheinkman, J. A. (1976), "On Optimal Steady States of N-Sector Growth Models when Utility is Discounted," *Journal of Economic Theory*, 12, 11–20.

Scheinkman, J. (1984), General Equilibrium Models of Business Fluctuations, mimeo, University of Chicago.

Shrinivasan, T. N. (1964), "Optimal Savings in a Two-Sector Model of Growth," *Econometrica*, 32, 358–373.

Sorger, G. (1992a), Minimum Impatience Theorems for Recursive Economic Models, Springer-Verlag, Heidelberg.

Sorger, G. (1992b), "On the Minimum Rate of Impatience for Complicated Optimal Growth Paths," *Journal of Economic Theory*, 56, 160–179.

Sorger, G. (1994a), "On the Structure of Ramsey Equilibrium: Cycles, Indeterminacy, and Sunspots," *Economic Theory*, 4, 745–764.

Sorger, G. (1994b), "Policy Functions of Strictly Concave Optimal Growth Models," *Ricerche Economeiche*, 48, 195–212.

Sorger, G. (1995), "On the Sensitivity of Optimal Growth Paths," *Journal of Mathematical Economics*, 24, 353–369.

Spear, S. E. (1991), "Growth, Externalities, and Sunspots," *Journal of Economic Theory*, 54, 215–223.

Stokey, N. L. and R. E. Lucas (1989), Recursive Methods in Economic Dynamics, Harvard University Press, Cambridge.

Sutherland, W. A. (1970), "On Optimal Development in a Multi-Sectoral Economy: The Discounted Case," *Review of Economic Studies*, 37, 585–589.

Uzawa, H. (1964), "Optimal Growth in a Two-Sector Model of Capital Accumulation," *Review of Economic Studies*, 31, 1–24.

Vial, J. P. (1983), "Strong and Weak Convexity of Sets and Functions," *Mathematics of Operations Research*, 8, 231–259.

Woodford, M. (1986), "Stationary Sunspot Equilibria in a Finance Constrained Economy," *Journal of Economic Theory*, 40, 128–137.

Part I
Optimal Growth and Endogenous Cycles

Chapter 3
The Hopf Bifurcation and Existence and Stability of Closed Orbits in Multisector Models of Optimal Economic Growth*

Jess Benhabib and Kazuo Nishimura**

3.1 Introduction

The local and global stability of multisector optimal growth models has been extensively studied in the recent literature. Brock and Scheinkman (1976), Cass and Shell (1976), McKenzie (1976), and Scheinkman (1976) have established strong results about global stability that require a small rate of discount. Burmeister and Graham (1973), Araujo and Scheinkman (1977), Magill (1977), and Scheinkman (1978) have established conditions that yield stability conditions independently of the rate of discount. Properties of unstable systems and of optimal paths that do no converge to steady states, however, have not been fully investigated. In this paper we will use bifurcation theory and treat the rate of discount as a parameter to study the properties of optimal paths. In particular, we will establish that, under certain assumptions, optimal paths become closed orbits as the steady state loses stability.

Bifurcations for differential equation systems arise when, for some value of a parameter, the Jacobian of the function describing the motion acquires eigenvalues with zero parts at a stationary point. If a real eigenvalue becomes zero an odd

*Journal of Economic Theory 21, 421–444, 1979

**We would like to thank Professor K.Lancaster and Professor L.McKenzie for their help and guidance at various stages of the paper. We are indebted to Professors W. A. Brock, E. Burmeister, D. Cass and K. J. P. Magill for valuable comments. Comments of an anonymous referee us helped us improve the paper and correct many mistakes.

J. Benhabib (✉)
Department of Economics, New York University, USA
e-mail: jess.benhabib@nyu.edu

K. Nishimura
Institute of Economic Research, Kyoto University, Japan
e-mail: nishimura@kier.kyoto-u.ac.jp

J. Stachurski et al. (eds.), *Nonlinear Dynamics in Equilibrium Models*,
DOI 10.1007/978-3-642-22397-6_3, © Published by Springer-Verlag

number of additional stationary solutions bifurcate from the original one for values of the parameter in the neighborhood of the bifurcation value. In our context this results in the multiplicity of steady states (see Brock 1973; Benhabib and Nishimura 1979, 1978).[1] In this paper we are concerned with a bifurcation that results from pure imaginary roots: the Hopf Bifurcation. In contrast to the bifurcation from a real eigenvalue, no additional stationary points arise out of a Hopf Bifurcation. Instead there emerge closed orbits around the stationary point. Since we would like to study orbits we have assumed nonjoint production and a single, possibly composite, consumption good to assure the uniqueness of the steady state. Under some additional assumption we show that the Jacobian of the functions describing the motion of the system cannot vanish at a steady state; thus only the Hopf Bifurcation is possible. In fact it can occur under the most general circumstances: in Sect. 3.3 we give an example of a Cobb-Douglas Technology with a totally unstable steady state which gives rise to optimal paths that are closed orbits (see Theorem in Sect. 3.4). In Sect. 3.5 we discuss the stability of orbits that arise from our optimal growth problem.

Optimal paths in growth theory can be regarded as competitive dynamical systems arising in descriptive models with perfect foresight.[2] The existence of closed orbits that arise under very general circumstances and which cannot be eliminated by intertemporal arbitrage may form a basis for the study of price and output fluctuations.

3.2 Saddle-Point Instability of Steady States

We define the optimal growth problem as follows:

$$\max \int_0^\infty e^{-(r-g)t} U\,(T\,(y,\,k))\,dt$$

$$\text{s.t. } \dot{k}_i = dk_i/dt = y_t - gk_{i,} \quad i = 1, \ldots, n. \tag{3.1}$$

Consumption is given by $c = T(y,\,k)$, where the vectors y and k represent per capita outputs and stocks of capital goods, respectively. $U(T(y,\,k))$ is the utility derived from consumption, g ($\geq$0) is the rate of populations growth, $r-g$ ($\geq$0) is the discount rate. r can be interpreted as the rate of interest. We assume the following:

[1] It is well known that one-sector optimal growth models with unique steady states are in general saddle-point stable. An expectation is an example due to Arrow, reported in Kurz (1968), where there is a unique unstable steady state if the discount rate exceeds the marginal product of capital and utility saturates. The optimal path either goes to the origin or to infinity unless it starts at the steady state. Saddle-point instability also arises in certain one-sector models where multiple steady states exist. See Kurz (1978) or Liviatan and Samuelson (1969).

[2] See Cass and Shell (1976) for the relationship between "optimal" and "descriptive" models of economic growth.

(A1) *All goods are produced nonjointly with production functions homogeneous of degree 1 strictly quasi-concave for nonnegative inputs, and twice differentiable for positive inputs.*

The social production function $T(y, k)$, which is derived from individual production functions, is twice differentiable if it is assumed that the vector of all input coefficients that maximize consumption for given (y, k) is strictly positive. In general, Inada-type conditions on production functions rule out corner solutions. We will use an alternative assumption which allows some inputs not to be used in the production of some goods:

(A2) *Let* $\left(K^j_{j_1} \cdots K^j_{j_{m_j}}\right)$ *be the set of inputs used in the production of good* j*. Then good* j *cannot be produced without* $K^j_{j_i}$, $i = 1, \ldots, m_j$.

In Appendix (AII) we show that $T(y, k)$ is twice differentiable if (A1) and (A2) hold. The static optimization conditions imply

$$\partial c/\partial y_i = \partial T/\partial y_i = -p_i, \quad \partial c/\partial k_i = \partial T/\partial k_i = w_i, \quad i = 1, \ldots, n$$

where p_i and w_i are the price and rental of the ith good in terms of the price of the consumption good.

To solve (3.1) we write the Hamiltonian

$$H = e^{-(r-g)t}\{U(T(y, k)) + q(y - gk)\}.$$

The Maximum Principle yields

$$\begin{aligned} \dot{k}_i &= y_i - gk_i, \\ \dot{q}_i &= -U'w_i + rq_i, \\ q_i &= U'p_i, \quad i = 1, \ldots, n \end{aligned} \tag{3.2}$$

Definition 1. $\left(\hat{k}, \hat{q}\right)$ is a steady-state equilibrium of the system (2) if $\left(\hat{k}, \hat{q}\right) \geqslant 0$ and $\dot{k}_i = \dot{q}_i = 0$ for all i.

Let the variable input coefficient matrix with nonnegative elements be given by

$$\hat{A} = \left[\begin{array}{c|c} a_{00} & a_{0\cdot} \\ \hline a_{\cdot 0} & A \end{array}\right],$$

where $a_{0\cdot}$ is the row vector of labor inputs, A is the capital input matrix to produce capital goods, $a_{\cdot 0}$ is the capital input vector for the consumption good, and a_{00} is the labor input coefficient for consumption good. We make the following assumptions:

(A3) *At the steady state under consideration, the capital coefficient matrix* A *is indecomposable.*

(A4) *At the steady state under consideration, direct labor and at least one capital input is required in the production of the consumption good and direct labor is required in the production of at least one capital good:* $a_{00} \neq 0,\ a'_{\cdot 0},\ a_{0\cdot} \neq 0.$

(A5) *In the neighborhood of the steady state the marginal utility of consumption is constant:* $U'' = d^2U(c)/dc^2 = 0,\ U' = dU(c)/dc = 1.$

(A6) *At the steady state under consideration the input coefficient matrix* $\hat{A}$ *is nonsingular.*

We shall begin by discussing the steady-state properties of the model described above. Under (A5) we observe from (3.2) that at the steady state, $q_i = p_i$, $i = 1, \ldots, n$. Under assumptions somewhat weaker than (A3) and (A4), Burmeister and Dobell (1970, Chap. 9) have shown that steady states are uniquely determined for values of $r \in (g, \bar{r})$,[3] where steady-state values $c(r)$, $p(r)$, $k(r)$, and $y(r)$ are all positive (steady states are interior) and continuous functions of r.[4]

The social production function $T(y, k)$ is concave in (y, k). When (A6) holds $c = T(y, k)$ is strictly concave in y for fixed k at the steady-state values (y, k) (see Lancaster 1968, Chap. 8 and the comment by Kelly (1969)). This reflects the strict concavity of the output production possibility surface at the steady state. Since $U'' = 0$ and $U' = 1$ by assumption (A5) it is easily seen from the above comments that the Hamiltonian

$$H = e^{-(r-g)t}\left[U(T(y, k)) + q(y - gk)\right]$$

is strictly concave in y at the steady state. Maximizing the Hamiltonian with respect to y yields a unique optimal vector. Since $U' = 1$ and $q = p$ in the neighborhood of the steady state, (3.2) are given by

$$\begin{aligned} \dot{k}_i &= y_i(k, p) - gk_i, \\ \dot{p}_i &= -w_i(k, p) + rp_i, \quad i = 1, \ldots, n. \end{aligned} \tag{3.3}$$

In Appendix (AIII) we show that $y(k, p)$ and $w(k, p)$ are differentiable under assumptions (A1) and (A6). If we linearize (3.3) around the steady state, the Jacobian of the linearized system will be given by

$$J = \begin{bmatrix} [\partial y/\partial k] - gI & [\partial y/\partial p] \\ -[\partial w/\partial k] & -[\partial w/\partial p] + rI \end{bmatrix}.$$

We will now consider the matrices $[\partial y/\partial k]$ and $[\partial w/\partial p]$ appearing in J. We have seen above that at the steady state all goods are produced. Then the following must hold:

---

[3]The results easily follow since A is indecomposable and $a'_{\cdot 0}$, $a_{0\cdot} \neq 0$. Burmeister and Dobell (1970) have weakened the indecomposability of A, but we will require it for some further results.

[4]$\bar{r}$ can be ∞: see Burmeister and Kuga (1970).

$$[w_0, w]\begin{bmatrix} a_{00} & a_{0\cdot} \\ a_{\cdot 0} & A \end{bmatrix} = [1, p] \quad \text{(price equals cost),} \tag{3.4a}$$

$$\begin{bmatrix} a_{00} & a_{0\cdot} \\ a_{\cdot 0} & A \end{bmatrix}\begin{bmatrix} c \\ y \end{bmatrix} = \begin{bmatrix} 1 \\ k \end{bmatrix} \quad \text{(full employment).} \tag{3.4b}$$

Consider the matrix of partials $[\partial y/\partial k]$. Since prices are fixed, the input–output coefficients remain fixed. Solving (3.4a) by eliminating c, we obtain

$$B^{-1} = [\partial y/\partial k] = [A - (1/a_{00})\, a_{\cdot 0} a_{0\cdot}]^{-1}$$

provided that B is nonsingular. The nonsingularity of B can established from the following observation. We have $\det \hat{A} = (a_{00}) \det [A - (1/a_{00})\, a_{\cdot 0} a_{0\cdot}]$ (see Gantmacher 1960, Vol. 1, pp. 45–46). Since $\det \hat{A} \neq 0$ by assumption (A6) and $a_{00} \neq 0$ by assumption (A5), we must have $\det B \equiv \det (A - (1/a_{00})\, a_{\cdot 0} a_{0\cdot}) \neq 0$. In evaluating $[\partial w/\partial p]$, we know that input-output coefficients change with prices. However, the well-known envelope theorem assures us that

$$[w_0, w]\begin{bmatrix} da_{00} & da_{0\cdot} \\ da_{\cdot 0} & dA \end{bmatrix} = 0.$$

(For an explicit discussion of this point, see Samuelson 1953–1954).

Thus we obtain

$$[\partial y/\partial k] = [\partial w/\partial p]' = [A - (1/a_{00})\, a_{\cdot 0} a_{0\cdot}]^{-1}.$$

(This is the duality between the Rybczynski and Stolper-Samuelson effects well known in trade theory).

Consider the matrix $[\partial w/\partial k]$ that appears in the Jacobian J. In general prices do not determine factor rentals uniquely, as is known from the debate on the factor price equalization. When capital stocks are given, however, the diversification cone where the economy operates is determined. If prices are fixed, a small change in capital stocks causes only outputs to adjust if the economy is not specialized to start with. The economy remains in the same diversification cone where factor rentals remain constant along with prices. Thus for a small change in factor endowments, there is no change in factor prices as is to be expected from the factor price equalization theorem.[5] We obtain, therefore, $[\partial w/\partial k] = 0$.

We can now write the Jacobian of (3.3) as

$$J = \begin{bmatrix} B^{-1} - gI & [\partial y/\partial p] \\ 0 & -B'^{-1} + rI \end{bmatrix},$$

[5]For a discussion and proof, see McKenzie (1955).

where, as shown before, $[\partial y/\partial k] = [\partial w/\partial p]' = B$ and $B = [A - (1/a_{00})\, a_{\cdot 0} a_{0\cdot}]$. None that J is "quasi-triangular"; it is easily shown that its roots will be the roots of its upper and lower diagonal matrices, $\left[B^{-1} - gI\right]$ and $\left[-B'^{-1} + rI\right]$. To further investigate the roots of J when $r > g$, we will need to establish some additional results.

It is well known that for a steady state to exist the rate of discount must be contained in some interval which depends on technology.[6] This requirement takes the form $r\lambda_A < 1$, where λ_A is the dominant root of the capital input coefficient matrix A at the steady state. For the sake of completeness, before establishing the main result in Proposition 2, we prove the following.

Proposition 1. *Let r be the rate of interest and A the associated capital input coefficient matrix defined above. Let λ_A be the dominant root of A. If A is indecomposable, then $r\lambda_A < 1$.*

Proof. Steady-state prices of capital goods are given by

$$p = a_{\cdot 0} w_0 + rpA,$$

which implies

$$\frac{1}{r} p \geqslant pA.$$

Debreu and Hernstein (1953) have shown that if A is indecomposable, $p \geqslant 0$, $p \neq 0$, then,

$$\frac{1}{r} \geqslant \lambda_A; \quad r\lambda_A \leqslant 1.$$

□

We will now establish a result which is of fundamental importance for stability, as the subsequent discussion will show. Note that the hypotheses of the second part of the following proposition are satisfied under assumptions (A3) and (A5).

Proposition 2. *Let $\begin{bmatrix} a_{00} & a_{0\cdot} \\ a_{\cdot 0} & A \end{bmatrix}$ be a nonnegative square matrix with $a'_{\cdot 0}, a_{0\cdot} \geqslant 0$, $a_{00} > 0$, partitioned such that a_{00} is a scalar. Let $B = [A - (1/a_{00})\, a_{\cdot 0} a_{0\cdot}]$. Let λ_A be the dominant root of A. Then $\lambda_{jB} \leqslant \lambda_A$, where λ_{jB} is any real root of B. If A is indecomposable and the vectors $a'_{\cdot 0}, a_{0\cdot}$ each have at least one positive element, then $\lambda_{jB} < \lambda_A$.*

Proof. Any root of B must satisfy the equation $\det\left[A - (1/a_{00})\, a_{\cdot 0} a_{0\cdot} - \lambda_{jB} I\right] = 0$. Assume then that $\lambda_{jB} > \lambda_A$. This implies that $\det\left(A - \lambda_{jB} I\right) \neq 0$. Furthermore from the properties of nonnegative matrices we have $\left(A - \lambda_{jB} I\right)^{-1} \leqslant 0$ (see Debreu and Hernstein 1953). Using the formula on the determinants of partitioned matrices

[6] We rule out utility saturation in (A5). If utility saturation is allowed some perverse cases may arise. See the example by Arrow in Kurz (1968).

that we already used above, we obtain,

$$\det\left[\begin{array}{c|c} 1 & a_{0\cdot} \\ \hline \frac{1}{a_{00}}a_{\cdot 0} & A-\lambda_{jB}I \end{array}\right] = (1)\det\left[A-\lambda_{jB}I-\frac{1}{a_{00}}a_{\cdot 0}a_{0\cdot}\right]$$

and also, since $(A-\lambda_{jB}I)$ is nonsingular,

$$\det\left[\begin{array}{c|c} 1 & a_{0\cdot} \\ \hline \frac{1}{a_{00}}a_{\cdot 0} & A-\lambda_{jB}I \end{array}\right] = \det\left[A-\lambda_{jB}I\right]\left(1-a_{0\cdot}\left(A-\lambda_{jB}I\right)^{-1}a_{\cdot 0}\left(1/a_{00}\right)\right). \tag{3.5}$$

Thus the right-hand sides of the above two equations are equal. The scalar quantity $\left(1-(1/a_{00})\,a_{0\cdot}\left(A-\lambda_{jB}I\right)^{-1}a_{\cdot 0}\right)\neq 0$ since $a'_{\cdot 0}$, $a_{0\cdot}\geqslant 0$ and $a_{00}\neq 0$ by hypothesis. But then $\det\left[A-(1/a_{00})\,a_{\cdot 0}a_{0\cdot}-\lambda_{jB}I\right]\neq 0$ and we have a contradiction. This establishes that $\lambda_{jB}\leqslant\lambda_A$.

To rule out $\lambda_{jB}=\lambda_A$, we assume that A is indecomposable. The righthand side of (3.5) can be written as

$$\begin{aligned}&\det\left[A-\lambda_{jB}I\right]-(1/a_{00})\,a_{0\cdot}\left(\det\left[A-\lambda_{jB}I\right]\left(A-\lambda_{jB}I\right)^{-1}\right)a_{0\cdot}\\&\quad=(-1)^n\left[\det\left[\lambda_{jB}I-A\right]+(1/a_{00})\,a_{0\cdot}\left(\det\left[\lambda_{jB}I-A\right]\left(\lambda_{jB}I-A\right)^{-1}\right)a_{0\cdot}\right].\end{aligned} \tag{3.6}$$

Define the adjoint matrix $D\left(\lambda_{jB}\right)=\det\left[\lambda_{jB}I-A\right]\left(\lambda_{jB}-A\right)^{-1}$.

If $\lambda_{jB}=\lambda_A$, $\det\left[\lambda_{jB}I-A\right]=0$ and $\left(\lambda_{jB}I-A\right)^{-1}$ is undefined, but for indecomposable, nonnegative A, $D\left(\lambda_{jB}\right)=\det\left[\lambda_{jB}I-A\right]\left(\lambda_{jB}I-A\right)^{-1}$ exists and moreover, $D\left(\lambda_{jB}\right)>0$ (see Gantmacher 1960, Vol. 2, Chap. 13, Theorem 3, p. 66, and Proposition 2, p. 69). Thus (3.6) reduces to

$$(-1)^n\,(1/a_{00})\,a_{0\cdot}D\left(\lambda_{jB}\right)a_{\cdot 0}\neq 0$$

since $a'_{\cdot 0}$, $a_{0\cdot}\neq 0$, $a_{00}\neq 0$. But then (3.5) cannot be zero for $\lambda_{jB}=\lambda_A$. Thus, if λ_{jB} is a real root of $\det\left[A-(1/a_{00})\,a_{0\cdot}a_{\cdot 0}\right]$, then $\lambda_{jB}<\lambda_A$. □

Proposition 3. *If* (A1) – (A6) *hold at the steady state, then for* $r\geqslant g$*, the real roots of* J *come in nonzero pairs of opposite sign.*

Proof. Let (A1)–(A6) holds. Under (A1), (A5), and (A6) the roots of J are given by the roots of $B^{-1}-gI$ and $-B^{-1}+rI$. We have shown above that B is nonsingular if $a_{00}\neq 0$ and (A6) holds. Consider the roots of J, $\left(1/\left(-\lambda_{jB}\right)\right)+r$, and $\left(1/\lambda_{jB}\right)-g$, $j=1,\ldots,n$. By Proposition 2, $r\lambda_{jB}<1$ and since $g\leqslant r$, $g\lambda_{jB}<1$. This implies that there are no zero real roots and sign $\left(\left(1/\left(-\lambda_{jB}\right)\right)+r\right)=$ sign $\left(1/\left(-\lambda_{jB}\right)\right)$, sign $\left(\left(1/\lambda_{jB}\right)-g\right)=$ sign $\left(1/\lambda_{jB}\right)$. Thus the real roots of J come in pairs of opposite sign: $\left(\left(1/\left(-\lambda_{jB}\right)\right)+r\right)$, $\left(\left(1/\lambda_{jB}\right)-g\right)$. □

Corollary *Let (A1) – (A6) hold at the steady state* $\left(\hat{k},\ \hat{p}\right)$ *and let the roots of* $B = (A - (1/a_{00})\, a_{0\cdot} a_{\cdot 0})$ *be real. Then the steady state is locally saddle-point stable.*

Proof. By Proposition 3, the real roots of J are nonzero and come in pairs of opposite sign. Since all roots of J are real if those of B are, the Corollary follows. □

A special case of the above corollary arises when the input coefficient matrix $\hat{A}$ is symmetric since in that case B is also symmetric and has only real roots. When the roots of B are not real, however, saddle-point stability need not hold. In the next section we give an example of a saddle-point unstable steady state and in Sect. 3.4 we show that the unstable steady state gives rise to an optimal path that is a periodic orbit.

3.3 An Example of a Totally Unstable Steady State

Consider the following steady-state technology matrix (At least two capital goods are necessary for out results).

$$\hat{A} = \begin{bmatrix} a_{00} & a_{\cdot 0} \\ a_{0\cdot} & A \end{bmatrix} = \begin{bmatrix} 0.100 & 0.700 & 0.100 \\ 0.001 & 0.107 & 0.801 \\ 0.070 & 0.010 & 0.170 \end{bmatrix}.$$

Row sums are less than unity, reflecting a productive technology. That the lower (2×2) principal minor has a dominant root less than unity is easily seen since both its row and column sums are less than unity. Thus for $r \leqslant 1$, $r\lambda_A < 1$ is satisfied. The roots of the corresponding Jacobian J are given by the roots of $\left[B^{-1} - gI\right]$ and $\left[-B^{-1} + rI\right]$, as shown in the previous section. The numerical values are $0.248 - g \pm 1.567i$ and $-0.248 + r \pm 1.567i$.

Finding a Cobb-Douglas technology that yields the above input–output coefficients for a given r is tedious but straightforward. If we set the wage rate equal to unity, prices will be determined by $p = a_{\cdot 0}\,[I - rA]^{-1}$. Steady-state factor rentals will be given by $w = rp$. The coefficients of each Cobb-Douglas production function that yield the above unit input coefficients at precisely the steady-state factor rentals can then be calculated from the first-order conditions of profit maximization. Below we give two examples of Cobb-Douglas technologies that yield the input coefficient matrix $\hat{A}$ at the steady state.

Let the Cobb-Douglas production functions be given as follows:

$$Y_i = b_i \prod_{J=0}^{2} K_{ji}^{\alpha_{ji}}, \tag{3.7}$$

where

$$\sum_{j=0}^{2} \alpha_{ji} = 1, \quad i = 0, 1, 2.$$

Y_0 is the output of the consumption good and K_{0j} is the labor input to good j.

(i) Let $r = 0.248$, $g = 0.1$. The Cobb-Douglas Technology (3.7) with the coefficients given below yields the input coefficient matrix $\hat{A}$ at the steady state.

$$\begin{array}{llll} b_0 = 10.2425 & \alpha_{00} = 0.9524 & \alpha_{01} = 0.9723 & \alpha_{02} = 0.3941 \\ b_1 = \ \ 1.5084 & \alpha_{10} = 0.0017 & \alpha_{11} = 0.0265 & \alpha_{12} = 0.5635 \\ b_2 = \ \ 3.0266 & \alpha_{20} = 0.0459 & \alpha_{21} = 0.0012 & \alpha_{22} = 0.0423 \end{array}$$

The roots of the Jacobian matrix for this technology will then be $0 \pm 1.567i$, $0.148 \pm 1.567i$. In (A IV) of the Appendix it is shown that the real part of the roots $u(r) \pm v(r)i = 0 \pm 1.576i$ increases with $r : du(r)/dr > 0$. Thus as r passes through the value 0.248, the steady state changes from saddle-point stable to totally unstable. The steady state quantities and prices at $r = 0.248$ are given below:

$$(c,\ k_1,\ k_2) = (9.285,\ 0.092,\ 0.649),$$
$$(p_0,\ p_1,\ p_2) = (1.000,\ 6.888,\ 2.433),$$
$$(w_0,\ w_1,\ w_2) = (2.480,\ 1.908,\ 0.605).$$

(ii) Let $r = 0.5$, $g = 0.1$. The Cobb-Douglas Technology (3.7) yields the steady-state input coefficient matrix $\hat{A}$ for the coefficients given below:

$$\begin{array}{llll} b_0 = 10.6462 & \alpha_{00} = 0.8627 & \alpha_{10} = 0.0032 & \alpha_{20} = 0.1341 \\ b_1 = \ \ 1.5878 & \alpha_{01} = 0.9437 & \alpha_{11} = 0.0534 & \alpha_{21} = 0.0029 \\ b_2 = \ \ 3.0555 & \alpha_{02} = 0.2304 & \alpha_{12} = 0.6848 & \alpha_{22} = 0.0858 \end{array}$$

the steady state values, for $g = 0.1$, are given by

$$\left(\hat{c},\ \hat{k}_1,\ \hat{k}_2\right) = (9.9169,\ 0.06272,\ 0.7061),$$
$$(w_0,\ w_1,\ w_2) = (8.6539,\ 3.2097,\ 1.8770),$$
$$(p_0,\ p_1,\ p_2) = (1.000,\ 6.4194,\ 3.7540).$$

Note that for $r = 0.5$, $g = 0.1$, the Jacobian J has four complex roots with positive real parts and the steady state is totally unstable.

In the next section we apply the Hopf Bifurcation Theorem to the problem given by (3.1) to generate optimal paths that are closed orbits. We then show that the Cobb-Douglas Technology (3.7) with the coefficients given in (i) gives rise to optimal paths that form closed periodic orbits

3.4 Bifurcation of Closed Orbits from Steady States

In this section we give Hopf Bifurcation Theorem and we use it to establish the existence of a closed orbit for the optimal growth problem.

Theorem 1. (*The Hopf Bifurcation Theorem*). *Let* $\dot{x} = F(x, \mu)$, $x = (x_1, \ldots, x_n)$ *be a real system of differential equations with real parameter* μ. *Let* $F(x, \mu)$ *be* C^s *in* x *and* μ *for* x *in a domain* G *and* $|\mu| < c$. *For* $|\mu| < c$ *let* $F(x, \mu)$ *possess a* C^r *family if stationary solutions* $\tilde{x} = \tilde{x}(\mu)$ *lying in* G :

$$F(\tilde{x}(\mu), \mu) = 0.$$

For $\mu = 0$,[7] *let the matrix* $F_x(\tilde{x}(0), 0)$ *have one pair of pure imaginary roots,* $\alpha(\mu) \pm \beta(\mu) i$; $\alpha(0) = 0$, $\beta(0) \neq 0$, *and* $\alpha(0)/d\mu \neq 0$. *Then there exists a family of real periodic solutions* $x = x(t, \epsilon)$, $\mu = (\epsilon)$ *which has properties* $\mu(0) = 0$, *and* $x(t, 0) = \tilde{x}(0)$, *but* $x(t, \epsilon) \neq \tilde{x}(\mu(\epsilon))$ *for sufficiently small* $\epsilon \neq 0$. $x(t, \epsilon)$ *is* C^r. $\mu(\epsilon)$ *and* $T(\epsilon)$, *where* $T(\epsilon) = 2\pi/|\beta(0)|$ *is the period of the orbit, are* C^{r-1}.

Proof. See Hopf (1976) and Sect. 3.2, pp. 197–198, of "Editorial Comments" by Hopf's translators, N.L. Howard and N. Koppel, in Marsden and McCracken (1969). Hopf states and proves the theorem for $F(x, \mu)$ analytic. Howard and Koppel revise Hopf's proof and provide the C^s version given above. □

Theorem 2A. *Let the optimal problem (3.1) satisfy assumptions (A1)–(A6) and the right-hand side of differential equations (3.3) be* C^s, $s \geqslant 1$.[8] *For* $r = r_0$ *let the Jacobian of (3.3),* J, *have one pair of pure imaginary roots* $\alpha(r_0) \pm \beta(r_0) i$, *where* $\alpha(r_0) = 0$, $\beta(r_0) \neq 0$, $d\alpha(r_0)/dr \neq 0$. *Then for the optimal growth problem given by (3.1) there exist a* C^{s-1} *function* $r = r(\epsilon)$, $r_0 = r(0)$, *and a* C^s *family of optimal paths* $(k(t, r(\epsilon)), p(t, r(\epsilon)))$ *that are nonconstant*[9] *closed orbits in the positive orthant for sufficiently small* $\epsilon \neq 0$.

Proof. By Proposition 3, J has no zero real roots. Thus we can apply the implicit function theorem to (3.3) by setting the left-hand side equal to zero and $r = r_0$. This yields steady-state values of k and p as C^s functions $k(r)$ and $p(r)$ at $r = r_0$. The Hopf Bifurcation Theorem immediately applies and we obtain closed orbits.

It is easy to show that for r close to r_0, the periodic orbits remain in the positive orthant. In Sect. 3.2 we showed that under our assumptions the steady state is interior, that is, steady-state values of $c(r)$, $y(r)$, $k(r)$, and $p(r)$ are positive. By Theorem 1, the orbits collapse into the stationary point as ϵ and $\mu(\epsilon) \to 0$, or

---

[7]There is no loss of generality in assuming that pure imaginary roots appear for $\mu = 0$

[8]Using the results and methods of Appendices (AII) and (AIII), we can show that for the Cobb-Douglas Technology used in the example of the previous section, $s = \infty$.

[9]Nonconstant only means that the closed orbit is not a stationary point.

in terms of our problem as $r \to r_0$ (see also Schmidt 1976). Then for sufficiently small ϵ the orbits $(k(t, r(\epsilon)), p(t, r(\epsilon)))$ remain in the positive orthant.

Finally, the optimality of a path that forms or approaches the orbit is assured by the transversality conditions. Let $\tilde{p}(t, r)$ be the price path approaching the orbit. Then $\lim_{t\to\infty} e^{-(r-g)t}\tilde{p}_i k_i \geqslant 0$ for any feasible nonnegative path of $k(t)$ and transversality conditions are satisfied (See Pitchford and Turnovsky 1977, pp. 28–30). □

Theorem 2B. *For the optimal growth problem (3.1) let the technology be defined by (3.7) with coefficients given in (i) of Sect. 3.3. Let the marginal utility of consumption be constant in the neighborhood of steady state corresponding to* $r = 0.248$. *Then there exists a continuous function*[10] $r = \hat{r}(\epsilon)$, $r_0 = r(0)$, *and a continuous family of optimal paths* $(k(t, r(\epsilon)), p(t, r(\epsilon)))$ *that are nonconstant closed orbits in the positive orthant for sufficiently small* $\epsilon \neq 0$.

Proof. We only have to show that the hypotheses of Theorem are satisfied for the technology (3.7) with coefficients given in (i) of Sect. 3.3. The steady-state input coefficient matrix $\hat{A}$, given in Sect. 3.3, satisfies assumptions (A3), (A4) and (A6). Furthermore, a Cobb-Douglas Technology satisfies (A1) and (A2). For $r_0 = 0.248$, the Jacobian J has one pair of pure imaginary roots, $u(r_0) \pm v(r_0)i$, $u(r_0) = 0$, $v(r_0) = 1.567$. In Appendix (AIV) it is shown that $du(r_0)/dr = 1.3046$. Thus all hypotheses Theorem are satisfied. □

Remark 1. It can be shown that the input coefficient matrix $\widehat{A}$ for a Cobb-Douglas Technology has a determinant of the same sign as that of the matrix formed by using the exponents of the inputs in the production functions as rows. In terms of the technology described in Sect. 3.3, sign det $\widehat{A}$ = sign det$[\alpha_{ij}]$. If $[\alpha_{ij}]$ is nonsingular assumption (A6) holds for all steady states in the interval $r \in [g, \overline{r}]$, where $\overline{r}$ is the upper bound of r for which steady states exist (As noted in footnote 5, $\overline{r}$ can be infinity). Thus B^{-1} is well defined on $r \in [g, \overline{r}]$. The following corollary applies to example (ii) of Sect. 3.3.

Corollary *Let* $\hat{A}(r)$ *be nonsingular for all* $r \in [g, \bar{r}]$. *Let the steady state be unstable in the saddle-point sense for some* $\hat{r}$ *in* $[g, \bar{r}]$. *Then there exists an* r^*, $r^* \leqslant \hat{r}$ *for which the Jacobian* J *has a pair of pure imaginary roots* $\alpha(r^*) \pm \beta(r^*)i$, $\alpha(r^*) = 0$, $\beta(r^*) \neq 0$. *If* $d\alpha(r^*)/dr \neq 0$, *there is a Hopf Bifurcation at* r^*.

Proof. For $(r - g)$ sufficiently small all the roots of J come in pairs of opposite sign and we have saddle-point stability. The corollary follows from the continuity of the roots of B^{-1} in r since B is nonsingular if $\hat{A}$ is nonsingular. □

Remark 2. An alternative approach to get a global, though limited, result can be attempted as follows. Using the results of Benveniste and Scheinkman (1979),

[10] See footnote 8 and Theorem 1.

one can express the price variables as functions of the stocks. Under certain assumptions,

$$p_i = \frac{\partial R(k)}{\partial k}, \quad i = 1, \ldots, n.$$

where R is the value of the optimal program from given initial values for k. The dynamic equations then reduce to

$$\dot{k}_i = y_i(p(k), k) - gk_i, \quad i = 1, \ldots, n.$$

The example in Sect. 3.3 assures a unique, totally unstable steady state. The stocks are bounded above by technology. If the path could be prevented from going to the origin by bending the linear utility function at low values of consumption (and therefore of the capital stocks), a limit cycle can be generated by using the Poincaré-Bendixon theorem for two-dimensional systems. Note that our example in Sect. 3.3 is two dimensional with stocks k_1 and k_2. The difficulty lies in establishing the existence of an optimal path with a utility function that effectively prevents the stocks from going to the origin.

An interesting point to consider is the possibility of a second bifurcation after the first one. Increasing r further may result in another pair of pure imaginary roots. If certain degeneracies are ruled out and the orbit is still in existence at the new bifurcation point, it will be transformed into a torus (see Ruelle and Tackens 1971).

3.5 The Stability of Closed Orbits

In the previous sections we studied orbits that bifurcate from steady states as the steady states become unstable in the saddle-point sense. In this section we study the stability of orbits in the sense defined below:

Definition 2. Let the $dz/dt = h(z)$ possess a periodic orbit solution $z = \gamma(t)$ of least period p, $p > 0$, where $h(z)$ is class C^1 on an open set. Let $\Im : z = \gamma(t)$, $0 \leqslant t \leqslant p$. Let the points z in the neighborhood of $\Im$, on solutions $z = z(t)$ of $dz/dt = h(z)$, and which satisfy $\text{dist}(\Im, z(t)) \to 0$ as $t \to \infty$ constitute a d-dimensional manifold. Then the periodic orbit is said to have a d-dimensional locally stable manifold.

The stability of a closed orbit can be studied by means of the roots of the Poincaré map called characteristic multipliers. The Poincaré map is the locus of points generated by the flow of the system crossing a $(2n - 1)$-dimensional submanifold transversal to the closed orbit at some point. The closed orbit, then, is a fixed point of the Poincaré map. Initial points close to the orbit generate points of the Poincaré map at time intervals close to the period of the orbit (see Hartman 1964, p. 25). Thus the Poincaré map is a discrete time map and the stability of its fixed point, the closed orbit, depends on whether its roots are within the unit circle. There are $2n$ characteristic multipliers, $2n - 1$ corresponding to the roots of the Poincaré map and one multiplier always equal to one, corresponding to the time-invariant fixed point (see Hartman 1964, p. 255). To study the stability of the orbit we will use characteristic exponents, λ_i, which are defined by characteristic multipliers σ_i as follows (see Hartman 1964, pp. 6, 252).

$$\text{Re}\,\lambda_i = 1/\hat{t}\ \text{Re}\ln|\sigma_i|,$$

where $\hat{t}$ is the period of the closed orbit. The following theorem from Hartman (1964) establishes stability conditions:

Theorem 3 *(Hartman). Let $\dot{x} = f(x)$ be m dimensional and of class C^1 on an open set containing $x = 0$. Let $x = \eta(t, x_0)$ be a solution, satisfying $\eta(0, x_0) = x_0$. Suppose $\gamma(t)$ is a periodic solution with positive period, p. Let $\lambda_1, \ldots, \lambda_n$ be characteristic exponents of $\gamma(t)$. Then one of these is 0. If exactly d of the remaining exponents have negative real parts, then there exists a $(d + 1)$-dimensional stable C^1-manifold in the neighborhood of the closed orbit.*

Proof. See Hartman (1964, pp. 251–256, Theorem 11.2). □

We now prove a theorem for the Hamiltonian system associated with our optimal growth problem (3.1) that relates the dimension of the stable manifold around the closed orbit to the rate of discount.

Theorem 4 *Let the optimal growth problem (3.1) give rise to a closed orbit. Then if σ_i is a characteristic multiplier of the closed orbit so is $e^{(r-g)\hat{t}}(1/\sigma_i)$, where $\hat{t}$ is the period of the closed orbit.*

Remark 3. When $r - g = 0$, the multipliers come in pairs inside and outside the unit circle. The result of Theorem is analogous to the one obtained by Kurz (1968) for a stationary point. For a closed orbit we work with characteristic multipliers and the proof must be different. Our proof modifies that the Whittaker (1944), who deals with the case where $(r - g) = 0$ (see Whittaker 1944, p. 402).

Proof of Theorem 4 The differential equations corresponding to (3.1) are given by

$$\begin{aligned}\dot{k}_i &= \partial H_0/\partial q_i,\\ \dot{q}_i &= -\partial H_0/\partial k_i + (r-g)\,q_i, \quad i = 1, \ldots, n,\end{aligned}$$

where $H_0 = e^{(r-g)t}H$. We define a small deviation $\left(\tilde{k}, \tilde{p}\right)$ from the closed orbit by

$$\begin{aligned}k(t) &= \phi_k(t) + \tilde{k},\\ p(t) &= \phi_p(t) + \tilde{p},\end{aligned}$$

where $(\phi_k(t), \phi_p(t))$ define the periodic solution for (k, p). The corresponding linearized variational equations are given, in matrix form, by

$$\begin{bmatrix} \frac{d\tilde{k}}{dt} \\ \frac{d\tilde{p}}{dt} \end{bmatrix} = \left[\begin{array}{c|c} \frac{\partial^2 H}{\partial p \partial k} - gI & \frac{\partial^2 H}{\partial p^2} \\ \hline -\frac{\partial^2 H}{\partial k^2} & -\frac{\partial^2 H}{\partial k \partial p} + rI \end{array} \right] \begin{bmatrix} \tilde{k} \\ \tilde{p} \end{bmatrix}. \tag{3.8}$$

(see Whittaker 1944, p. 397, and also p. 268.) Let $\left(\tilde{k}', \tilde{p}'\right)$ represent another deviation from the closed orbit so that

$$\begin{bmatrix} \frac{d\tilde{k}'}{dt} \\ \frac{d\tilde{p}'}{dt} \end{bmatrix} = \left[\begin{array}{c|c} \frac{\partial^2 H}{\partial p \partial k} - gI & \frac{\partial^2 H}{\partial p^2} \\ \hline -\frac{\partial^2 H}{\partial k^2} & -\frac{\partial^2 H}{\partial k \partial p} + rI \end{array} \right] \begin{bmatrix} \tilde{k}' \\ \tilde{p}' \end{bmatrix}. \tag{3.9}$$

Multiplying (3.8) by $\left(-\tilde{p}', \tilde{k}'\right)$ and (3.9) by $\left(-\tilde{p}, \tilde{k}\right)$, and subtracting, we obtain[11]

$$\frac{d\left(\tilde{p}'\tilde{k} - \tilde{p}\tilde{k}'\right)}{dt} = (r - g)\left(\tilde{p}'\tilde{k} - \tilde{p}\tilde{k}'\right). \tag{3.10}$$

The right-hand side follows because all matrix terms, $[\partial^2 H/\partial p \partial k]$, $[\partial^2 H/\partial p^2]$, $[\partial^2 H/\partial p \partial k]$, and $[\partial^2 H/\partial p^2]$ cancel out, leaving terms corresponding to gI and rI. It follows that

$$\left(\tilde{p}'\tilde{k} - \tilde{p}\tilde{k}'\right)(t) = \left(\tilde{p}'\tilde{k} - \tilde{p}\tilde{k}'\right)(0)\, e^{(r-g)t}. \tag{3.11}$$

We now have to consider how values of (k, p) are transformed after a complete period. This is where the Poincaré map plays a role. Linearizing the Poincaré map around its fixed point corresponding to the closed orbit we obtain

$$\begin{bmatrix} \approx \\ k \\ \approx \\ p \end{bmatrix} = R \begin{bmatrix} \tilde{k} \\ \tilde{p} \end{bmatrix},$$

where $(\overset{\approx}{k}, \overset{\approx}{p})$ correspond to the values that $\left(\tilde{k}, \tilde{p}\right)$ acquire after the lapse of one period. The matrix R is the transformation matrix and it can be shown that one of its roots is unity (see Hartman 1964, Lemma 10.2, p. 251)). Combining the above transformation and the fact that it must satisfy (3.11), we can obtain information about the roots of R. Specifically, we have

[11](3.10) is where our proof has to differ from the proof given by Whittaker (1944). In our terminology, $r = g$ in the case considered by Whittaker and the right-hand side is zero.

$$\left(\tilde{\tilde{p}}'\tilde{\tilde{k}} - \tilde{\tilde{p}}\tilde{\tilde{k}}'\right) = \left[\tilde{p}'\tilde{k}'\right] R^T \left[\begin{array}{c|c} 0 & I \\ \hline -I & 0 \end{array}\right] R \begin{bmatrix} \tilde{p} \\ \tilde{k} \end{bmatrix}$$

$$= \left[\tilde{p}'\tilde{k}'\right] \left[\begin{array}{c|c} 0 & I \\ \hline -I & 0 \end{array}\right] \begin{bmatrix} \tilde{p} \\ \tilde{k} \end{bmatrix} e^{(r-g)\hat{t}},$$

where $\hat{t}$ denotes the period of the orbit and R^T is the transpose of R. Since the above analysis holds for any small deviation in $(k,\ p)$, we have the matrix equation

$$e^{(r-g)\hat{t}} \left[\begin{array}{c|c} 0 & I \\ \hline -I & 0 \end{array}\right] = R^T \left[\begin{array}{c|c} 0 & I \\ \hline -I & 0 \end{array}\right] R$$

or, if

$$\left[\begin{array}{c|c} 0 & I \\ \hline -I & 0 \end{array}\right] = S,$$

$$e^{(r-g)\hat{t}} S R^{-1} S^{-1} = R^T.$$

From this we see that R and $e^{(r-g)\hat{t}} R^{-1}$ have the same roots. If σ is a root of R, so is $e^{(r-g)\hat{t}}/\sigma$. Since the roots of R are the characteristic multipliers Theorem is proved. □

In terms of characteristic exponents the result of Theorem can be expressed as follows: If λ is a characteristic exponent so is $-\lambda + (r - g)$.

We can establish a stronger result if we constrain ourselves to the first bifurcation at $\hat{r}$. Let $\dot{x} = F(x,\ r)$ and for $r = \hat{r}$ let the Jacobian of $F(x,\ \hat{r})$, F_x, evaluated at $x(\hat{r})$ such that $F(x(\hat{r}),\ \hat{r}) = 0$, have $n - 2$ roots with negative real parts, n roots with positive real parts, and two pure imaginary roots giving rise to a Hopf Bifurcation. For $r < \hat{r}$ let $F_x(r)$ have n roots with positive and n roots with negative real parts. Periodic orbits will exist for values of $r(\epsilon)$, $\epsilon > 0$, where ϵ is defined in Theorem 1. The characteristic exponents of the orbit depend continuously on r and at $r = \hat{r}$ there are two characteristic exponents with zero real parts corresponding to the pure imaginary roots of F_x, $(n - 2)$ characteristic exponents corresponding to the $(n - 2)$ roots of F_x with negative real parts, and n characteristic exponents with positive real parts corresponding to the n characteristic roots of F_x with positive real pars (see Marsden 1976, p. 110 or Iooss 1972, remark on p. 319). For values of r in the neighborhood of $\hat{r}$, by the continuity of the exponents in r, $(n - 2)$ exponents have roots with negative and n exponents have roots with positive real parts. One exponent must always remain zero, as discussed earlier. The sign of the remaining exponent is given by $-2\mu_2 (d\alpha(\hat{r})/dr)$, where $\alpha(\hat{r}) = 0$ is the real part of the pair of pure imaginary roots of F_x at $\hat{r}$. μ_2 is defined in terms of the higher-order terms in the expansion of $F(x,\ \hat{r})$ around $x(\hat{r})$ such that $F(x(\hat{r}),\ \hat{r}) = 0$ (see Hopf 1976).

If $\mu_2 > 0$ ($\mu_2 < 0$) orbits will exist in a right (left) neighborhood of $\hat{r}$.[12] Thus for $d\alpha(r)/dr > 0$ and $\mu_2 > 0$ there will exist, by Theorem, an n-dimensional stable manifold around the orbit. For $d\alpha(r)/dr > 0$ and $\mu_2 < 0$ an $(n-1)$-dimensional stable manifold will exist around the orbit. If $d\alpha(r)/dr < 0$ we can redefine r as $(-r)$ and the same results follow.

Let the bifurcation of orbits for $r > \hat{r}$ ($r < \hat{r}$) be called supercritical (subcritical). Then we can formalize the above discussion in the following theorem.

Theorem 5 *Let the stationary solutions of $\dot{x} = F(x, r)$ be defined by $x(r)$. For $r = \hat{r}$ let the Jacobian $\partial F(x(\hat{r}), \hat{r})/\partial x$ have $(n-2)$ roots with negative real parts, n roots with positive real parts, and two pure imaginary roots giving rise to a Hopf Bifurcation. For $r < \hat{r}$ let $\partial F(x(\hat{r}), \hat{r})/\partial x$ have n roots with negative and n roots with positive real parts. Then supercritical orbits have an n-dimensional stable manifold while subcritical orbits have an $(n-1)$-dimensional stable manifold.*

Figures 3.1 and 3.2 illustrate how supercritical and subcritical Hopf Bifurcations can occur in two dimensions as a stationary point loses stability.

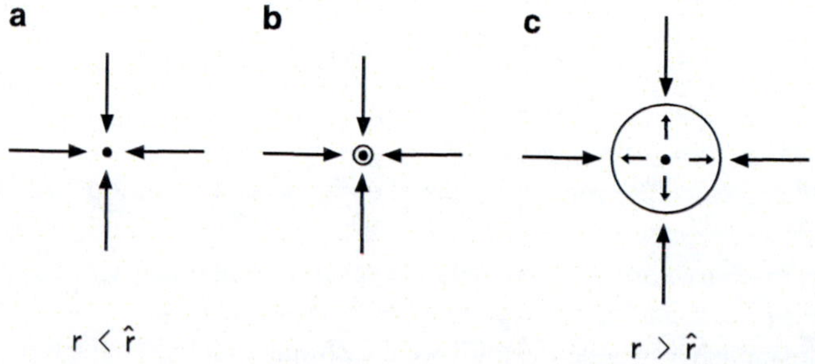

Fig. 3.1 Supercritical Hopf Bifurcation: Stable orbit bifurcates as the stationary point loses stability for $r > \hat{r}$

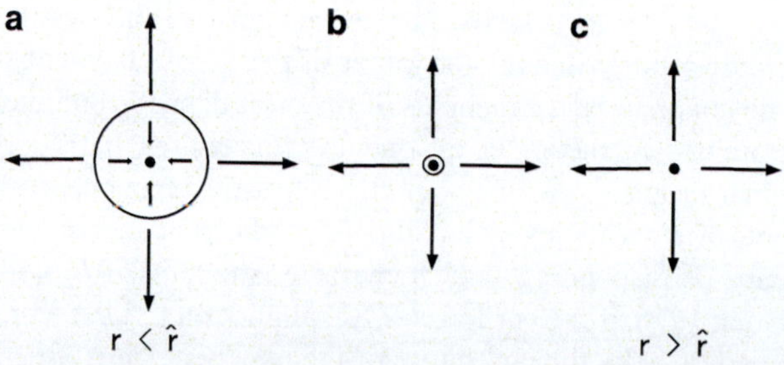

Fig. 3.2 Subcritical Hopf Bifurcation: Unstable orbit collapses onto the stationary point as the stationary point loses stability for $r > \hat{r}$

[12]For the case where $\mu_2 = 0$ see Hopf (1976).

Applying Theorem to our optimal growth problem we see that for supercritical orbits we can always find initial prices p, given values of k in the neighborhood of the orbit, to generate an optimal path $k(t)$, $p(t)$ converging to the orbit.

Finally, let us note that the orbits resulting from the Hopf Bifurcation are generic. Since only one of the characteristic multipliers is equal to unity for $|r-\hat{r}|<\epsilon$, $r \neq \hat{r}$, and $\mathrm{Re}\,(d\alpha(\hat{r})/dr) \neq 0$, orbits will persist under small perturbations of the vector field (see Theorem 1, p. 309, in Hirsch and Smale 1976). This establishes the structural stability of the system.

3.6 Appendix

(AI) Let Y_i and K_{ij} be the ith output and the jth input to produce the ith output. Y_0 is the consumption good and K_{i0} is the labor input to the ith good. Consider the following neoclassical technology:

$$Y_i = f^i(K_{io}, K_{i1}, \ldots, K_{in}), \quad i = 0, 1, \ldots, n. \tag{B1}$$

To obtain the social production function $c = T(y_1, \ldots, y_n, k_1, \ldots, k_n)$, we set

$$c = \max f^0(K_{00}, K_{01}, \ldots, K_{0n})$$

subject to

$$y_i = f^i(K_{i0}, K_{i1}, \ldots, K_{in}), \tag{B2}$$

$$1 = \sum_{1i=0}^{n} K_{i0},$$

$$k_i = \sum_{i=0}^{n} K_{ij}.$$

Lowercase letters represent aggregate quantities normalized by the total amount of labor. We restrict $(y_1, \ldots, y_n, k_1, \ldots, k_n)$ to be nonnegative.

(AII) *$T(y, k)$ is differentiable in (y, k) for $(y, k) > 0$, $c > 0$.*

Proof. To solve (B2) we write the Lagrangian:

$$L = f^0(K_{00} \ldots K_{0n}) + \sum_{i=0}^{n} p_i\left(f^i(K_{i0} \ldots K_{in}) - y_i\right)$$
$$+ w_0\left(1 - \sum_{i=0}^{n} K_{i0}\right) + \sum_{i=0}^{n} w_j\left(K_j - \sum_{i=0}^{n} K_{ij}\right).$$

To simplify notation and without loss of generality, we assume that all inputs are required for the production of every good. Otherwise we would appeal to assumption (A2) and a reduction in the number of first-order conditions would be required.

First-order conditions for the above problem are giving as follows:

$$
\begin{aligned}
p_t f_i^t - w_{ti} &= 0, \quad i, t = 0, 1, \ldots, n, \qquad \text{(B3)}\\
f^t - y_t &= 0, \quad t = 0, 1, \ldots, n,\\
1 - \sum_{i=0}^{n} K_{i0} &= 0,\\
k_i - \sum_{i=0}^{n} K_{ij} &= 0, \quad j = 1, \ldots, n,
\end{aligned}
$$

where $p_0 = 1$. The Jacobian with respect to $\left(K_{ij},\ p,\ w\right)$ is given by

$$
Q = \left[\begin{array}{ccccc|ccc}
p_0 f_{ij}^0 & \cdots & 0 & \cdots & 0 & 0 & \cdots & -I\\
\vdots & \ddots & \vdots & & \vdots & \vdots & & \vdots\\
\vdots & \cdots & p_1 f_{ij}^1 & \cdots & 0 & F_1 & \cdots & -I\\
\vdots & & \vdots & \ddots & \vdots & \vdots & & \vdots\\
0 & \cdots & 0 & \cdots & p_n f_{ij}^n & F_n & \cdots & -I\\
\hline
0 & \cdots & F_1' & \cdots & F_n' & 0 & \cdots & 0\\
\vdots & & \vdots & & \vdots & \vdots & & \vdots\\
-I & \cdots & -I & \cdots & I & 0 & \cdots & 0
\end{array}\right]
= \left[\begin{array}{c|c} R & P \\ \hline P' & 0 \end{array}\right],
$$

where $\left[f_{ij}^t\right]$ is the Hessian matrix of the production function f^t and

$$
F_i = \left[\begin{array}{ccc}
\underbrace{\begin{matrix} 0 \cdots 0 \\ \vdots \quad \vdots \\ 0 \cdots 0 \end{matrix}}_{i-1} &
\begin{matrix} f_1^i \\ \vdots \\ f_n^i \end{matrix} &
\underbrace{\begin{matrix} 0 \cdots 0 \\ \vdots \quad \vdots \\ 0 \cdots 0 \end{matrix}}_{n-i}
\end{array}\right].
$$

If Q is nonsingular, by the implicit function theorem, $K_{ij}(y, k)$, $p_i(y, k)$, and $w_i(y, k)$, $i, j = 0, 1, \ldots, n$, are locally differentiable functions (see Hirsch and Smale 1976). Then $c = f^0(K_{00}(y, k), K_{01}(y, k), \ldots, K_{0n}(y, k))$ must be differentiable in y and k. The proof of the nonsingularity of Q is due to Hirota and Kuga (1971). We sketch it below.

Assume Q is singular. Then there exists a vector x such that $Qx = 0$. Define $x = [x_1, x_{21}, x_{22}]$. Then $x'Qx = x_1 R x_1$, where R is negative semidefinite since it contains the Hessians $[f^t_{ij}]$ along the diagonal and zeros everywhere else. If $x_1 R x_1 = 0$ for $x_1 \neq 0$, then

$$x_1 = \begin{pmatrix} \lambda_0 K^0 \\ \vdots \\ \lambda_n K^n \end{pmatrix}$$

where K^i is the input vector for f^i. But then $P'x_1 = \sum \lambda_i F_i' K^i$. But by Euler's theorem $F_i' K^i \neq 0$ since one of the elements of $F_i' K_i = \sum_{j=0}^n f^i_j K_{ij} = f^i \neq 0$ by hypothesis. Then $\lambda_i = 0,\ i = 0, 1, \ldots, n$, and $x_1 = 0$. But if $x_1 = 0$, the first row of Q implies $-I x_{22} = 0$ and $x_{22} = 0$. If $x_{22} = 0$ and $x_1 = 0$, from Qx we get $F_i x_{21} = 0$ for $i = 1, \ldots, n$. Then $x_{21} = 0$ unless all $f^i_j = 0$ for $i = 1, \ldots, n$. But if $f^i_j = 0$, by Euler's theorem $\sum_j f^i_j K_{ij} = f^i = 0$, which contradicts the hypothesis $f^i > 0,\ i = 0, 1, \ldots, n$. Thus $x_1 = 0,\ x_{21},\ x_{22} = 0$. But then Q is nonsingular. □

(AIII) *Let (A6) hold. Then* $y(k, p)$ *and* $w(k, p)$ *are twice differentiable functions in* $(k, p) > 0$ *for* $c > 0,\ y > 0$.

Proof. As in the proof of (AII) we assume without loss of generality that all inputs are required to produce every output. This simplifies notation. We differentiate (B3) with respect to K_{ij}, y, and w. This yields the Jacobian

$$M = \left[\begin{array}{c|cc} R & 0 & -I \\ & 0 & -I \\ \hline P' & I & 0 \\ & 0 & 0 \end{array}\right],$$

where R and P are as in the proof of (AII). If M is singular

$$M \begin{pmatrix} x_1 \\ x_2 \\ x_3 \end{pmatrix} = 0$$

and

$$\text{(i)} \qquad R x_1 - \begin{bmatrix} I \\ \vdots \\ I \end{bmatrix} x_3 = 0,$$

$$\text{(ii)} \qquad [0,\ F_1, \ldots,\ F_n] x_1 + x_2 = 0,$$

$$\text{(iii)} \qquad [\,I\ , \ldots,\ I\,]\, x_1 = 0.$$

Hence $x_1 R x_1 = 0$. By (A1) the Hessians of each production function, occurring in R, are negative-semidefinite. Thus $x_1' R x_1 = 0$ implies that

$$x_1 = \begin{pmatrix} \lambda_0 K^0 \\ \vdots \\ \lambda_n K^n \end{pmatrix}.$$

Then (iii) implies $[I \cdots I] x_1 = \left(\sum_{i=0}^{n} \lambda_i K_j^i\right) = 0$. By assumption (A6) the input coefficient matrix is nonsingular. Thus $\lambda_1, \ldots, \lambda_n = 0$ and $x_1 = 0$. From (i), $x_1 = 0$ implies $x_3 = 0$ and from (ii), $x_1 = 0$ implies $x_2 = 0$. But then M is nonsingular. The implicit function then implies that $y(k, p)$ and $w(k, p)$ are differentiable in (k, p) for $k, p, y > 0$ and $c > 0$. □

(AIV) *At the steady state corresponding to* $r = 0.248$ *for the technology given in Sect. 3.3, the pure imaginary roots of the matrix* J, *defined in Sect. 3.2, have real parts that are not stationary with respect to* r.

Proof. Let the pure imaginary roots of Jacobian matrix J corresponding to our Cobb-Douglas Technology be given by $u(r) \pm v(r) i = 0 + 1.56i$. They are the roots of $-B^{-1} + rI$, where $B = [A - (1/a_{00})\, a_{\cdot 0} a_{0 \cdot}]$ is defined in Sect. 3.3. Let the roots of B be $\alpha \pm \beta i$. The roots of B^{-1} are then

$$\frac{\alpha}{\alpha^2 + \beta^2} \pm \frac{\beta}{\alpha^2 + \beta^2} i.$$

The real parts of the imaginary roots of $-B^{-1} + rI$ are

$$u(r) = \frac{\alpha(r)}{\alpha^2(r) + \beta^2(r)} + r$$

and we would like to show that this quantity is not stationary with respect to r. Differentiating, we obtain

$$\frac{du(r)}{dr} = \left(\frac{-\alpha'(r)(\alpha^2(r)) + \beta^2(r) + 2\alpha(r)^2 \alpha'(r) + 2\beta(r)\alpha(r)\beta'(r)}{(\alpha^2(r) + \beta^2(r)^2)} + 1\right). \tag{B4}$$

To evaluate, we have to calculate $\alpha'(r) = d\alpha(r)/dr$ and $\beta'(r) = d\beta(r)/dr$.

First note that $\alpha(r) = 1/2$ Trace B and $\beta^2(r) = \det B - \alpha^2(r)$. Then

$$\alpha'(r) = \frac{1}{2} \frac{d\,(\text{Trace } B)}{dr} \quad \text{and} \quad \beta'(r) = \frac{1}{2\beta} \frac{d\,(\det B)}{dr} - 2\alpha(r)\alpha'(r).$$

We also have

$$\frac{d\,(\text{Trace } B)}{dr} = \frac{db_{11}}{dr} + \frac{db_{22}}{dr}$$

and

$$\frac{d(\det B)}{dr} = \frac{db_{11}}{dr}b_{22} + b_{11}\frac{db_{22}}{dr} - b_{21}\frac{db_{12}}{dr} - b_{12}\frac{db_{21}}{dr}.$$

We have to calculate the elements of B from the elements of $\begin{bmatrix} a_{00} & a_{\cdot 0} \\ a_{0\cdot} & A \end{bmatrix}$.

The change in the elements of the input coefficient matrix can be calculated using the Allen partial elasticity of substitution coefficients σ_{ij} (see Allen 1968, p. 509):

$$\frac{da_{ij}}{dr} = \sum_k \frac{a_{ij}}{w_k}\frac{a_{kj}w_k}{p_j}\sigma_{ik}\frac{dw_k}{dr} = a_{ij}\sum_k \frac{a_{kj}}{p_j}\sigma_{ik}\frac{dw_k}{dr}.$$

For Cobb-Douglas production functions $\sigma_{ik} = 1,\ i \neq k$. Furthermore $\sum_{i\neq k} m_i\sigma_{kk}$, where m_i is the proportion of the cost of factor i to total cost. Thus $\sum_i m_i = 0$. This yields $\sum_{i\neq k} m_i = 1 - m_k = m_k\sigma_{kk}$ since $\sigma_{ik} = 1$ for $i \neq k$. We obtain $\sigma_{kk} = (1 - m_k)/m_k$. For the Cobb-Douglas case $m_i = \alpha_i$, where α_i is the exponent of factor i appearing in the production function.

To calculate dw/dr we first find dp/dr. This is given as $dp/dr = pA[I - rA]^{-1}$, where p is the price vector of capital goods in terms of the wage rate and A is the capital input coefficient matrix (see Burmeister and Dobell 1970, Chap. 9, Theorem 3). Since $w = rp$ at the steady state $dw/dr = p + rdp/dr$ (Note that we normalized by the wage rate w_0 so that $dw_0 = 0$).

The rates of change of the elements of the matrix B is given as follows:

$$\frac{db_{ij}}{dr} = \frac{da_{ij}}{dr} - \frac{a_{0j}}{a_{00}}\frac{da_{10}}{dr} - \frac{a_{10}}{a_{00}}\frac{da_{0j}}{dr} + \frac{a_{i0}a_{0j}}{a_{00}^2}\frac{da_{00}}{dr}.$$

We have expressed all elements of db_{ij}/dr in terms of known quantities. Tedious calculations yield

$$\frac{d\left[b_{ij}\right]}{dr} = \begin{bmatrix} -0.4036 & -1.2216 \\ 2.8907 & -0.4365 \end{bmatrix}$$

and

$$\frac{d\alpha(r)}{dr} = \frac{1}{2}\text{Trace}\left[\frac{d\left[b_{ij}\right]}{dr}\right] = -0.420,$$

$$\frac{d\beta(r)}{dr} = \frac{1}{2\beta}\left(\frac{d\left(\det\left[b_{ij}\right]\right)}{dr} - 2\alpha(r)\frac{d\alpha(r)}{dr}\right) = 1.249.$$

Substituting into expression (B4) we obtain

$$\frac{du(r)}{dr} = \frac{d\left(\frac{\alpha(r)}{\alpha^2(r)+\beta(r)^2} + r\right)}{dr} = 1.311.$$

This shows that for our example the real parts of the imaginary roots of J are not stationary with respect to the parameter r. □

References

Allen, R. G. D. (1968), Mathematical Analysis for Economists, St. Martin's Press, New York.

Araujo, A. and J. A. Scheinkman (1977), "Smoothness, Comparative Dynamics and the Turnpike Property," *Econometrica*, 45, 601–620.

Benhabib, J. and K. Nishimura (1979), "On the Uniqueness of Steady States in an Economy with Heterogeneous Capital Goods," *International Economic Review*, 20, 59–82.

Benhabib, J. and K. Nishimura (1978), Multiple Equilibria and Stability in Growth Models, working paper.

Benveniste, L. M. and J. A. Scheinkman (1979), "Differentiable Value Functions in Concave Dynamic Optimization Problems," *Econometrica*, 47, 727–732.

Brock, W. (1973), "Some Results on the Uniqueness of Steady States in Multisector Models of Optimum Growth when Future Utilities are Discounted," *International Economic Review*, 14, 535–559.

Brock, W. A. and J. A. Scheinkman (1976), "Global Asymptotic Stability of Optimal Control Systems with Applications to the Theory of Economic Growth," *Journal of Economic Theory*, 12, 164–190.

Burmeister, E. and A. R. Dobell (1970), Mathematical Theories of Economic Growth, Macmillan & Co., London.

Burmeister, E. and D. Graham (1973), Price Expectations and Stability in Descriptive and Optimally Controlled Macro-Economic Models, *J. E. E. Conference Publication No. 101*, Institute of Electrical Engineers, London.

Burmeister, E. and K. Kuga (1970), "The Factor Price Frontier in a Neoclassical Multi-sector Model," *International Economic Review*, 11, 163–176.

Cass, D. and K. Shell (1976), "The Structure and Stability of Competitive Dynamical Systems," *Journal of Economic Theory*, 12, 31–70.

Debreu, G. and I. N. Hernstein (1953), "Non-negative Square Matrices," *Econometrica*, 21, 597–607.

Gantmacher, F. R. (1960), The Theory of Matrices, Vol. 1, 2, Chelsea, New York.

Hartman, P. (1964), Ordinary Differential Equations, Wiley, New York.

Hirota, M. and K. Kuga (1971), "On an Intrinsic Joint Production," *International Economic Review*, 12, 87–105.

Hirsch, M. W. and S. Smale (1976), Differential Equations, Dynamic Systems, and Linear Algebra, Academic Press, New York.

Hopf, E. (1976), Bifurcation of a Periodic Solution from a Stationary Solution of a System of Differential Equations, translated by L. N. Howard and N. Kopell, in *The Hopf Bifurcation and Its Applications*, ed. by Marsden, J. E. and M. McCracken, Springer-Verlag, New York, 163–194.

Iooss, G. (1972), "Existence et stabilité de la solution périodique secondaire intervenant dans les problèmes d'évolution du type Navier-Stokes," *Archive for Rational Mechanics and Analysis*, 49, 301–329.

Joseph, D. D. and D. H. Sattinger (1972), "Bifurcating Time–periodic Solutions and their Stability," *Archive for Rational Mechanics and Analysis*, 45, 79–109.

Kelly, J. S. (1969), "Lancaster vs Samuelson on the Shape of the Neoclassical Transformation Surface," *Journal of Economic Theory*, 1, 347–351.

Kurz, M. (1968), "The General Instability of a Class of Competitive Growth Processes," *Review of Economic Studies*, 102, 155–174.

Kurz, M. (1978), "Optimal Economic Growth and Wealth Effects," *International Economic Review*, 9, 348–357.

Lancaster, K. (1968), Mathematical Economics, Macmillian & Co., London.

Liviatan, N. and P. A. Samuelson (1969), "Notes on Turnpikes: Stable and Unstable," *Journal of Economic Theory*, 1, 454–475.

Magill, M. J. P. (1977), "Some New Results on the Local Stability of the Process of Capital Accumulation," *Journal of Economic Theory*, 15, 174–210.

Marsden, E. J. (1976), and M. McCracken, The Hopf Bifurcation and its Applications, Springer-Verlag, New York.

McKenzie, L. W. (1955), "Equality of Factor Prices in World Trade," *Econometrica*, 23, 239–257.

McKenzie, L. W. (1976), "Turnpike Theory," *Econometrica*, 44, 841–865.

Ruelle, D. and F. Tackens (1971), "On the Nature of Turbulence," *Communications in Mathematical Physics*, 20, 167–192.

Pitchford, J. D. and S. J. Turnovsky (1977), Applications of Control Theory to Economic Analysis, North-Holland, New York.

Samuelson, P. A. (1953–1954), "Prices of Factors and Goods in General Equilibrium," *Review of Economic Studies*, 21, 1–20.

Sattinger, D. H. (1973), Topics in Stability and Bifurcation Theory, Springer-Verlag, New York.

Schmidt, D. S. (1976), Hopf Bifurcation Theorem and the Center Theorem of Liapunov, in *The Hopf Bifurcation and Its Applications*, ed. by Marsden, J. E. and M. McCracken, Springer-Verlag, New York, 95–103.

Scheinkman, J. A. (1976), "On Optimal Steady States of n-Sector Growth Models when Utility is Discounted," *Journal of Economic Theory*, 12, 11–30.

Scheinkman, J. A. (1978), "Stability of Separable Hamiltonians and Investment Theory," *Review of Economic Studies*, 45, 559–570.

Whittaker, E. T. (1944), A Treatise on the Analytical Dynamics of Particles and Rigid Bodies, Dover, New York.

Chapter 4
Competitive Equilibrium Cycles*

Jess Benhabib and Kazuo Nishimura**

4.1 Introduction

Recently there has been a surge of interest in endogenous business cycles that arise in competitive laissez-faire economies. In the context of standard overlapping generations economies, conditions for the existence of equilibrium cycles have been given by Grandmont (1983) and by Benhabib and Day (1982). Models of the economy with extrinsic uncertainty or "sunspots" that have been developed by Shell (1977) and by Cass and Shell (1983) (see also Balasko (1983)) can also lead to equilibrium cycles. The relation between sunspot equilibria and the existence of deterministic cycles has been explored in a recent paper by Azariadis and Guesnerie (1983). A search model where beliefs of agents also play a role in generating endogenous cycles is given in a paper by Diamond and Fudenberg (1983). Although the equilibrium cycles in many of the works cited above arise in a competitive framework, they are not necessarily Pareto-efficient and they allow the possibility of policy intervention. In contrast, works by Kydland and Prescott (1982) and Long

*Journal of Economic Theory 35, 284–306, 1985.

**Jess Benhabib's research was supported in part by the National Science Foundation Grant SES 8308225, and by the C. V. Starr Center for Applied Economics at New York University. We would like to thank the referees for pointing out errors as well as making very valuable suggestions. We are also grateful to Clive Bull and Chuck Wilson for enlightening discussions. For any remaining errors, we are entirely responsible.

J. Benhabib (✉)
Department of Economics, New York University, USA
e-mail: jess.benhabib@nyu.edu

K. Nishimura
Institute of Economic Research, Kyoto University, Japan
e-mail: nishimura@kier.kyoto-u.ac.jp

J. Stachurski et al. (eds.), *Nonlinear Dynamics in Equilibrium Models*,
DOI 10.1007/978-3-642-22397-6_4,

and Plosser (1983) develop models with a single infinitely lived representative agent that generate Pareto-optimal equilibria. They then explore the possibility of business fluctuations in these models by using simulation methods.

The purpose of this paper is to provide economically interpretable sufficient conditions under which equilibrium cycles arise in a deterministic, perfect foresight model with an infinitely lived representative agent and a neoclassical technology. One of our main results, obtained for a standard two-sector model of production, gives sufficient conditions for generating "robust" (see the Remark following the proof of Corollary 1) periodic cycles in outputs, stocks, and relative prices. These sufficient conditions are expressed in terms of the discount rate and differences in the relative factor-intensities of the two industries. The main contribution is to give an insight into the structure of general classes of neoclassical technologies that lead to equilibrium cycles in a fully stationary environment.

While we obtain sufficient conditions for persistent cycles in a two-sector discrete-time model, similar results hold in multisector continuous time models. Benhabib and Nishimura (1979b) provide a general method for constructing periodic equilibrium trajectories with a multisector Cobb-Douglas technology in a continuous time framework. The structure of their examples suggests that the existence of cycles can be linked to generalized capital intensity conditions when the rate of time preference is not too close to zero. So far, however, there have not been any sufficient conditions for the existence of cycles given in the literature.

In the next section Theorem 1 gives sufficient conditions for the existence of optimal periodic trajectories in an abstract and general setting. Section 4.3 discusses existing examples of cycles due to Sutherland (1979) and due to Weitzman as reported in Samuelson (1973), especially as they relate to Theorem 1.[1] In Sect. 4.4 Theorem 2 gives general conditions under which the equilibrium trajectory is globally monotonic or oscillatory and Theorem 3 gives conditions under which the optimal equilibrium trajectory converges either to a stationary point or to a cycle of period two. We obtain our main results in Sect. 4.5 by applying Theorem 1 to a two-sector neoclassical technology. We give capital-intensity conditions that lead to persistently cyclical, efficient equilibrium trajectories of outputs, stocks, and relative prices. We also discuss an application to the adjustment-cost model of investment. Finally, in Sect. 4.6 we summarize the intuitive explanation for the existence of cyclical equilibria that arise in neoclassical technologies.

4.2 The Existence and Stability of Periodic Cycles

A technology set D is defined as a closed convex subset of $R^2_+ = \{x \in R^2 \mid x \geqslant 0\}$, where $(x_1, x_2) \in D$ represents the input stock x_1 and a feasible output stock x_2.

---

[1] Although we explore the existence of cyclical trajectories, our framework of analysis is also in the spirit of turnpike theory. For an excellent survey of turnpike results see McKenzie (1981a, 1981b).

If $x_2 > x_1$ and $(x_1, y_1) \in D$, we assume that there exists a $y_2 > y_1$ such that $(x_2, y_2) \in D$. Furthermore, assume that there exists stock level $\overline{x} > 0$ such that if $x > \overline{x}$ and $(x, y) \in D$, then $y \leqslant x$, and if $x > 0$ and $x < \overline{x}$, then there exists some y such that $(x, y) \in D$ and $y > x$. Also assume that $(x, y) \in D$ implies $(x, y') \in D$ for $0 \leqslant y' \leqslant y$. Let $\overline{D} = \{(x, y) \in D | 0 \leqslant x \leqslant \overline{x}\}$. $\overline{D}$ is clearly a compact and convex subset of R^2_+.

We consider the following problem:

$$\max W(k_0) = \sup \sum_{t=0}^{\infty} \delta^t V(k_t, k_{t+1})$$

subject to

$$(k_t, k_{t+1}) \in \overline{D} \tag{4.1}$$

$$0 \leqslant k_0 \leqslant \overline{x} \quad \text{given } t \geq 0,$$

$$0 < \delta < 1.$$

(A1) *V is continuous and concave on $\overline{D}$, of class C^2 on the interior of $\overline{D}$, with $V_{11}, V_{22} < 0$, $V_{11}V_{22} - V_{12}^2 \geqslant 0$ and $|V(x, y)| < \infty$ for $(x, y) \in \overline{D}$.*

It can be shown that under (A1), problem (4.1) has an optimal solution (see Brock (1970) and Majumdar (1975)). A path is an interior Euler path if it satisfies $V_2(k_{t-1}, k_t) + \delta V_1(k_t, k_{t+1}) = 0$ and $\overline{x} > k_t > 0$ for all t. A steady state $\overline{k}(\delta)$ is a solution of $V_2(k, k) + \delta V_1(k, k) = 0$.

(A2) *A steady state $\overline{k}(\delta)$ exists on $(0, \overline{x})$ for $\delta \in [\underline{\delta}, 1]$, where $1 > \underline{\delta} > 0$.*
(A3) *$V_{12}(x_1, x_2) < 0$ for $(x_1, x_2) \in$ interior $\overline{D}$.*[2]

Lemma 1. *The steady state $\overline{k}(\delta)$ is a differentiable function on $[\underline{\delta}, 1]$.*

Proof. Consider $H(k, k; \delta) = V_2(k, k) + \delta V_1(k, k)$ on $\overline{D} \times (\underline{\delta}, 1)$. For any δ, $dH/dk < 0$ by (A1)–(A3) and thus for any $\delta \in (\underline{\delta}, 1)$ there is a unique $\overline{k}(\delta)$ such that $H(\overline{k}(\delta), \delta) = 0$. The differentiability follows from the implicit function theorem.

Assumptions (A3) and (A4) (which follows) will be interpreted in Sect. 4.5 in terms of capital intensity conditions for a neoclassical economy.

[2]This assumption can be relaxed to hold only on some relevant subset of $\overline{D}$, but the additional notation need would be cumbersome. For a class of models where $(0, y) \in \overline{D}$ implies $y = 0$ (the impossibility of the Land of Cockaigne), $V_{12} < 0$ may hold only in the interior of $\overline{D}$ as in the two-sector model of Sect. 4.5 below, or in other models only on some subset of $\overline{D}$. Note that for (k_t, k_{t+1}) in the neighborhood of $(\overline{k}(\delta^-), \overline{k}(\delta^-))$, (A4) implies $V_{12} < 0$ and the steady state is locally unique. If we choose the set B in the proof of Theorem 1 small enough to exclude other steady states that may exist when (A3) does not hold, we can dispense with (A3) in proving Theorem 1.

(A4) *There exists* δ^-, δ^+ *with* $[\delta^-, \delta^+] \subset [\delta, 1]$ *such that*

(i) $[V_{22}\overline{k}(\delta^-), \overline{k}(\delta^-)) + \delta^- V_{11}(\overline{k}(\delta^-), \overline{k}(\delta^-))] > (1+\delta^-)V_{12}(\overline{k}(\delta^-), \overline{k}(\delta^-))$
(ii) $[V_{22}\overline{k}(\delta^+), \overline{k}(\delta^+)) + \delta^+ V_{11}(\overline{k}(\delta^+), \overline{k}(\delta^+))] < (1+\delta^+)V_{12}(\overline{k}(\delta^+), \overline{k}(\delta^+))$.

Remark. Requirement (A4)(ii) that a δ^+ exists is only mildly restrictive. If $V(k_{t+1}, k_t)$ is strongly concave, $V_{11} + V_{22} < 2V_{12}$ and setting δ^+ sufficiently close or equal to 1 satisfies (A4(ii)). If $V(k_{t+1}, k_t)$ is only concave δ^+ exists if $V_{11} \neq V_{22}$. To see this, note that if $V_{11} + V_{22} = 2V_{12}$, the concavity of V implies $V_{11}V_{22} \geqslant V_{12}^2 = \frac{1}{4}(V_{11}^2 + 2V_{11}V_{22} + V_{22}^2)$, which in turn gives $0 \geqslant (V_{11} - V_{22})^2$. This holds only if $V_{11} = V_{22}$. Thus the main restriction imposed by (A4) is that a δ^- exists. We use (A4) in establishing the properties of the roots of (4.3) (see Lemma 2).

Remark. In our one capital good model (A4(ii)) is the same as the dominant diagonal conditions that Araujo and Scheinkman (1977) use to obtain global stability results in a multisector economy. In this paper, however, we require (A4(i)) and (A4(ii)) to hold only at the steady state, rather than along the optimal path. The result (ii) of Lemma 2 shows that the dominant diagonal condition (A4(ii)) implies the local asymptotic stability of the steady state. In a multisector context, Dasgupta and McKenzie (1984) have shown that if $[V_{12}]$ is symmetric, then the dominant diagonal condition is both necessary and sufficient for local asymptotic stability.

For later use we also define the following sets:

$$P^- = \left\{\delta | V_{22} + \delta V_{11} > (1+\delta)V_{12} \text{ at } \overline{k}(\delta); \overline{k}(\delta) \in \overline{D}\right\}$$

$$P^+ = \left\{\delta | V_{22} + \delta V_{11} < (1+\delta)V_{12} \text{ at } \overline{k}(\delta); \overline{k}(\delta) \in \overline{D}\right\} \tag{4.2}$$

$$P^0 = \left\{\delta | V_{22} + \delta V_{11} = (1+\delta)V_{12} \text{ at } \overline{k}(\delta); \overline{k}(\delta) \in \overline{D}\right\}.$$

The roots of the Jacobian of an Euler path evaluated at a steady state will be the solutions of

$$\delta V_{21}\lambda^2 + (V_{22} + \delta V_{11})\lambda + V_{12} = 0. \tag{4.3}$$

It is easily shown that both roots of (4.3) are real and that $V_{12} < 0$ implies that both are negative. We will need the following lemmas.

Lemma 2. *Let (A1)–(A3) hold and* λ_1 *and* λ_2 *be the roots of (4.3).*

(i) If $V_{22} + \delta V_{11} > (1+\delta)V_{12}, \lambda_1, \lambda_2 < -1$.

(ii) If $V_{22} + \delta V_{11} < (1+\delta)V_{12}, \lambda_1 \in (-\infty, -1), \lambda_2 \in (-1, 0)$.
(iii) If $V_{22} + \delta V_{11} = (1+\delta)V_{12}, \lambda_1 = -1/\delta,\ \lambda_2 = -1$.[3]

Proof. Dividing (4.3) by δV_{21} we note that $\lambda_1 + \lambda_2 = -(V_{22} + \delta V_{11})(\delta V_{12})^{-1}$ and $\lambda_1\lambda_2 = \delta^{-1}$. Therefore, by the concavity of V, and remembering that V_{11}, $V_{22} < 0$,

$$\begin{aligned} \lambda_1 + \lambda_2 + 2 &= (\delta V_{12})^{-1}\left[-V_{22} - \delta V_{11} + 2\delta V_{12}\right] \\ &\leqslant (\delta V_{12})^{-1}\left[-V_{22} - \delta V_{11} - 2(\delta V_{11}V_{22})^{1/2}\right] \\ &= (\delta V_{12})^{-1}\left[(|V_{22}|)^{1/2} - (|\delta V_{11}|)^{1/2}\right]^2 \leqslant 0. \end{aligned} \tag{4.4}$$

Also, we have

$$\begin{aligned} (\lambda_1 + 1)(\lambda_2 + 1) &= 1 + \lambda_1\lambda_2 + \lambda_1 + \lambda_2 \\ &= \delta^{-1}\left[1 + \delta - (V_{12})^{-1}(V_{22} + \delta V_{11})\right]. \end{aligned} \tag{4.5}$$

Combining (4.4) and (4.5) the results of the lemma follow immediately.

Lemma 3. *There exists a differentiable function* F: *interior* $\overline{D} \rightarrow R$ *which satisfies* $k_{1+2} = F(k_t, k_{t+1}; \delta)$ *along an interior Euler path.*

Proof. Let $\{k_t\}$ be an interior Euler path. Then it satisfies $0 = S(k_{t+2}, k_{t+1}, k_t; \delta) = V_2(k_t, k_{t+1}) + \delta V_1(k_{t+1}, k_{t+2})$. Since $dS/dk_{t+2} < 0$ by (A1)–(A3), along an Euler path k_{t+2} is unique for each (k_{t+1}, k_t) and $\delta \in (\boldsymbol{\delta}, 1)$ such that $S = 0$. The differentiability follows from the implicit function theorem.

Remark. Note that $F_1 \equiv \partial F(k_t, k_{t+1}; \delta)/\partial k_t = -V_{21}(k_t, k_{t+1})(\delta V_{12}(k_{t+1}, k_{t+2}))^{-1}$, $F_2 \equiv \partial F(k_t, k_{t+1}; \delta)/\partial k_{t+1} = -(V_{22}(k_t, k_{t+1}) + \delta V_{11}(k_{t+1}, k_{t+2}))(\delta V_{12}(k_{t+1}, k_{t+2}))^{-1}$.

Theorem 1. *Under (A1)–(A4), there exists a nonempty set* $L \subset [\delta^-, \delta^+]$ *such that for* $\delta \in L$, *problem (4.1) has optimal solution trajectories which are cycles of period two.*

Proof. We can convert F in Lemma 3 to a first-order system as follows:

$$\begin{aligned} k_{t+2} &= F(y_{t+1}, k_{t+1}; \delta) \\ y_{t+2} &= k_{t+1}. \end{aligned} \tag{4.6}$$

Consider also the following system obtained by expressing (y_{t+1}, k_{t+1}) in the terms of (y_t, k_t), using (4.7) and substituting back in (4.6):

[3] For higher order systems similar conditions can be obtained by exploiting Schur-Cohn conditions. See Marden (1949).

$$k_{t+2} = F(k_t, F(y_t, k_t; \delta), \delta) \tag{4.7}$$
$$y_{t+2} = F(y_t, k_t; \delta).$$

Clearly the fixed points of (4.6) are fixed points of (4.7) but not vice versa. The fixed points of (4.7) that are not fixed points of (4.6) correspond to periodic points of (4.6) of period two. We will prove that such fixed points of (4.7) exist.

Consider the vector function on the interior of $\overline{D}$:

$$M(\delta, k, y) = \begin{pmatrix} k - F(k, F(y, k; \delta)) \\ y - F(y, k; \delta). \end{pmatrix}. \tag{4.8}$$

A stationary point of (4.7) is given by $M(\delta, k, y) = 0$. A simple calculation shows that if $[G]$ is the Jacobian of the right side of (4.6), evaluated at a stationary point of (4.6), $[G]^2$ is the Jacobian of the right side of (4.7). Thus if λ_1 and λ_2 are the roots of the right side of (4.6) evaluated at the steady state $\overline{k}(\delta)$, the roots of the right side of (4.8) will be given by $(1 - \lambda_1^2)$ and $(1 - \lambda_2^2)$. Note that λ_1 and λ_2 are also the roots of (4.3). Consider an interval $[\delta^-, \delta^+]$, where δ^- and δ^+ are defined by (A4). We can consider $M(\delta, k, y)$ as a homotopy on the interior of $\overline{D}$ over $[\delta^-, \delta^+]$.

(i) Consider first the case where the set $Z(\delta) = \{(k, y) | M(\delta, k, y) = 0\}$ does not, for any $\delta \in [\delta^-, \delta^+]$, intersect the boundary of a nonempty, convex, open subset B of $\overline{D}$ which contains the steady state $\overline{k}(\delta)$ in its interior. By construction and by Lemma 2, for any $\delta \in [\delta^-, \delta^+] \cap P^-$ (where P^- is defined by (4.2)), the Jacobian determinant of the right side of (4.8), that is, of $M(\delta, \overline{k}(\delta), \overline{k}(\delta))$, evaluated at the steady state of (4.6), will be $(1 - \lambda_1^2(\delta))(1 - \lambda_2^2(\delta)) > 0$. Also, for any $\delta \in [\delta^-, \delta^+] \cap P^+$, the Jacobian determinant of $M(\delta, \overline{k}(\delta), \overline{k}(\delta))$, evaluated at the steady state of (4.6) corresponding to δ, will be $(1 - \lambda_1(\delta)^2)(1 - \lambda_2(\delta)^2) < 0$. But the topological degree of $M(\delta, k, y)$ over the boundary of B is homotopy invariant. This implies that if the Jacobian of $M(\delta, k, y)$ is $[J(\delta)]$, then $\sum_{(k,y) \in Z(\delta)}$ sign $Det[J(\delta)]$ is constant over $\delta \in [\delta^-, \delta^+]$ (see Milnor (1965)). But since $Det[J(\delta)]$ evaluated at the steady state of (4.6) changes sign as δ crosses from $[\delta^-, \delta^+] \cap P^-$ to $[\delta^-, \delta^+] \cap P^+$, either $M(\delta, k, y) = 0$ has at least two solutions with $Det[J(\delta)] < 0$ (i.e., solutions other than the steady state of (4.6)) in B for all $\delta \in [\delta^-, \delta^+] \cap P^-$, or $M(k, y; \delta) = 0$ has at least two solutions with $Det[J(\delta)] > 0$ (i.e., solutions other than the steady state of (4.6)) in B for all $\delta \in [\delta^-, \delta^+] \cap P^+$. Since such solutions are stationary points of (4.7) and periodic paths of (4.6), the set L in Theorem 1 is either $[\delta^-, \delta^+] \cap P^-$ or $[\delta^-, \delta^+] \cap P^+$.

(ii) Now consider the case where no matter how we choose a convex open subset of B in $\overline{D}$ containing the steady state $\overline{k}(\delta)$ in its interior, $Z(\delta)$ intersects the boundary of B for some $\delta \in [\delta^-, \delta^+]$ for any feasible choice of the interval $[\delta^-, \delta^+]$. Since $(\overline{k}(\delta), \overline{k}(\delta))$ stays in the interior of $\overline{D}$ by (A2) and $(\overline{k}(\delta), \overline{k}(\delta))$ changes continuously with δ by Lemma 1, it must be the case that $Z(\delta)$ contains other points than $(\bar{k}(\delta), \bar{k}(\delta))$ for some $\delta \in [\delta^-, \delta^+]$, which by construction are periodic solutions of (4.6).

We note that (4.6) gives the first order conditions for the problem given by (4.1). However, since $V(k_t, k_{t+1})$ in problem (4.1) is concave and bounded in $\overline{D}$ and since the periodic cycles are bounded by B, standard arguments assure the sufficiency of transversality conditions for the periodic paths to be optimal (see Weitzman (1973) or the proof of Lemma 16 in Scheinkman (1976)).

We can sharpen Theorem 1 to obtain the corollary below, provided we add the following assumptions:

(A5) *P^0 (defined by (4.2)) is a discrete set.*
(A6) *The stationary points of (4.7) are isolated for each $\delta \in [\delta^-, \delta^+]$.*

Corollary 1. *(i) Under (A1)–(A6) the interval L in Theorem 1 is of positive length.*
(ii) Either there exist periodic cycles of (4.6) (i.e., stationary points of (4.7)) for $\delta \in [\widehat{\delta^-}, \widehat{\delta^+}] \cap P^+$ which are locally unstable (i.e., both roots of the Jacobian of the right side of (4.7), evaluated at the stationary point of (4.7), are outside the unit circle), or there exist periodic cycles of (4.6) for $\delta \in [\widehat{\delta^-}, \widehat{\delta^+}] \cap P^-$, which locally are saddle points (i.e., one root of the Jacobian of the right side of (4.7), evaluated at the stationary point of (4.7), is inside, the other outside the unit circle).

Remark. Part (ii) of the above corollary implies that if the periodic cycles of (4.6) exist for $[\widehat{\delta^-}, \widehat{\delta^+}] \cap P^+$, they are locally repelling and if they exist for $[\widehat{\delta^-}, \widehat{\delta^+}] \cap P^-$, locally they have a 1-dimensional stable manifold, such that initial conditions on this manifold lead to convergence to the cycle.

Proof. Under (A5) P^0 is discrete. So for any $\delta^0 \in P^0$ we can choose an interval $[\widehat{\delta^-}, \widehat{\delta^+}] \subset [\delta^-, \delta^+]$ such that $\delta^0 \in \{\delta | [\widehat{\delta^-}, \widehat{\delta^+}] \cap P^0\}$. Moreover, as the stationary points of (4.7) are isolated by (A6) we can choose a convex neighborhood B of $(\bar{k}(\delta), \bar{k}(\delta))$ in the interior of $\overline{D}$ such that stationary points of (4.7) do not lie on the boundary of B as long as $[\widehat{\delta^-}, \widehat{\delta^+}]$ is sufficiently small. Therefore, the arguments used in case (i) of the proof of Theorem 1 apply using $[\widehat{\delta^-}, \widehat{\delta^+}]$ and B. Thus, for every $\delta \in [\widehat{\delta^-}, \widehat{\delta^+}] \cap P^-$ or for every $\delta \in [\widehat{\delta^-}, \widehat{\delta^+}] \cap P^+$, B contains periodic paths of (4.6). $L = [\widehat{\delta^-}, \widehat{\delta^+}] \cap P^-$ or $[\widehat{\delta^-}, \widehat{\delta^+}] \cap P^+$, both of which are of positive length. This proves part (i) of the corollary. By the proof of Theorem 1 either there exist at least two nonstationary periodic cycles of (4.6) for $\delta \in [\widehat{\delta^-}, \widehat{\delta^+}] \cap P^-$ with $Det[J(\delta)] < 0$ (where $[J(\delta)]$ is the Jacobian of the right side of (4.8)) or there exist at least two nonstationary periodic cycles of (4.6) for $\delta \in [\widehat{\delta^-}, \widehat{\delta^+}] \cap P^+$ with $Det[J(\delta)] > 0$, where $Det[J(\delta)]$ is evaluated at the corresponding nonstationary periodic points. Consider first the case of nonstationary cycles for $\delta \in [\delta^-, \delta^+] \cap P^-$. By proof of Theorem 1, $Det[J(\delta)] = (1 - \mu_1)(1 - \mu_2) < 0$, where μ_1 and μ_2 are the roots of the Jacobian of the right side of (4.7) evaluated at these periodic points.

Calculating the determinant and trace of the Jacobian of the right side of (4.7), say $[H(\delta)]$, and using the fact that $k_{t+2} = k_t$, $k_{t+3} = k_{t+1}$ at the periodic points, we obtain

$$Det\,[H(\delta)] = \delta^{-2} > 0.$$

$$\begin{aligned}
Trace\,[H(\delta)] &= -\left[\frac{V_{21}(k_t,k_{t+1})}{\delta V_{12}(k_{t+1},k_{t+2})} + \frac{V_{21}(k_{t+1},k_{t+2})}{\delta V_{12}(k_t,k_{t+1})}\right] \\
&\quad + \left[\frac{V_{22}(k_t,k_{t+1}) + \delta V_{11}(k_{t+1},k_{t+2})}{\delta V_{12}(k_{t+1},k_{t+2})}\right] \\
&\quad \times \left[\frac{V_{22}(k_{t+1},k_{t+2}) + \delta V_{11}(k_t,k_{t+1})}{\delta V_{12}(k_t,k_{t+1})}\right] \\
&= \frac{\begin{pmatrix}\delta(V_{11}(k_t,k_{t+1})V_{22}(k_t,k_{t+1}) - V_{12}(k_t,k_{t+1})^2) \\ + \delta(V_{11}(k_{t+1},k_{t+2})V_{22}(k_{t+1},k_{t+2}) - V_{12}(k_{t+1},k_{t+2})^2)\end{pmatrix}}{\delta^2 V_{12}(k_{t+1},k_{t+2})V_{12}(k_t,k_{t+1})} \\
&\quad + \frac{V_{22}(k_t,k_{t+1})V_{22}(k_{t+1},k_{t+2}) + \delta^2 V_{11}(k_{t+1},k_{t+2})V_{11}(k_t,k_{t+1})}{\delta^2 V_{12}(k_{t+1},k_{t+2})V_{12}(k_t,k_{t+1})} \\
&> 0
\end{aligned}$$

by (A1).

Since $Trace\,[H(\delta)] > 0$ and $Det\,[H(\delta)] > 0$, $\mu_1, \mu_2 > 0$. Thus, $Det[J(\delta)] = (1-\mu_1)(1-\mu_2) < 0$ implies that $\mu_1 < 1$, $\mu_2 > 1$. Therefore, the fixed point of (4.7) are saddle points. For $\delta \in [\widehat{\delta}^-, \widehat{\delta}^+] \cap P^+$, $Det\,[J(\delta)] > 0$ at periodic points of (4.7). This implies that $\mu_1 > 1$, $\mu_2 > 1$ since $\mu_1\mu_2 = \delta^{-2} \geqslant 1$. Therefore, the periodic points of (4.7) are unstable.

Remark. We can point out without getting into too many technical details that the periodic cycles of the corollary are generic. Let $\mathfrak{F}$ be the class of all C^r maps $F : P \times M \to M$, where P is a 1-dimensional C^r manifold and M is an n-dimensional C^r manifold, $r \geqslant 2$, such that for every $p \in P$, $F_p : M \to M$, is a diffeomorphism. The function F in (4.6) satisfies these requirements. Then for every F from a residual subset $\mathfrak{F}_1 \subset \mathfrak{F}$ (*i.e.*, $\mathfrak{F}_1$ is the countable intersection of open sets, each of which is dense in $\mathfrak{F}$), (A5) and (A6) are satisfied (see Brunovsky (1970) [Theorem 1, (ii) and (iii)]). Furthermore, if P^0 is discrete as in (A5), under the corollary above cycles will exist for intervals of δ for which $\delta \notin P^0$. In addition, for any such δ, the Jacobian of (4.6) does not have roots on the unit circle by Lemma 3. For any such δ consider any C^2-perturbation of $F(k, y; \delta)$. Then we can show that cycles will persist in the perturbed system and will be "close" to the cycles of the unperturbed system (For a proof see Hirsch and Smale (1974), Proposition, p.305, Chap. 16).[4]

[4]Using the same techniques, it is possible to generalize the results of Theorem 1 to higher dimlensional cases. We will pursue this in further work.

4.3 Discussion of Examples

There are several examples in the literature of periodic cycles in the setting of problem (4.1). The example given by Sutherland (1979) fits precisely the conditions of our Theorem 1. In the example, $V(k_{t-1}, k_t) = 9k_{t-1}^2 - 11k_tk_{t-1} - 4k_t^2 + 43k_t$, where $V_{12} = -11$, $V_{11} = -18$, and $V_{22} = -8$ for any (k_{t-1}, k_t) since $V(k_{t-1}, k_t)$ is quadratic. Sutherland obtains cycles for $\delta = \frac{1}{3}$. For $\delta = \frac{1}{3}$, $(V_{22} + \delta V_{11})/V_{12} = \frac{14}{11} < 1 + \frac{1}{3}$, but for $\delta = 1$, $(V_{22} + \delta V_{11})/V_{12} = \frac{24}{11} > 2$. Furthermore, there is a unique $\delta^0 = \frac{3}{7}$ such that $(V_{22} + \delta^0 V_{11})/V_{12} = 1 + \delta^0$. Thus the set P^0 in (A5) is discrete. Theorem 1 and its corollary immediately apply.

Another example due to Weitzman is reported in Samuelson (1973). The example is $\max \sum_{t=0}^{\infty} \delta^t k_t^{\alpha}(1 - k_{t+1})^{\beta}$, for $\alpha = \beta = \frac{1}{2}$. The roots of the associated linear system for this example are -1 and $-\delta^1$, so one roots is always on the unit circle. Given δ any pair (x, y) satisfying $(1-x)/x = \delta^2(y/(1-y))$ qualifies as a periodic cycle. Since for any $\delta \in (0, 1]$ none of the roots of the associated linear system is inside the unit circle, our (A4) fails and Theorem 1 cannot be applied. Note that (A5) also fails since one root is always -1 and the set P^0 is a continuum. This suggests that the cycles may not persist under small C^2-perturbations (see the Remark following Corollary 1). In fact, if we set $\alpha, \beta > 0, \alpha + \beta < 1$, then $k_t^{\alpha}(1 - k_{t+1})^{\beta}$ is strictly concave at a steady state and periodic solutions will disappear for δ sufficiently close to 1 (see Scheinkman (1976) and McKenzie (1981b)). It is, however, possible to use Theorem 1 to obtain cycles if δ is not close to 1. Consider the expression $q(\delta) = (V_{22} + \delta V_{11})/V_{12} - (1+\delta) = ((1-\beta)/\alpha)(\bar{k}(\delta)/(1-\bar{k}(\delta))) + \delta((1-\alpha)/\beta)((1-\bar{k}(\delta)/\bar{k}(\delta)) - (1+\delta)$ for the above model where the steady state $\bar{k}(\delta)$ is the solution to $(1-\bar{k}(\delta))/\bar{k}(\delta) = \beta/\delta\alpha$. Consider the cases $\alpha + \beta \leqslant 1$, $\frac{1}{2} < \alpha < 1$, $0 < \beta < \frac{1}{2}$. Then $\lim_{\delta \to 0} q(\delta) = (1-2\alpha)/\alpha < 0$. Also, $q(1) = (1-\beta)/\beta + (1-\alpha)/\alpha - 2 = \alpha^{-1} + \beta^{-1} - 4 = \left(\alpha^{-1/2} - \beta^{-1/2}\right)^2 + 2(\alpha\beta)^{-1/2} - 4 > 2(\alpha\beta)^{-1/2} - 4 = 4\left((\alpha+\beta) - \left(\alpha^{-1/2} - \beta^{-1/2}\right)^2\right)^{-1} - 4 \geqslant 4(\alpha+\beta)^{-1} - 1 \geqslant 0$. Thus, $q(1) > 0$ and (A4) is satisfied. Furthermore, $dq(\delta)/d\delta = (1-2\beta)/\beta > 0$ and there is a unique $\delta^0 = \beta(2\alpha - 1)/\alpha(1-2\beta) > 0$ such that $q(\delta^0) = 0$. Thus Theorem 1 applies to the above cases.

4.4 Conditions for Monotonic and Oscillatory Trajectories

Let $W(k_0) = \max \sum_{t=0}^{\infty} \delta^t V(k_t, k_{t+1})$ for $(k_t, k_{t+1}) \in \overline{D}$ for all t. We can show that under (A3) the optimal path from k_0 is unique. We will use the following lemma in proving Theorem 2.

Lemma 4. *Let (A1)–(A3) hold. Then the optimal path $\{k_t\}$ from given k_0 is unique.*

Proof. Since $V(k_t, k_{t+1})$ is concave, $W(k_0)$ is concave. Consider $W(k_0) = \max_{k_1} V(k_0, k_1) + \delta W(k_1)$. Since $V_{22} < 0$, $V(k_0, k_1) + \delta W(k_1)$ is strictly concave in k_1 and the choice of k_1 that maximizes it is unique. This follows because if there were two optimizing k_1's their convex combinations would be feasible and yield a higher value for $V(k_0, k_1) + \delta W(k_1)$. Since the argument holds for any k_{t+1} given k_t, the optimal path is unique.

Theorem 2. *Let $\{k_t\}$ be an optimal path. Let $(k_t, k_{t+1}) \in$ interior $\overline{D}$ and let (A1), (A2) hold. Then*

(i) If $V_{12}(x, y) > 0$ for all $(x, y) \in$ interior $\overline{D}$, $k_t < k_{t+1}$ implies $k_{t+1} \leqslant k_{t+2}$. If $(k_{t+1}, k_{t+2}) \in$ interior $\overline{D}$, $k_t < k_{t+1}$ implies $k_{t+1} < k_{t+2}$ (i.e., any interior segment of an optimal path is monotonic).
(ii) If $V_{12}(x, y) < 0$ for all $(x, y) \in$ interior $\overline{D}$, $k_t < k_{t+1}$ implies $k_{t+1} \geqslant k_{t+2}$. If $(k_{t+1}, k_{t+2}) \in$ interior $\overline{D}$, $k_t < k_{t+1}$ implies $k_{t+1} > k_{t+2}$ (i.e., any interior segment of an optimal path is oscillatory).

Proof. Let $W(k_0) = \max \sum_{t=0}^{\infty} \delta^t V(k_t, k_{t+1})$ with $(k_t, k_{t+1}) \in \overline{D}$ for all $t \geqslant 0$. Let $\{k_t\}$ and $\{k_t'\}$ be optimal paths from k_0 and k_0', where $k_0' > k_0$. These paths are unique from Lemma 4. Suppose that $(k_0, k_1) \in int\overline{D}$. Then for k_0' sufficiently close to k_0, $(k_0, k_1') \in intD$. This follows from the continuity of optimal stock in initial stocks. For the same reason we have $(k_0', k_1) \in int\overline{D}$. By the principle of optimality

$$W(k_0) = V(k_0, k) + \delta W(k_1) \geqslant V(k_0, k_1') + \delta W(k_1')$$
$$W(k_0') = V(k_0', k_1') + \delta W(k_1') \geqslant V(k_0', k_1) + \delta W(k_1).$$

Adding, we obtain

$$V(k_0, k) + V(k_0', k_1') \geqslant V(k_0, k_1') + V(k_0', k_1).$$

By the convexity of $\overline{D}$, $\left[k_0, k_0'\right] \times \left[k_1', k_1\right] \subset \overline{D}$. Then we have

$$V(k_0, k_1') = \int_{k_1}^{k_1'} (V_2(k_0, s_1) + V(k_0, k_1))ds_1$$

and

$$V(k_0', k_1) = \int_{k_1'}^{k_1} (V_2(k_0', s_1) + V(k_0', k_1'))ds_1.$$

Substituting these above, we obtain

$$0 \geqslant \int_{k_1'}^{k_1} [V_2(k_0', s_1) - V_2(k_0, s_1)]ds_1$$
$$= \int_{k_1'}^{k_1} \int_{k_0}^{k_0'} V_{21}(s_0, s_1)ds_0 ds_1.$$

The above inequality implies $k_1^{'} \leqslant k_1$ if $V_{12} < 0$ and $k_1^{'} \geqslant k_1$ if $V_{12} > 0$, provided $k_0^{'}$ is sufficiently close to k_0. But the choice of k_0 was arbitrary. Hence this local property may be extended to the whole domain $(0, \overline{x})$. It follows that for any $k_0^{'}$ and k_0 in $(0, \overline{x})$, $k_0^{'} > k_0$ implies $k_1^{'} \leqslant k_1$ if $V_{12} < 0$ and $k_1^{'} \geqslant k_1$ if $V_{12} > 0$.

Now consider an optimal path $\{k_t\}$ and let $k_0^{'} = k_1$ and $k_1^{'} = k_2$. The above result shows that if $k_0 > k_1$, $V_{12} < 0$ implies $k_2 \leqslant k_1$ and $V_{12} > 0$ implies $k_2 \geqslant k_1$. This proves the first parts of (i) and (ii) in Theorem 2.

To obtain the results with strict inequalities we use the first order conditions or the Euler equations. We prove the results by contradiction. Suppose $k_1^{'} = k_1$. Then $\{k_0, k_1, k_2 \ldots\}$ and $\{k_0^{'}, k_1, k_2 \ldots\}$ are both optimal paths. If (k_0, k_1) and (k_1, k_2) are in the interior of $\overline{D}$ the Euler equation holds with equality and

$$V_2(x, k_1) + \delta V_1(k_1, k_2) = 0$$

holds for $x = k_0, k_0^{'}$. But if $V_{12} \neq 0$ on the interior of $\overline{D}$ this cannot be true and a contradiction follows. Thus $k_0 \neq k_0^{'}$ implies $k_1 \neq k_1^{'}$. This proves the second parts of (i) and (ii) of Theorem 2.

Remark. We used Lemma 4 in proving Theorem 2 where we assumed that $V_{22} < 0$. If V_{22} is allowed to be zero the optimal choice k_1 from k_0 may not be unique but will be a closed convex set. We can define $S(k_0) = \{k_1 | k_1 = \arg\max_{k_1} V(k_0, k_1) + \delta W(k_1)\}$. Theorem 2 still goes through as is, provided we assume that $k_t \times S(k_t)$ and $S(k_t) \times S(k_{t+1})$ are in the interior of $\overline{D}$ for all $k_{t+1} \in S(k_t)$.

Part (i) of Theorem 2 (monotonicity) can be proved under a weaker set of assumptions. We can drop the assumption of the convexity of $\overline{D}$ and the concavity of V and replace (A1) and (A3) with the following:

(A1′) V is continuous on $\overline{D}$ and of class C^2 in the interior of $\overline{D}$ with $V_1 > 0$ and $V_2 < 0$.

(A3′) $V_{12}(x_1, x_2) \geqslant 0$ for $(x_1, x_2) \in$ *interior*$\overline{D}$ and $V_{12} = 0$ at most for a countable infinity of points.

The following theorem generalizes a theorem by Dechert and Nishimura (1983). For a slightly different formulation proving the monotonicity of the optimal path see Majumdar (1982).

Theorem 2′ *Assume that (A1′) and (A3′) hold. Then for an optimal path* $\{k_t\}$, $k_t \underset{(<)}{>} k_{t+1}$ *implies* $k_{t+1} \underset{(\leqslant)}{\geqslant} k_{t+2}$. *Further, if* $(k_t, k_{t+1}), (k_{t+1}, k_{t+2}) \in int\overline{D}$, *then*

$k_t \underset{(<)}{>} k_{t+1}$ *implies* $k_{t+1} \underset{(<)}{>} k_{t+2}$.

Proof. Let $W(k_0) = \max \sum_{t=0}^{\infty} \delta^t V(k_t, k_{t+1})$ and define $k_t' = k_{t+1}$, $t \geqslant 0$. Then k_0' is an optimal path from $k_0' = k_1$. Assume that $k_0' > k_0$ and $k_1 > k_1'$. Then $(k_0, k_1) \in \overline{D}$ implies that $(k_0, k_1') \in \overline{D}$ for $k_1' < k_1$ and $(k_0', k_1) \in \overline{D}$ for $k_0' > k_0$. Hence again using the principle of optimality, as in the proof of Theorem 2, we obtain

$$V(k_0, k_1) + V(k_0', k_1') \geqslant V(k_0, k_1') + V(k_0', k_1).$$

Since $\left[k_0, k_0'\right] \times \left[k_1', k_1\right] \in D$ by assumption and the above hypotheses,

$$0 \geqslant \int_{k_1'}^{k_1} \int_{k_0}^{k_0'} V_{21}(s_0, s_1) ds_0 ds_1.$$

The above inequality implies that $k_1' \geqslant k_1$ under (A3′). To get the strict inequality, we use the Euler equation as in the proof of Theorem 2.

The intuitive idea behind Theorems 2 and 2′ is not hard to grasp from the structure of the proof. If an increase in k_t decreases the marginal benefit of k_{t+1}, that is, if V_{12} is negative, then the optimal adjustment is to decrease k_{t+1}. As the optimal choice of k in any period depends only on the value of k in the preceding period, if k_{t+1} is greater than k_t, k_{t+2} will then be less than k_{t+1}. We will discuss this point further in the next section in terms of a neoclassical two-sector model.

The above theorems raise the question of whether cycles of periodicity higher than two or more complicated chaotic dynamics is possible in our model. Unlike overlapping generation models (see Benhabib and Day (1982) or Grandmont (1983)) it is very difficult to construct examples which generate chaotic dynamics for infinite-horizon models that have concave utility functions. The difficulty in constructing such examples arises because the range of parameter values that lead to chaotic behavior also violate the concavity of the function $V(x, y)$.[5,6] For nonconcave functions it should be possible to construct chaotic optimal paths although we do not know of any particular economic example that has been worked out.

In the context of our model we can rule out complicated dynamics or periodic cycles of order greater than two. In Theorem 3 below we show that if $V_{12}(x, y)$ is of uniform sign over $\overline{D}$, any interior optimal path must converge either to a steady state or to a cycle of period two. In the next section (see Theorem 6) we will show that for a standard two-sector neoclassical technology, a uniform sign for $V_{12}(x, y)$

[5]Ray Deneckere of Northwestern University and S. Pelikan of the University of Cincinnati have recently constructed a particular example that does not violate concavity.

[6]The well-known Hénon map, for example, generates chaotic dynamics for parameter ranges that violate the concavity of $V(x, y)$.

restricts differences in the relative factor intensities of the two industries and rules out factor intensity reversals.

Theorem 3. *Let $\{k_t\}$ be an interior optimal path so that $(k_t, k_{t+1}) \in$ interior $\overline{D}$ and let (A1),(A2) hold. If $V_{12}(x, y) \neq 0$ for all $(x, y) \in$ interior $\overline{D}$, $\{k_t\}$ converges either to a stationary point or to a cycles of period two.*

Proof. If $V_{12}(x, y) > 0$ for all $(x, y) \in interior\overline{D}$, by Theorem 2 the trajectory $\{k_t\}$ is monotonic and bounded and there can be no cycles, $\{k_t\}$ must converge to a point in $\overline{D}$. So consider the case $V_{12}(x, y) < 0$ for all $(x, y) \in$ interior $\overline{D}$. In the proof of Theorem 2 we showed that $k_0^{'} > k_0$ implies $k_1^{'} < k_1$ if $V_{12} < 0$. It then followed that $k_1 > k_0$ implies $k_2 < k_1$. Suppose $k_2 > k_0$. If we define $k_2 = k_0^{'}$, it follows by the same logic that the optimal choice from k_2, that is k_3, is less than the optimal choice from k_0, that is k_1. But if $k_3 < k_1$, then by the same argument $k_4 > k_2$. Therefore $k_2 > k_0$ implies $k_4 > k_2$. Similarly, the converse also holds so that $k_2 < k_0$ implies $k_4 < k_2$. Thus the even iterates are monotonic and since the optimal path is bounded they must converge to a point in $\overline{D}$, say $\overline{z}$. The same applies to odd iterates and they must also converge to a point in $\overline{D}$, say z. Thus any interior path either converges to a period two cycles $(\overline{z}, z)$ or to a single point if $\overline{z} = z$.

4.5 Competitive Cycles

We will now apply the results of the previous sections to a model in which a representative agent maximizes the sum of discounted consumption and where production is characterized by a two-sector neoclassical nonjoint technology. Before applying the results of the previous section we will start by giving a detailed exposition and discussion of the model.

Let the per capita stock at time t be k_t. The per capita output of the capital good, y_t, represents the gross accumulation of capital, given as $y_t = k_{t+1} - (1 - g)k_t$, where g is the rate of depreciation with $g > 0$. The efficient per capita output of the consumption good can be obtained as a function of the per capita stock and the output of the capital good as follows. For $Y, Y, L \geqslant 0$ consider

$$\max_{(l,k)} C(l, k)$$

subject to

$$F(L - l, K - k) \geqslant Y \quad 0 \leqslant l \leqslant L, 0 \leqslant k \leqslant K,$$

where $C(\cdot, \cdot)$, $F(\cdot, \cdot)$ are standard neoclassical production functions relating the inputs of capital and labor to the outputs of the consumption good and of the capital good, respectively, and where Y is the minimum output of the capital good, K is the stock of capital, and L is the stock of labor. It is easily shown that there is a function $\tau : R_+^3 \to R_+$ such that $C = \tau(Y, K, L) \geqslant 0$. Assuming constant returns to scale in both industries, we can normalize by labor. Define $c = C/L$, $f = F/L$, $k = K/L$, and $y = Y/L$. Expressing all quantities in per capita terms we have $c_t = \tau(y_t, k_t, 1) = T(y_t, k_t) = T(k_{t+1} - (1-g)k_t, k_t) \geqslant 0$. Under standard differentiability and quasi-concavity assumptions on the underlying production functions, it can also be shown that T is of class C^2 and concave (see Benhabib and Nishimura (1979b) [Appendix AI]).

For given k_t, $T(y_t, k_t)$ describes the production possibility frontier between c_t and y_t. We will assume that there exists a $\overline{k} > 0$ such that for y satisfying $T(y,k) = 0$, $y + (1-g)k < k$ if $k > \overline{k}$ and $y + (1-g)k > k$ if $k < \overline{k}$. Thus it is not possible to maintain stocks above $\overline{k}$. The technology set $\overline{D}$ given in Sect. 4.2 here will be given by $\overline{D} = \{(k_t, k_{t+1}) | 0 \leqslant k_t \leqslant \overline{k}, 0 \leqslant k_{t+1} \leqslant \overline{k}, y_t = k_{t+1} - (1-g)k_t \geqslant 0, T(y_t, k_t) \geqslant 0\}$ for $t = 0, 1, \ldots$. The consumer's problem is to maximize $\sum_{t=0}^{\infty} \delta^t T(k_{t+1} - (1-g)k_t, k_t)$ given k_0, where δ is the discount factor. First-order conditions for an interior solution are given by

$$\begin{aligned} &T_1(k_{t+1} - (1-g)k_t, k_t) + \delta T_2(k_{t+2} - (1-g)k_{t+1}, k_{t+1}) \\ &\quad - \delta(1-g)T_1(k_{t+2} - (1-g)k_{t+1}, k_{t+1}) = 0, \end{aligned} \tag{4.9}$$

where $T_1 = \partial T(y,k)/\partial y$ and $T_2 = \partial T(y,k)/\partial k$. It is well known that the solution to this problem can be viewed as a competitive intertemporal equilibrium, where the rate of return on capital is equal to the discount rate, where factors earn their marginal products and the production of outputs is determined according to relative (shadow) prices. Furthermore, we can specify that $T_1(y,k) = -p$ and $T_2(y,k) = w$, where p can be taken as the price (or the slope of the production possibility frontier) and w as the rental of the capital good expressed in terms of the price of the consumption good. Setting $\delta = 1/(1+r)$, where r is the discount rate, the first order conditions (4.9) then can be written as

$$1 + r = \frac{w_{t+1}}{p_t} + (1-g)\frac{p_{t+1}}{p_t}, \tag{4.9'}$$

which states that one plus the discount rate equals the rental return plus the capital gain on the depreciated stock as a percentage of the cost. It is of course possible to start with the function $c = T(y,k)$ describing efficient production and allocation of resources. The first-order conditions (4.9) then describe competitive intertemporal allocation without using the representative agent. The additional restriction that the representative agent imposes (via the transversality conditions) is to rule out the nonoptimal explosive or implosive ("bubble") paths of accumulation. While the analysis of Sect. 4.2 and Theorem 1 operates on the first-order conditions directly, a bounded cyclical solution or the paths converging to it will automatically satisfy the transversality conditions.

To apply the theorems of the previous two sections and give them economic interpretations in terms of our competitive model we now turn to establish some results concerning the technology and the function $T(y,k)$.

The details and proofs of the relationships and results used below are given for the general multi-sector case in Benhabib and Nishimura (1979a). For completeness we give a heuristic as well as a diagrammatic exposition. First we note that we can express prices as functions of factor prices alone, as $p = p(w^0, w)$, where w^0 is wage rate for labor, the primary input. Note that we an express w^0 in terms of the technology as well. We set

$$w^0 = \frac{\partial\left[L \cdot \tau\left(\frac{Y}{L}, \frac{K}{L}, 1\right)\right]}{\partial L} = \frac{\partial\left[L \cdot T\left(\frac{Y}{L}, \frac{K}{L}\right)\right]}{\partial L}$$
$$= T(y, k) + p(y, k) \cdot y - w(y, k) \cdot k,$$

where w^0 is the marginal product of labor in terms of the price of the consumption good and can be expressed as an function of per capita variables (y, k). Since $T_{12} = T_{21}$, we obtain $T_{22} = \partial w/\partial k$, $T_{12} = -(\partial p/\partial k) = -(\partial p/\partial w)(\partial w/\partial k) = -b(\partial w/\partial k)$, and $T_{11} = -(\partial p/\partial w)(\partial w/\partial y) = -(\partial p/\partial w)(T_{21}) = b^2(\partial w/\partial k)$, where $b = (\partial p/\partial w)$ is a weighted relative capital intensity difference (invariant to redefinitions of units) given by

$$b = a_{01}\left(\frac{a_{11}}{a_{01}} - \frac{a_{10}}{a_{00}}\right) < 1,$$

and where $a_{11}(\delta)$ and $a_{01}(\delta)$ are steady state per unit capital and labor inputs to the capital good and $a_{10}(\delta)$ and $a_{00}(\delta)$ are the steady state per unit capital and labor inputs to the consumption good. This result follows directly from the envelope theorem applied to cost minimization under a neoclassical technology and is also known as Shepard's lemma. The reason b is not simply an input coefficient is that p is a relative price. To get b, note that from Shepard's lemma $(dp, 0) = (dw^0, dw)A$, where A is the input coefficient matrix for factor prices (w^0, w). The 0-element in $(dp, 0)$ is due to the fact that the price of the consumption good is fixed at unity. Eliminating dw_0 we can solve for $dp/dw = b$. Note that b is fully determined by relative factor prices. At a steady state, however, we have from the non-substitution theorem that all relative prices as well as all the input coefficients a_{ij} are functions of the discount factor δ.

We can now illustrate diagrammatically how p responds to a change in k. Since $T_{22} = (\partial w/\partial k)$ is negative the sign of $T_{21} = (\partial p/\partial k) = -b(\partial w/\partial k)$ depends on b.[7] Since $T_1 = -p$ we see from Fig. 4.1 that the change in p when the capital-labor ratio k changes and y remains fixed depends on how the production possibility surface shifts in response to an increase in k. If the consumption good is capital

[7]Concavity implies $T_{22} \leqslant 0$. However, it can be shown that strictly diminishing marginal products imply $T_{22} < 0$. For a discussion in the case of a multisector model see Benhabib and Nishimura (1979b).

Fig. 4.1

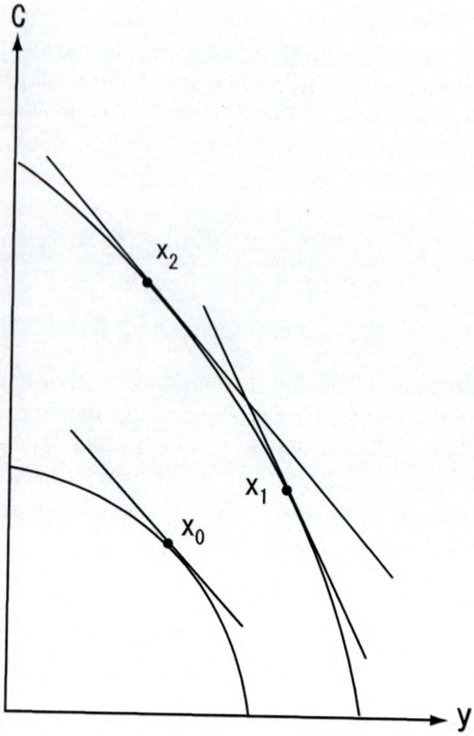

intensive, that is, if b is negative, the surface shifts outward favoring the production of c. This implies that at a given y, p gets steeper and $(\partial p/\partial k)$ is negative.

This is illustrated in Fig. 4.1, as the production point shifts from x_0 to x_1. Alternatively, if k declines at constant relative prices, the output of y increases. To maintain a fixed y, p must decline, implying that $(\partial p/\partial k)$ is negative. Note that a decline in k results in a higher y at constant prices because the production of c declines sufficiently and factors released from the production of c flow into the production of y. This is possible for a marginal change in k as long as c is produced by a positive amount, that is, as long as $(k_t, k_{t+1}) \in$ *interior* $\overline{D}$.

We now adapt the consumer's optimization problem for our model to problem (4.1) of Sect. 4.2. Define $V(k_t, k_{t+1}) \equiv T(k_{t+1} - (1-g)k_t, k_t)$, where $V(k_t, k_{t+1})$ was defined in Sect. 4.2. To apply Theorems 1 and 2 we have to establish the sign of V_{12} on the interior of $\overline{D}$. We have $V_{12} = -(1-g)T_{11} + T_{12} = b(\partial w/\partial k)$ $[-1 - (1-g)b]$, provided y, k, $T(y,k) > 0$. Since $T_{22} = \partial w/\partial k < 0$ by the concavity of T (see footnote 7 as well), we have $V_{12} < 0$ if $b \in (-(1-g)^{-1}, 0)$ and $(k_t, k_{t+1}) \in$ *interior* $\overline{D}$. The latter implies that y, k, $T(y,k) > 0$. As discussed above, under constant returns to scale in both industries input coefficients and therefore b depends only on relative factor prices (w/w^0). Furthermore, as shown above, $T_2(y_t, k_t) = w_t$ and $T(y_t, k_t) + p(y_t, k_t) - w(y_t, k_t)k_t = w_t^0$ for $(k_t, k_{t+1}) \in$ *interior* $\overline{D}$ where, for all t, $y_1 = k_{t+1} - (1-g)k_t$. Then relative factor prices (w/w^0)

and therefore b depends on (k_t, k_{t+1}). We can now state an assumption which restricts b, or the differences between the capital labor ratios in the two industries:

(B1) *For the two-sector technology described above, for all* $(k_1, k_0) \in$ *interior* $\overline{D}$, $b \in (-(1-g)^{-1}, 0)$.

Of course if for all factor-price ratios (w/w_0), cost minimizing factor intensities are such that $b \in (-(1-g)^{-1}, 0)$, (B1) will be automatically satisfied. Note that for the special technology of this section, (A1) holds and (B1) implies that (A2) and (A3) of Sect. 4.2 also hold.

Theorem 4. [8]*Let (B1) hold for the two-sector technology described above. Then along an optimal path* $(k_{t+2} - k_{t+1})(k_{t+1} - k_t) < 0$ *if* $k_t \neq k_{t+1}$ *and* $(k_t, k_{t+1}) \in$ *interior* $\overline{D}$*; that is, the optimal path oscillates.*

Proof. Theorem 3 follows directly from Theorem 2.

As noted at the end of Sect. 4.4, a negative V_{12} implies that the marginal benefit of k_{t+1} declines with k_t and the optimal sequence of k's can then be shown to be oscillatory. We can interpret this result in the context of our competitive equilibrium model. An increase in k_t in our two-sector model results in two opposing forces. When the consumption good is capital intensive the trade-off in production becomes more favorable to consumption goods. For a given y_t, this is reflected in the steeper price line at x_t than at x_0 in Fig. 4.1. This tends, like a substitution effect, to lower investment and the capital stock in the next period. On the other hand, the capital intensity differences affect not only the shift but the shape of the production possibility frontier. For a given k_{t+1}, an increase in k_t reduces the required gross investment $y_t = k_{t+1} - (1-g)k_t$, and, like a wealth effect, makes it possible to produce more c. However, y cannot be transformed into c at a constant rate. When the consumption good is substantially more capital intensive than the investment good the slope of the production possibility is more sensitive to changes in the composition of output. A reduction in y_t changes the trade-off in production in favor of the investment good. The combined change in p due to a change in k can be seen in Fig. 4.1 as a switch from x_0 to x_2. Thus $V_{12} < 0$ if the slope $-p$ is steeper at x_2 than at x_0. This requires that $b \in (-(1-g)^{-1}, 0)$, that is, the consumption good to be more capital intensive than the investment good but not too strongly so. However, with full depreciation $g = 1$ and oscillations occur if $b < 0$. In this case k_t, fully depreciates and $y_t = k_{t+1}$. This implies that the second effect described above, the movement from x_1 to x_2 along the production possibility frontier, is absent.

So far the above discussion concerns the existence of oscillations but not that of persistent cycles. Persistent cycles require a further restriction of the capital-intensity differences between consumption and investment goods. For cycles to be sustained, the oscillations in relative prices must not present intertemporal arbitrage

[8]For a generalization of this theorem to a stochastic economy with an endogenous labor supply see Benhabib and Nishimura (1984).

opportunities. Thus possible gains from postponing consumption from periods when the marginal rate of transformation between consumption and investment is high to periods when it is low must not be worth it. Whether this is the case or not depends on the discount rate as well as the slope of the production possibility frontier. Thus the existence of cycles depends not only on b but also on δ, the discount factor. As shown rigorously below, b must lie in an interval defined by the discount factor and the rate of depreciation.

To establish cycles we have to investigate the conditions of Theorem 1, so that the set P^- defined in Theorem 1 is not empty. That is, for some δ, the condition $V_{22} + \delta V_{11} > (1 + \delta)V_{12}$ has to be satisfied. In terms of technology, this reduces to $Ab^2 + Hb + \delta < 0$, where $A = (1+\delta(1-g)^2+(1+\delta)(-g))$, $H = (1+\delta(3-2g))$. This inequality is satisfied if $b \in M = (-(1 + (1 - g))^{-1}, -\delta(1 + \delta(1 - g))^{-1})$, which is empty if $\delta = 1$. At the same time, for $\delta = 1$, it is easily shown that $V_{11} + V_{22} - 2V_{12} = (1 - b(2 - g))^2 T_{22} < 0$, which implies that $V_{11} + V_{22} < 2V_{12}$. Therefore, there exists some $\delta < 1$ such that $\delta \in (\overline{\delta}, 1] \subset P^+$. Thus, P^+ is not empty and there exists a δ^+ as required in (A4). Then the following assumption covers (A4) of Theorem 1.

(B2) *There exist a δ^* such that at the steady state value of b at δ^*, $b(\delta^*) \in M$, where $M = (-(1 + (1 - g))^{-1}, -\delta^*(1 + \delta^*(1 - g))^{-1}) \subset (-1, 0)$.*

Note that $\delta = 0$ implies $M = (-(1+(1-g))^{-1}, 0)$. (B2) assure that there exists a δ^- as required by (A4). Note also that with $g = 1$ (full depreciation) we have $M = (-1, -\delta^*)$ and (B2) reduces to $b(\delta^*) \in (-1, -\delta^*)$.

Theorem 5. *Let (B1) and (B2) hold. The consumer's intertemporal optimization problem with the two-sector technology T described above has optimal solution trajectories of the capital stock k_t, outputs y_t, c_t and prices p_t and w_t, which are cycles of period 2 for δ in some nonempty set $[\delta^-, \delta^+]$.*

Proof. Theorem 4 follows directly from Theorem 1

The results of Corollary 1 also apply to the two-sector technology of this section. Applying Theorem 3, we can rule out periodic cycles of higher order and more complicated dynamic behavior in our two-sector model if we impose conditions on the relative factor intensities. We have shown that $V_{12} = b(\partial w/\partial k)(-1 - (1 - g)b)$ and that b and $(\partial w/\partial k)$ can be expressed in terms of (k_t, k_{t+1}). The following assumption rules out $V_{12} = 0$, as required by Theorem 3.

(B3) *For all $(k_t, k_{t+1}) \in$ interior $\overline{D}$, $b \neq 0$ and $b \neq -(1 - g)^{-1}$.*

Since the sign of b determines whether the consumption or the capital good is capital intensive, $b \neq 0$ implies that there are no factor intensity reversals for factor price ratios which correspond to $(k_t, k_{t+1}) \in \overline{D}$. If the consumption good is capital intensive, that is, if $b < 0$, the requirement of (B3) that $b \neq -(1-g)^{-1}$ prevents the consumption good from becoming too capital-intensive relative to the capital good.

Note that as g approaches 1 (full depreciation), this last requirement becomes less and less binding and disappears in the limit.

Theorem 6. *Let (B3) hold for the two-sector technology described above. Then an interior optimal path converges either to a steady state or to a cycle of period two.*

Proof. Since (B3) assures that $V_{12} \neq 0$ for $(k_t, k_{t+1}) \in$ *interior* $\overline{D}$, the result follows immediately from Theorem 3.

Remark. The above theorems can be extended to include an endogenous labor supply which depends on the wage rate. The technology function can be defined as $T(y, k, l)$, where l is the labor input. The model can then be closed by specifying an inverse labor supply function $w_0(l)$, since labor demand is given by $\partial T(y, k, l)/\partial l = w_0(y, k, l)$, where w_0 is the real wage in terms of the price of the consumption good. Solving for l as a function of (y, k), we have $T(y, k, l(y, k))$.[9] Then the above theorems can be directly applies to this function and, if the assumptions hold, employment will behave in a cyclical manner as well. Of course, there is no involuntary unemployment in such a model. However, if the equilibrium trajectory is cyclical, it may be plausible to introduce frictions preventing the continuous flow of factors between industries and obtain some unemployment of factors.

Remark. The results of this section merely illustrate how the structure of technology can lead to efficient cycles. The capital-intensity relations will be more complicated in a multi-sector model. The initial accumulation of a particular type of machine can lead to its subsequent decumulation if it causes a shift in production which favors other capital good and thereby generates cycles. Furthermore, it can be seen from the examples of Benhabib and Nishimura (1979b) that generalized capital-intensity conditions play a role in producing cycles in their continuous time model as well. In a continuous time framework, however, at least two distinct capital goods are required for cycling. As one capital stock crosses its steady state level the other capital stock, also on its optimal path, will be away from its steady state value.

Remark. It is also possible to interpret the technology function $T(k_{t+1} - (1 - g)k_t, k_t)$ as the output of a firm subject to adjustment costs, as in the standard adjustment-cost literature. The first argument of $T(\cdot, \cdot)$ can represent gross investment, which affects output directly. By Theorem 2, the optimal path would be oscillatory if $-T_{11}(1 - g) + T_{12} < 0$. To show that periodic cycles exist we have to show that (A4) holds and the crucial condition is the existence of some δ^* such that $(V_{22} + \delta^* V_{11})/V_{12} < 1 + \delta^*$ at a stationary point. This will hold for some δ^* if at the stationary point corresponding to δ^*, $AT_{11} - HT_{12} + \delta^* T_{22} > 0$,

---

[9]Equivalently, we could specify a utility function $V(k_t, k_{t+1}, l_t) = T(y_t, k_t, l_t) - v(l_t)$, where $v(l_t)$ is the disutility of work. The first order conditions would imply $\partial T(y_t, k_t, l_t)/\partial l_t = \partial v(l_t)/\partial l_t$ for all t.

where $A = (1 + \delta^*(1 - g)^2 + (1 + \delta^*)(1 - g))$ and $H = (1 + \delta^*(3 - 2g))$. Since T_{11}, $T_{12} < 0$, the inequality is likely to hold for small δ provided T_{12} is sufficiently negative. A negative value of T_{12} implies an adjustment cost function for which investment has a negative effect on the marginal productivity of the existing stock. Thus, a firm with a large enough discount rate and negative T_{12} is likely to have an alternating level of capital stock, with gross investment falling below and overshooting[10] depreciation in alternating periods. If a secondary market for capital goods exists we could also observe the firm buying and selling capital equipment in alternating periods, even though it incurs positive adjustment costs whenever it is buying or selling the equipment.

4.6 Conclusion

The basic reason why the oscillatory behavior of economic variables can persist in the stationary environment of our model is because the accumulation of capital assets changes the trade-offs in the production of different goods. If the process of accumulation of a certain capital good changes the slope of the production possibility frontier in favor of other goods or of consumption goods, that particular capital good may then be decumulated. Then as this capital good is decumulated the process will be reversed. As such a process results in oscillations of relative prices in a competitive equilibrium context, the possibility of intertemporal arbitrage may arise. Thus persistent cycles in prices require some degree of discounting of the future. These anticipated relative price changes will not necessarily be dampened or eliminated by arbitrage since when the future is discounted relative prices can change across periods without generating profitable trading or storing opportunities. As has been shown in Sect. 4.5, it is the relation between the capital intensity differences in the production of different goods (which determine the curvature of the production possibility frontier) and the rate of time preference (or discount rate) which determines whether or not the competitive solution will exhibit persistent and "robust" cycles in outputs, stocks, relative prices, and employment. A potentially interesting issue that we do not address here is the possibility of friction and resistance (possibly due to retraining costs or informational difficulties) in the movement of factors across industries in response to the cyclical changes in the equilibrium relative prices.[11]

---

[10]For some empirical evidence of such overshooting behavior see Nadiri and Rosen (1973).

[11]According to Von Hayek, cyclical changes in the relative prices of consumption and investment goods lead to business cycles by generating a tendency to the reallocation of factors of production that cannot easily retrain or move across industries. In "Prices and Production" (1931) Von Hayek argues that a Wicksellian elastic money supply may be responsible for cyclical changes in relative prices. In "Profits, Interest and Investment" (1939) he argues that relative prices cycle because an expansion in the demand for and production of consumer goods at the expense of capital goods lowers real wages and changes the tradeoff in the production of consumer and investment goods

References

Araujo, A. and L. A. Scheinkman (1977), "Smoothness, Comparative Dynamics, and the Turnpike Property," *Econometrica*, 45, 601–620.

Azariadis, C. and R. Guesnerie (1983), Sunspots and Cycles, CARESS Working Paper, University of Pennsylvania.

Balasko, Y. (1983), "Extrinsic Uncertainty Revisited," *Journal of Economic Theory*, 31, 203–220.

Benhabib, J. and R. H. Day (1982), "A Characterization of Erratic Dynamics in the Overlapping Generations Model," *Journal of Economic Dynamics and Control*, 4, 37–55.

Benhabib, J. and K. Nishimura (1984), Stochastic Equilibrium Cycles, C. V. Starr Center for Applied Economics, New York University.

Benhabib, J. and K. Nishimura (1979a), "On the Uniqueness of Steady States in an Economy with Heterogeneous Capital Goods," *International Economic Review*, 20, 59–82.

Benhabib, J. and K. Nishimura (1979b), "The Hopf Bifurcation and the Existence and Stability of Closed Orbits in Multisector Models of Optimal Economic Growth," *Journal of Economic Theory*, 21, 421–444.

Bewley, T. (1982), "An Integration of General Equilibrium Theory and Turnpike Theory," *Journal of Mathematical Economics*, 10, 233–267.

Brock, W. (1970), "On Existence of Weakly Maximal Programs in a Multisector Economy," *Review of Economic Studies*, 37, 275–280.

Brunovsky, P. (1970), "On One-Parameter Families of Diffeomorphisms," *Commentationes Mathematicae Universitatis Carolinae*, 3, 559–582.

Cass, D. and K. Shell (1983), "Do Sunspots Matter?" *Journal of Political Economy*, 91, 193–227.

Dasgupta, S. and L. W. McKenzie (1984), The Comparative Statics and Dynamics of Stationary State, working paper, Department of Economics, University of Rochester.

Dechert, W. D. and K. Nishimura (1983), "A Complete Characterization of Optimal Growth Paths in an Aggregate Model with a Non-Concave Production Function," *Journal of Economic Theory*, 31, 332–354.

Diamond, P. and D. Fudenberg (1983), An Example of Rational Expectations Business Cycles in Search Equilibrium, working paper, Massachusetts Institute of Technology.

Grandmont, J. M. (1983), On Endogenous Competitive Business Cycles, CEPREMAP DP No.8316.

Von Hayek, F. A. (1931), Prices and Production, Routledge, London.

Von Hayek, F. A. (1939), Profits, Interest and Investment, Routledge, London.

Hirsch, M. W. and S. Smale, (1974), Differential Equations, Dynamical Systems, and Linear Algebra, Academic Press, New York.

Iooss, G. (1979), Bifurcation of Maps and Applications, North-Holland, Amsterdam.

Kydland, F. E. and E. C. Prescott (1982), "Time to Build and Aggregate Fluctuations," *Econometrica*, 50, 1345–1370.

Long, J. B. and C. I. Plosser (1983), "Real Business Cycles," *Journal of Political Economy*, 91, 39–69.

Majumdar, M. (1975), "Some Remarks on Optimal Growth with Intertemporally Dependent Preferences in the Neo-Classical Model," *Review of Economic Studies*, 42, 147–153.

Majumdar, M. (1982), Notes on Discounted Dynamic Optimization when the Felicity Function Depends on both Consumption and Input, mimeo.

Marden, M. (1949), The Geometry of the Zero's of Polynomials, American Mathematical Society, New York.

---

in favor of investment. This reverses the initial expansion of consumption and the contraction of investment.

McKenzie, L. W. (1981a), Optimal Growth and Turnpike Theorems, in *Handbook of Mathematical Economics*, ed. by Arrow, K. and M. Inrilligator, North-Holland, New York.

McKenzie, L. W. (1981b), Turnpike Theory, Discounted Utility and the Von Neumann Facet, Discussion paper No. 833, University of Rochester, revised October 1981.

Milnor, J. (1965), Topology from the Differentiable Viewpoint, Univ. of Virginia Press, Charlottesville.

Nadiri, M. I. and S. Rosen (1973), A Disequilibrium Model of Demand for Factors of Production, National Bureau of Economic Research, New York.

Samuelson, P. A. (1973), "Optimality of Profit, Including Prices under Ideal Planning," *Proceedings of the National Academy of Sciences of the United States of America*, 70, 2109–2111.

Scheinkman, J. A. (1976), "On Optimal Steady States of *n*-Sector Growth Models when Utility is Discounted," *Journal of Economic Theory*, 12, 11–30.

Shell, K. (1977), Monnaie et allocation intertemporelle, CNRS Seminaire D'Econometrie Roy-Malinvaud, Paris, mimeo.

Sutherland, W. A. (1979), "On Optimal Development in Multi-Sectoral Economy: The Discounted Case," *Review of Economic Studies*, 46, 585–589.

Weitzman, M. L. (1973), "Duality Theory for Infinite Horizon Convex Models," *Management Science*, 19, 783–789.

Chapter 5
Interlinkage in the Endogenous Real Business Cycles of International Economies*

Kazuo Nishimura and Makoto Yano**

5.1 Introduction

International trade interlinks the business cycles of trading countries, as it relates economic activities of agents in one country to those in another. In some cases, as a result of trade, a country's business cycles may be spread throughout the world. In other cases, the same country's business cycles may be erased. The present study explains this function of international trade, which has not been treated in the literature, from the viewpoint of endogenous real business cycles. For this purpose, we consider a perfect foresight equilibrium model with two countries and characterize the interlinkage of the two countries' business cycles.

In a deterministic growth model, it has been revealed, an optimal path may be cyclic or even totally irregular (chaotic); see Benhabib and Nishimura (1979, 1985), Boldrin and Montrucchio (1986) and Deneckere and Pelikan (1986). (A similar possibility is demonstrated in overlapping generation models as well; see Benhabib and Day 1982, and Grandmont 1985). This implies that business cycles may appear as a result of optimization even in a fully adjustable deterministic economy. Such business cycles are called endogenous real business cycles. It has been demonstrated, moreover, that in a perfect foresight model with many consumers, a competitive equilibrium path behaves like an optimal growth path; see Becker (1980), Bewley (1982), Yano (1984a), Lucas and Stokey (1984), Coles (1985), Epstein (1987a), Benhabib et al. (1988), and Marimon (1989). As these results

*Economic Theory 3, 151–168, 1993

**We are grateful to Takashi Negishi for useful conversations.

K. Nishimura (✉) · M. Yano
Institute of Economic Research, Kyoto University, Japan
e-mail: nishimura@kier.kyoto-u.ac.jp; yano@kier.kyoto-u.ac.jp

J. Stachurski et al. (eds.), *Nonlinear Dynamics in Equilibrium Models*,
DOI 10.1007/978-3-642-22397-6_5, © Springer-Verlag Berlin Heidelberg 1993

suggest, a perfect foresight equilibrium path may exhibit business cycles even in a many-consumer model such as a large-country trade model. Endogenous real business cycles in such a setting, however, have not been characterized in the existing literature.[1]

This study does not attempt simply to characterize endogenous real business cycles in a large-country trade model. Rather, it attempts to analyze interlinkage in the business cycles of large-country economies in a free-trade equilibrium. For this purpose, we identify the determinants of each country's global accumulation pattern in an autarky equilibrium. By means of those determinants, we characterize fluctuant and monotone free-trade equilibrium paths.

As determinants of individual countries' autarky accumulation patterns, we focus on production technologies. We build a simple two-country trade model in which the countries have independent technologies. By using a result of Benhabib and Nishimura (1985), we identify the characteristic of a production function that determines a country's autarky accumulation pattern. By means of those characteristics, we explain global accumulation patterns in the free-trade case.

Our comparison between free-trade and autarky equilibria is a practice of comparative dynamics. In general, as the work of Yano (1984b, 1990), Epstein (1987b) and Kehoe et al. (1990) demonstrates, comparative dynamics is subject to an extra complication in a multi-consumer model, relative to a single-consumer model. This is because in order to find an equilibrium in a multi-consumer model, it is necessary to find an equilibrium price path that equates the demand and supply derived from the multiple consumers' optimization problems. In a single-consumer model, in contrast, an equilibrium can be found by solving a single optimal growth problem. In our analysis, therefore, we face with this extra difficulty.

In the trade literature, this study is not the first attempt analytically to characterize a perfect foresight equilibrium in a large-country model. Wan (1971) considers capital accumulation in a two-country trade model with one good; Grossman and Helpman (1989) considers technology transfer in a model with non-convex technology; Yano (1990) analyzes the transfer problem in a two-by-two model; Fukao and Hamada (1990) considers the case of heterogeneous discount factors.

In what follows, Sect. 5.2 presents our model. Section 5.3 reviews the results in the existing literature that are used in our analysis. Section 5.4 characterizes the global pattern of the aggregate capital accumulation in the free-trade equilibrium in relation to the individual countries' accumulation patterns in the autarky equilibrium. Section 5.5 demonstrates that trade may be super-stabilizing in the sense that an economy may jump to a stationary state as soon as trade opens, no matter how the autarky equilibrium paths behave in individual countries. Sections 5.6 and 5.7 relate the post-trade aggregate accumulation pattern to each country's post-trade accumulation pattern. Section 5.8 summarizes our results.

---

[1] Kemp et al. (1990) considers a similar issue in a different context; that work considers the interaction between the optimization of the government and that of the private sector.

5.2 Model

We consider a simple perfect foresight trade model with two goods. As seen below, fluctuations along an equilibrium path can be attributed to the fact that, in our model, the relative price between the goods can vary and is endogenously determined. The way in which different countries' business cycles are interlinked depends on various elements. Among them, we focus on production technologies in order to keep our characterization tractable. The model's consumption side is kept as simple as possible.

There are two countries α and β. Call X and Y the two goods. Good X is a pure consumption good, and good Y is a produced good that is used for production (produced good input). Good Y is not consumed.

In production processes, good Y input is used in combination with primary good input, such as labor. In order to produce output in a certain period, produced-good input (Y) must be put into production processes one period prior to that period. In this sense, we call good Y the capital good. Primary good input is put into production processes in the same period as output is produced. In each country, primary factors are supplied inelastically and time-independently. They are immobile across countries.

We focus on perfect foresight equilibria; in a perfect foresight equilibrium, the present value prices of all future goods and factors are known in the present period. Denote by q_{ht} and p_{ht}, respectively, the present value prices of goods X and Y in country h in period t. Goods X and Y are assumed to be freely mobile between countries once trade opens, whereas labor is internationally immobile both before and after the opening of trade. This implies that in free-trade equilibrium, $q_{\alpha t} = q_{\beta t}$ and $p_{\alpha t} = p_{\beta t}$ must hold.

We do not restrict the number of primary production factors, denoted by n; this enables us to cover various models of production in the literature (see footnote 1). Denote by $w_{ht} \in R^n$ the present value prices of primary factors, where R is the set of real numbers and where R^n is its n-copies. Because the primary factors are immobile across countries, w_{ht} differs between countries, in general, even in the free-trade case.

Let R_+ and R_{++}, respectively, be the sets of non-negative and positive real numbers. Let $\overline{\ell}_h \in R^n_{++}$ be country h's primary good endowments that does not vary over time. Let $\overline{k}_{h0} > 0$ country h's capital good endowment in the initial period. Country h's wealth is, therefore, $p_{h0}\overline{k}_{h0} + \sum_{t=1}^{\infty} w_{ht}\overline{\ell}_h$.

Denote by c_{ht} h's aggregate consumption in period t and by $u_h(c_{ht})$ h's period-wise utility function. In order to keep the consumption side as simple as possible, we assume that the utility functions are linear.

Assumption 1 $u_h : R_+ \to R$ *satisfies* $u_h(c_{ht}) = c_{ht}$.

Country h's consumers' optimization is summarized as follows:

$$(c_{h1}, c_{h2}, \ldots) = \arg \max_{(\zeta_{h1}, \zeta_{h2}, \ldots)} \sum_{t=1}^{\infty} \rho^t u_h(\zeta_{ht})$$

$$\text{subject to} \sum_{t=1}^{\infty} q_{ht}\zeta_{ht} = p_{h0}\overline{k}_{h0} + \sum_{t=1}^{\infty} w_{ht}\overline{\ell}_h, \quad h = \alpha, \beta. \tag{5.1}$$

(For the moment, we proceed our discussions with general utility function $u_h(c_h)$, which will be useful for explaining the idea of our analysis.)

Denote by x_{ht} and y_{ht} country h's output of goods X and Y, respectively, in period t. Denote by k_{ht-1} and ℓ_{ht}, respectively, the capital good and primary factors that are used in order to produce x_{ht} and y_{ht}. Recall that the capital good (Y) is put into production processes in period $t-1$ whereas the primary factors ℓ_{ht} in period t. Thus, country h's production technology can be expressed by function $x_{ht} = f_h(k_{ht-1}, y_{ht}, \ell_{ht})$.

Assumption 2 *(a)* $f_h : D_h^f \to R$ *is concave and linearly homogeneous.*
(b) D_h^f *is a closed and convex subset of* $R_+^2 \times R_+^n$.
(c) If $(k, y, \ell) \in D_h^f$, $0 \le y' \le y$, $(k', \ell') \ge (k, \ell)$, *then* $(k', y', \ell') \in D_h^f$.

Assumption 2a implies constant returns to scale. Assumption 2c implies free disposal. Because the primary factors are inelastically supplied, it is more convenient to work with function $g_h(k, y) = f_h(k, y, \overline{\ell}_h)$.[2] Denote the domain of g_h by $D_h = \{(k, y) | (k, y, \overline{\ell}_h) \in D_h^f\}$.

Assumption 3 *(a) There are* $\theta > 0$ *and* $0 < \sigma < 1$ *such that if* $k > \theta$ *and* $(k, y) \in D_h$, $y < \sigma k$.
(b) $g_h : D_h \to R$ *is continuously twice differentiable on interior* D_h.
(c) On the interior of D_h, $\partial g_h(k, y)/\partial k > 0$, $\partial g_h(k, y)/\partial y < 0$, $\partial^2 g_h(k, y)/\partial k^2 < 0$ *and* $\partial^2 g_h(k, y)/\partial y^2 < 0$.
(d) There is $y > \overline{k}_{h0}$ *such that* $(\overline{k}_{h0}, y) \in D_h$.

Recall that under Assumption 2, the average productivity of capital input decreases as capital input increases. Assumption 3a, however, implies that too large capital input cannot be reproduced even if all the productive resources are used for

---

[2]The "production" function $x_h = g_h(k_h, y_h)(= f_h(k_h, y_h, \overline{\ell}_h))$ captures the production possibility frontier if k_h is given. It can be derived from the production process in which goods X and Y are produced by independent industries, which use capital input and primary good input to produce their respective products. The case in which there is only one primary factor, i.e., $\overline{\ell}_h$ is one-dimensional, coincides with the standard Heckscher-Ohlin-Samuelson model (see Jones 1965, for example). The model that Kemp and Khang (1974) considers coincides with the case in which there are two primary factors that are freely mobile between the industries. The specific factor model (see Jones 1971) coincides with the case in which $\overline{\ell}_h = (\overline{\ell}_{hX}, \overline{\ell}_{hY})$ is two-dimensional and in which $\overline{\ell}_{hX}$ and $\overline{\ell}_{hY}$ are used exclusively by industries X and Y, respectively.

production of the capital good (i.e., $g_h(k, y) = 0$). Behind this assumption, it is implicit that the primary-goods endowments work as a bottle-neck in production processes. Assumption 3c implies a trade-off between goods X and Y and positive marginal productivity of capital input in either X or Y. Moreover, the marginal rate of transformation between X and Y and the marginal productivity of good X is decreasing.[3] Assumption 3d implies expansibility of each country's initial stock. These assumptions are borrowed from the standard literature on optimal growth (see McKenzie 1986).

Optimization on the production side is captured as follows:

$$(x_{ht}, k_{ht-1}, y_{ht}, \ell_{ht}) = \arg \max_{(x,k,y,\ell)} q_{ht}x + p_{ht}y - p_{ht-1}k - w_{ht}\ell$$

$$s.t.\ \ x = f_h(k, y, \ell) \qquad h = \alpha, \beta \text{ and } t = 1, 2, \ldots. \tag{5.2}$$

In the case in which free trade is allowed between α and β (*free-trade case*), output prices must be equated between countries. Denote by p_t and q_t the world prices of X and Y in t.

$$p_t = p_{ht}, \qquad h = \alpha, \beta \text{ and } t = 0, 1, \ldots; \tag{5.3}$$

$$q_t = q_{ht}, \qquad h = \alpha, \beta \text{ and } t = 1, 2, \ldots. \tag{5.4}$$

After trade opens, the demand and supply of produced goods must be met in the world market. The market clearing conditions for goods X and Y are, therefore, as follows:

$$\sum_h c_{ht} = \sum_h x_{ht}, \qquad t = 1, 2, \ldots; \tag{5.5}$$

$$\sum_h k_{ht} = \sum_h y_{ht}, \qquad t = 1, 2, \ldots. \tag{5.6}$$

In the case in which the countries do not trade with each other (*autarky case*), the output prices are generally different between the countries. Demand and supply must be met domestically. In this case, therefore, the market clearing conditions for goods X and Y are expressed as follows:

$$c_{ht} = x_{ht}, \qquad h = \alpha, \beta \text{ and } t = 1, 2, \ldots; \tag{5.7}$$

$$k_{ht} = y_{ht}, \qquad h = \alpha, \beta \text{ and } t = 1, 2, \ldots. \tag{5.8}$$

[3] Because of the concavity of g_h (see Assumption 2a), the Hessian of g_h is non-positive definite. At this stage, however, we do not assume the negative definiteness for the Hessian in order to keep g_h consistent with the Heckscher-Ohlin-Samuelson model, which results in the determinant of the Hessian of g_h equal to zero (see Jones 1965, for an explanation).

Labor is immobile between countries both before and after trade opens. Thus, both before and after trade opens, each country's labor market clearing condition is as follows:

$$\ell_{ht} = \bar{\ell}_h, \quad h = \alpha, \beta \text{ and } t = 1, 2, \ldots. \tag{5.9}$$

Definition 1. Path $e_t = (c_{ht}, x_{ht}, k_{ht-1}, y_{ht}, \ell_{ht}, q_t, p_{t-1}, w_{ht}; h = \alpha, \beta)$, $t = 1, 2, \ldots$, is in a *free-trade equilibrium* if it satisfies conditions (5.1)–(5.6) and (5.9).

Definition 2. Path $e_{ht} = (c_{ht}, x_{ht}, k_{ht-1}, y_{ht}, \ell_{ht}, q_{ht}, p_{ht-1}, w_{ht})$, $h = \alpha, \beta$ and $t = 1, 2, \ldots$, is in an *autarky equilibrium* if it satisfies conditions (5.1), (5.2) and (5.7)–(5.9).

5.3 Preliminary Results

This section states several existing results that we will use in our analysis. As is shown below, a competitive equilibrium path may be thought of as an optimal path in an optimal growth model (Bewley 1982; Negishi 1960). We will use this fact together with the characterization of the global dynamics of an optimal growth path in Benhabib and Nishimura (1985). We start with stating this characterization.

Consider a *reduced form optimal growth model* $\sum_{t=1}^{\infty} \rho^t v(\kappa_{t-1}, \kappa_t)$; function $v(\kappa_{t-1}, \kappa_t)$ denotes the maximum social utility that is attainable in period t when the economy's capital stock is κ_{t-1} at the end of period $t-1$ and κ_t at the end of period t.

Path κ_t, $t = 0, 1, 2, \ldots$, is an *optimal path* if it maximizes $\sum_{t=1}^{\infty} \rho^t v(\kappa'_{t-1}, \kappa'_t)$ subject to $\kappa'_0 = \kappa_0$, where $0 < \rho < 1$. It is a *stationary optimal path* if it is optimal and stationary (i.e., $\kappa_t = \kappa$, $t = 0, 1, 2, \ldots$). It is often said that a stationary optimal path is in a *modified golden rule state*. Take a sufficiently large rectangular in the interior of the domain of v, and denote it by $\mathfrak{S}$. Under the assumption that v is continuously twice differentiable on $\mathfrak{S}$, the following lemma holds (see Benhabib and Nishimura 1985, for a proof).

Lemma 1. *Suppose that path κ_t is an optimal path that is not stationary and that $(\kappa_{t-1}, \kappa_t) \in \mathfrak{S}$ for $t = 1, 2, \ldots$. The following hold:*

(a) If $\partial^2 v(\kappa, \eta)/\partial\kappa\partial\eta > 0$ on $\mathfrak{S}$, $(\kappa_{t-1} - \kappa_t)(\kappa_t - \kappa_{t+1}) > 0$;
(b) If $\partial^2 v(\kappa, \eta)/\partial\kappa\partial\eta = 0$ on $\mathfrak{S}$, $\kappa_1 \neq \kappa_0$ and $\kappa_t = \kappa_1$ for $t = 2, 3, \ldots$;
(c) If $\partial^2 v(\kappa, \eta)/\partial\kappa\partial\eta < 0$ on $\mathfrak{S}$, $(\kappa_{t-1} - \kappa_t)(\kappa_t - \kappa_{t+1}) < 0$.

The existence of a free-trade equilibrium path can be proved by following Bewley (1972); a proof is not presented here.

Lemma 2. *(a) From initial stocks $\bar{k}_{\alpha 0}$ and $\bar{k}_{\beta 0}$, there is a free-trade equilibrium path, e_t, $t = 1, 2, \ldots$, such that the associated price paths, q_t, p_{t-1} and w_{ht}, $t = 1, 2, \ldots$, satisfy $q_t \geq 0$, $p_{t-1} \geq 0$, $w_{ht} \geq 0$, $\sum_{t=1}^{\infty} |q_t| < \infty$, $\sum_{t=1}^{\infty} |p_{t-1}| < \infty$, and $\sum_{t=1}^{\infty} |w_{ht}| < \infty$.*

(b) There is also an autarky equilibrium path, e_{ht}, $h = \alpha, \beta$ and $t = 1, 2, \ldots$, such that the associated price paths, q_{ht}, p_{ht-1} and w_{ht}, $h = \alpha, \beta$ and $t = 1, 2, \ldots$, satisfy $q_t \geq 0$, $p_{t-1} \geq 0$, $w_{ht} \geq 0$, $\sum_{t=1}^{\infty} |q_{ht}| < \infty$, $\sum_{t=1}^{\infty} |p_{ht-1}| < \infty$ and $\sum_{t=1}^{\infty} |w_{ht}| < \infty$.

The following lemma implies that an equilibrium consumption path of a country can be associated with a *marginal utility of wealth*. (A proof is routine and omitted.)

Lemma 3. *If c_{ht} and q_{ht}, $t = 1, 2, \ldots$, and h's consumption path and the price path of good X in h in an equilibrium (either free-trade or autarky). Then, there is $\lambda_h > 0$ such that*

$$\sum_{t=1}^{\infty} \rho^t u_h(c_{ht}) - \lambda_h \sum_{t=1}^{\infty} q_{ht} c_{ht} \geq \sum_{t=1}^{\infty} \rho^t u_h(\zeta_{ht}) - \lambda_h \sum_{t=1}^{\infty} q_{ht} \zeta_{ht} \tag{5.10}$$

for any $\zeta_{ht} \in R_+$, $t = 1, 2, \ldots$.

Take a free-trade equilibrium and an autarky equilibrium. In order to avoid unnecessary confusion, denote the free-trade equilibrium path by

$$e_t^F = (c_{ht}^F, x_{ht}^F, y_{ht}^F, k_{ht-1}^F, \ell_{ht}^F, q_t^F, p_{t-1}^F, w_{ht}^F; h = \alpha, \beta), \quad t = 1, 2, \ldots, \tag{5.11}$$

and the autarky equilibrium path by

$$e_{ht}^A = (c_{ht}^A, x_{ht}^A, y_{ht}^A, k_{ht-1}^A, \ell_{ht}^A, q_{ht}^A, p_{ht-1}^A, w_{ht}^A), \quad h = \alpha, \beta \text{ and } t = 1, 2, \ldots. \tag{5.12}$$

Denote by λ_h^F $(h = \alpha, \beta)$ country h's marginal utility of wealth associated with the free-trade equilibrium, e_t^F. By following Bewley (1982), we normalize the price paths in the free-trade equilibrium so that the marginal utilities of wealth add up to a constant value. For the sake of the analysis below, it is convenient to make the price normalization so that

$$\lambda_\alpha^F + \lambda_\beta^F = 2. \tag{5.13}$$

Given $(\lambda_\alpha^F, \lambda_\beta^F)$, define

$$v^F(k, y; \lambda_\alpha^F, \lambda_\beta^F) = \max_{(c_h, x_h, y_h, k_h; h=\alpha,\beta)} \sum_h u_h(c_h)/\lambda_h^F$$

$$\text{subject to} \begin{cases} \sum_h c_h \leq \sum_h g_h(k_h, y_h) \\ \sum_h (k_h, y_h) = (k, y). \end{cases} \tag{5.14}$$

Denote by D^F the domain of function $v^F(\cdot, \cdot; \lambda_\alpha^F, \lambda_\beta^F)$, i.e.,

$$D^F = \{(k, y) \in R_+^2 | (k, y) = \sum_h (k_h, y_h), (k_h, y_h) \in D_h, h = \alpha, \beta\}. \tag{5.15}$$

Set D^F determines feasible activities. In the case in which outputs are freely traded between the countries, path k_t, $t = 1, 2, \ldots$, is feasible if $(k_{t-1}, k_t) \in D^F$ for $t = 1, 2, \ldots$, and $k_0 = \sum_h \overline{k}_0^h$.

The next lemma states that the aggregate accumulation path in a free-trade equilibrium is optimal with respect to the reduced form optimal growth model $\sum_{t=1}^{\infty} \rho^t v^F(\kappa_{t-1}, \kappa_t; \lambda_\alpha^F, \lambda_\beta^F)$. We state a proof, on which we build proofs of the main results in the subsequent sections.

Lemma 4. *Let* $k_t^F = \sum_h k_{ht}^F$ *be the capital accumulation path associated with the free-trade equilibrium path,* e_t^F. *Then,*

$$(k_0^F, k_1^F, k_2^F, \ldots) = \arg \max_{(\kappa_0, \kappa_1, \kappa_2, \ldots)} \sum_{t=1}^{\infty} \rho^t v^F(\kappa_{t-1}, \kappa_t; \lambda_\alpha^F, \lambda_\beta^F)$$

$$\textit{subject to} \quad \kappa_0 = \sum_h \overline{k}_{h0}. \tag{5.16}$$

Proof. Note that (5.10) holds for $q_{ht} = q_t^F$ and $\lambda_h = \lambda_h^F$. Therefore, by setting $\zeta_\tau = c_{h\tau}$ for $\tau \neq t$,

$$\rho^t u_h(c_{ht}^F) - \lambda_h^F q_t^F c_{ht}^F \geq \rho^t u_h(\zeta_{ht}) - \lambda_h^F q_t^F \zeta_{ht} \tag{5.17}$$

for any $\zeta_{ht} \in R_+$. This implies

$$\rho^t \sum_h u_h(c_{ht}^F)/\lambda_h^F - q_t^F \sum_h c_{ht}^F \geq \rho^t \sum_h u_h(\zeta_{ht})/\lambda_h^F - q_t^F \sum_h \zeta_{ht} \tag{5.18}$$

for any $\zeta_{ht} \in R_+$, $h = \alpha, \beta$. Since $p_{ht} = p_t^F$, condition (5.2) implies

$$q_t^F g_h(k_{ht-1}^F, k_{ht}^F) + p_t^F k_{ht}^F - p_{t-1}^F k_{ht-1}^F \geq q_t^F g_h(\kappa_{ht-1}, \kappa_{ht}) + p_t^F \kappa_{ht} - p_{t-1}^F \kappa_{ht-1} \tag{5.19}$$

for any $(\kappa_{ht-1}, \kappa_{ht}) \in D_h$. Since $\sum_h g_h(k_{ht-1}^F, k_{ht}^F) = \sum_h c_{ht}^F$, by (5.18) and (5.19),

$$\rho^t \sum_h u_h(c_{ht}^F)/\lambda_h^F + p_t^F k_t^F - p_{t-1}^F k_{t-1}^F \geq \rho^t \sum_h u_h(\zeta_{ht})/\lambda_h^F + p_t^F \kappa_t - p_{t-1}^F \kappa_{t-1} \tag{5.20}$$

for any $(\kappa_{t-1}, \kappa_t) \in D^F$ and for any $\zeta_{ht} \geq 0$ such that $\sum_h g_h(\kappa_{ht-1}, \kappa_{ht}) = \sum_h \zeta_{ht}$, where $k_t^F = \sum_h k_{ht}^F$. Thus, by the definition of v^F, (5.20) implies

$$\begin{aligned} &\rho^t v^F(k_{t-1}^F, k_t^F; \lambda_\alpha^F, \lambda_\beta^F) + p_t^F k_t^F - p_{t-1}^F k_{t-1}^F \\ &\geq \rho^t v^F(\kappa_{t-1}, \kappa_t; \lambda_\alpha^F, \lambda_\beta^F) + p_t^F \kappa_t - p_{t-1}^F \kappa_{t-1} \end{aligned} \tag{5.21}$$

for any $(\kappa_{t-1}, \kappa_t) \in D^F$. Since Assumption 3a implies that $|k_t|$ are bounded uniformly in t, and since $p_t^F \to 0$ by Lemma 2, the lemma can be established by adding (5.21) up through t and setting $\kappa_0 = \sum_h \overline{k}_{h0}$. □

A similar but simpler result holds in the autarky case. To state that result, define $v_h^A : D_h \to R$,

$$v_h^A(k, y) = u_h(g_h(k, y)). \tag{5.22}$$

The following lemma can be proved in a manner similar to Lemma 4 (a proof is omitted).

Lemma 5. *Let k_{ht}^A be the capital accumulation path in country h that is associated with the autarky equilibrium e_{ht}^A. Then,*

$$(k_{h0}^A, k_{h1}^A, k_{h2}^A, \dots) = \arg \max_{(\kappa_0, \kappa_1, \kappa_2, \dots)} \sum_{t=1}^{\infty} \rho^t v_h^A(\kappa_{t-1}, \kappa_t)$$
$$\textit{subject to} \quad \kappa_0 = \overline{k}_{h0}. \tag{5.23}$$

5.4 Post-trade Aggregate Accumulation

We compare the global dynamics of a free-trade equilibrium path e_t^F, $t = 1, 2, \dots$, with those of autarky equilibrium paths, e_{ht}^A, $t = 1, 2, \dots$. To this end, we characterize the dynamics of each of these paths in terms of the fundamental structure of a model, i.e., production technologies in the present case. In our characterization, we focus on accumulation patterns, i.e., fluctuant and monotone capital accumulation paths, in order to capture interlinkage in the accumulation paths of countries.

In the trade-theoretic terminology, we are dealing with the large-country case, in which each country's economic behavior affects the world equilibrium prices. In order to find a free-trade equilibrium, therefore, it is necessary to find market clearing prices. In order to find an autarky equilibrium, in contrast, it suffices to solve a single optimization problem.

This difference is captured by Lemmas 4 and 5 above. Lemma 5 indicates that an autarky equilibrium path is optimal with respect to $v^A(k_h, y_h)$, which is determined solely by the fundamental structure of a model. In contrast, Lemma 4 indicates that a free-trade equilibrium path is optimal with respect to $v^F(k, y; \lambda_\alpha^F, \lambda_\beta^F)$. Its dynamics, therefore, depends on marginal utilities of wealth, λ_α^F and λ_β^F, which are determined endogenously in the market. These facts explain the difficulty in comparative dynamics in a many-consumer model, to which Bewley (1982), Yano (1984b) and Epstein (1987b) point. The present study faces with the same difficulty in characterizing the global dynamics of a free-trade equilibrium. The next lemma enables us to overcome this difficulty. Define

$$\mathbf{v}^F(k, y) = \max_{(k_h, y_h; h=\alpha,\beta)} \sum_h g_h(k_h, y_h) \text{ subject to } \sum_h (k_h, y_h) = (k, y). \tag{5.24}$$

Lemma 6.

$$\mathbf{v}^F(k, y) = v^F(k, y; \lambda^F_\alpha, \lambda^F_\beta).$$

Proof. First, we prove

$$p^F_0 > 0. \tag{5.25}$$

By Lemma 1, $p^F_0 \geq 0$. Suppose $p^F_0 = 0$. Then, by (5.21), $\sum_{t=1}^{\infty} \rho^t v^F(k^F_{t-1}, k^F_t; \lambda^F_\alpha, \lambda^F_\beta) \geq \sum_{t=1}^{\infty} \rho^t v^F(\kappa_{t-1}, \kappa_t; \lambda^F_\alpha, \lambda^F_\beta)$ for any path $(\kappa_{t-1}, \kappa_t) \in D^F$, $t = 1, 2, \ldots$, from an arbitrary initial stocks κ_0. This contradicts the fact that utilities cannot be saturated (see Assumption 1).

Next, we prove

$$\lambda^F_\alpha = \lambda^F_\beta. \tag{5.26}$$

By Assumption 1, (5.18) implies

$$\rho^t(c^F_{\alpha t}/\lambda^F_{\alpha t} + c^F_{\beta t}/\lambda^F_\beta) - q^F_t(c^F_{\alpha t} + c^F_{\beta t}) \geq \rho^t(\zeta_{\alpha t}/\lambda^F_\alpha + \zeta_{\beta t}/\lambda^F_\beta) - q^F_t(\zeta_{\alpha t} + \zeta_{\beta t}) \tag{5.27}$$

for any $(\zeta_{\alpha t}, \zeta_{\beta t}) \in R^2_+$, $t = 1, 2, \ldots$. Suppose $\lambda^F_\alpha > \lambda^F_\beta$. Then, (5.27) implies that for any t such that $c^F_{\alpha t} + c^F_{\beta t} > 0$, $c^F_{\beta t} > c^F_{\alpha t} = 0$. Note that for any t such that $c^F_{\alpha t} + c^F_{\beta t} = 0$, $c^F_{\alpha t} = 0$ by $c_{ht} \geq 0$ (Assumption 1). By these facts, $c^F_{\alpha t} = 0$ for $t = 1, 2, \ldots$. Thus, $0 = \sum_{t=1}^{\infty} q^F_t c^F_{\alpha t} = p^F_0 \overline{k}_{\alpha 0} + \sum_{t=1}^{\infty} w^F_t \overline{\ell}_F$, which contradicts $p^F_0 > 0$ and $\overline{k}_{\alpha 0} > 0$, since $w^F_t \geq 0$ and $\overline{\ell}_F \geq 0$. In a similar manner, we may establish a contradiction in the case of $\lambda^F_\alpha < \lambda^F_\beta$.

By (5.26) and (5.13), $\lambda^F_\alpha = \lambda^F_\beta = 1$. By Assumption 1, the lemma follows from the definitions of v^F and $\mathbf{v}^F$. □

For the sake of simplicity, introduce the following notation with respect to function $g_h(k_h, y_h)$: $g^1_h = \partial g_h/\partial k_h$, $g^2_h = \partial g_h/\partial y_h$, $g^{11}_h = \partial^2 g_h/\partial k^2_h$, $g^{12}_h = \partial^2 g_h/\partial k_h \partial y_h$ and $g^{22}_h = \partial^2 g_h/\partial y^2_h$. Denote by G_h the Hessian of g_h; i.e., $G_h = \left[g^{ij}_h\right]$. Moreover, denote by $|A|$ the determinant of matrix A.

Assumption 4 *If* $g^{12}_\alpha g^{12}_\beta \geq 0$, $\sum_h |G_h| > 0$.

Assumption 4 implies that if the cross partial derivatives of g_α and g_β have opposite signs, then at least one country's production function is strictly concave.

Lemma 7. *Suppose* $\mathbf{v}^F(k, y) = \sum_h g_h(k_h, y_h)$, $\sum_h (k_h, y_h) = (k, y)$ *and* $(k_h, y_h) \in$ *interior* D_h. *Then,* $\mathbf{v}^F$ *is twice continuously differentiable around* (k, y). *The second partial derivatives of* v^F *satisfy the following:*

$$\partial^2 \mathbf{v}^F/\partial k^2 = \frac{1}{\Delta}\left[g_\alpha^{11}\left|G_\beta\right| + g_\beta^{11}\left|G_\alpha\right|\right] \tag{5.28}$$

$$\partial^2 \mathbf{v}^F/\partial y^2 = \frac{1}{\Delta}\left[g_\alpha^{22}\left|G_\beta\right| + g_\beta^{22}\left|G_\alpha\right|\right] \tag{5.29}$$

$$\partial^2 \mathbf{v}^F/\partial k\partial y = \frac{1}{\Delta}\left[g_\alpha^{12}\left|G_\beta\right| + g_\beta^{12}\left|G_\alpha\right|\right] \tag{5.30}$$

where

$$\Delta = (g_\alpha^{11} g_\beta^{22} - g_\alpha^{12} g_\beta^{12}) + (g_\beta^{11} g_\alpha^{22} - g_\beta^{12} g_\alpha^{12}) + \sum_h |G_h| > 0. \tag{5.31}$$

Proof. Write the Lagrangean associate with the maximization problem in (5.24) as follows:

$$\mathcal{L} = \sum_h g_h(k_h, y_h) + \mu_1(k - \sum_h k_h) + \mu_2(y - \sum_h y_h). \tag{5.32}$$

By the hypothesis of Lemma 7, $(k_h, y_h;\ h = \alpha, \beta)$ satisfies the first order condition associated with (5.33), i.e.,

$$g_h^j(k_h, y_h) - \mu_j = 0, \quad j = 1, 2 \text{ and } h = \alpha, \beta. \tag{5.33}$$

Let G_h be the Hessian of g_h, I be the 2×2 identity matrix and 0 be the 2×2 0 matrix. Moreover, denote by ${}^tB$ the transpose of matrix B. Totally differentiating (5.33) and the constraints, we have

$$A d\pi = d\omega, \tag{5.34}$$

where A, $d\pi$, and $d\omega$ are defined as follows:

$$A = \begin{bmatrix} G_\alpha & 0 & -I \\ 0 & G_\beta & -I \\ -I & -I & 0 \end{bmatrix}; \tag{5.35}$$

$$d\pi =^t [dk_\alpha, dy_\alpha, dk_\beta, dy_\beta, d\mu_1, d\mu_2] \in R^6; \tag{5.36}$$

$$d\omega =^t [0, 0, 0, 0, -dk, -dy] \in R^6. \tag{5.37}$$

By the Implicit Function Theorem, it suffices to prove that A is invertible. To this end, note

$$|A| = \Delta \tag{5.38}$$

where $|A|$ denotes the determinant of matrix A. Since, by concavity of g_h (Assumption 2a),

$$|G_h| \geq 0, \tag{5.39}$$

$$\sqrt{g_h^{11} g_h^{22}} \geq \begin{cases} g_h^{12} & \text{if } g_h^{12} \geq 0 \\ -g_h^{12} & \text{if } g_h^{12} < 0. \end{cases} \tag{5.40}$$

Take the case of $g_\alpha^{12} g_\beta^{12} \geq 0$. Since $\sqrt{g_\alpha^{11} g_\alpha^{22} g_\beta^{11} g_\beta^{22}} \geq g_\beta^{12} g_\alpha^{12}$ by (5.40), by Assumption 4 and (5.31),

$$\Delta > g_\alpha^{11} g_\beta^{22} + g_\beta^{11} g_\alpha^{22} - 2\sqrt{g_\alpha^{11} g_\alpha^{22} g_\beta^{11} g_\beta^{22}} = \left(\sqrt{g_\alpha^{11} g_\beta^{22}} - \sqrt{g_\beta^{11} g_\alpha^{22}}\right)^2 \geq 0. \tag{5.41}$$

Take the case of $g_\alpha^{12} g_\beta^{12} < 0$. Then, by (5.31), (5.39) and Assumption 3c,

$$\Delta \geq g_\alpha^{11} g_\beta^{22} + g_\beta^{11} g_\alpha^{22} > 0. \tag{5.42}$$

Thus, in either case, by (5.38),

$$|A| > 0. \tag{5.43}$$

Note that $\mu_1 = \partial \mathbf{v}^F / \partial_k$ and $\mu_2 = \partial \mathbf{v}^F / \partial y$. Thus, in order to derive (5.30), for example, it suffices to find $d\mu_1/dy|_{dk=0} = \partial^2 \mathbf{v}^F / \partial k \partial y$ in system (5.34). By (5.35),

$$d\mu_1/dy|_{dk=0} = \frac{1}{|A|} \begin{vmatrix} \multicolumn{2}{c}{G_\alpha} & \multicolumn{2}{c}{0} & -1 & 0 \\ & & & & 0 & 0 \\ \multicolumn{2}{c}{0} & \multicolumn{2}{c}{G_\beta} & -1 & 0 \\ & & & & 0 & 0 \\ -1 & 0 & -1 & 0 & 0 & 0 \\ 0 & -1 & 0 & -1 & 0 & -dy \end{vmatrix}, \tag{5.44}$$

which implies (5.30). Expressions (5.28) and (5.29) can be derived in a similar manner. □

For the sake of analysis in the subsequent sections, we note the following lemma:

Lemma 8. *There is a function* $(\mathbf{k}_h(k, y), \mathbf{y}_h(k, y); h = \alpha, \beta)$ *such that*

$$(\mathbf{k}_h(k, y), \mathbf{y}_h(k, y); h = \alpha, \beta) = \arg \max_{(k_h, y_h; h=\alpha,\beta)} \sum_h g_h(k_h, y_h)$$

$$\textit{subject to} \sum_h (k_h, y_h) = (k, y). \tag{5.45}$$

If $(\mathbf{k}_h(k,y), \mathbf{y}_h(k,y)) \in$ interior D^h for $h = \alpha, \beta$, it is continuously differentiable in (k,y). In particular, $\mathbf{k}_h(k,y)$ satisfies the following:

$$\frac{\partial}{\partial k}\mathbf{k}_h(k,y) = \frac{1}{\Delta}(g_h^{22}g_{h'}^{11} - g_h^{12}g_{h'}^{12} + |G_{h'}|); \tag{5.46}$$

$$\frac{\partial}{\partial y}\mathbf{k}_h(k,y) = \frac{1}{\Delta}(g_h^{22}g_{h'}^{12} - g_h^{12}g_{h'}^{22}), \tag{5.47}$$

where $\{h, h'\} = \{\alpha, \beta\}$.

Proof. This readily follows from (5.34) to (5.37). □

Define φ: interior $(D_\alpha \times D_\beta) \to R$ as

$$\varphi(k_\alpha, y_\alpha, k_\beta, y_\beta) = g_\alpha^{12}(k_\alpha, y_\alpha)|G_\beta(k_\beta, y_\beta)| + g_\beta^{12}(k_\beta, y_\beta)|G_\alpha(k_\alpha, y_\alpha)|. \tag{5.48}$$

In order to apply Lemma 1, take a large rectangular in the interior of D_h, $h = \alpha, \beta$, and denote it by S; i.e., $S \subset interior\ D_h$ for $h = \alpha, \beta$.

Assumption 5 *The free-trade (post-trade) equilibrium path, e_t^F, $t = 1, 2, \ldots$, satisfies $(k_{ht-1}^F, y_{ht}^F) \in S$ for $h = \alpha, \beta$ and $t = 1, 2, \ldots$, where $e_t^F = (c_{ht}^F, x_{ht}^F, k_{ht-1}^F, y_{ht}^F, \ell_{ht}^F, q_t^F, p_{t-1}^F, w_{ht}^F; h = \alpha, \beta)$ and $k_t^F = \sum_h k_{ht}^F$.*

Our first main result is as follows:

Theorem 1. *Suppose that the post-trade aggregate accumulation path, k_t^F, is not stationary. The following holds:*

(a) If $\varphi > 0$ on S^2, $\left(k_{t-1}^F - k_t^F\right)\left(k_t^F - k_{t+1}^F\right) > 0$;
(b) If $\varphi = 0$ on S^2, $k_1^F \neq \sum_h \overline{k}_{h0}$ and $k_t^F = k_1^F$ for $t = 1, 2, \ldots$;
(c) If $\varphi < 0$ on S^2, $\left(k_{t-1}^F - k_t^F\right)\left(k_t^F - k_{t+1}^F\right) < 0$.

Proof. The theorem readily follows from Lemmas 1 and 7. □

The aggregate dynamics of an autarky equilibrium path in country h, e_{ht}^A, $t = 1, 2, \ldots$, can be characterized in a much simpler way.

Assumption 6 *An autarky (pre-trade) equilibrium path in country h, e_{ht}^A, $t = 1, 2, \ldots$, satisfies $(y_{ht}^A, k_{ht-1}^A) \in S$ for $t = 1, 2, \ldots$, where $e_{ht}^A = (c_{ht}^A, x_{ht}^A, y_{ht}^A, k_{ht-1}^A, \ell_{ht}^A, q_{ht}^A, p_{ht-1}^A, w_{ht}^A)$.*

The following holds with respect to autarky equilibrium path e_{ht}^A:

Theorem 2. *Suppose that each country's pre-trade accumulation path, k_{ht}^A, is not stationary. The following folds:*

(a) If $g_h^{12} > 0$ on S, $(k_{ht-1}^A - k_{ht}^A)(k_{ht}^A - k_{ht+1}^A) > 0$;
(b) If $g_h^{12} = 0$ on S, $k_{h1}^A \neq \overline{k}_{h0}$ and $k_{ht}^A = k_{h1}^A$ for $t = 1, 2, \ldots$;
(c) If $g_h^{12} < 0$ on S, $(k_{ht-1}^A - k_{ht}^A)(k_{ht}^A - k_{ht+1}^A) < 0$.

Proof. Under Assumption 1, $v_h^A(y_h, k_h) = g_h(y_h, k_h)$. Thus, the theorem follows from Lemmas 1 and 5. □

Theorems 1 and 2 demonstrate that the pattern of each country's pre-trade (autarky) capital accumulation path and that of the post-trade (free-trade) aggregate capital accumulation path depend on g_h^{12} and φ. It is therefore important to consider their economic interpretations. The next lemma characterizes the sign of g_h^{12}.

Lemma 9. *The y_h satisfying*

$$y_h = \arg\max_{\eta_h} g_h(k_h, \eta_h) + p\eta_h \tag{5.49}$$

can be written as a continuously twice differentiable function of p and k_h, $y_h = \boldsymbol{\eta}_h(p, k_h)$. Moreover,

$$\partial \boldsymbol{\eta}_h / \partial k_h = -g_h^{12} / g_h^{22}. \tag{5.50}$$

Proof. By the first order condition, $g_h^2(k_h, \eta_h) = -p$. By totally differentiating this, we have (5.50). □

Lemma 9 captures the response in good-Y output to an exogenous increase in capital input, given constant output prices. In Yano (1990), the effect on output that an exogenous increase in capital input causes, given output prices, is called the *capital expansion effect*. The capital expansion effect is a concept that parallels to the income effect in consumption theory; as is well-known, the income effect can be measured by the change in compensated demand that follows an exogenous increase in utility level, given the prices of goods.

As (5.50) indicates, the capital expansion effect can work in either positive or negative direction, just like the income effect[4]:

Remark 1. The capital expansion effect works on sector Y in the positive direction (i.e., increases sector Y's output) if $g_h^{12} > 0$ and in the negative direction if $g_h^{12} < 0$. The magnitude of a capital expansion effect on sector Y is captured by $|g_h^{12}|/|g_h^{22}|$.

The next lemma characterizes the sign of φ.

Lemma 10. *Suppose $|G_h| \neq 0$. Then, $|G_h| > 0$. Moreover,*

$$\varphi = |G_\alpha||G_\beta|(g_\alpha^{12}/|G_\alpha| + g_\beta^{12}/|G_\beta|). \tag{5.51}$$

Moreover, the (k_h, y_h) satisfying

$$(k_h, y_h) = \arg\max_{(\kappa_h, \eta_h)} g_h(\kappa_h, \eta_h) + p\eta_h - r\kappa_h \tag{5.52}$$

[4] Sector Y may be called a normal sector if the capital expansion effect works in the positive direction and an inferior sector if it works in the negative direction. The term sector Y originates from the case in which independent industries are producing goods X and Y and in which production function f_h merely represents the aggregate relationship between inputs and outputs in the economy as a whole, as is discussed in footnote 2.

can be written as a continuously twice differentiable function of q *and* r, $(k_h, y_h) = (\kappa_h(p,r), \eta_h(p,r))$. *Moreover,*

$$\partial \kappa_h / \partial_r = g_h^{22} / |G_h|. \tag{5.53}$$

$$\partial \eta_h / \partial_r = -g_h^{12} / |G_h|. \tag{5.54}$$

Proof. By the first order condition, $g_h^1(\kappa_h, \eta_h) = r$ and $g_h^2(\kappa_h, \eta_h) = -p$. By totally differentiating this system and setting $dq = 0$, we have $G_h \begin{bmatrix} d\kappa_h \\ d\eta_h \end{bmatrix} = \begin{bmatrix} dr \\ 0 \end{bmatrix}$. This implies (5.53) and (5.54). □

Lemma 10 captures the responses in capital input and in the capital good output (Y) to an exogenous change in the price of capital input. As in Yano (1990), we call this effect the *capital cost effect*. Note that $|G_h| \neq 0$ implies $|G_h| > 0$. Thus, by (5.54), we have the following:

Remark 2. The capital cost effect works on sector Y in the positive direction (i.e., increases sector Y's output) if $g_h^{12} < 0$ and in the negative direction if $g_h^{12} < 0$. The magnitude of a capital cost effect is captured by $|g_h^{12}| / |G_h|$.

Remarks 1 and 2 imply that *the capital expansion effect and the capital cost effect works on sector Y in oppositive directions*. This is because the capital cost effect on sector Y can be decomposed into the capital cost effect on sector Y's capital input and the capital expansion effect on Y's output. For the sake of explanation, suppose that the price of capital input, r, rises. By $g_h^{22} < 0$ (Assumption 3c) and $|G_h| > 0$, (5.50) implies that sector Y's capital input unambiguously falls. If this fall increases sector Y's output (i.e., if the capital expansion effect works on sector Y in the negative direction), then the capital cost effect works in the positive direction.

Several conclusions can be derived from Theorems 1 and 2 with respect to the relationship between the dynamics of each country's pre-trade capital accumulation path, k_{ht}^A, and that of the post-trade aggregate accumulation path, k_t^F. Before discussing those conclusions, two technical points should be noted. First, our characterizations do not depend on whether or not a modified golden rule state is stable. This feature distinguishes our study from the early literature on optimal growth (Brock and Scheinkman 1976; Cass and Shell 1976; McKenzie 1976; Scheinkman 1976) and the literature on equilibrium turnpike (Bewley 1982; Coles 1985; Marimon 1989; Yano 1984b), which are concerned with the stability of a modified golden rule state.

Second, because we are interested in the global dynamics of an equilibrium path, the conditions that we impose on partial derivatives of g_h are all concerned with the global (as oppose to local) relationships of and between the two countries' production technologies. In other words, those conditions are always made globally.

The first conclusion, derived from Theorems 1 and 2, is that each country's pre-trade accumulation pattern is determined by the direction of a capital expansion effect on sector Y. That is to say, by Theorem 2 and Remark 1, h's pre-trade accumulation is monotone if the capital expansion effect on sector Y works in the positive direction ($g_h^{12} > 0$) and fluctuant if it works in the negative direction ($g_h^{12} < 0$).

Second, if the capital expansion effect on sector Y works in the same direction in one country as in the other, the post-trade aggregate accumulation pattern and the countries' pre-trade accumulation patterns coincide with one another. This follows from Theorems 1 and 2. If the capital expansion effect on sector Y works in the positive direction in both countries, then, by Remark 1, $g_{\alpha}^{12} > 0$ and $g_{\beta}^{12} > 0$. Moreover, by Lemma 10, $\varphi > 0$. Thus, by Theorems 1 and 2, not only pre-trade capital accumulation paths are monotone in both countries but also the post-trade aggregate capital accumulation path is monotone. If, in contrast, the capital expansion effect on sector Y works in the negative direction in both countries ($g_{\alpha}^{12} < 0$ and $g_{\beta}^{12} < 0$), not only pre-trade capital accumulation paths are fluctuant in both countries but also the post-trade aggregate accumulation path is fluctuant.

Third, if the capital expansion effect on sector Y works in opposite directions between the countries (g_{α}^{12} and g_{β}^{12} have opposite signs), the post-trade aggregate accumulation pattern depends on magnitudes of the capital cost effects on sector Y in the two countries. Suppose that, for example, the capital expansion effect on sector Y works in the positive direction in country α and in the negative direction in country β. Then, by Remark 1, $g_{\alpha}^{12} > 0$ and $g_{\beta}^{12} < 0$. Then, the post-trade aggregate accumulation is monotone if $|g_{\alpha}^{12}|/|G_{\alpha}|>|g_{\beta}^{12}|/|G_{\beta}|$ and fluctuant if $|g_{\alpha}^{12}|/|G_{\alpha}|<|g_{\beta}^{12}|/|G_{\beta}|$. This implies that, by Remark 2, the post-trade aggregate accumulation pattern is the same as the pre-trade accumulation pattern of the country in which the capital cost effect on sector Y has a larger magnitude than in the other country.

These results may be summarized as follows:

Proposition 1. *A country's pre-trade accumulation is monotone if the capital expansion effect on the country's capital good production sector works in the positive direction and fluctuant if it works in the negative direction.*

Proposition 2. *Suppose that the capital expansion effect on the capital good production sector works in the same direction in one country as in the other country. Then the post-trade aggregate accumulation pattern coincides with each country's pre-trade accumulation pattern.*

Proposition 3. *Suppose that the capital expansion effect on the capital good production sector works in the positive direction in one country and in the negative direction in the other country. If the magnitude of a capital cost effect on the capital good production sector is larger in country h than in country h', the post-trade aggregate accumulation pattern coincides with the pre-trade accumulation pattern of country h.*

5.5 Super-Stabilization Effect of Trade

One of the most striking implications of Theorems 1 and 2 is that trade may sometimes be "extremely" stabilizing. If, as Lemma 1 indicates, the cross partial derivative of a reduced form utility function is zero, an optimal growth path reaches a stationary state in one period. We say that such a stationary state is "*super-stable.*"

Theorems 1 and 2 imply that the stationary state to which the post-trade equilibrium path will converge is super-stable if $|G_h| = 0$ for $h = \alpha, \beta$.[5] Therefore, trade may "super-stabilize" an equilibrium path, i.e., that even if it takes infinitely many periods for pre-trade equilibrium paths to reach stationary states, the post-trade equilibrium path may reach a stationary state within one period.

The super stabilization of trade can be observed in the case in which the production process can be described by the standard Heckscher-Ohlin-Samuelson model (see footnote 1). In that case, $|G_h| = 0$ (see footnote 2). Moreover, by the Rybczynski theorem (see Jones 1965), the capital expansion effect on the labor intensive sector works in the negative direction . Therefore, if sector Y is capital intensive in one country and labor intensive in the other country, g_α^{12} and g_β^{12} have different signs. In this case, by Theorem 1, trade is super-stabilizing.

Proposition 4. *Suppose that the social production processes can be described by the Heckscher-Ohlin-Samuelson model and that the factor intensity ranking of industries is reversed between the countries. Trade may be super-stabilizing, i.e., the post-trade equilibrium path may reach a stationary state one period after opening trade, no matter how each-country's pre-trade equilibrium path behaves.*

5.6 Post-trade Accumulation in the Dominant Country

In this and the next sections, we consider the relationship between the aggregate capital accumulation in a free-trade equilibrium (post-trade aggregate accumulation) and that of each country's capital accumulation in that equilibrium (an individual country's post-trade accumulation).

Suppose that the capital expansion effect on sector Y works in opposite directions between the countries. Then, by Remark 1, $g_\alpha^{12} g_\beta^{12} < 0$. Proposition 3 implies that if the capital cost effect on sector Y has the larger magnitude in country h^* than in country h^{**} (i.e., by Remark 1, $|g_{h^*}^{12}|/|G_{h^*}| > |g_{h^{**}}^{12}|/|G_{h^{**}}|$), the *post-trade* aggregate accumulation pattern coincides with the *pre-trade* accumulation pattern of h^* and is reversed from that of h^{**}. In this case, therefore, we call countries h^* and h^{**} *the dominant and subordinate countries,* respectively, in the determination of post-trade accumulation patterns. Given this terminology, as is shown in Sect. 5.6.1, the dominant country's *post-trade* accumulation pattern coincides with the *post-trade* aggregate accumulation pattern. As is seen in Sect. 5.7, moreover, the subordinate country's post-trade accumulation pattern is ambiguous.

If the capital expansion effect on sector Y works in the same direction in one country as in the other country ($g_\alpha^{12} g_\beta^{12} > 0$), the *post-trade* aggregate

[5] In this case, $g_\alpha^{12} g_\beta^{12} < 0$ must hold under Assumption 4.

accumulation pattern is the same as both countries' *pre-trade* accumulation patterns (see Proposition 2). Therefore, the countries cannot be distinguished between dominant and subordinate countries in the same sense as above. In this case ($g_\alpha^{12} g_\beta^{12} > 0$), we call country h^* the dominant country if the capital cost effect on sector Y has a larger magnitude in the country than in the other country (i.e., if $|g_{h^*}^{12}|/|g_{h^*}^{22}| > |g_{h^{**}}^{12}|/|g_{h^{**}}^{22}|$). Given this terminology, just like in the above case, we may conclude that the dominant country's *post-trade* accumulation pattern coincides with the *post-trade* aggregate accumulation pattern (see Sect. 5.6.2). As is seen in Sect. 5.7, moreover, the subordinate country's post-trade accumulation pattern is ambiguous.

If $|G_h| = 0$ for $h = \alpha, \beta$, as is discussed above, the super-stabilization effect of trade is at work. In this case, as Lemma 8 indicates, each country's post-trade accumulation path as well reaches its stationary level in period 2. In what following, therefore, we exclude this trivial case:

Assumption 4′ $|G_h| \neq 0$ *for* $h = \alpha, \beta$.

5.6.1 *Case I* (g_α^{12} *and* g_β^{12} *with Opposite Signs)*

The next theorem implies that the dominant country's post-trade accumulation pattern is the same as the post-trade aggregate accumulation pattern.

Theorem 3. *Let* $h' \in \{\alpha, \beta\}$ *and* $h'' \in \{\alpha, \beta\} \backslash \{h'\}$. *If* $g_{h'}^{12} \geq 0$ *and* $g_{h''}^{12} \leq 0$, *the post-trade aggregate accumulation path,* k_t^F, *and the individual countries' post-trade accumulation paths,* $k_{h' t}^F$ *and* $k_{h'' t}^F$, *satisfy the following:*

(a) If $k_t^F \lesseqgtr k_{t+1}^F \lesseqgtr k_{t+2}^F$, *then* $k_{h' t}^F \lesseqgtr k_{h' t+1}^F$;
(b) If $k_t^F = k_{t+1}^F = k_{t+2}^F$, *then* $k_{h' t}^F = k_{h' t+1}^F$ *and* $k_{h'' t}^F = k_{h'' t+1}^F$;
(c) If $k_t^F < k_{t+1}^F$ *and* $k_{t+1}^F > k_{t+2}^F$, *then* $k_{h'' t}^F < k_{h'' t+1}^F$; *moreover, if* $k_{t+1}^F > k_{t+2}^F$ *and* $k_{t+2}^F < k_{t+3}^F$, *then* $k_{h'' t+1}^F > k_{h'' t+2}^F$.

Proof. Since $g_{h'}^{12} \geq 0$ and $g_{h''}^{12} \leq 0$, $\partial \mathbf{k}_{h'}(k, y)/\partial k > 0$ and $\partial \mathbf{k}_{h'}(k, y)/\partial y \geq 0$, $\partial \mathbf{k}_{h''}(k, y)/\partial k > 0$ and $\partial \mathbf{k}_{h''}(k, y)/\partial y \leq 0$. If k_t^F is monotone increasing, by $\partial \mathbf{k}_{h'}/\partial k > 0$ and $\partial \mathbf{k}_{h'}/\partial y > 0$,

$$k_{h' t}^F = \mathbf{k}_{h'}(k_t^F, k_{t+1}^F) < \mathbf{k}_{h'}(k_{t+1}^F, k_{t+1}^F) \leq \mathbf{k}_{h'}(k_{t+1}^F, k_{t+2}^F) = k_{h' t+1}^F. \tag{5.55}$$

If it is monotone decreasing, $k_{h' t}^F > k_{h' t+1}^F$ can be proved in a similar way. Thus, (a) is proved. In order to prove (c), let $k_t^F < k_{t+1}^F$ and $k_{t+1}^F > k_{t+2}^F$. Then, by $\partial \mathbf{k}_{h''}/\partial k > 0$ and $\partial \mathbf{k}_{h''}/\partial y < 0$,

$$k_{h'' t}^F = \mathbf{k}_{h''}(k_t^F, k_{t+1}^F) < \mathbf{k}_{h''}(k_{t+1}^F, k_{t+1}^F) \leq \mathbf{k}_{h''}(k_{t+1}^F, k_{t+2}^F) = k_{h'' t+1}^F. \tag{5.56}$$

If $k^F_{t+1} > k^F_{t+2}$ and $k^F_{t+2} < k^F_{t+3}$, similarly,

$$k^F_{h''t+1} = \mathbf{k}_{h''}(k^F_{t+1}, k^F_{t+2}) > \mathbf{k}_{h''}(k^F_{t+2}, k^F_{t+2}) \geq \mathbf{k}_{h''}(k^F_{t+2}, k^F_{t+3}) = k^F_{h''t+2}. \tag{5.57}$$

From (5.56) and (5.57), (c) follows. Let $k^F_t = k^F_{t+1} = k^F_{t+2} = k$. Then, $k^F_{h't} = k^F_{h't+1} = \mathbf{k}_{h'}(k,k)$, and $k^F_{h''t} = k^F_{h''t+1} = \mathbf{k}_{h''}(k,k)$. Thus, (c) holds. □

Suppose that $g^{12}_{h^*}/|G_{h^*}| > -g^{12}_{h^{**}}/|G_{h^{**}}| > 0$. Then, by (5.48), $\varphi > 0$. Therefore, by Theorems 1 and 3, both the post-trade aggregate accumulation and the post-trade accumulation of country h^* are monotone. Suppose, instead, that $g^{12}_{h^*}/|G_{h^*}| < -g^{12}_{h^{**}}/|G_{h^{**}}| < 0$. Then, $\varphi < 0$. Therefore, by Theorems 1 and 3, both the post-trade aggregate accumulation and the post-trade accumulation in country h^* are fluctuant. Because $|g^{12}_h/|G_h|$ measures the magnitude of a capital cost effect (see Remark 2), this may be summarized as follows:

Proposition 5. *Suppose that the capital expansion effect on the capital good production sector works in the positive direction in one country and in the negative direction in the other country. If the magnitude of a capital cost effect on the capital good production sector is larger in country h^* than in country h^{**}, the post-trade accumulation pattern in country h^* (the dominant country) coincides with the post-trade aggregate accumulation pattern.*

5.6.2 Case II (g^{12}_α and g^{12}_β with the Same Signs)

In order to simplify our discussion, we assume the following:

Assumption 7 *On S^2, the following holds:*

$$g^{22}_\alpha g^{11}_\beta + |G_\beta| > g^{12}_\alpha g^{12}_\beta; \tag{5.58}$$

$$g^{22}_\beta g^{11}_\alpha + |G_\alpha| > g^{12}_\alpha g^{12}_\beta. \tag{5.59}$$

By Lemma 8, this assumption implies that each country's capital input increases as the aggregate capital input increases while the aggregate capital output is kept constant, i.e., $\partial \mathbf{k}_h/\partial k > 0$. Because $\sum_h \partial \mathbf{k}_h/\partial k = 1$, an increase in aggregate capital input increases at least one country's capital input but may decrease the other country's capital input, in general, which is excluded under Assumption 7. Note that Assumption 7 is always satisfied in Case I ($g^{12}_\alpha g^{12}_\beta < 0$).

Theorem 4 *Suppose that for $h' \in \{\alpha, \beta\}$ and $h'' \in \{\alpha, \beta\} \backslash \{h'\}$,*

$$\frac{\left|g^{12}_{h'}\right|}{\left|g^{22}_{h'}\right|} > \frac{\left|g^{12}_{h''}\right|}{\left|g^{22}_{h''}\right|}. \tag{5.60}$$

Then, with respect to the capital accumulation path of country h', $k^F_{h'\,t}$, the following holds:

(a) In the case in which $g^{12}_\alpha > 0$ and $g^{12}_\beta > 0$,

$$k^F_{h'\,t} \lesseqgtr k^F_{h'\,t+1} \text{ if } k^F_t \lesseqgtr k^F_{t+1} \lesseqgtr k^F_{t+2};$$

(b) In the case in which $g^{12}_\alpha < 0$ and $g^{12}_\beta < 0$, the following holds:

$$k^F_{h'\,t} < k^F_{h'\,t+1} \text{ if } k^F_t < k^F_{t+1} \text{ and } k^F_{t+1} > k^F_{t+2};$$
$$k^F_{h'\,t+1} > k^F_{h'\,t+2} \text{ if } k^F_{t+1} > k^F_{t+2} \text{ and } k^F_{t+2} < k^F_{t+3}.$$

Proof. By Assumption 5, $\partial \mathbf{k}_{h'}/\partial k > 0$. Suppose $g^{12}_{h'} > 0$ for $h = \alpha, \beta$. By Lemma 8, (5.60) implies $\partial \mathbf{k}_{h'}/\partial y > 0$. Thus, if k^F_t is monotone increasing,

$$k^F_{h'\,t} = \mathbf{k}_{h'}(k^F_t, k^F_{t+1}) < \mathbf{k}_{h'}(k^F_{t+1}, k^F_{t+1}) < \mathbf{k}_{h'}(k^F_{t+1}, k^F_{t+2}) = k^F_{h'\,t+1}. \quad (5.61)$$

If it is monotone decreasing, $k^F_{h'\,t} > k^F_{h'\,t+1}$ can be proved in a similar way. Thus, (a) is proved. In order to prove (b), suppose $g^{12}_{h'} < 0$ for $h = \alpha, \beta$. Then, by Lemma 8, (5.60) implies $\partial \mathbf{k}_{h'}/\partial y < 0$. Let $k^F_t < k^F_{t+1}$ and $k^F_{t+1} > k^F_{t+2}$. Then, by $\partial \mathbf{k}_{h'}/\partial k > 0$ and $\partial \mathbf{k}_{h'}/\partial y < 0$,

$$k^F_{h'\,t} = \mathbf{k}_{h'}(k^F_t, k^F_{t+1}) < \mathbf{k}_{h'}(k^F_{t+1}, k^F_{t+1}) \leq \mathbf{k}_{h'}(k^F_{t+1}, k^F_{t+2}) = k^F_{h'\,t+1}. \quad (5.62)$$

If $k^F_{t+1} < k^F_{t+2}$ and $k^F_{t+2} > k^F_{t+3}$, similarly,

$$k^F_{h'\,t+1} = \mathbf{k}_{h'}(k^F_{t+1}, k^F_{t+2}) > \mathbf{k}_{h'}(k^F_{t+2}, k^F_{t+2}) \geq \mathbf{k}_{h'}(k^F_{t+2}, k^F_{t+3}) = k^F_{h'\,t+2} \quad (5.63)$$

from (5.62) and (5.63), (b) follows. □

Take the case in which $g^{12}_{h^*}/|g^{22}_{h^*}| > g^{12}_{h^{**}}/|g^{22}_{h^{**}}| > 0$. Then, by (5.48), $\varphi > 0$. Therefore, by Theorems 1 and 3, both the post-trade aggregate accumulation and the post-trade accumulation of country h^* are monotone. Take, instead, the case in which $g^{12}_{h^*}/|g^{22}_{h^*}| < g^{12}_{h^{**}}/|g^{22}_{h^{**}}| < 0$. Then, $\varphi < 0$. Therefore, by Theorems 1 and 3, both the post-trade aggregate accumulation and the post-trade accumulation in country h^* are fluctuant. Because $|g^{12}_{h^{**}}|/|g^{22}_{h^{**}}|$ measures the magnitude of a capital expansion effect on sector Y (see Remark 1), the above discussion can be summarized as follows:

Proposition 6. *Suppose that the capital expansion effect works in the same direction in one country as in the other and that the capital expansion effect on the capital good production sector is stronger in country h^* than in country h^{**}. Then, the post-trade accumulation pattern in country h^* (the dominant country) coincides with the post-trade aggregate accumulation pattern.*

5.7 Post-trade Accumulation in the Subordinate Country

Recall that the country that is not dominant is the *subordinate country*. As is seen below, the relationship between the subordinate country's post-trade accumulation pattern and the post-trade aggregate accumulation pattern is generally ambiguous. Following the above treatment, we denote by h^* the dominant country and by h^{**} the subordinate country.

Diagrammatic expositions are convenient in explaining the post-trade dynamics in the subordinate country. To this end, define the following value function,

$$\mathbf{V}^F(y) = \max_{(k_0,k_1,\ldots)} \sum_{t=1}^{\infty} \mathbf{v}^F(k_{t-1}, k_t) \text{ subject to } k_0 = y. \tag{5.64}$$

Because, by Lemma 4, the post-trade aggregate accumulation path is optimal with respect to $\sum_{t=1}^{\infty} \rho^t \mathbf{v}^F(k_{t-1}, k_t)$, $y = k_t^F$ must satisfy, given $k = k_{t-1}^F$,

$$y = \arg\max_h \mathbf{v}^F(k, h) + \rho \mathbf{V}^F(h) \tag{5.65}$$

(the optimal principle). This indicates that, in equilibrium, the stock at the end of a period, y, may be written as a function of the stock at the period's beginning, k. If this function, which we call an *optimal choice function*, is denoted by $y = \mathbf{O}(k)$, it must hold that

$$k_t^F = \mathbf{O}(k_{t-1}^F). \tag{5.66}$$

Following Benhabib and Nishimura (1985), we may prove the existence of $\mathbf{O}(k)$ and that for any $\Delta k > 0$.

$$\mathbf{O}(k + \Delta k) \gtrless \mathbf{O}(k) \text{ if } \varphi \gtrless 0. \tag{5.67}$$

The curves $\mathbf{O}$ in Figs. 5.1 and 5.2, respectively, illustrate the optimal choice functions for the cases of $\varphi > 0$ and $\varphi < 0$. The sequence $S, S', S'', \ldots$ indicates a post-trade aggregate accumulation path. Recall that the relationship between aggregate world activity (k, y) and country h's capital input k_h is given by $\mathbf{k}_h(k, y) = k_h$ in Lemma 8; i.e., by $y_t^F = k_t^F$,

$$k_{ht-1}^F = \mathbf{k}_h(k_{t-1}^F, k_t^F). \tag{5.68}$$

By Lemma 8 and Assumption 5, the curve that depicts $\mathbf{k}_h(k, y) = k_h$ on the k-y plane for a given k_h is downward-sloping for one country and upward-sloping for the other. Moreover, as k_h increases, the curve shifts to the right. This implies that the country of which curve $\mathbf{k}_h(k, y) = k_h$ is sloped in the direction opposite to optimal choice curve $\mathbf{O}$ is the dominant country. In Figs. 5.1 and 5.2, curves

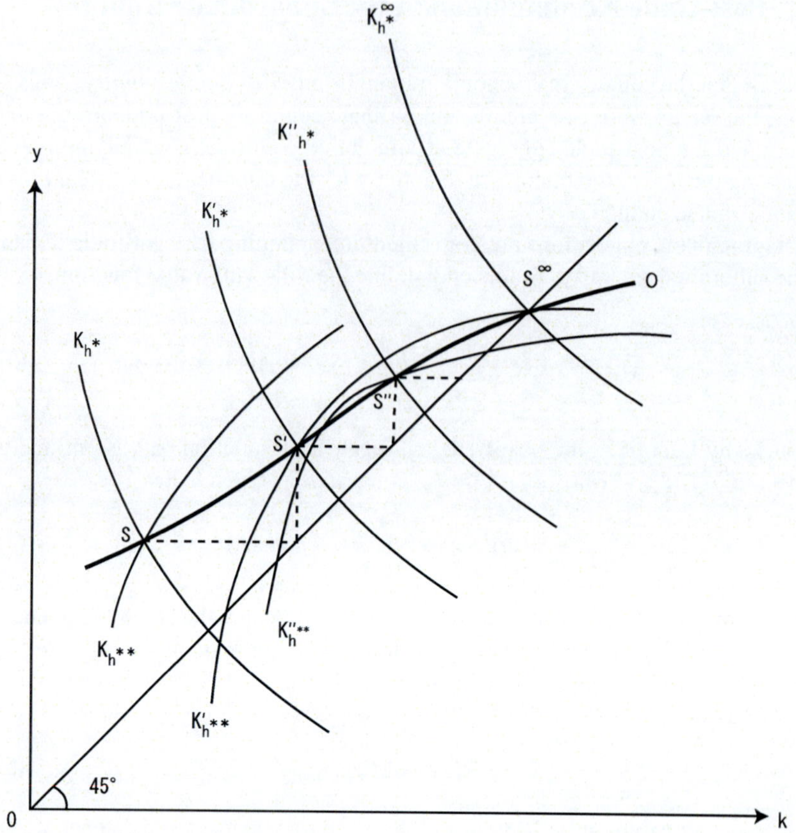

Fig. 5.1

K_{h^*}, K'_{h^*} and K''_{h^*}, which depict curves $\mathbf{k}_{h^*}(k, y) = k_{h^*}$, $\mathbf{k}_{h^*}(k, y) = k'_{h^*}$ and $\mathbf{k}_{h^*}(k, y) = k''_{h^*}$, respectively, indicate capital accumulation in the dominant country.

The optimal choice curve, **O**, is sloped in the same direction as the subordinate country's curve that depicts $\mathbf{k}_{h^{**}}(k, y) = k_{h^{**}}$ on the k-y plane for a given $k_{h^{**}}$. This suggests that *the relationship between the post-trade aggregate accumulation pattern and the subordinate country's post-trade accumulation pattern is generally ambiguous.*

Figure 5.1 illustrates the case in which the subordinate country's accumulation pattern does not coincide with the aggregate accumulation pattern. In Fig. 5.1, the aggregate accumulation path is monotone increasing, as it shifts from S to S' and to S'' and, finally, converges to S^{∞}. The subordinate country's capital stocks at points S, S' and to S'' are, respectively, indicated by curves $K_{h^{**}}$, $K'_{h^{**}}$ and $K''_{h^{**}}$ that, respectively, illustrate $\mathbf{k}_{h^{**}}(k, y) = k_{h^{**}}$, $\mathbf{k}_{h^{**}}(k, y) = k'_{h^{**}}$ and $\mathbf{k}_{h^{**}}(k, y) = k''_{h^{**}}$;

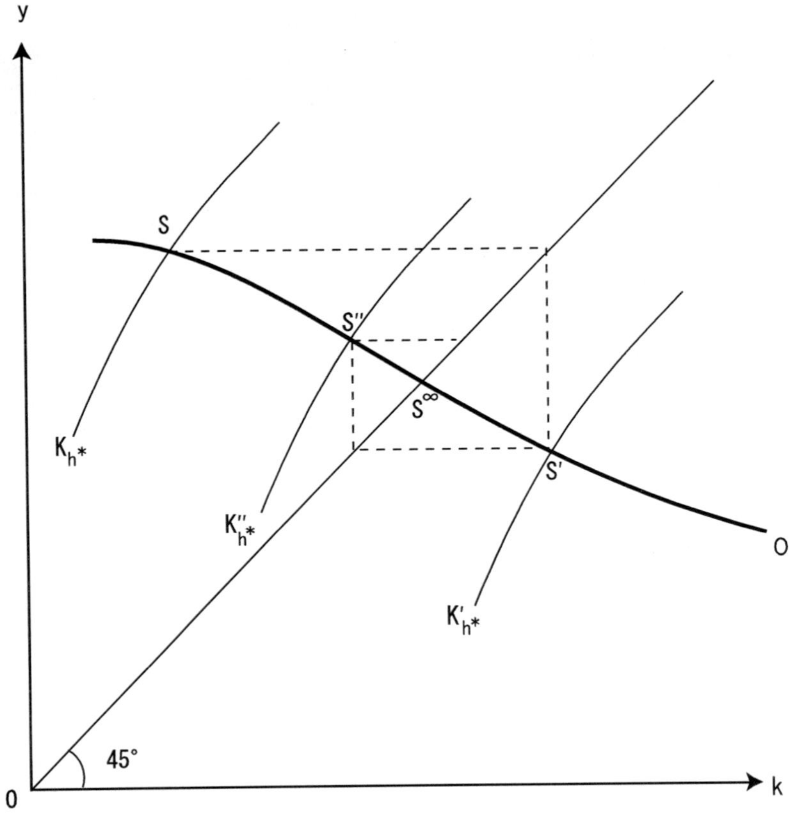

Fig. 5.2

thus, the stock at point S^∞ is indicated by curve $K'_{h^{**}}$. As the figure indicates, the subordinate country's stock at S is smaller than at S^∞; that at S' is the same as at S^∞; that at S'' is larger than at S^∞. After the aggregate stocks passes point S'', the subordinate country's stock decreases to the level indicated at point S^∞ (suggesting the possibility of an overshooting).

Although the subordinate country's post-trade accumulation pattern is generally ambiguous, we may provide sufficient conditions under which it becomes the same as the post-trade aggregate accumulation pattern. The next two theorems provide such conditions:

Theorem 5 *Suppose* $\varphi > 0$ *and*

$$\frac{g_{h^{**}}^{22} g_{h^*}^{11} - g_{h^{**}}^{12} g_{h^*}^{12} + |G_{h^*}|}{g_{h^{**}}^{12} g_{h^*}^{22} - g_{h^{**}}^{22} g_{h^*}^{12}} > \frac{g_{h^{**}}^{12} |G_{h^*}| + g_{h^*}^{12} |G_{h^{**}}|}{|g_{h^{**}}^{22}||G_{h^*}| + |g_{h^*}^{22}||G_{h^{**}}|}. \tag{5.69}$$

Then, h^{**} *is the subordinate country. Moreover,* $k_{h^{**}t}^F \lesseqgtr k_{h^{**}t+1}^F$ *if* $k_t^F \lesseqgtr k_{t+1}^F$.

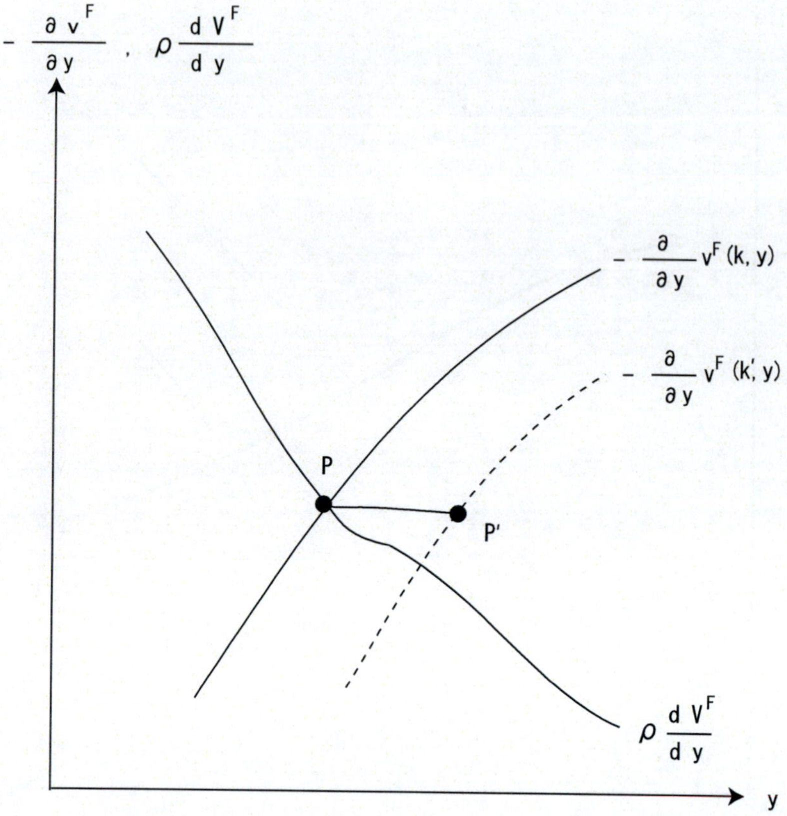

Fig. 5.3

Proof. By (5.48), $\varphi > 0$ implies the right-hand side of (5.69) is positive since $g_h^{22} < 0$ (Assumption 3c) and $|G_h| > 0$ (Assumption 2a and 4′). Thus, by Assumption 7, $g_{h^{**}}^{12} g_{h^*}^{22} - g_{h^{**}}^{22} g_{h^*}^{12} > 0$. Thus, h^{**} is the dominant country.

See Fig. 5.1. The above discussion implies that if, unlike in Fig. 5.1, curve $K_{h^{**}}$ is always steeper than curve **O** at their intersection, the subordinate country's stock always moves in the same direction as the aggregate stock. Thus, it suffices to demonstrate that (5.69) implies that curve $K_{h^{**}}$ is steeper than curve **O** at their intersection.

By Assumptions 2 and 3, the value function $\mathbf{V}^F(k)$ is concave and monotone increasing. Thus, $\rho d\mathbf{V}^F/dk$ exists almost everywhere and is positive and downward-sloping, as shown in Fig. 5.3. Note that $-\partial \mathbf{v}^F(k, y)/\partial y > 0$ is upward-sloping in y and that, by (5.65), $y = \mathbf{O}(k)$ is determined by the intersection between curve $\rho d\mathbf{V}^F/dk$ and $-\partial \mathbf{v}^F/\partial y$.

By $\varphi > 0$, curve $-\partial \mathbf{v}^F/\partial y$ shifts downwards as k increases. Thus, as k increases, the value of y at the intersection can increase at most by PP'. Because, by totally differentiating $\partial \mathbf{v}^F(k, y)/\partial y$ = constant, PP' can be approximated by $dy = -\left(\frac{\partial^2 \mathbf{v}^F}{\partial k \partial y} / \frac{\partial^2 \mathbf{v}^F}{\partial y^2}\right) dk$, Lemma 9 implies

$$0 \leq \lim_{\Delta k \to 0} \sup \frac{\mathbf{O}^F(k + \Delta k)}{\Delta k} \leq -\frac{g_\alpha^{12}|G_\beta| + g_\beta^{12}|G_\alpha|}{g_\alpha^{22}|G_\beta| + g_\beta^{22}|G_\beta|}, \tag{5.70}$$

which gives the upper-bound of the slope of curve $\mathbf{O}$ in Fig. 5.1.

Recall that curve $K_{h^{**}}$ in Fig. 5.1 depicts $\mathbf{k}_{h^{**}}(k, y) = k_{h^{**}}$ for a given $k_{h^{**}}$. Thus, By Lemma 8,

$$\left.\frac{dy}{dk}\right|_{\mathbf{k}_{h^{**}}(k,y)=\text{constant}} = -\frac{g_{h^{**}}^{22} g_{h^*}^{11} - g_{h^{**}}^{12} g_{h^*}^{12} + |G_{h^*}|}{g_{h^{**}}^{22} g_{h^*}^{12} - g_{h^{**}}^{12} g_{h^*}^{22}}. \tag{5.71}$$

By (5.70) and (5.71), curve $K_{h^{**}}$ is steeper than curve $\mathbf{O}$ at their intersection if condition (5.69) is satisfied. □

A similar result can be proved for the case of $\varphi < 0$ (a proof is omitted).

Theorem 5′ *Suppose $\varphi < 0$ and*

$$\frac{g_{h^{**}}^{22} g_{h^*}^{11} - g_{h^{**}}^{12} g_{h^*}^{12} + |G_{h^*}|}{g_{h^{**}}^{12} g_{h^*}^{22} - g_{h^{**}}^{22} g_{h^*}^{12}} < \frac{g_{h^{**}}^{12}|G_{h^*}| + g_{h^*}^{12}|G_{h^{**}}|}{|g_{h^{**}}^{22}||G_{h^*}| + |g_{h^*}^{22}||G_{h^{**}}|} (< 0). \tag{5.72}$$

Then, h^{**} is the subordinate country. Moreover, $k_{h',t}^F \lesseqgtr k_{h',t+1}^F$ if $k_t^F \lesseqgtr k_{t+1}^F$.

If, by Lemma 9, the capital expansion effect on sector Y in one country is similar to that in the other country, the denominators of the left-hand sides of (5.69) and (5.72) are small. This implies the following conclusion:

Proposition 7. *If the capital expansion effects on sector Y are sufficiently similar between the countries, the subordinate country's post-trade accumulation pattern as well as that of the dominant country is the same as the post-trade aggregate accumulation pattern.*

5.8 Summary

In a perfect foresight model with two countries, we have investigated the way in which international trade interlinks the endogenous real business cycles of countries. We have demonstrated that in some cases a country's business cycles are spread throughout the world as a result of trade and that they are erased in other cases. In a certain special case, moreover, an economy jumps to a stationary state as soon as trade opens, no matter how the autarky equilibrium paths behave in individual countries (the super-stabilization effect of trade).

As main determinants for the interlinkage of business cycles, we have identified what Yano (1990) calls the *capital expansion effect* and the *capital cost effect* on the capital good production sector (call it sector Y). The effect that an exogenous increase in a country's aggregate capital input has on a sector's output is thought of as a capital expansion effect; the effect that an exogenous increase in the price of capital input has on a sector's output is thought of as a capital cost effect. The capital expansion effect and the capital cost effect work on sector Y in opposite directions.

The *pre-trade* accumulation pattern in a country depends on directions in which the capital expansion effect works on sector Y. A country's *pre-trade* accumulation pattern is monotone if, in that country, the capital expansion effect on sector Y works in the positive direction and fluctuant if it works in the negative direction.

If the capital expansion effect on sector Y works in the same direction in one country as in the other country, the *post-trade* aggregate accumulation pattern coincides with the countries' *pre-trade* accumulation patterns, which do not differ between the countries. That is to say, the *post-trade* aggregate accumulation pattern is monotone if the *pre-trade* accumulation is monotone in the countries and fluctuant if it is fluctuant.

Suppose that the capital expansion effect on sector Y works in opposite directions between the countries, The *post-trade* aggregate accumulation pattern coincides with the *pre-trade* accumulation pattern in the country in which the capital cost effect on sector Y has the larger magnitude between the two countries. We have called that country the dominant country in the determination of a *post-trade* aggregate accumulation pattern.

In the case in which the capital expansion effect on sector Y works in the same direction, we call the country in which the capital expansion effect is stronger the dominant country. We have called the country that is not dominant the subordinate country.

With this terminology, we have demonstrated that the dominant country's *post-trade* accumulation pattern coincides with the *post-trade* aggregate accumulation pattern. If the technologies of the two countries are sufficiently similar, the subordinate country's *post-trade* accumulation pattern also coincides with the *post-trade* aggregate accumulation pattern. In general, however, the relationship between the subordinate country's *post-trade* accumulation pattern and the *post-trade* aggregate accumulation pattern is ambiguous.

References

Becker, R. (1980), "On the Long-Run Steady State in a Simple Dynamic Model of Equilibrium with Heterogeneous Households," *Quarterly Journal of Economics,* 95, 375–82.

Benhabib, J. and R. Day (1982), "A Characterization of Erratic Dynamics in the Overlapping Generation Model," *Journal of Economic Dynamics and Control,*4, 34–55.

Benhabib, J. and K. Nishimura (1979), "The Hopf Bifurcation and the Existence and Stability of Closed Orbits in Multisector Models of Optimal Economic Growth," *Journal of Economic Theory,* 21, 421–444.

Benhabib, J. and K. Nishimura (1985), "Competitive Equilibrium Cycles," *Journal of Economic Theory*, 35, 284–306.

Benhabib, J., Jafarey, S. and K. Nishimura (1988), "The Dynamics of Efficient Intertemporal Allocations with Many Agents, Recursive Preferences, and Production," *Journal of Economic Theory*, 44, 301–320.

Bewley, T. (1972), "Existence of Equilibria in Economies with Infinitely Many Commodities," *Journal of Economic Theory*, 4, 514–40.

Bewley, T. (1982), "An Integration of Equilibrium Theory and Turnpike Theory," *Journal of Mathematical Economics*, 10, 233–67.

Boldrin, M. and L. Montrucchio (1986), "On the Indeterminacy of Capital Accumulation Paths," *Journal of Economic Theory*, 40, 26–39.

Brock, W. and J. Scheinkman (1976), "Global Asymptotic Stability of Optimal Growth with Applications to the Theory of Economic Growth," *Journal of Economic Theory*, 12, 164–90.

Cass, D. and K. Shell (1976), "The Structure and Stability of Competitive Dynamical Systems," *Journal of Economic Theory*, 12, 31–70.

Coles, J. (1985), "Equilibrium Turnpike Theory with Constant Returns to Scale and Possibly Heterogeneous Discount Factors," *International Economic Review*, 26, 671–80.

Deneckere, R. and S. Pelikan (1986), "Competitive Chaos," *Journal of Economic Theory*, 40, 13–25.

Epstein, L. (1987a), "The Global Stability of Efficient Intertemporal Allocations," *Econometrica*, 55, 329–355.

Epstein, L. (1987b), "A Simple Dynamic General Equilibrium Model," *Journal of Economic Theory*, 41, 68–95.

Fukao, K. and K. Hamada (1990), International Trade and Investment under Different Rates of Time Preferences, mimeo, Hitotsubashi University.

Grandmont, J. (1985), "On Endogenous Competitive Business Cycles," *Econometrica*, 53, 995–1045.

Grossman, G. and E. Helpman (1989), "Product Development and International Trade," *Journal of Political Economy*, 97, 1262–1283.

Jones, R. (1965), "The Structure of Simple General Equilibrium Models," *Journal of Political Economy*, 74, 557–72.

Jones, R. (1971), A Tree-Factor Model in Theory, Trade and History, in *Trade, the Balance of Payments, and Growth*, ed. by Bhagwati, J. et al., North-Holland.

Kehoe, T., Levine, D. and P. Romer (1990), "Determinacy of Equilibria in Dynamic Models with Finitely Many Consumers," *Journal of Economics Theory*, 50, 1–21.

Kemp, M. and C. Khang (1974), "A Note on Steady-State Price: Output Relationships," *Journal of International Economics*, 2, 187–97.

Kemp, M., Long, N. and K. Shimomura (1990), Cyclical and Non-Cyclical Redistributive Taxation, mimeo, University of New Southwales.

Lucas, R. and N. Stokey (1984), "Optimal Growth with Many Consumers," *Journal of Economic Theory*, 31, 139–71.

Marimon, R. (1989), "Stochastic Turnpike Property and Stationary Equilibrium," *Journal of Economic Theory*, 47, 282–306.

McKenzie, L. (1976), "Turnpike Theory," *Econometrica*, 44, 841–865.

McKenzie, L. (1986), Optimal Economic Growth and Turnpike Theorems, in *Handbook of Mathematical Economics*, ed. by Arrow, K. and M. Intriligator, North-Holland, Amsterdam.

Negishi, T. (1960), "Welfare Economics and Existence of an Equilibrium for Competitive Economy," *Metroeconomica*, 12, 92–97.

Scheinkman, J. (1976), "On Optimal Steady Stages of *n*-Sector Growth Models when Utility is Discounted," *Journal of Economic Theory*, 12, 11–20.

Wan, H. (1971), A Simultaneous Variational Model for International Capital Movement, in *Trade, Balance of Payments, and Growth*, ed. by Bhagwati, J. et al., North-Holland.

Yano, M. (1984a), "Competitive Equilibria on Turnpikes in a McKenzie Economy, I: A Neighborhood Turnpike Theorem," *International Economic Review*, 25, 695–717.

Yano, M. (1984b), "The Turnpike of Dynamic General Equilibrium Paths and Its Insensitivity to Initial Conditions," *Journal of Mathematical Economics*, 13, 235–254.

Yano, M. (1990), Comparative Dynamics in a Simple Trade Model, mimeo, Yokohama National University.

Part II
Optimal Growth and Chaotic Dynamics

Chapter 6
Non-linear Dynamics and Chaos in Optimal Growth: An Example*

Kazuo Nishimura and Makoto Yano**

6.1 Introduction

In the recent literature, it has been demonstrated that optimal capital accumulation may be chaotic; see Boldrin and Montrucchio (1986) and Deneckere and Pelikan (1986).[1] This finding indicates, as Scheinkman (1990) discusses, that the deterministic equilibrium model of a dynamic economy may explain various complex dynamic behaviors of economic variables, and, in fact, search for such explanations has already begun (see Brock 1986; Scheinkman and LeBaron 1989, for example). In the existing literature, however, not much has yet been revealed with respect to the circumstances under which optimal accumulation exhibits complex nonlinear dynamics. In particular, it has not yet been determined whether or not chaotic optimal accumulation may appear in the case in which future utilities are discounted not so strongly.[2]

***Econometrica 63**, 981–1001, 1995

**We would like to thank Nobuo Aoki, Michele Boldrin, William Brock, Dick Day, Toru Koda, Andreu Mas-Colell, Lionel McKenzie, Luigi Montrucchio, Jose Scheinkman, and Yoichiro Takahashi for useful conversations and comments and anonymous referees for detailed comments and suggestions.

[1]These studies are followed by the recent work of Boldrin and Deneckere (1990) and Majumdar and Mitra (1994). For earlier treatments of chaotic economic behaviors, see Benhabib and Day (1982), Day (1982), and Grandmont (1985), the models of which are not of optimal growth.

[2]It has been shown that if, given a technology and a period-wise preference, the discounting of future utilities is sufficiently weak, the classical turnpike result holds, i.e., the optimal paths converge to a stationary state; see Cass and Shell (1976), Scheinkman (1976), Brock and

K. Nishimura (✉) · M. Yano
Institute of Economic Research, Kyoto University, Japan
e-mail: nishimura@kier.kyoto-u.ac.jp; yano@kier.kyoto-u.ac.jp

J. Stachurski et al. (eds.), *Nonlinear Dynamics in Equilibrium Models*,
DOI 10.1007/978-3-642-22397-6_6, © Econometric Society 1995. Published by
Springer-Verlag Berlin Heidelberg 2012. All Rights Reserved

The main purpose of this study is to demonstrate the possibility of ergodically chaotic optimal accumulation for the case in which future utilities are discounted arbitrarily weakly. As in the standard literature, we consider a discrete time model with the objective function $\sum_{t=1}^{\infty} \rho^t u_t$. In such a model, the discount factor of future utilities, ρ, and the length of the model's individual period, say n years, determine the long-run annual real interest rate, say r, by $(1+r)^n = \rho^{-1}$.[3] Once, therefore, we choose a long-run annual real interest rate inside of the range existing in the real world, the closer to 1 discount factor ρ is, the shorter the span of periods in which chaotic behavior, if it exists, can be observed. We demonstrate the possibility of ergodically chaotic optimal accumulation for ρ arbitrarily close to 1 while the existing studies assume ρ around 0.01.[4]

We employ a simple two-sector model with common Leontief production functions and a linear utility function.[5] Despite the simplicity, however, it is technically difficult to demonstrate the existence of a chaotic policy function; because we deal with production and utility functions that are not strictly concave, even the uniqueness of a policy function is not a priori guaranteed. In order to deal with this difficulty, we first construct a condition under which the optimal dynamical system is unimodal and expansive; as is well known, a unimodal and expansive dynamical system is ergodically chaotic (Collet and Eckmann 1980; Lasota and Yorke 1973; Li and Yorke 1978).[6] We then demonstrate that the set of parameter values satisfying that condition is nonempty no matter how close ρ is to 1.

Our result is based on the assumption that the consumption good sector is capital intensive.[7] If, under this assumption, capital is small at the beginning of a

---

Scheinkman (1976), and McKenzie (1983). Sorger (1992), moreover, demonstrates the existence of an upper bound of discount factors with which a given trajectory is a solution to an optimal growth model. These results, however, do not exclude the possibility that for any given degree of the discounting of future utilities, there may be technology and a period-wise preference that result in chaotic optimal dynamics.

[3]Suppose that the long-run annual real interest rate is 5%. Then, the length of the model's individual period is 1 year if $\rho = 0.952$, 10 years if $\rho = 0.61$, and 20 years if $\rho = 0.377$.

[4]Both Boldrin and Montrucchio (1986, p. 37) and Deneckere and Pelikan (1986, p. 22) obtain the ergodically chaotic policy function $h(x) = 4x(1-x)$. In a two-sector model in which the consumption sector is CES and the investment sector is Leontief, Boldrin and Deneckere (1990, p. 641) report chaos at discount factors as high as $\rho = 0.25$. However, the latter paper is concerned with the existence of topological chaos (Li and Yorke 1975), which is weaker than the notion of ergodic chaos.

[5]As is shown in Sect. 6.2, this model gives rise to a model with nontrivial von Neumann facets. See McKenzie (1968) for the possibility of cyclical optimal paths in such a model. See also Chung (1986), which considers such a possibility in a model similar to that of this study.

[6]For economic applications of an expansive map, see Boldrin and Woodford (1990), Day and Pianigiani (1991), and Deneckere and Judd (1992). Because the graph of an expansive and unimodal map is tent-shaped, the policy function of this study is nondifferentiable; on the differentiability of an optimal transition function, see Araujo (1991), Boldrin and Montrucchio (1989), and Santos (1991).

[7]Benhabib and Nishimura (1985) demonstrate that if and only if the sign of the cross partial derivative of a reduced form utility function is negative, the optimal transition function is

period, it may be optimal in that period to accumulate capital as much as possible by sacrificing consumption.[8] Once capital is accumulated sufficiently, it may be optimal to consume to the extent that the existing capital is decumulated. We demonstrate that optimal chaotic accumulation results from this alternation between capital accumulation and decumulation.

In what follows, we introduce our model in Sect. 6.2 and discuss our main result in Sect. 6.3. We provide preliminary lemmas in Sect. 6.4 and prove the main result in Sect. 6.5. Section 6.6 is for concluding remarks.

6.2 Model

The model we consider is of two sectors, each of which has a Leontief production function. The model is simple and standard but suffices to bring out chaos in optimal growth. Take two goods C and K. Good C is a pure consumption good. Each sector uses both good K and labor L as inputs. The input of good K must be carried out one period prior to the period in which output is produced; in this sense, we call good K a capital good. Labor input is made in the same period as output is produced. Good K is not consumed. Sectors C and K have Leontief production functions as follows:

$$c_t = \min\{K_{Ct-1}, L_{Ct}/\alpha\}; \tag{6.1}$$

$$y_t = \mu \min\{K_{Kt-1}, L_{Kt}/\beta\}. \tag{6.2}$$

For $i = C, K$, $K_{it-1} \geqslant 0$ is sector i's capital input in period $t-1$, and $L_{it} \geqslant 0$ is sector i's labor input in period t. Moreover, $c_t \geqslant 0$ and $y_t \geqslant 0$ are outputs in period t. Denote by k_{t-1} the aggregate capital input,

$$k_{t-1} = K_{Ct-1} + K_{Kt-1} \geqslant 0. \tag{6.3}$$

Denote by n_t the aggregate labor input,

$$n_t = L_{Ct} + L_{Kt} \geqslant 0. \tag{6.4}$$

Assume that labor supply is inelastic and time independent. Goods are normalized so that the labor endowment $\overline{n}$ is equal to 1. Thus,

downward sloping (This result can be derived as a special case of the results obtained by Topkis (1978) and Marshall and Olkin (1979)). Because the existence of a capital intensive consumption good sector implies a negative cross partial derivative, the transition function is monotone decreasing. Because Benhabib and Nishimura (1985) assume differentiability, the result of that study does not directly apply to our case, which is nondifferentiable. However, a more or less similar relationship appears in our setting, as is shown below.

[8] In our model, more precisely speaking, the optimal choice is to choose an activity on the boundary of the feasible set. A similar role of the boundary is suggested by Boldrin and Woodford (1990), who consider the boundary imposed by the limit of capital depreciation.

$$n_t \leqslant \bar{n} = 1. \tag{6.5}$$

The consumers' preference is represented by a linear utility function

$$u(c_t) = c_t \geqslant 0 \tag{6.6}$$

for $c_t \geqslant 0$ in each period. Denote the set of feasible input/output combinations as

$$F = \{(k, n, y, c) \in R_+^4 | (1) \text{ through } (6) \text{ are satisfied}\}. \tag{6.7}$$

The optimal growth model is described by

$$\max_{(k_{t-1}, n_t, k_t, c_t)} \sum_{t=1}^{\infty} \rho^t u(c_t) \text{ subject to}$$

$$k_0 \leqslant \bar{k},\ k_t \leqslant y_t \text{ and } (k_{t-1}, n_t, y_t, c_t) \in F \text{ for } t = 1, 2, \ldots, \tag{6.8}$$

where discount factor ρ satisfies

$$0 < \rho < 1. \tag{6.9}$$

We treat ρ, μ, β, and α as the parameters of the model and denote the model by $M(\rho, \mu, \beta, \alpha)$. Assume the following:

$$\rho\mu > 1; \tag{6.10}$$

$$\beta > \alpha > 0. \tag{6.11}$$

Conditions similar to (6.10) are often employed in order to guarantee the existence of a nontrivial stationary equilibrium (see McKenzie 1986). In the present setting, μ is the marginal product of capital in the case in which labor is not binding. Moreover, it is known that $\rho^{-1} - 1$ is equal to the real interest rate in the stationary state (see McKenzie 1986). Thus, (6.10) implies that the system is productive to the extent to which the marginal product of capital can cover principal and interest. Condition (6.11) implies that the consumption good sector is capital intensive.

An optimal growth model can be expressed with a reduced form utility function

$$v(k, y) = \max_{(n,c)} c \quad \text{subject to } (k, n, y, c) \in F. \tag{6.12}$$

By using this function, define the value function

$$V(y) = \max_{(k_0, k_1, \ldots)} \sum_{t=1}^{\infty} \rho^t v(k_{t-1}, k_t) \text{ subject to}$$

$$k_0 = y \text{ and } (k_{t-1}, k_t) \in D \qquad (t = 1, 2, \ldots,) \tag{6.13}$$

where D is the domain of v. Define the *optimal transition correspondence*

$$H(k) = \{\eta \in R | v(k, \eta) + V(\eta) \geqslant v(k, y) + V(y) \tag{6.14}$$
$$\text{for any } y \geqslant 0 \text{ such that } v(k, y) \text{ is well defined}\}.$$[9]

In the case in which $H(k)$ is a singleton for every k, we call $H(k)$ an *optimal transition function.*

Let $H^t(k) = H(H^{t-1}(k))$, where $H^0(k) = k$. It is easy to demonstrate that if (k_{t-1}, n_t, y_t, c_t), $t = 1, 2, \ldots$, is an optimal path from $\overline{k}$ solving the optimization problem (6.8), then

$$k_t \in H^t(\overline{k}) \tag{6.15}$$

for any t. We may prove that for each $k \in R$, the optimal transition correspondence, $H(k)$, is nonempty (McKenzie 1983).

We are concerned with "ergodic chaos" that results from an expansive and unimodal transition function (Collet and Eckmann 1980).[10] Such a transition function is "chaotic" in several senses. First, it is chaotic in the sense of ergodic oscillations; that is, there is a unique absolutely continuous probability measure ν that satisfies that for almost every $x \in I$, $\lim_{t\to\infty} t^{-1} \sum_{\tau=0}^{t-1} \chi_B(f^\tau(x)) = \nu(B)$ for any Borel set $B \subset I$, where $\chi_B(x) = 1$ if $x \in B$ and $= 0$ if $x \notin B$.[11] Second, it is chaotic in the sense of geometric sensitivity; that is, there is a $g > 0$ such that, for any t, there is an $\varepsilon > 0$ such that $|\Delta x| < \varepsilon$ implies $|f^\tau(x + \Delta x) - f^\tau(x)| \geqslant (1 + g)^\tau |\Delta x|$ for $\tau = 1, \ldots, t$.[12]

6.3 Main Result

In order to explain our main result, it is useful to have the actual form of the reduced form utility function, which is

$$v(k, y) = \begin{cases} k - \frac{1}{\mu} y & \text{if } y \leqslant -\frac{\alpha}{\beta-\alpha}\mu k + \frac{1}{\beta-\alpha}\mu, \\ \frac{1}{\alpha} - \frac{\beta/\alpha}{\mu} y & \text{if } y \geqslant -\frac{\alpha}{\beta-\alpha}\mu k + \frac{1}{\beta-\alpha}\mu. \end{cases} \tag{6.16}$$

[9]Note that $H(k)$ is the solution set to the maximization of $[v(k, \eta) + V(\eta)]$ but not that of $[v(k, \eta) + \rho V(\eta)]$. This is because, following McKenzie (1986), we adopt the objective function in which the first period utility is discounted by ρ.

[10]A function f on an interval I is expansive if it is continuous and piecewise twice continuously differentiable and if there is $\gamma > 0$ such that : $f'(x) \geq 1 + \gamma$ for any x at which f has a derivative. Moreover, a function f on an interval I is unimodal if it continuous and if there is c in the interior of I such that f is strictly increasing for any $x < c$ and that f is strictly decreasing for any $x > c$. (See Collet and Eckmann 1980 for details on these definitions.)

[11]An expansive and unimodal transition function is chaotic in the sense of ergodic oscillations (Lasota and Yorke 1973; Li and Yorke 1978).

[12]This is a direct consequence of expansiveness.

By (6.16) and the nonnegativity conditions, the domain of v is

$$D = \{(k, y) \in R_+^2 | y \leqslant \mu k \text{ if } k \leqslant 1/\beta \text{ and } y \leqslant \mu/\beta \text{ if } k \geqslant 1/\beta\}. \tag{6.17}$$

In Fig. 6.1, D is the region on and below segment OP and half line PZ. The kinked line OPZ is the indifference curve with $v(k, y) = 0$. Any indifference curve is parallel to the kinked line OPZ with the kink on line PQ.

In order to construct an expansive and unimodal optimal transition function, we focus on the graph consisting of segments OP and PQ. This kinked segment may be thought of as the graph of an expansive and unimodal transition function under the following conditions:

$$\mu/\beta < 1/\alpha; \tag{6.18}$$

$$\frac{\alpha}{\beta - \alpha}\mu > 1. \tag{6.19}$$

Condition (6.19) guarantees that the slope of segment PQ is less than -1; condition (6.18) guarantees that the graph consisting of OP and PQ in fact describes a dynamical system; the function corresponding to the graph does not map the interval between O and P into itself if $1/\alpha < \mu/\beta$. Because, under (6.10), the slope of segment OP is larger than 1, the graph becomes expansive and unimodal. Because the range of this graph is the interval between 0 and μ/β, we may restrict its domain to the same interval, i.e.,

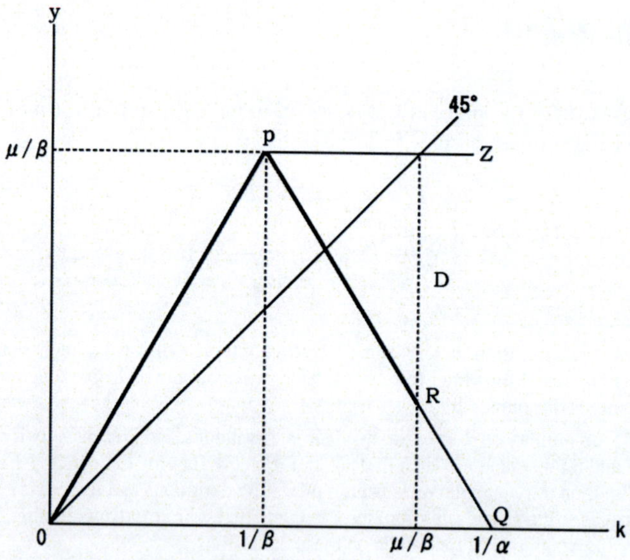

Fig. 6.1

$$I \equiv \{k \in R | 0 \leqslant k \leqslant \mu/\beta\}. \tag{6.20}$$

Our main result implies that, under a certain choice of parameter values, OPR is the graph of the optimal transition function for ρ arbitrarily close to 1. In order to make a formal statement, on domain I, define

$$h(k;\mu,\beta,\alpha) = \begin{cases} \mu k & \text{if } 0 \leqslant k \leqslant 1/\beta, \\ -\frac{\alpha}{\beta-\alpha}\mu k + \frac{1}{\beta-\alpha}\mu & \text{if } 1/\beta \leqslant k \leqslant \mu/\beta. \end{cases} \tag{6.21}$$

Then, the graph of $h(k)$ is OPQ; we often drop (μ, β, α) in the expression of h and denote $h(k) = h(k;\mu,\beta,\alpha)$. Denote by (x, y) the open interval between x and y. The main result is as follows.

MAIN RESULT: *For any $\rho' < 1$, there are μ, α, β and an open interval $U \subset (\rho', 1)$ such that if $\rho \in U$, h is the expansive and unimodal optimal transition function of model $M(\rho, \mu, \beta, \alpha)$ (i.e., ergodically chaotic optimal accumulation occurs).*[13]

In order to explain an intuition behind this result, note that if and only if (k, y) is on segment PQ, both labor and capital are fully employed. Labor is not fully employed below line PQ while capital is not fully employed above line PQ. Line PQ is downward-sloping because $\beta > \alpha$, i.e., the consumption good sector is capital intensive (see Jones 1965).

Take the case of $k > 1/\beta$. Because, in this case, it is possible fully to employ all the resources that the economy has, it is reasonable that the optimal activity is the full employment, i.e., to choose (k, y) on PQ. If $k < 1/\beta$, instead, the insufficiency of capital input k serves as a bottleneck preventing the full employment of labor. In this case, it is reasonable that the optimal activity is to produce as much capital as possible, i.e., to choose (k, y) on OP, so as to clear the bottleneck as quickly as possible.

6.4 Ergodically Chaotic Optimal Transition Functions

It is not a priori the case that $h(k)$ is an optimal transition function: in fact, the optimal transition correspondence, $H(k)$, may not even be a function, for production and utility functions are not strictly concave. Therefore, we directly derive a condition under which h is an optimal transition function. This is, however, technically difficult in the general setting of our model, for it involves solving an infinite-horizon optimization problem. We handle this difficulty by focusing on the

[13]The existence of an open subset of the discount factor with which optimal paths follow chaotic dynamics is demonstrated also by Deneckere and Pelikan (1986). However, their result focuses on topological chaos while our result is concerned with ergodic chaos.

case in which the top of the graph of h (i.e., point P in Fig. 6.1) corresponds to a cyclical point of h or, more precisely, the case in which $1/\beta$ is a cyclical point of h.

Lemma 1. *Suppose that $1/\beta$ is a cyclical point of transition function h; i.e., there is $\tau > 1$ such that $h^\tau(1/\beta) = 1/\beta$. Then, h is the optimal transition function if and only if the cyclical path from $1/\beta$ that h generates is the unique optimal path from $1/\beta$, i.e., $H^t(1/\beta) = \{h^t(1/\beta)\}$ for any $t = 1, \ldots, \tau$.*

Instead of a formal proof, we provide a diagrammatical explanation for this lemma. Denote by ∂f the subgradient of a convex function f in the standard sense. Because we deal with concave functions, for a concave function f, denote $-\partial[-f]$ as ∂f.

In Figs. 6.2 and 6.3, curve MV illustrates the graph of $\partial V(y)$. By the concavity of V, curve MV is downward sloping or, more precisely, satisfies that if $r \in \partial V(y)$, $r' \in \partial V(y')$ and $y < y'$, then $r \geqslant r'$. Although it may have flat and vertical segments, it does not matter for the following discussion. Note that, for any fixed k, $-v(k, y)$ may be thought of as a convex function of y, the subgradient of which we denote by $\partial_2[-v(k, y)]$. Given k, curve $MC|_k$ illustrates the graph of $\partial_2[-v(k, y)]$. By the principle of optimality, the optimal transition correspondence $H(k)$ appears at the intersection of MV and $MC|_k$ curves. That is, by (6.14),

$$H(k) = \{y \geqslant 0 | \partial_2[-v(k, y)] \cap [\partial V(y)] \neq \varnothing\}. \tag{6.22}$$

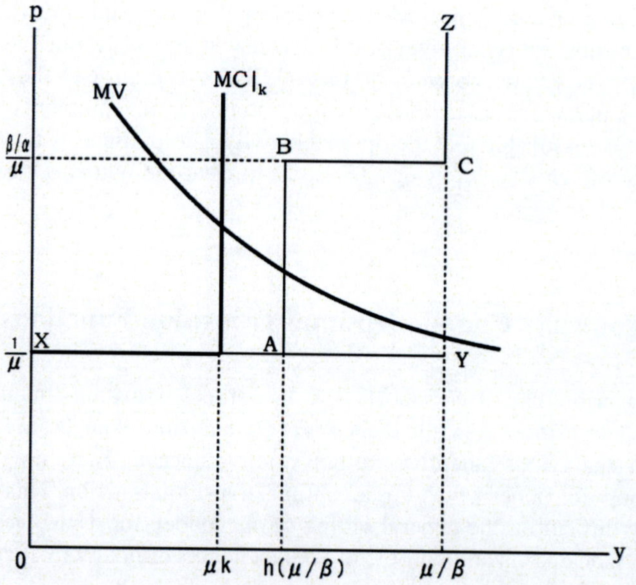

Fig. 6.2

Fig. 6.3

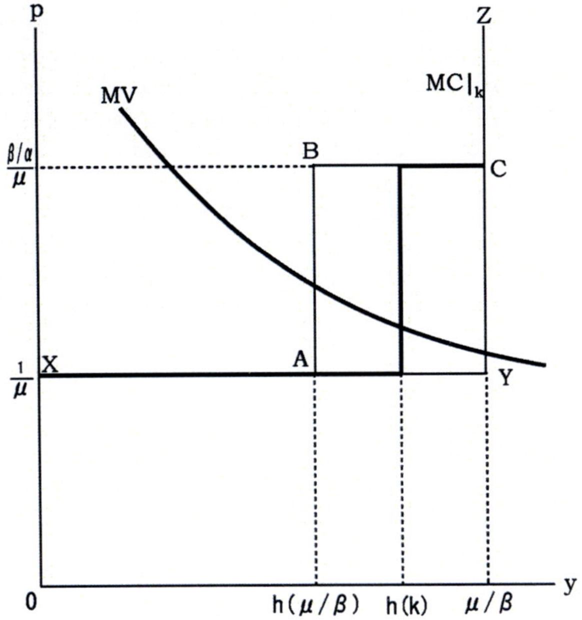

By (6.16), the shape of curve $MC|_k$ differs between the cases of $k < 1/\beta$ and $k > 1/\beta$. That is, if $0 < k \leqslant 1/\beta$,

$$\partial_2[-v(k,y)] = \begin{cases} \{1/\mu\} & \text{if } y < \mu k, \\ \{r|r \geqslant 1/\mu\} & \text{if } y = \mu k, \end{cases} \tag{6.23}$$

which is illustrated in Fig. 6.2. If $1/\beta < k \leqslant \mu/\beta$,

$$\partial_2[-v(k,y)] = \begin{cases} \{1/\mu\} & \text{if } y < -\frac{\alpha}{\beta-\alpha}\mu k + \frac{1}{\beta-\alpha}\mu, \\ \{r|1/\mu \leqslant r \leqslant (\beta/\alpha)/\mu\} & \text{if } y = -\frac{\alpha}{\beta-\alpha}\mu k + \frac{1}{\beta-\alpha}\mu, \\ \{(\beta/\alpha)/\mu\} & \text{if } -\frac{\alpha}{\beta-\alpha}\mu k + \frac{1}{\beta-\alpha}\mu < y < \mu/\beta, \\ \{r|r \geqslant (\beta/\alpha)/\mu\} & \text{if } y = \mu/\beta, \end{cases} \tag{6.24}$$

which is illustrated in Fig. 6.3.

In order to demonstrate the lemma, it suffices to prove that if and only if $H^t(1/\beta) = \{h^t(1/\beta)\}$ for any $t = 1, \ldots, \tau$,

$$H(k) = \{h(k)\} \tag{6.25}$$

for any k, $0 \leqslant k \leqslant \mu/\beta$. To this end, assume $H^t(1/\beta) = \{h^t(1/\beta)\}$, $t = 1, \ldots, \tau$. Since $H(1/\beta) = \{h(1/\beta)\}$, by (6.22), curves MV and $MC|_{1/\beta}$ must intersect each other on the half line YZ in Fig. 6.2; this is because curve $MC|_{1/\beta}$ coincides with the kinked line XYZ in the figure. Since, moreover, $H^2(1/\beta) = \{h^2(1/\beta)\} = \{h(\mu/\beta)\}$, by (6.22), curves MV and $MC|_{\mu/\beta}$ must intersect each other on segment AB in Fig. 6.2; this is because curve $MC|_{\mu/\beta}$ coincides with the kinked line $XABCZ$ in the figure. In summary, curve MV intersects segments AB and YC in Figs. 6.2 and 6.3.

In order to prove (6.25), increase k from 0 to μ/β. First, note that if $k = 0$, $H(k) = \{h(k)\} = 0$. Next, note that, by (6.23), curve $MC|_k$ in Fig. 6.2 illustrates a typical graph of $\partial_2[-v(k, y)]$ for the case of $0 < k \leqslant 1/\beta$. In this case, as Fig. 6.2 illustrates, MV and $MC|_k$ have a unique intersection at $y = \mu/k$. This implies that (6.25) holds for any k such that $0 < k \leqslant 1/\beta$. Finally, note that, by (6.24), curve $MC|_k$ in Fig. 6.3 illustrates a typical graph of $\partial_2[v(k, y)]$ for the case of $1/\beta < k \leqslant \mu/\beta$. In this case, as Fig. 6.3 illustrates, curves MV and $MC|_k$ have a unique intersection at $y = h(k)$. This implies that (6.25) holds for any k such that $1/\beta < k \leqslant \mu/\beta$. This establishes (6.25) for any k such that $0 \leqslant k \leqslant \mu/\beta$. Because the converse is obvious, the lemma is established.

Given Lemma 1, we need to derive a condition under which the cyclical trajectory from $1/\beta$ is, in fact, the unique optimal path from $1/\beta$.[14] For this purpose, we use the value loss method (McKenzie 1986) by constructing a support price path. Let

$$\gamma = (\beta - \alpha)/\alpha. \tag{6.26}$$

We will prove the following in Appendix A.

Lemma 2. *The trajectory that h generates from $1/\beta$ is a unique optimal path from $1/\beta$ if the following is satisfied.*

A: $1/\beta$ is a cyclical point of transition function h with the order of periodicity $N \geqslant 3$.

B: There are prices $q_0, q_1, \ldots, q_N$ such that the following are satisfied:

(i) $q_N = q_0$;

(ii) $$\begin{cases} q_0 > 0, \\ -q_0 + \rho\mu q_1 > 0, \\ \gamma q_0 + \rho\mu q_1 > \rho(1+\gamma); \end{cases}$$

(iii) let $k_t = h^t(1/\beta)$ *and* $t = 2, 3, \ldots, N$;

[14]The existence of cyclical optimal paths plays important roles in Deneckere and Pelikan (1986), Deneckere (1988), and Boldrin and Woodford (1990) as well. Those studies characterize a cyclical optimal path by the Euler equation. In contrast, we characterize it by support prices. This difference is due to the fact we focus on noninterior solutions in the nondifferentiable model while the existing studies consider interior solutions in a differentiable model.

$$(iii.a)\quad q_t = \begin{cases} q_{t-1}/(\rho\mu) & \textit{if } k_{t-1} < 1/\beta, \\ [-\gamma q_{t-1} + \rho(1+\gamma)]/(\rho\mu) & \textit{if } k_{t-1} > 1/\beta; \end{cases}$$

$$(iii.b)\quad \begin{cases} q_{t-1} > \rho & \textit{if } k_{t-1} < 1/\beta, \\ 0 < q_{t-1} < \rho & \textit{if } k_{t-1} > 1/\beta. \end{cases}$$

(Note that, given condition A, $k_{t-1} \neq 1/\beta$ for $t = 2, 3, \ldots, N$.)

6.5 Ergodic Chaos for the Discount Factor Arbitrarily Close to 1

In order to establish our main result, stated in Sect. 6.3, we will characterize conditions A and B of Lemma 2. This characterization will reveal the fact that, given a periodicity of point $1/\beta$, N, an upper bound of ρ exists under which μ and γ satisfy conditions A and B; if $N = 3$, for example, the upper bound is about 0.36. Finally, we will demonstrate that this upper bound of ρ increases to 1 if periodicity $N = 3 \times 2^n$ increases with $n = 0, 1, \ldots$.

At the outset, assume

$$\rho\mu \leqslant \sqrt{\gamma} \tag{6.27}$$

(this assumption will be used in order to establish (B.10) in the Appendix). Define the following expressions:

$$m[n] = [2^n - (-1)^n]/3, \tag{6.28}$$

$$f(\mu;\gamma,n) = (\mu^{2^n})^2 - \gamma^{m[n+1]}(\mu^{2^n}) - \gamma^{m[n+2]}. \tag{6.29}$$

Next, define

$$g(\theta;\gamma,n) = (\theta^{2^n})^2 + \gamma^{2m[n]}(\theta^{2^n}) - \gamma^{m[n+2]}. \tag{6.30}$$

With these preparations, we will prove the following in Appendix B.

Lemma 3. *Let $n = 0, 1, \ldots$. Suppose $f(\mu;\gamma,n) = 0$. If $g(\rho\mu;\gamma,n) < 0$ and if $g(\rho\mu;\gamma,n)$ is sufficiently close to 0, conditions A and B of Lemma 2 are satisfied for $N = 3 \times 2^n$.*

We explain Lemma 3 for $n = 0$ by following the proof of the general case in Appendix B. To this end, let $\widetilde{k}_t = h^t(1/\beta)$. Then, by (6.21),

$$\widetilde{k}_3 - \frac{1}{\beta} = \frac{1}{\beta}(\mu - 1)\left(1 + \mu\left(1 - \frac{\mu}{\gamma}\right)\right). \tag{6.31}$$

Since $\mu > 1$ by (6.10), (6.31) implies that $1/\beta$ is a cyclical point of h with periodicity $N = 3 \times 2^0 = 3$ (i.e., $\widetilde{k}_3 = h^3(1/\beta) = 1/\beta) = 1/\beta$) if and only if

$$f(\mu;\gamma,0) = \mu^2 - \gamma\mu - \gamma = 0. \tag{6.32}$$

Suppose, as in Lemma 3, that (6.32) is satisfied by (μ, γ). In order to construct sequence q_0, q_1, q_2, q_3 of Lemma 2, set $k_t = \widetilde{k}_t$ in condition B (iii.a). First, pick an arbitrary q_1, and use conditions B (iii.a) and B (i) to construct q_2, q_3, and q_0. Then, by construction, (q_0, q_1) satisfies

$$(q_0 - \rho) + \frac{\gamma}{(\rho\mu)^2}\left(q_1 - \frac{1}{\mu}\right) = \rho\left(\frac{1}{\rho\mu} - 1\right)\left(1 + \frac{1}{\rho\mu}\left(1 - \frac{\gamma}{\rho\mu}\right)\right). \tag{6.33}$$

In particular, denote the sequence that ends up with $q_0 = \rho$ by $q_1^\rho, q_2^\rho, q_3^\rho, q_0^\rho$; that is, by condition B (iii.a), $q_1^\rho = \rho + \rho[1 - (\rho\mu)^2]/\gamma$, $q_2^\rho = \rho^2\mu$, and $q_3^\rho = q_0^\rho = \rho$. We will demonstrate that $q_t = q_t^\rho$, $t = 0, 1, 2, 3$, is the sequence satisfying the conditions of Lemma 2 with $k_t = \widetilde{k}_t$. By construction, it satisfies conditions B (iii.a) and B (i) of Lemma 2. Moreover, since the above expressions imply $q_1^\rho < \rho$ and $q_2^\rho > \rho$ by (6.10), sequence $q_t = q_t^\rho$, $t = 0, 1, 2, 3$, satisfies condition B(iii.b) with $k_t = \widetilde{k}_t$. Therefore, it suffices to demonstrate that the sequence satisfies condition B(ii).

To this end, in Fig. 6.4, line L illustrates (6.33); region Γ depicts the set of (q_0, q_1) at which condition B (ii) of Lemma 2 is satisfied. Note that line L is negatively sloped, that (q_0^ρ, q_1^ρ) lies on line L, and that $q_0^\rho = \rho$. Thus, as Fig. 6.4 illustrates, (q_0^ρ, q_1^ρ) satisfies condition B(ii) if and only if line L intersects region Γ. With this consideration, we will next derive a condition under which line L intersects region Γ.

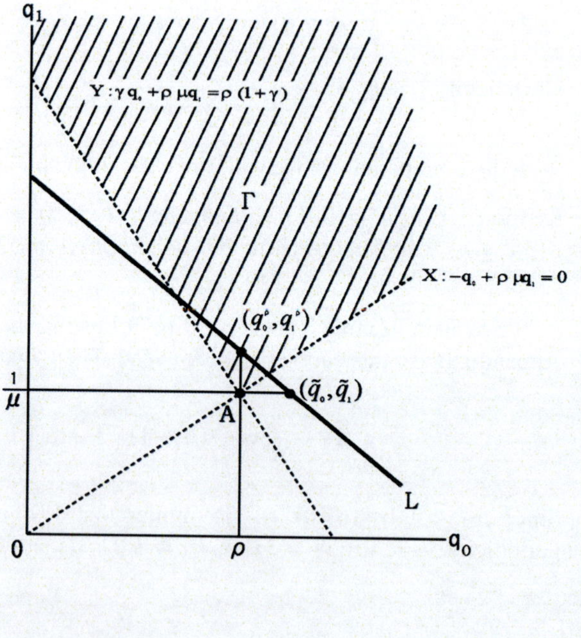

Fig. 6.4

Denote by $(\widetilde{q}_1, \widetilde{q}_2, \widetilde{q}_3, \widetilde{q}_0)$ the sequence q_1, q_2, q_3, q_0 constructed from $\widetilde{q}_1 = 1/\mu$ in the same way. Since line L is negatively sloped, and since $(\widetilde{q}_0, \widetilde{q}_1)$ lies on line L, line L intersects Γ if any only if $\widetilde{q}_0 > \rho$. By (6.33), this condition means

$$\widetilde{q}_0 - \rho = \rho\left(\frac{1}{\rho\mu} - 1\right)\left(1 + \frac{1}{\rho\mu}\left(1 - \frac{\gamma}{\rho\mu}\right)\right) > 0. \tag{6.34}$$

By (6.10), this holds if and only if the last parenthesis on the right-hand side of (6.34) is negative, i.e., $g(\rho\mu; \gamma, 0) < 0$ by (6.30).

In short, if $g(\rho\mu; \gamma, 0) < 0$, given $f(\mu; \gamma, 0) = 0, q_t = q_t^\rho, t = 0, 1, 2, 3,$ satisfies condition B of Lemma 2. (In the above discussion, we did not use the condition that $g(\rho\mu; \gamma, 0) < 0$ is sufficiently close to 0. We will, however, use this condition in order to guarantee condition B (iii.b) for the general case of n.)

The restrictions that we have made on parameters can be restated as follows:

$$(\mu, \gamma) \in \Pi = \{(\mu, \gamma) | 0 < \gamma < \mu < \gamma + 1\}, \tag{6.35}$$

$$\rho \in R(\mu, \gamma) = \{\rho | 1/\mu < \rho < \sqrt{\gamma}/\mu\}. \tag{6.36}$$

(In the above expressions, $\gamma > 0$ follows from (6.11); $\mu > \gamma$ follows from (6.19); $\gamma + 1 > \mu$ follows from (6.18); $\rho > 1/\mu$ follows from (6.10); $\sqrt{\gamma}/\mu > \rho$ follows from (6.27)). Think of $f(\mu; \gamma, n) = 0$ and $g(\theta; \gamma, n) = 0$ as equations of μ and θ, respectively. Each of these equations has only one positive solution. Denote by $\mu(\gamma, n)$ the positive solution to $f(\mu; \gamma, n) = 0$ and by $\theta(\gamma, n)$ that to $g(\theta; \gamma, n) = 0$;

$$\mu(\gamma, n) = \left(\frac{\gamma^{m[n+1]} + \sqrt{\gamma^{2m[n+1]} + 4\gamma^{m[n+2]}}}{2}\right)^{1/2^n}, \tag{6.37}$$

$$\theta(\gamma, n) = \left(\frac{-\gamma^{2m[n]} + \sqrt{\gamma^{4m[n]} + 4\gamma^{m[n+2]}}}{2}\right)^{1/2^n}. \tag{6.38}$$

Theorem 1. *Let $n = 0, 1, 2, \ldots$. If $(\mu, \gamma) \in \Pi$ satisfies $\mu = \mu(\gamma, n)$ and if $\rho \in R(\mu, \gamma)$ is smaller than and sufficiently close to $\theta(\gamma, n)/\mu(\gamma, n)$, transition function h is unimodal, expansive, and optimal.*

Proof. $\mu = \mu(\gamma, n)$, $f(\mu; \gamma, n) = 0$. If $\rho < \theta(\gamma, n)/\mu(\gamma, n)$ is sufficiently close to $\theta(\gamma, n)/\mu(\gamma, n)$, $g(\rho\mu; \gamma, n) < 0$. Thus, by Lemma 3, conditions A and B of Lemma 2 are satisfied for $N = 3 \times 2^n$. Since this implies that the trajectory that h generates from $1/\beta$ is a unique optimal path from $1/\beta$, the theorem follows from Lemma 1. □

Let $\rho(n)$ be the upper bound of ρ under which h is an expansive and unimodal optimal transition function and for which the periodicity of $1/\beta$ is $N = 3 \times 2^n$. By Theorem 1, we may find this upper bound for each n. when $n = 0$ (that is, the periodicity of $1/\beta$ is $3 = 3 \times 2^0$), by (6.31) and (6.34), we may prove that $\rho(n)$

is just above $\rho = 0.363$. Moreover, as n increases, $\rho(n)$ is approximately "square rooted"; that is, $\rho(n) \approx (0.363)^{n/2}$. The next theorem implies that, following this regularity, $\rho(n)$ actually increases to 1 as $n \to 1$, in other words, that our main result holds.[15] Let

$$\Gamma(n) = \left\{ \gamma \in R | \left(\frac{1+\sqrt{5}}{2}\right)^{1/2} < \gamma^{2^n/3} < \frac{1+\sqrt{5}}{2} \right\}. \tag{6.39}$$

Theorem 2. *For any $\varepsilon > 0$, there is an $\overline{n}$ such that for any $n \geqslant \overline{n}$ and $\gamma \in \Gamma(n)$, it holds that $(\mu(\gamma, n), \gamma) \in \Pi$, that $\theta(\gamma, n)/\mu(\gamma, n) \in R(\mu(\gamma, n), \gamma)$, and that $1 - \varepsilon < \theta(\gamma, n)/\mu(\gamma, n) < 1$.*

Proof. First, we demonstrate that $(\mu(\gamma, n), \gamma) \in \Pi$ for any sufficiently large n. Since, by (6.39),

$$\left(\frac{1+\sqrt{5}}{2}\right)^{3/2^{n+1}} < \gamma < \left(\frac{1+\sqrt{5}}{2}\right)^{3/2^n},$$

$\gamma \to 1$ as $n \to \infty$. Moreover, since $\gamma^{2^n/3}$ satisfies (6.39), it follows from (6.37) and the definition of $m[n]$ that $\mu(\gamma, n) \to 1$ as $n \to \infty$. Thus, $\mu(\gamma, n) < \gamma + 1$ for n sufficiently large. Since $\gamma > 0$ by (6.11) and (6.27), it suffices to prove $\mu(\gamma, n) > \gamma$. By (6.37) and the definition of $m[n]$, this is equivalent to

$$\gamma^{(-1)^{n-1} \cdot 1/3} \left(\gamma^{2^n/3}\right)^2 - \left(\gamma^{2^n/3}\right) - \gamma^{(-1)^{n-1} \cdot 2/3} < 0. \tag{6.40}$$

Recall $\gamma \to 1$ as $n \to \infty$. Thus, for any constant number ξ,

$$\gamma^{(-1)^n \cdot \xi/3} \to 1 \text{ as } n \to \infty. \tag{6.41}$$

Thus, by setting $x = \gamma^{2^n/3}$, (6.40) becomes $x^2 - x - 1 < 0$ in the limit. Since, given (6.39), this inequality is satisfied, (6.40) is satisfied if n is sufficiently large.

Next, we demonstrate that for any sufficiently large n, $\theta(\gamma, n)/\mu(\gamma, n) \in R(\mu(\gamma, n), \gamma)$. By (6.36), this is equivalent to $1 < \theta(\gamma, n) < \sqrt{\gamma}$. This condition is, by (6.38) and the definition of $m[n]$, equivalent to the following:

---

[15]When $n = 0$ and $\rho = 0.363$, the corresponding parameter values are $\beta = 3.1$ and $\mu = \mu(2.1, 0)$. If $n = 1$, $\rho(n)$ is just above $\rho = 0.61$; the parameter values are $\beta = 2.5$ and $\mu = \mu(1.5, 1)$. If $n = 2$ and $\rho(n)$ is just above $\rho = 0.78$, the parameter values are $\beta = 2.22$ and $\mu = \mu(1.22, 2)$. If $n = 3$, $\rho(n)$ is just above $\rho = 0.886$; the parameter values are $\beta = 2.1$ and $\mu(1.1, 3)$. In short, $\rho(0) \approx 0.36$, $\rho(1) \approx 0.61 \approx \sqrt{0.36}$, $\rho(2) \approx 0.78 \approx \sqrt{0.61}$, and $\rho(3) \approx 0.886 \approx \sqrt{0.78}$. The recognition of this systematic structure led us to Theorem 1.

$$\gamma^{(-1)^{n-1}\cdot 1/3}\left(\sqrt{\gamma^{2^n/3}}\right)^2 - \gamma^{(-1)^{n-1}\cdot 2/3}\left(\sqrt{\gamma^{2^n/3}}\right) - 1 < 0, \tag{6.42}$$

$$\gamma^{(-1)^{n-1}\cdot 1/3}\left(\gamma^{2^n/3}\right)^4 - \gamma^{(-1)^{n-1}\cdot 2/3}\left(\gamma^{2^n/3}\right)^2 - 1 > 0. \tag{6.43}$$

Thus, by (6.41)–(6.43) become $x - \sqrt{x} - 1 < 0$ and $x^4 - x^2 - 1 > 0$. Since, given (6.39), these inequalities are satisfied, (6.42) and (6.43) are satisfied if n is sufficiently large.

In order to complete the proof, it suffices to prove $\theta(\gamma, n)/\mu(\gamma, n) \to 1$ as $n \to 1$. For any sufficiently large n, as is noted above, $\theta(\gamma, n)/\mu(\gamma, n) \in R(\mu(\gamma, n), \gamma)$. Since $\gamma \in \Gamma(n)$, as $n \to \infty$, $\gamma \to 1$ by (6.39). Moreover, as $n \to \infty$, $\mu(\gamma, n) \to 1$, as is proved above. Since $\gamma \to 1$ and $\mu(\gamma, n) \to 1$, it follows from the definition of $R(\mu, \gamma)$ that $R(\mu(\gamma, n), n) \to \{1\}$. Thus, $\theta(\gamma, n)/\mu(\gamma, n) \to 1$, since $\theta(\gamma, n)/\mu(\gamma, n) \in R(\mu(\gamma, n), n)$.

In order to establish the Main Result, stated in Sect. 3, we increase the periodicity of $1/\beta$, $N = 3 \times 2^n$, by taking $n \to \infty$.[16] For each $\varepsilon > 0$, by Theorem 2, choose n in such a way that $(\mu(\gamma, n), \gamma) \in \Pi$, $\theta(\gamma, n)/\mu(\gamma, n) \in R(\mu(\gamma, n), \gamma)$ and $1 - \varepsilon < \theta(\gamma, n)/\mu(\gamma, n) < 1$. Let $\mu = \mu(\gamma, n)$, and choose an open interval $U \subset \{x | 1 - \varepsilon < x < \theta(\gamma, n)/\mu(\gamma, n)\}$ in such a way that the lower bound of U is sufficiently close to $\theta(\gamma, n)/\mu(\gamma, n)$. Then, $(\mu, \gamma) \in \Pi$ and $\rho \in R(\gamma)$. Thus, by Theorem 1, h is expansive, unimodal, and optimal. Since $\rho \to 1$ as $\varepsilon \to 0$, the Main Result is proved.

6.6 Some Remarks

In our model, the von Neumann facets are not trivial. That is to say, we deal with a case in which the modified golden rule state lies on a flat of the graph of the non-strictly concave reduced-form utility function. McKenzie (1983) studies a similar case and demonstrates that for any small $\varepsilon > 0$, it is possible to find a lower bound of the values of the discount factor with which any optimal paths converge into the ε–neighborhood of the von Neumann facet containing the modified golden rule state. Whereas McKenzie assumes differentiability for the reduced-form utility function, we deal with the nondifferentiability case.

A Leontief production function does not allow for substitutability between inputs. As is well-known, however, it can be written as the limit of a CES production

[16] Theorems 1 and 2 indicate that, for each given n, there are both upper and lower bounds for ρ with which values of the other parameters can be chosen in such a way that h may be an ergodically chaotic optimal transition function. The existence of such lower and upper bounds is comparable to the results obtained in Deneckere and Pelikan (1986), Boldrin and Montrucchio (1986), and Boldrin and Deneckere (1990).

function, which allows for the substitutability. One conjecture is, therefore, that our results may extend to the substitutable case by using CES production functions. Moreover, it might be the case that the higher the substitutability between capital and labor, the less likely chaotic optimal accumulation. It is of interest to investigate these points.

In this study, we have focused on the case in which optimal accumulation can be expressed by a one-dimensional transition function. It may be possible, however, to extend part of our results to the case of a higher dimension. In the model with two capital good sectors, for example, the graph of an optimal transition function may consist of several flat faced in the four-dimensional space of beginning- and end-of-a-period stocks. Thus, it may be possible to characterize those faces of the graph in a way similar to that of this study. Such a characterization may reveal, in the multiple-dimensional case, the existence of optimal chaos with respect to a suitably extended concept of chaos; it is straight-forward to extend, for example, expansiveness and geometric sensitivity in the multiple-dimension case.

6.7 Appendix A: Proof of Lemma 2

Call a vector (q, r) a support price vector of activity (k, y) if

$$v(k, y) + ry - \rho^{-1}qk \geqslant v(\zeta, \xi) + r\xi - \rho^{-1}q\zeta \tag{A.1}$$

for all $(\zeta, \xi) \in D$. For an activity vector $(k, y) \in D$ and its support price vector (q, r), define a value loss (McKenzie (1986)):

$$\Delta(\zeta, \xi; q, r; k, y) = v(k, y) + ry - \rho^{-1}qk - [v(\zeta, \xi) + r\xi - \rho^{-1}q\zeta] \geqslant 0. \tag{A.2}$$

The following sublemma characterizes the support price vector of $P = (1/\beta, \mu/\beta)$ in Fig. 6.1.

Sublemma 1. *Suppose that (q_0, q_1) satisfies condition B(ii) of Lemma 2. Then, it is a support price vector of activity $(1/\beta, \mu/\beta)$. Moreover, it holds that for $(k, y) \in D$,*

$$\Delta(k, y; q_0, q_1; 1/\beta, \mu/\beta) > 0 \text{ if and only if } (k, y) \neq (1/\beta, \mu/\beta). \tag{A.3}$$

Proof. Note that v is concave, since the model is a convex model, and that D is the region on and below OPZ in Fig. 6.1. By (6.16), $v(1/\beta, \mu/\beta) = 0$. Thus, point $(v, k, y) = (0, 1/\beta, \mu/\beta)$ lies on the graph of function $v = v(k, y)$ at $(k, y) = (1/\beta, \mu/\beta)$. A plane through point $(0, 1/\beta, \mu/\beta)$ is

$$v = \rho^{-1}q_0(k - 1/\beta) - q_1(y - \mu/\beta) \equiv f(k, y). \tag{A.4}$$

In Fig. 6.1, $P = (1/\beta, \mu/\beta)$ and $Q = (1/\alpha, 0)$. Take $Z = (1 + 1/\beta, \mu/\beta)$ and $K = (1 + 1/\beta, 0)$. The corresponding utility levels are $v(1/\beta, \mu/\beta) = 0$ at P, $v(1/\alpha, 0) = 1/\alpha$ at Q, $v(1/\beta, \mu/\beta) = 0$ at Z, and $v(1 + 1/\beta, 0) = 1/\alpha$ at K. Let $O' = (0, 0, 0)$, $P' = (0, 1/\beta, \mu/\beta)$, $Q' = (1/\alpha, 1/\alpha, 0)$, $Z' = (0, \mu/\beta, \mu/\beta)$, and $K' = (1/\alpha, 1 + 1/\beta, 0)$ in the $v - k - y$ space. Then, by (6.16), the graph of $v = v(k, y)$ consists of triangle $O'P'Q'$ and the face surrounded by segment $P'Q'$ and half lines $P'Z'$ and $Q'K'$. Given condition B(ii) of Lemma 2, thus, the rays from P' through O', Q' and Z' all lie strictly below the plane defined by (A.4) but point P'. Thus, by the concavity of $v = v(k, y)$, that plane supports the graph of $v(k, y)$ at P'. In particular, $\Delta(k, y; q_0, q_1; 1/\beta, \mu/\beta) = 0$ if and only if $(k, y) = (1/\beta, \mu/\beta)$.

The next sublemma characterizes the support price vector of a point on segment PQ of Fig. 6.1 but not at the endpoints.

Sublemma 2. *Suppose that (q_{t-1}, q_t) satisfies condition B (iii) of Lemma 2. Suppose, in addition, that $k_t = h(k_{t-1})$ and $k_{t-1} > 1/\beta$. Then, (q_{t-1}, q_t) is a support price vector of (k_{t-1}, k_t). Moreover, for $(k, y) \in D$,*

$$\Delta(k, y; q_{t-1}, q_t; k_{t-1}, k_t) > 0 \textit{ if and only if } y \neq h(k) \textit{ or } k < 1/\beta. \tag{A.5}$$

Proof. Let $\lambda = \rho^{-1} q_{t-1}$. Under the hypothesis of the sublemma, by condition B (iii) of Lemma 2, it holds that $0 < \lambda < 1$ and that

$$(\rho^{-1} q_{t-1}, q_t) = \lambda \left(1, -\frac{1}{\mu}\right) + (1 - \lambda) \left(0, -\frac{\beta/\alpha}{\mu}\right). \tag{A.6}$$

Since (k_{t-1}, k_t) lies on segment PQ but not at the endpoints, $(v, k, y) = (v(k_{t-1}, k_t), k_{t-1}, k_t)$ lies on the plane defined by $v = k - (1/\mu)y$ as well as that defined by $v = (1/\alpha) - ((\beta/\alpha)/\mu)y$. Thus, by the concavity of $v(k, y)$, it holds that

$$v(k_{t-1}, k_t) + \frac{1}{\mu} k_t - k_{t-1} \geqslant v(k, y) + \frac{1}{\mu} y - k \tag{A.7}$$

for any $(k, y) \in D$, and that

$$v(k_{t-1}, k_t) + \frac{\beta/\alpha}{\mu} k_t - 0 \cdot k_{t-1} \geqslant v(k, y) + \frac{\beta/\alpha}{\mu} y - 0 \cdot k \tag{A.8}$$

for all $(k, y) \in D$. Therefore, it follows from (A.6)–(A.8) that (q_{t-1}, q_t) is a support price vector of (k_{t-1}, k_t). The rest can be proved by following an argument similar to the proof of Sublemma 1.

The next sublemma characterizes the support price vector of a point on segment OP of Fig. 6.1 but not at the endpoints.

Sublemma 3. *Suppose that (q_{t-1}, q_t) satisfies condition B(iii) of Lemma 2. Suppose, in addition, that $k_t = h(k_{t-1})$ and $0 < k_{t-1} < 1/\beta$. Then, (q_{t-1}, q_t) is a support price vector of (k_{t-1}, k_t). Moreover, for $(k, y) \in D$,*

$$\Delta(k, y; q_{t-1}, q_t; k_{t-1}, k_t) > 0 \text{ if and only if } y \neq h(k) \text{ or } k > 1/\beta. \qquad \text{(A.9)}$$

Proof. Let $1 + \delta = \rho^{-1} q_{t-1}$. Under the hypothesis of the sublemma, by condition B (iii) of Lemma 2, it holds that $\delta > 0$ and that

$$(\rho^{-1} q_{t-1}, -q_t) = (1 + \delta)\left(1, -\frac{1}{\mu}\right). \qquad \text{(A.10)}$$

Since $k_t = \mu k_{t-1}$ by the hypothesis of the sublemma, by (6.16),

$$\frac{1}{\mu} k_t - k_{t-1} \geqslant \frac{1}{\mu} y - k \qquad \text{(A.11)}$$

for all $(k, y) \in D$. Since $(v, k, y) = (0, k_{t-1}, k_t)$ is on the plane defined by $v = k - (1/\mu) y$, the concavity of $v(k, y)$ implies that

$$v(k_{t-1}, k_t) + \frac{1}{\mu} k_t - k_{t-1} \geqslant v(k, y) + \frac{1}{\mu} y - k \qquad \text{(A.12)}$$

for any $(k, y) \in D$. It follows from (A.10)–(A.12) that (q_{t-1}, q_t) is a support price vector of (k_{t-1}, k_t). The rest can be proved by following an argument similar to the proof of Sublemma 1.

We now prove Lemma 2. Let $k_t = h^t(1/\beta)$. Since $1/\beta$ is a cyclical point,by Lemma 1, it suffices to prove that path k_t is the unique optimal path from $1/\beta$. By Sublemma 1–3, (q_{t-1}, q_t) is a support price vector of activity (k_{t-1}, k_t) for $t = 1, 2, \ldots$. Define

$$\Delta_t(k, y) = \Delta(k, y; q_{t-1}, q_t; k_{t-1}, k_t). \qquad \text{(A.13)}$$

Then, for any $(k, y) \in D$,

$$\Delta_t(k, y) \geqslant 0 \qquad (t = 1, 2, \ldots). \qquad \text{(A.14)}$$

In order to prove the optimality of path k_t, take an arbitrary alternative path h_t, $t = 0, 1, \ldots$, such that $(h_{t-1}, h_t) \in D$ and $h_0 = 1/\beta$. Then, by the definition of value losses, for any T,

$$\sum_{t=1}^{T} \rho^t (v(k_{t-1}, k_t) - v(h_{t-1}, h_t)) = \sum_{t=1}^{T} \rho^t \Delta_t(h_{t-1}, h_t) - \rho^T q_T (h_T - k_T). \qquad \text{(A.15)}$$

Since $q_T \in \{q_0, q_1, \ldots, q_{N-1}\}$ by definition, by taking $T \to \infty$,

$$\sum_{t=1}^{\infty} \rho^t v(k_{t-1}, k_t) - \sum_{t=1}^{\infty} \rho^t v(h_{t-1}, h_t) = \sum_{t=1}^{\infty} \rho^t \Delta_t(h_{t-1}, h_t) \geqslant 0 \tag{A.16}$$

for any path h_t, $t = 0, 1, \ldots$, such that $(h_{t-1}, h_t) \in D$ and $h_0 = 1/\beta$. This implies that path k_t is optimal.

In order to prove the uniqueness of path k_t, suppose that there is another optimal path h_t, $t = 0, 1, \ldots$, from $1/\beta$. Then, there is t^* such that $h_{t^*} \neq h(h_{t^*-1})$. Therefore, by (A.3), (A.5), and (A.9), $h_{t^*} \neq h(h_{t^*-1})$ implies

$$\Delta_t(h_{t^*-1}, h_{t^*}) > 0. \tag{A.17}$$

This implies, however, by (A.14), (A.16), and (A.17),

$$\sum_{t=1}^{\infty} \rho^t v(k_{t-1}, k_t) - \sum_{t=1}^{\infty} \rho^t v(h_{t-1}, h_t) \geqslant \Delta_t(h_{t^*-1}, h_{t^*}) > 0, \tag{A.18}$$

which contradicts the optimality of path h_t. This proves Lemma 2.

6.8 Appendix B: Proof of Lemma 3

In order to prove Lemma 3, we use Lemma 2. For this purpose, we first characterize condition A of Lemma 2 for $N = 3 \times 2^n$, where $n = 0, 1, \ldots$. Set

$$\alpha = 1. \tag{B.1}$$

Let $\widetilde{k}_1 = \mu/\beta$ and $\widetilde{k}_t = h(\widetilde{k}_{t-1})$ for $t = 2, 3\ldots$. The following holds:

Sublemma 4. *Let $n = 0, 1, \ldots$. Point $k = 1/\beta$ is a cyclical point of h with the order of periodicity $N = 3 \times 2^n$ if and only if $f(\mu; \gamma, n) = 0$.*

Proof. Note $\widetilde{k}_1 = h(1/\beta)$. As is shown in Sect. 6.4, Lemma 3 holds for $n = 0$. For the sake of convenience, shift the origin of $h(k)$ to $(1/\beta, 1/\beta)$ by defining $\kappa = k - 1/\beta$ and $\zeta = y - 1/\beta$. By (6.21), $h(k)$ is equivalent to

$$\eta(\kappa) = \begin{cases} \mu\kappa + \frac{1}{\beta}(\mu - 1) & \text{if } -1/\beta < \kappa \leqslant 0, \\ -\frac{\mu}{\gamma}\kappa + \frac{1}{\beta}(\mu - 1) & \text{if } 0 < \kappa < \frac{1}{\beta}(\mu - 1). \end{cases} \tag{B.2}$$

We provide an heuristic method, which is more intuitive than a formal induction. In Sect. 6.4, we have proved the sublemma for $n = 0$. Let $n = 1$. In order to characterize the condition that $1/\beta$ is a cyclical point of h with periodicity $6 = 3 \times 2^1$, we first construct $\eta^2(\kappa)$ and characterize the condition that 0 is a cyclical point of η^2 with periodicity 3, i.e., that of η with periodicity $6 = 3 \times 2^1$.

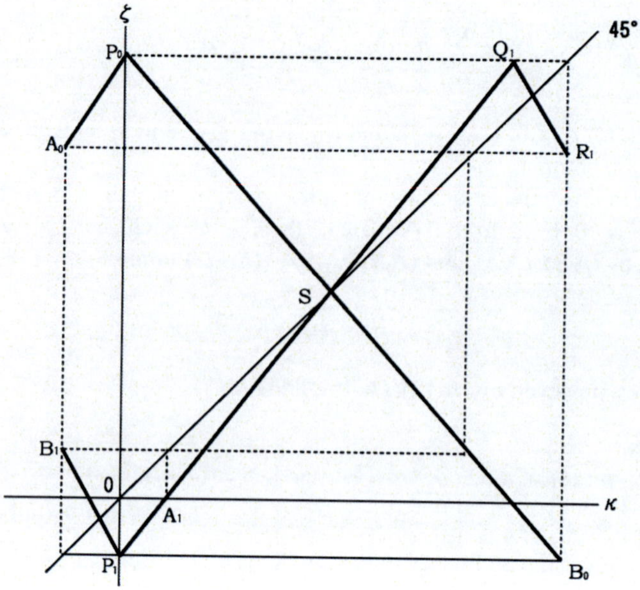

Fig. 6.5

Figure 6.5 depicts $\eta(\kappa)$ by kinked segment $A_0P_0B_0$. As the diagram indicates, P_0 indicates $(0, \eta(0))$, B_0 indicates $(\eta(0), \eta^2(0))$, and A_0 indicates $(\eta^2(0), \eta^3(0))$. By construction, $(\eta^{t-1}(0), \eta^t(0))$ must lie on kinked segment $A_0P_0B_0$ for any $t = 1, 2, \ldots$. Curve $B_1P_1Q_1R_1$ illustrates function $\eta^2(\kappa) = \eta \circ \eta(\kappa)$. If 0 is a cyclical point of η^2 with the order of periodicity equal to 3, then $(0, \eta^2(0))$ is at P_1 and $(\eta^2(0), \eta^{2\times2}(0))$ is at B_1. Moreover, $A_1 = (\eta^{2\times2}(0), \eta^{2\times3}(0))$ must lie on the open segment between P_1 and S; otherwise, it is impossible to construct a period-3 path from 0. In order to construct a condition under which 0 is a cyclical point of η^2 with order 3, it suffices to focus on function η^2 restricted to the interval between κ_{B_1} and κ_{A_1}, where κ_{A_1} and κ_{B_1} are, respectively, the κ-coordinate of points A_1 and B_1. On this interval, $\eta^2(\kappa)$ can be expressed as follows:

$$\eta^2(\kappa) = \begin{cases} -\frac{\mu^2}{\gamma}\kappa - \frac{1}{\beta}(\mu-1)\left(\frac{\mu}{\gamma}-1\right) & \text{if } \kappa_{B_1} \leqslant \kappa \leqslant 0 \\ \frac{\mu^2}{\gamma^2}\kappa - \frac{1}{\beta}(\mu-1)\left(\frac{\mu}{\gamma}-1\right) & \text{if } 0 < \kappa \leqslant \kappa_{A_1}. \end{cases} \tag{B.3}$$

Therefore, 0 is a cyclical point of η^2 with periodicity 3 if and only if

$$\widetilde{\kappa}_6 - 1/\beta = \widetilde{\kappa}_6 = \frac{1}{\beta}(\mu-1)\left(1-\frac{\mu}{\gamma}\right)\left(1+\frac{\mu^2}{\gamma^2}\left\{1-\frac{\mu^2}{\gamma}\right\}\right) = 0, \tag{B.4}$$

which is, by $\mu > \gamma > 1$, equivalent to $f(\mu; \gamma, n) = 0$.

For the general case of $N = 3 \times 2^n$, we use an argument similar to this. That is, we first construct $\eta^{2n}(\kappa)$ inductively and characterize the condition that 0 is a cyclical point of η^{2^n} with periodicity 3, i.e., that of η with periodicity 3×2^n. To this end, find kinked segment $A_n P_n B_n$ such that $(0, \eta^{2^n}(0))$ is at P_n, $(\eta^{2^n}(0), \eta^{2^n \times 2}(0))$ is at B_n, and $(\eta^{2^n \times 2}(0), \eta^{2^n \times 3}(0))$ is at A_n. Moreover, denoting by κ_{A_n} and κ_{B_n}, respectively, the κ-coordinates of points A_n and B_n, function $\eta^{2^n}(\kappa)$ restricted to the interval between κ_{A_n} and κ_{B_n} is as follows:

If n is even,

$$\eta^{2^n}(\kappa) = \begin{cases} \frac{\mu^{2^n}}{\gamma^{2m[n]}}\kappa + \frac{1}{\beta}\left(1-\frac{\mu}{\gamma}\right)\left(1-\frac{\mu^2}{\gamma}\right)\left(1-\frac{\mu^4}{\gamma^2}\right)\ldots\left(1-\frac{\mu^{2^n-1}}{\gamma^{m[n]}}\right) \\ \qquad \text{if } \kappa_{A_n} \leqslant \kappa \leqslant 0, \\ -\frac{\mu^2}{\gamma^{m[n+1]}}\kappa + \frac{1}{\beta}\left(1-\frac{\mu}{\gamma}\right)\left(1-\frac{\mu^2}{\gamma}\right)\left(1-\frac{\mu^4}{\gamma^2}\right)\ldots\left(1-\frac{\mu^{2^n-1}}{\gamma^{m[n]}}\right) \\ \qquad \text{if } 0 \leqslant \kappa \leqslant \kappa_{B_n}. \end{cases} \tag{B.5}$$

If n is odd,

$$\eta^{2^n}(\kappa) = \begin{cases} -\frac{\mu^{2^n}}{\gamma^{m[n+1]}}\kappa + \frac{1}{\beta}\left(1-\frac{\mu}{\gamma}\right)\left(1-\frac{\mu^2}{\gamma}\right)\left(1-\frac{\mu^4}{\gamma^2}\right)\ldots\left(1-\frac{\mu^{2^n-1}}{\gamma^{m[n]}}\right) \\ \qquad \text{if } \kappa_{B_n} \leqslant \kappa \leqslant 0, \\ \frac{\mu^{2n}}{\gamma^{2m[n]}}\kappa + \frac{1}{\beta}\left(1-\frac{\mu}{\gamma}\right)\left(1-\frac{\mu^2}{\gamma}\right)\left(1-\frac{\mu^4}{\gamma^2}\right)\ldots\left(1-\frac{\mu^{2^n-1}}{\gamma^{m[n]}}\right) \\ \qquad \text{if } 0 \leqslant \kappa \leqslant \kappa_{A_n}. \end{cases} \tag{B.6}$$

(Note: By $\mu > \gamma > 1$, the constant term, i.e., the term that does not depend on κ, in (B.5) and (B.6) is positive if n is even and negative if n is odd). Therefore, it follows from (B.5) and (B.6) that

$$\widetilde{\kappa}_N - 1/\beta = \frac{1}{\beta}(\mu - 1)\left(1-\frac{\mu}{\gamma}\right)\left(1-\frac{\mu^2}{\gamma}\right)\cdots\left(1-\frac{\mu^{2^n-1}}{\gamma^{m[n]}}\right) \\ \times\left(1+\frac{\mu^{2^n}}{\gamma^{2m[n]}}\left(1-\frac{\mu^{2^n}}{\gamma^{m[n+1]}}\right)\right). \tag{B.7}$$

Since $\mu > \gamma > 1$, the first parenthesis on the right-hand side of (B.7) is positive, and the rest of the parameters are negative except the last. Therefore, $1/\beta$ is a cyclical point of h with periodicity $N = 3\times 2^n$ if and only if the last parenthesis on the right-hand side of (B.7) is 0. By (B.7), this condition is equivalent to $f(\mu; \gamma, n) = 0$.

By the hypothesis of Lemma 3, we may assume that $1/\beta$ is a cyclical point of h with periodicity $N = 3 \times 2^n$. Recall $\widetilde{k}_t = h^t(1/\beta)$ by definition. Given $k_t = \widetilde{k}_t$, define $q_1, \ldots, q_N, q_0$ by condition B (iii.a) and B(i) of Lemma 2 with an arbitrary q_1. Then, (q_0, q_1) may be represented by a line like line L in Fig. 6.4. In the case of $n = 1$ (i.e., $N = 6$), it is possible to demonstrate that line L is

$$(q_0-\rho)-\tfrac{\gamma^4}{(\rho\mu)^5}\left(q_1-\tfrac{1}{\mu}\right)\\ =\rho\left(\tfrac{1}{\rho\mu}-1\right)\left(1-\tfrac{\gamma}{\rho\mu}\right)\left(1+\tfrac{\gamma^2}{(\rho\mu)^2}\left(1-\tfrac{\gamma}{(\rho\mu)^2}\right)\right). \tag{B.8}$$

In particular, denote by $q_1^\rho, ..., q_N^\rho, q_0^\rho$ the sequence $q_1, ..., q_N, q_0$ that satisfies $q_0 = \rho$ in addition to conditions B (iii.a) and B(i).

Sublemma 5. *(q_0^ρ, q_1^ρ) satisfies condition B(ii) of Lemma 2 if and only if $g(\rho\mu;\gamma, n) < 0$.*

Proof. Denote by $\widetilde{q}_1, ..., \widetilde{q}_N, \widetilde{q}_0$ the $q_1, ..., q_N, q_0$ that satisfies $q_1 = 1/\mu$ in addition to conditions B(iii.a) and B(i). Since $\widetilde{k}_1$ is a cyclical point of order N, $\widetilde{k}_{t-1} \neq 1/\beta$ for any $t = 2, ..., N$. Recall $\alpha = 1$. Thus, whenever $\widetilde{k}_{t-1} > 1/\beta$, $\widetilde{k}_t = -(\mu/\gamma)\widetilde{k}_{t-1}+(\mu/\gamma)$ and $\widetilde{q}_t = -(\gamma/\rho\mu)\widetilde{q}_{t-1}+((1+\gamma)/\mu)$ by (6.21) and condition B(iii.a). In contrast, whenever $\widetilde{k}_{t-1} < 1/\beta$, $\widetilde{k}_t = \mu\widetilde{k}_{t-1}$ and $\widetilde{q}_t = (1/\rho\mu)\widetilde{q}_{t-1}$. This parallel relationship between $\widetilde{k}_t - 1/\beta$ and $\widetilde{q}_t - \rho$ together with (B.7) indicates

$$\widetilde{q}_0-\rho=\rho\left(\frac{1}{\rho\mu}-1\right)\left(1-\frac{\gamma}{\rho\mu}\right)\left(1-\frac{\gamma}{(\rho\mu)^2}\right)\cdots\left(1-\frac{\gamma^{m[n]}}{(\rho\mu)^{2^n-1}}\right)\\ \times\left(1+\frac{\gamma^{2m[n]}}{(\rho\mu)^{2^n}}\left(1-\frac{\gamma^{m[n+1]}}{(\rho\mu)^{2^n}}\right)\right). \tag{B.9}$$

Note that both (q_0^ρ, q_1^ρ) and $(\widetilde{q}_0, \widetilde{q}_1)$ lies on line L in Fig. 6.4. Thus, (q_0^ρ, q_1^ρ) satisfies condition B(ii) if and only if it lies in region Γ. In the case in which line L is upward-sloping, (q_0^ρ, q_1^ρ) lies in region Γ if and only if $\widetilde{q}_0 < \rho$. In the case in which line L is downward-sloping, (q_0^ρ, q_1^ρ) lies in region Γ if and only if $\widetilde{q}_0 > \rho$. Moreover, it is possible to prove that, as is indicated in (6.33) and (B.8), line L is downward-sloping if and only if n is an even number and upward-sloping if and only if n is an odd number. Since $N = 3\times 2^n$, therefore, line L cuts region Γ if and only if $(\widetilde{q}_0 - \rho)(-1)^n > 0$.

In order to determine the sign of $(\widetilde{q}_0 - \rho)(-1)^n$, we use (B.9). To this end, we prove, inductively for $n = 1, 2, ...,$

$$\gamma^{m[n]}/(\rho\mu)^{2^{n-1}} > 1. \tag{B.10}$$

Since $\sqrt{\gamma} > \rho\mu > 1$ by (6.10) and (6.27), $\gamma > \rho\mu$ and $\gamma > (\rho\mu)^2$. By $\gamma > \rho\mu$, (B.10) holds for $n = 1$. By $\gamma > (\rho\mu)^2$, (B.10) holds for $n = 2$. Suppose that (B.10) holds for $n = i$ and $i + 1$. Then, by the definition of $m[n]$,

$$\left[\gamma^{m[i+2]}/(\rho\mu)^{2^{(i+2)-1}}\right]=\left[\gamma^{m[i]}/(\rho\mu)^{2^{i-1}}\right]^2\times\left[\gamma^{m[i+1]}/(\rho\mu)^{2^{(i+2)-1}}\right]>1^2\times 1=1.$$

This implies (B.10) for $n = i + 2$; (B.10) is proved.

By $\rho\mu > 1$, (B.10) implies that the parentheses on the right-hand side of (B.9) are all negative except the last. Thus, $(\widetilde{q}_0 - \rho)(-1)^n > 0$ if and only if the last parenthesis is negative. By (6.30), this condition is equivalent to $g(\rho\mu; \gamma, n) < 0$.

In order to complete the proof, we prove that $q_1^\rho, ..., q_N^\rho, q_0^\rho$ can be chosen in such a way that condition B (iii,b) of Lemma 2 is satisfied. To this end, first take an arbitrary combination ρ^*, μ^* and γ^* that satisfy $f(\mu^*; \gamma^*, n) = 0$ and $g(\rho^*\mu^*; \gamma^*, n) = 0$; let β^* correspond to these values. Given these parameter values, construct $\widetilde{k}_t^*$ and $\widetilde{q}_t^*$ in the way specified before Sublemma 5. The structural similarity between the expression of $\widetilde{q}_0^* - \rho^*$, given by Sublemma 5, and that of $\widetilde{k}_0^* - 1/\beta^*$, given by Sublemma 4, implies the following: If $(\widetilde{k}_t^*, \widetilde{k}_{t+1}^*)$ lies on segment OP, but point P, of Fig. 6.1, $(\widetilde{q}_t^*, \widetilde{q}_{t+1}^*)$ lies on segment AX, but point A, of Fig. 6.4. If $(\widetilde{k}_t^*, \widetilde{k}_{t+1}^*)$ lies on segment PR, but point P, of Fig. 6.1, $(\widetilde{q}_t^*, \widetilde{q}_{t+1}^*)$ lies on segment AX, but point A, of Fig. 6.4. In short, for any $t = 1, 2, ..., N-1$,

$$\begin{cases} \widetilde{q}_t^* > \rho^* & \text{if } \widetilde{k}_t^* < 1/\beta^*, \\ 0 < \widetilde{q}_t^* < \rho^* & \text{if } \widetilde{k}_t^* > 1/\beta^*. \end{cases} \tag{B.11}$$

Note $\widetilde{q}_0^* = \rho^*$ by $g(\rho^*\mu^*; \gamma^*, n) = 0$ and (B.9) and $\widetilde{q}_1^* = 1/\mu^*$ by construction.

By the hypothesis of Lemma 3, $f(\mu; \gamma, n) = 0$. Moreover, $g(\rho\mu; \gamma, n) < 0$ is arbitrarily close to 0. With these parameters, construct $\widetilde{k}_t$ and $q_1^\rho, ..., q_N^\rho, q_0^\rho$ in the way specified above. Since $g(\rho\mu; \gamma, n) < 0$ is arbitrarily close to 0, by construction of $q_1^\rho, ..., q_N^\rho, q_0^\rho$ it is possible to find such ρ^*, μ^*, and γ^* as those in the previous paragraph and to make q_n^ρ and $\widetilde{k}_t$ arbitrarily close to $\widetilde{q}_t^*$ and $\widetilde{k}_t^*$. Thus, by (B.11), condition B(iii.b) holds. This establishes Lemma 3.

References

Araujo, A. (1991), "On the Differentiability of the Policy Function," *Econometrica*, 59, 1383–1394.

Benhabib, J. and R. Day (1982), "A Characterization of Erratic Dynamics in the Overlapping Generation Model," *Journal of Economic Dynamics and Control*, 4, 37–55.

Benhabib, J. and K. Nishimura (1985), "Competitive Equilibrium Cycles," *Journal of Economic Theory*, 35, 284–306.

Boldrin, M. and R. Deneckere (1990), "Sources of Complex Dynamics in Two-Sector Growth Models," *Journal of Economic Dynamics and Control*, 14, 627–653.

Boldrin, M. and L. Montrucchio (1986), "On the Indeterminacy of Capital Accumulation Paths," *Journal of Economic Theory*, 40, 26–39.

Boldrin, M. and L. Montrucchio (1989), On the Differentiability of the Policy Function, mimeo, UCLA.

Boldrin, M. and M. Woodford (1990), "Equilibrium Models Displaying Endogenous Fluctuations and Chaos," *Journal of Monetary Economics*, 25, 189–222.

Brock, W. (1986), Distinguishing Random and Deterministic Systems: An Expanded Version, mimeo, University of Wisconsin.

Brock, W. and J. Scheinkman (1976), "Global Asymptotic Stability of Optimal Control Systems with Applications to the Theorem of Economic Growth," *Journal of Economic Theory*, 12, 164–190.

Cass, D. and K. Shell (1976), "The Structure and Stability of Competitive Dynamic Systems," *Journal of Economic Theory*, 12, 31–70.

Chung, C. -S. (1986), "On the Stability of Invariant Capital Stock in a Two Dimensional Planing Model," *Mathematical Programming*, 34, 201–211.

Collet, P. and J. Eckmann (1980), Iterated Maps on the Interval as Dynamical Systems, Boston: Birkhauser.

Day, R. (1982), "Irregular Growth Cycles," *American Economic Review*, 72, 406–414.

Day, R. and G. Pianigiani (1991), "Statistical Dynamics and Economics," *Journal of Economic Behavior and Organizations*, 16, 37–83.

Deneckere, R. (1988), An Improved Sufficiency Condition for Complex Dynamics in Optimal Growth Models, mimeo, Northwestern University.

Deneckere, R. and K. Judd (1992), Cyclical and Chaotic Behavior in a Dynamic Equilibrium Model, with Implications for Fiscal Policy, in *Cycles and Chaos in Economic Equilibrium*, ed. by Benhabib, J., Princeton: Princeton University Press, 308–329.

Deneckere, R. and S. Pelikan (1986), "Competitive Chaos," *Journal of Economic Theory*, 40, 13–25.

Grandmont, J. (1985), "On Endogenous Competitive Business Cycles," *Econometrica*, 53, 995–1045.

Jones, R. (1965), "Duality in International Trade: A Geometrical Note," *Canadian Journal of Economics and Political Sciences*, 31, 390–393.

Lasota, A. and J. Yorke (1973), "On the Existence of Invariant Measures for Piecewise Monotonic Transformations," *Transaction of the American Mathematical Society*, 186, 481–488.

Li, T.-Y. and J. Yorke (1975), "Period Three Implies Chaos," *American Mathematical Monthly*, 82, 985–992.

Li, T.-Y. and J. Yorke (1978), "Ergodic Transformations from an Interval into Itself," *Transaction of the American Mathematical Society*, 235, 183–192.

Majumdar, M. and T. Mitra (1994), "Periodic and Chaotic Programs of Optimal Intertemporal Allocation in an Aggregative Model with Wealth Effects," *Economic Theory*, 4, 649–676.

Marshall, A. and I. Olkin (1979), Inequalities: Theory of Majorization and its Applications, New York: Academic Press.

McKenzie, L. (1968), Accumulation Programs of Maximum Utility and the Von Neumann Facet, in *Value, Capital and Growth*, ed. by Wolfe, J., Edinburgh: Edinburgh University Press, 353–383.

McKenzie, L. (1983), "Turnpike Theory, Discounted Utility and the Von Neumann Facet," *Journal of Economic Theory*, 27, 330–352.

McKenzie, L. (1986), Optimal Economic Growth, Turnpike Theorems and Comparative Dynamics, in *Handbook of Mathematical Economics: Vol III*, ed. by Arrow, K. and M. Intriligator, Amsterdam: North-Holland, 1281–1355.

Santos, M. (1991), "Smoothness of the Policy Function in Discrete Time Economic Models," *Econometrica*, 59, 1365–1382.

Scheinkman, J. (1976), "On Optimal Steady State of n-Sector Growth Models when Utility is Discounted," *Journal of Economic Theory*, 12, 11–20.

Scheinkman, J. (1990), "Nonlinearities in Economic Dynamics," *Economic Journal*, 33–48.

Scheinkman, J. and B. LeBaron (1989), "Nonlinear Dynamics and Stock Returns," *Journal of Business*, 62, 311–337.

Sorger, G. (1992), "On the Minimum Rate of Impatience for Complicated Optimal Growth Paths," *Journal of Economic Theory*, 56, 160–179.

Topkis, D. (1978), "Minimizing a Submodular Function on a Lattice," *Operation Research*, 26, 305–321.

Chapter 7
Chaotic Solutions in Dynamic Linear Programming*

Kazuo Nishimura and Mokoto Yano**

7.1 Introduction

Chaotic phenomena have been observed in various fields of sciences. We are concerned with linear programming (LP) and demonstrate that chaos may emerge as a solution to a dynamic LP problem. For this purpose, we work with an infinite time-horizon problem, for chaos appears in a dynamical system with no terminal date. As a result, it is not straightforward to find a solution, which cannot be derived from a simple repetition of arithmetics. In the finite time-horizon case, in contrast, a solution can be, at least in theory, obtained by such a method; the simplex method is one such procedure, repeating computations systematically.

It is well known that dynamic programming can be treated in the standard LP framework by adding a time structure. In their classic book, for example, Dorfman et al. (1958) dealt with an LP problem of the following type:

$$\begin{cases} \max \ \sum_{t=1}^{\infty} \rho^{t-1} p x_t \\ (x_1, x_2, \ldots) \geq 0 \\ \text{subject to} \ \begin{pmatrix} B & 0 & 0 & 0 & \cdots \\ -A & B & 0 & 0 & \cdots \\ 0 & -A & B & 0 & \cdots \\ 0 & 0 & -A & B & \cdots \\ \vdots & \vdots & \vdots & \ddots & \ddots \end{pmatrix} \begin{pmatrix} x_1 \\ x_2 \\ x_3 \\ x_4 \\ \vdots \end{pmatrix} \leq \begin{pmatrix} Ax+d \\ d \\ d \\ d \\ \vdots \end{pmatrix} \end{cases} \quad (7.1)$$

***Chaos, Solitons and Fractals* 7**, 1941–1953, 1996.

**An early part of this research was conducted during the workshop, 'Various Approaches to Complex Systems' held at the International Institute for Advanced Studies. We are gratefully to Richard Day, Gustav Feichtinger and Robert Kalaba for useful conversations and encouragement.

K. Nishimura (✉) · M. Yano
Institute of Economic Research, Kyoto University, Japan
e-mail: nishimura@kier.kyoto-u.ac.jp; yano@kier.kyoto-u.ac.jp

J. Stachurski et al. (eds.), *Nonlinear Dynamics in Equilibrium Models*,
DOI 10.1007/978-3-642-22397-6_7, © Elsevier Ltd. 1996. Published by Springer-Verlag
Berlin Heidelberg 2012. All Rights Reserved

(what Dorfman et al. actually studied was a finite time-horizon version of this problem). In the above problem, the discount factor ρ is a number between 0 and 1, A and B are $m \times n$ matrices of non-negative components, d and p are $n \times 1$ matrices of non-negative components, and p' is the transpose of p. The intended interpretation of this problem is to maximize the objective function $\sum_{t=1}^{\infty} \rho^{t-1} p' x_t$, which is the discounted sum of $p' x_t$ over the time periods $t = 1, 2, \ldots$, under the recursive constraints $Bx_t \leq Ax_{t-1} + d$ with the initial condition $x_0 = x$.

By the Bellman principle (Bellman 1957 and Bellman and Kalaba 1965), the solutions to problem (7.1) can be described by a binary relation F. That is to say, $(x_1^*, x_2^*, \ldots)$ solves (7.1) if and only if $x_0^* = x$ and

$$x_t^* \in F(x_{t-1}^*) \tag{7.2}$$

for $t = 1, 2, \ldots, \infty$. In this sense, we may call the binary relation F an *optimal program*. Note that because of its linear structure, the maximization problem (7.1) generally has multiple solutions. As a result, the binary relation describing the solutions, F, is a set-valued function.

The question that we face is whether or not such an optimal program can be a chaotic dynamical system. In order to deal with this issue, we need to answer the following two specific questions. (1) Under what condition is the optimal program (7.2), which is in fact a dynamical system of the standard sense, described by a single-valued function instead of the set-valued function F? (2) Under what condition is the resulting dynamical system chaotic? In order to answer these questions, we construct a simple LP problem with an infinite time-horizon. We then derive conditions under which the solutions are described by a chaotic dynamical system.

7.2 Chaotic Solutions to a Simple Dynamic LP Problem

Think of the following LP problem with parameters $a_{11} > 0$, $a_{12} > 0$, $a_{21} > 0$, $a_{22} > 0$. $k > 0$ and ρ, $0 < \rho < 1$.

$$\begin{cases} \max\limits_{(c_1,k_1,c_2,k_2,\ldots)\geq 0} \sum\limits_{t=1}^{\infty} \rho^{t-1} c_t \\ \text{subject to (i) } a_{11}c_t + a_{12}k_t \leq 1 \\ \qquad\qquad \text{(ii) } a_{21}c_t + a_{22}k_t \leq k_{t-1} \\ \qquad\qquad\qquad t = 1, 2, \ldots, \text{ and} \\ \qquad\qquad \text{(iii) } k_0 = k. \end{cases} \tag{7.3}$$

As noted above, the solutions to problem (7.3) can be described by a generalized dynamical system. To this end, for each $(k_{t-1}, k_t) \geq 0$, define $c(k_{t-1}, k_t)$ as the maximum value of $c_t \geq 0$ satisfying conditions (i) and (ii) of (7.3).

Proposition 1. *For each $k \geq 0$, there is a non-empty subset of R_+, $H(k)$, such that if $(c_1, k_1, c_2, k_2, \ldots)$ is a solution to (7.3), then it holds that*

$$k_t \in H(k_{t-1}), \quad t = 1, 2, \ldots, \tag{7.4}$$

with $k_0 = x$ *and that*

$$c_t = c(k_{t-1}, k_t). \tag{7.5}$$

We call system H a *generalized optimal dynamical system*. If, in particular, H is a function, we call it an *optimal dynamical system*. In what follows, we will demonstrate that H can in fact be a chaotic optimal dynamical system. For the characterization of chaotic motion, we will use the following result (see Lasota and Yorke (1974) and Li and Yorke (1978)).

Proposition 2. *Let f be a function on a closed interval I into itself satisfying that it is continuously twice differentiable everywhere except one point $b \in I$, and that there is an $\epsilon > 0$ such that $|f'(x)| > 1 + \epsilon$ for any x at which f' exists (expansive and unimodal). Then, there is a unique invariant measure on I, μ, that is ergodic with respect to f and absolutely continuous with respect to the Lebesgue measure.*

The above result implies that almost every trajectory following an expansive and unimodal dynamical system behaves as if it were stochastic. The tent map is a well known example of an expansive and unimodal system.

Our result implies that if parameter values are suitably chosen, the solutions to the LP problem (7.3) can be described by an optimal dynamical system that is expansive and unimodal. In order to explain this result, in Fig. 7.1 the kinked line OPR illustrates the maximum k_t that satisfies conditions (i) and (ii) of problem (7.3) as well as $(c_t, k_t) \geq 0$, given k_{t-1}. In order to obtain this kinked line, set $c_t = 0$ in condition (i) and (ii), and see that line OPR is

$$k_t = \min\{\mu k_{t-1}, \mu/(1+\gamma)\}, \tag{7.6}$$

where $\mu = 1/a_{22}$ and $\gamma = a_{12}/a_{22} - 1$. Denote by $\mathfrak{D}$ the region in the non-negative quadrant below and on the kinked line OPR. Then, if and only if $(k_{t-1}, k_t) \in \mathfrak{D}$, there is $c_t \geq 0$ such that conditions (i) and (ii) of problem (7.3) are satisfied.

If conditions (i) and (ii) of problem (7.3) are both satisfied with equality, the value of k_t (as well as that of c_t) is uniquely determined by k_{t-1}. By setting $a_{11}/a_{21} = 1$, this relationship between k_{t-1} and k_t is given by

$$k_t = -(\mu/\gamma)(k_{t-1} - 1). \tag{7.7}$$

If, in particular, $\gamma = a_{12}/a_{22} - 1 > 0$, the graph of (7.7) is a negatively sloping line through point P. In Fig. 7.1, this line is illustrated by line PQ.

The candidate for our chaotic optimal dynamical system is the function the graph of which coincides with the kinked line OPQ; i.e.

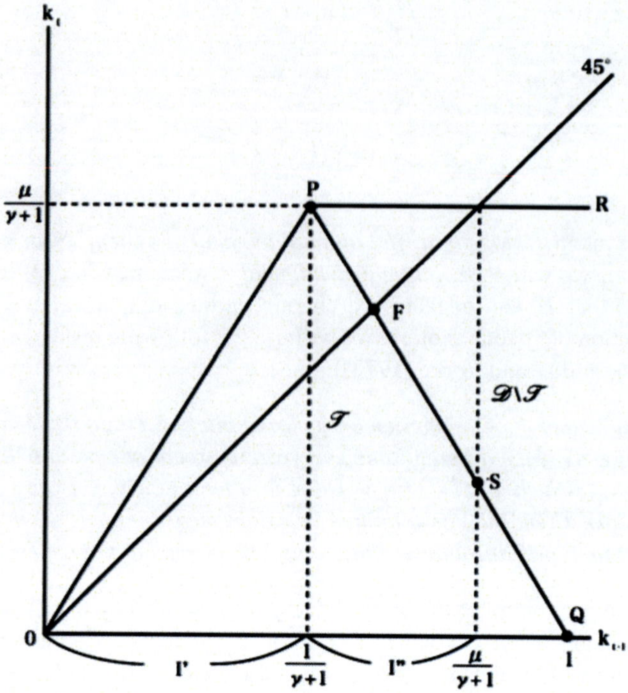

Fig. 7.1

$$h(k_{t-1}) = \begin{cases} \mu k_{t-1} & \text{if} \quad 0 \le k_{t-1} \le 1/(\gamma+1) \\ -(\mu/\gamma)(k_{t-1}-1) & \text{if} \quad 1/(\gamma+1) \le k_{t-1} \le 1. \end{cases} \tag{7.8}$$

Using the assumption $\mu/(1+\gamma) \le 1$, function h maps the unit interval [0, 1] into itself. For all the practical purposes, we may restrict h to the closed interval $I = [0, \mu/(1+\gamma)]$ and treat it as a function on I onto itself. Our main result is as follows:

Theorem 1. *Let $a_{11}/a_{21} = 1$, $\mu = 1/a_{22}$ and $\gamma = a_{12}/a_{22} - 1$. Moreover, let h_1 be the function (7.8) restricted to interval $I = [0, \mu/(1+\gamma)]$. Suppose that parameters μ, ρ and γ satisfy*

$$0 < \rho < 1, \quad \rho\mu > 1 \quad and \quad \mu \le \gamma + 1. \tag{7.9}$$

Then, on interval I, the generalized optimal dynamical system $H(k)$ coincides with function h_I if one of the following two conditions are satisfied.

Condition A

$$\mu \le \gamma;$$

Condition B

$$\gamma < \mu \leq \min\left\{\frac{\gamma + \surd(\gamma^2 + 4\gamma)}{2}, \frac{-1 + \surd(1 + 4\gamma)}{2\rho}\right\}.$$

Before proving the theorem, it is useful to explain the meaning of Conditions A and B. At the outset, note that γ satisfies

$$\gamma > 0 \tag{7.10}$$

in the setting of the theorem. As is noted above, therefore, line PQ is negatively sloped. Moreover, condition (7.9) implies that the slope of segment OP is strictly larger that 1.

Under Condition A, the slope of line PQ is larger than or equal to -1. If, therefore, Condition A is satisfied with strict inequality, the optimal dynamical system, OPS, is globally stable. As Fig. 7.2 illustrates, the optimal solution from any initial condition $k > 0$ converges to the fixed point F at the intersection between segment PS and the 45° line. If, instead, Condition A is satisfied with equality, any optimal solution from $k > 0$ converges to period-2 cycles that appear on segment PS (see Fig. 7.3) except for that going into fixed point F.

Under Condition B, the slope of line PQ is smaller than -1. In this case, h is expansive and unimodal. By Proposition 2, therefore, the optimal dynamical system

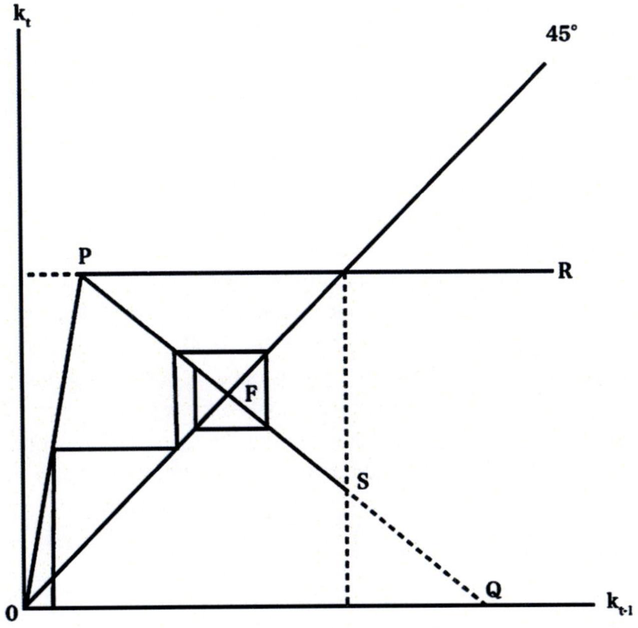

Fig. 7.2

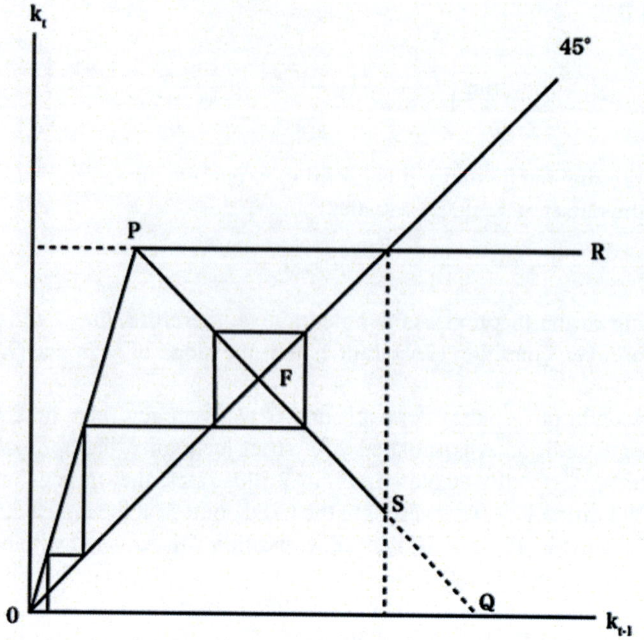

Fig. 7.3

is chaotic. Figure 7.4 illustrates a typical optimal path in this case; in Fig. 7.4, the parameters are set to $\mu = 3.5$, $\gamma = 3$, $\rho = 0.35$ and $k(= k_0) = 0.2$.

The upperbound μ in Condition B is determined by the interaction between the optimality of h and the emergence of cyclical paths from h. Inequality

$$\mu \leq \frac{\gamma + \surd(\gamma^2 + 4\gamma)}{2} \tag{7.11}$$

results from the condition under which dynamical system h generates cyclical paths of period 3. That is to say, if (7.11) is satisfied with strict inequality, given segment OP, the slope of line PQ is not steep enough to generate a cyclical path of period 3. If, in contrast, (7.11) is satisfied with equality, as Fig. 7.5 illustrates, the slope of line PQ becomes steep enough just to generate a cyclical path of period 3 (see Li and Yorke (1975) for the chaos based on the existence of period-3 cycles). If (7.11) is violated, the slope of line PQ becomes even steeper.

As the proof of the theorem indicates, the smaller (steeper) the slope of line PQ, the more stringent condition we need to impose on the relationship between μ, γ and ρ. In the case in which condition (7.11) is satisfied, inequality

$$\mu \leq \frac{-1 + \surd(1 + 4\gamma)}{2\rho} \tag{7.12}$$

can guarantee that function h is in fact the optimal dynamical system.

Our main result may be summarized as follows.

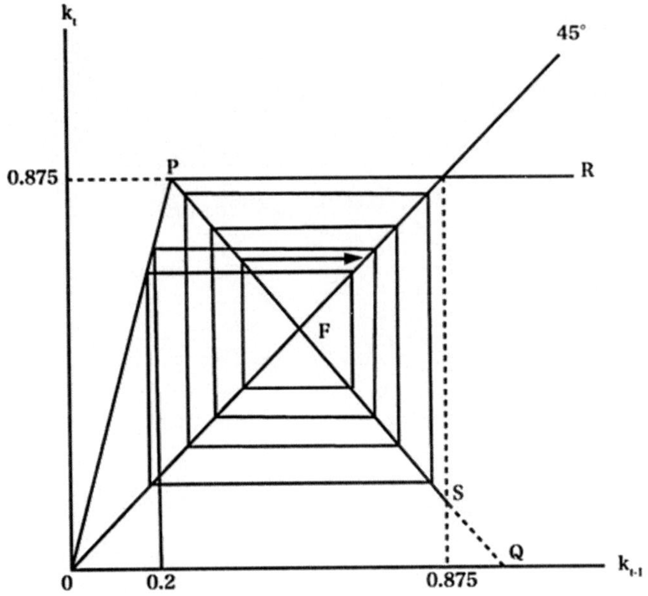

Fig. 7.4

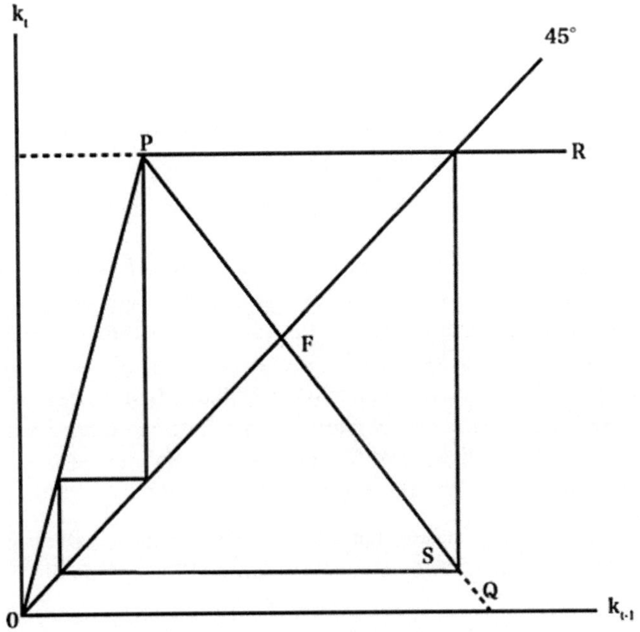

Fig. 7.5

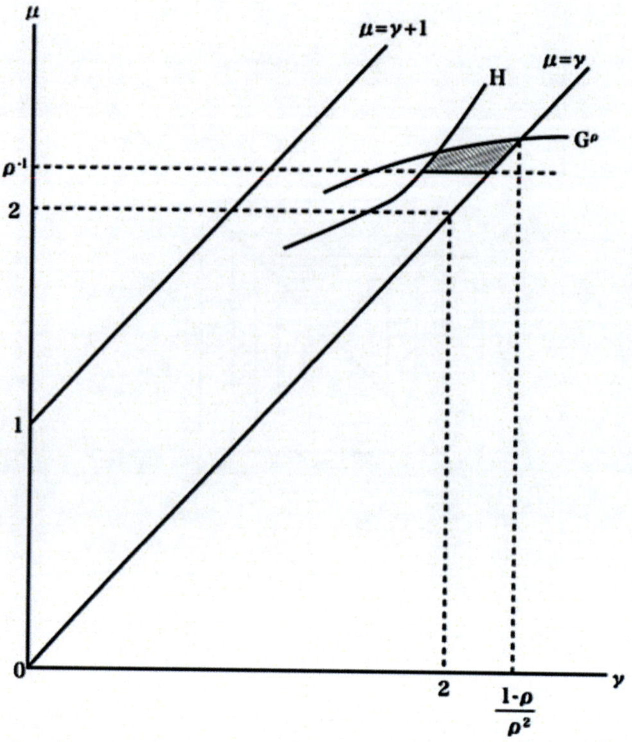

Fig. 7.6

Proposition 3. *Parameter values can be chosen in such a way that the solutions to the LP problem (7.3) follow a chaotic dynamical system.*

Proof. Given Theorem 1, it suffices to demonstrate that the set of (ρ, μ, γ) satisfying conditions B and (7.9) is non-empty. For this purpose, in Fig. 7.6, curve H illustrates $\mu = (\gamma + \surd(\gamma^2 + 4\gamma))/2$, which lies between lines $\mu = \gamma$ and $\mu = \gamma + 1$. Thus, conditions (7.9), (7.11) and $\mu > \gamma$ are satisfied at the same time in the region between curve H and the 45° line and above $\mu = 1/\rho$. Call this region Γ. Moreover, denote by the G^ρ curve, $\mu = (-1 + \surd(1 + 4\gamma))/(2\rho)$. Then, given ρ, Condition B is non-empty under (7.9) if and only if the region below curve G^ρ and region Γ have a non-empty intersection. This non-emptiness is guaranteed if and only if $\rho < 1/2$, since curve G^ρ intersects the 45° line at $\mu = (1 - \rho)/\rho^2$, and since this intersection lies above $\mu > 1/\rho$ if and only if $\rho < 1/2$.

The above proof demonstrates that it is possible to choose values of parameters μ and γ in such a way that the dynamical system OPS is optimal if

$$\rho < 1/2. \tag{7.13}$$

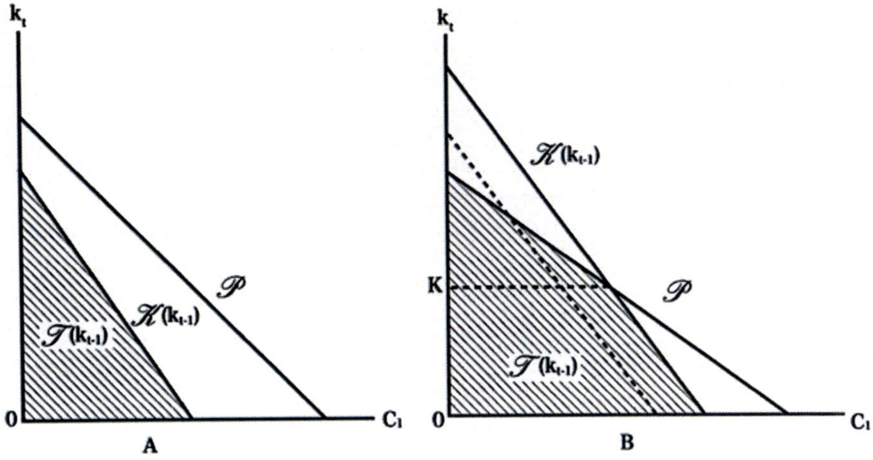

Fig. 7.7

7.3 Proof of the Theorem

For the sake of proof, it is convenient to illustrate the constraints of problem (7.3) diagrammatically. In Fig. 7.7, lines $\mathfrak{L}$ and $\mathfrak{K}(k_{t-1})$, respectively, correspond to the boundaries of conditions (i) and (ii) of problem (7.3). Moreover, denote by $\mathfrak{F}(k_{t-1})$ the region in which conditions (i) and (ii) as well as $(c_t, k_t) \geq 0$ are satisfied.

The definition of function c may be written as follows.

$$c(k_{t-1}, k_t) = \max c_t \ s.t. \ (c_t, k_t) \in \mathfrak{F}(k_{t-1}). \tag{7.14}$$

By using Fig. 7.7, it is easy to prove that c is

$$c(k_{t-1}, k_t) = \begin{cases} \frac{1}{\gamma+1}(\mu k_{t-1} - k_t) & \text{if } \ k_t \leq -\frac{\mu}{\gamma}(k_{t-1} - 1) \\ \frac{\mu}{\gamma+1} - k_t & \text{if } \ k_t \geq -\frac{\mu}{\gamma}(k_{t-1} - 1). \end{cases} \tag{7.15}$$

This is because the k_t corresponding to the intersection of lines $\mathfrak{L}$ and $\mathfrak{K}(k_{t-1})$ is given by $k_t = -(\mu/\gamma)(k_{t-1} - 1)$. If $k_t \leq -(\mu/\gamma)(k_{t-1} - 1)$, the maximum c_t is at (c_t, k_t) on line $\mathfrak{K}(k_{t-1})$; (c_t, k_t) does not lie on line $\mathfrak{L}$ if the inequality holds strictly. If $k_t \geq -(\mu/\gamma)(k_{t-1} - 1)$, the maximum c_t is at (c_t, k_t) on line $\mathfrak{L}$; (c_t, k_t) does not lie on line $\mathfrak{K}(k_{t-1})$ if the inequality holds strictly. Expression (7.15) follows from these facts.

It follows from a comparison between expressions (7.15) and (7.8) that the intersection of lines $\mathfrak{L}$ and $\mathfrak{K}(k_{t-1})$ corresponds to line PQ in Fig. 7.1. Define intervals I' and I'' as follows.

$$I' = [0, 1/(\gamma + 1)]; \tag{7.16}$$

$$I'' = [1/(\gamma + 1), \mu(\gamma + 1)]. \tag{7.17}$$

Recall that $1/(\gamma+1)$ is the k_{t-1}-coordinate of point P in Fig. 7.1. Thus, if $k_{t-1} \in I'$, $(k_{t-1}, k_t) \in \mathfrak{D}$ cannot lie on line PQ in Fig. 7.1. This case corresponds to Fig. 7.7a. The case of $k_{t-1} \in I''$ corresponds to Fig. 7.7b.

It is useful to note the basic facts concerning Figs. 7.1 and 7.7. For this purpose, in Fig. 7.1, let $\mathfrak{I}$ be the region surrounded by segments OP, PQ and the horizontal axis (including the boundary). Moreover, let $\mathfrak{I}^\circ$ be the interior of $\mathfrak{I}$ relative to the non-negative orthant of the k_{t-1}-k_t space. First, note the following fact, which is obvious.

Fact 1. *Vector* $(c_1, k_1, c_2, k_2, \ldots) \geq 0$ *satisfies conditions (i) and (ii) of problem (7.3) if and only if* $(c_t, k_t) \in \mathfrak{F}(k_{t-1})$ *for* $t = 1, 2, \ldots$ *and* $k_0 = k$.

The next fact follows immediately from (7.15).

Fact 2. *Let* $c_t = c(k_{t-1}, k_t)$ *and* $c_t + \Delta c_t = c(k_{t-1}, k_t + \Delta k_t)$.

(i) $\Delta k_t < 0$ *if and only if* $\Delta c_t > 0$
(ii) If $c(k_{t-1}, k_t) = 0$, (k_{t-1}, k_t) *lies on the kinked line* OPR.

Let $[O, P)$ be the half open interval between points O and P in Fig. 7.1 that includes point O and excludes point P. Also, let $[PQ]$ be the closed interval between points P and Q. The next fact follows from the explanation of (7.15) above.

Fact 3. *Let* $c_t = c(k_{t-1}, k_t)$.

(i) If $(k_{t-1}, k_t) \in [O, P) \cup \mathfrak{I}^\circ$, *then* $(c_t, k_t) \in \mathfrak{K}(k_{t-1})$ *and* $(c_t, k_t) \notin \mathfrak{L}$.
(ii) If $(k_{t-1}, k_t) \in \mathfrak{D} \backslash \mathfrak{I}$, *then* $(c_t, k_t) \in \mathfrak{L}$ *and* $(c_t, k_t) \notin \mathfrak{K}(k_{t-1})$.
(iii) If $(k_{t-1}, k_t) \in [PQ]$, *then* $(c_t, k_t) \in \mathfrak{K}(k_{t-1})$ *and* $(c_t, k_t) \in \mathfrak{L}$.

Denote by $(c_1^*, k_1^*, c_2^*, k_2^*, \ldots)$ a solution to problem (7.3). Moreover, denote by $\partial\mathfrak{F}(k_{t-1})$ the boundary of $\mathfrak{F}(k_{t-1})$ relative to the non-negative orthant of the c_t-k_t space. It follows from Proposition 1 that

$$c_t^* = c(k_{t-1}^*, k_t^*); \tag{7.18}$$

or in other words,

$$(c_t^*, k_t^*) \in \partial\mathfrak{F}(k_{t-1}^*). \tag{7.19}$$

Lemma 1. *For* $t = 2, 3, \ldots, (k_{t-1}^*, k_t^*) \notin \mathfrak{D} \backslash \mathfrak{I}$.

Proof. Suppose $(k_{t-1}^*, k_t^*) \in \mathfrak{D} \backslash \mathfrak{I}$. Then, by Fact 3, $(c_t^*, k_t^*) \in \mathfrak{L}$ but $(c_t^*, k_t^*) \notin \mathfrak{K}(k_{t-1}^*)$. Therefore, by $(c_t^*, k_t^*) \in \mathfrak{F}(k_{t-1}^*)$, there is a k_{t-1} such that $(c_t^*, k_t^*) \in \mathfrak{K}(k_{t-1})$ and $k_{t-1} < k_{t-1}^*$. Thus, $(c_{t-1}, k_{t-1}) \in \mathfrak{F}(k_{t-2}^*)$ for $c_{t-1} = c(k_{t-2}^*, k_{t-1})$. Since $c_{t-1}^* = c(k_{t-2}^*, k_{t-1}^*)$ and $c_{t-1} = c(k_{t-2}^*, k_{t-1})$, by Fact 2, $k_{t-1} < k_{t-1}^*$

implies $c_{t-1} > c^*_{t-1}$. Thus, $(c^*_1, k^*_1, \ldots, c_{t-1}, k_{t-1}, c^*_t, \ldots)$ satisfies the constraint of problem (7.3) and gives a value of the objective function larger than $\sum_{t=1}^{\infty} \rho^{t-1} c^*_t$, a contradiction.

In Fig. 7.1, define $[OP]$ as the closed segment between points O and P. Recall that $(c^*_1, k^*_1, c^*_2, k^*_2, \ldots)$ is a solution to problem (7.3), the initial condition of which is $k_0 = k; i.e. k^*_1 \in H(k)$.

Corollary 1. *Let* $k^*_1 \in H(k)$. *If* $(k, k^*_1) \in [OP]$ *for any* $k \in I'$, *then* $(k, k^*_1) \notin \mathfrak{D} \backslash \mathfrak{I}$ *for any* $k \in I''$.

Proof. Let $k \in I''$. Then, there is $k_0 \in I'$ such that $(k_0, k) \in [OP]$. Let $c_0 = c(k_0, k)$. Then, by the hypothesis of the corollary and Proposition 1, $(c_0, k, c^*_1, k^*_1, c^*_2, k^*_2, \ldots)$ is a solution to problem (7.3) when $k = k_0$. Thus, by Lemma 1, $(k, k^*_1) \notin \mathfrak{D}/\mathfrak{I}$.

In order to prove the theorem, it suffices to demonstrate $(k, k^*_1) \notin \mathfrak{I}^\circ$. This is because if $(k, k^*_1) \notin \mathfrak{I}^\circ$, by Proposition 1, the graph of $H(k)$ must coincide with segment OP on interval I'. Thus, by Lemma 1, the graph cannot lie above segment PQ on interval I''. Because $(k, k^*_1) \notin \mathfrak{I}^\circ$ implies that, on interval I'', the graph cannot lie below segment PQ either, it must lie on segment PQ on interval I''. Thus, the graph of H lies on the kinked line OPQ, which implies Theorem 1.

In what follows, we will first establish several lemmas under the hypothesis that $(k, k^*_1) \notin \mathfrak{I}^\circ$. At the end, by using those lemmas, we derive a contradiction, establishing $(k, k^*_1) \notin \mathfrak{I}^\circ$. Recall that, in Fig. 7.1, $\mathfrak{L}$ is the line determined by the boundary of condition (i) of problem (7.3).

Lemma 2. *Let* $k^*_1 \in H(k)$ *and* $(k, k^*_1) \in \mathfrak{I}^\circ$. *If* $\rho\mu > 1$, $(c^*_2, k^*_2) \in \mathfrak{L}$.

Proof. Since $(k, k^*_1) \in \mathfrak{I}^\circ$, by Fact 3, $(c^*_1, k^*_1) \notin \mathfrak{L}$ and $(c^*_1, k^*_1) \in \mathfrak{K}(k)$. Let $\Delta c_t < 0$, and define

$$\Delta k_t = -a_{21} \Delta c_1 / a_{22}. \tag{7.20}$$

Let $c_1 = c^*_1 + \Delta c_1$ and $k_1 = k^*_1 + \Delta k_1$. Then, by condition (ii) of problem (7.3), it follows from $(c^*_1, k^*_1) \in \mathfrak{K}(k)$ that $(c_1, k_1) \in \mathfrak{K}(k)$. Since it follows from $(c^*_1, k^*_1) \notin \mathfrak{L}$ and $(c^*_1, k^*_1) \in \mathfrak{F}(k)$ that $c^*_1 > 0$, we may choose Δc_1 sufficiently close to 0 so that

$$(c_1, k_1) \in \mathfrak{F}(k). \tag{7.21}$$

Let $\Delta c_2 = \Delta k_1 / a_{21}$ and $c_2 = c^*_2 + \Delta c_2$.

Suppose $(c^*_2, k^*_2) \notin \mathfrak{L}$. Since $\Delta k_1 > 0$ and $(c^*_2, k^*_2) \in \mathfrak{F}(k^*_1)$, $(c_2, k^*_2) \in \mathfrak{F}(k^*_1 + \Delta k_1)$. This together with (7.21) implies that $(c_1, k_1, c_2, k^*_2, c^*_3, \ldots)$ satisfies the constraints of problem (7.3) and gives rise to the value of the objective function $\sum_{t=1}^{\infty} \rho^{t-1} c^*_t + \rho \Delta c_2 + \Delta c_1$. Since $\Delta c_2 = \Delta k_1 / a_{21} = -\Delta c_1 / a_{22} = -\mu \Delta c_1$, $\rho \Delta c_2 + \Delta c_1 = -(\rho\mu - 1) \Delta c_1 > 0$ by (7.9) and $\Delta c_1 < 0$. This contradicts the fact that $(c^*_1, k^*_1, \ldots)$ is a solution to problem (7.3).

Lemma 3. *Let $k_1^* \in H(k)$, $(k, k_1^*) \in \mathfrak{I}^\circ$ and $k_2^* > 0$. If $\rho\mu > 1$ and $\rho\mu < \gamma$, $c_3^* = 0$.*

Proof. Keep c_1, Δc_1, k_1 and Δk_1 in the proof of Lemma 2. However, change the definition of Δc_2 as follows and that of $c_2 = c_2^* + \Delta c_2$ accordingly. Let $k_1' \geqslant 0$ be such that $(c_2^*, k_2^*) \in \mathfrak{K}(k_1')$. Since $(c_2^*, k_2^*) \in \mathfrak{F}(k_1^*)$, $k_1' \leqslant k_1^*$. Define $(\Delta c_2, \Delta k_2)$ as follows.

$$(\Delta c_2, \Delta k_2) = \left(\frac{a_{12}/a_{21}}{a_{12} - a_{22}}, -\frac{1}{a_{12} - a_{22}} \right) \Delta k_1. \tag{7.22}$$

Let $k_2 = k_2^* + \Delta k_2$.

Since $(k, k_1^*) \in \mathfrak{I}^\circ$ and $k_1^* \in H(k)$, by Lemma 2, it holds that $(c_2^*, k_2^*) \in \mathfrak{L}$. Thus, $(c_2^*, k_2^*) \in \mathfrak{L} \cap \mathfrak{K}(k_1')$. Thus, by the definition of c_2, Δc_2, k_2, and Δk_2, it follows from conditions (i) and (ii) of problem (7.3) that $(c_2, k_2) \in \mathfrak{L} \cap \mathfrak{K}(k_1' + \Delta k_1)$. Under the setting of Theorem 1, by (7.20), (7.22) implies

$$(\Delta c_2, \Delta k_2) = \left(-\frac{(\gamma + 1)\mu}{\gamma}, a_{21}\frac{\mu^2}{\gamma} \right) \Delta c_1. \tag{7.23}$$

Since $\gamma > 0$ by (7.10), and since $\Delta c_1 < 0$, $\Delta k_2 < 0$ and $\Delta c_2 > 0$. Since $k_2^* > 0$ by hypothesis, by choosing Δc_1 sufficiently close to 0, we have $(c_2^* + \Delta c_2, k_2 + \Delta k_2) \in \mathfrak{F}(k_1' + \Delta k_1)$. This implies, by $k_1' \leqslant k_1^*$ and $k_1' + \Delta k_1 < k_1^* + \Delta k_1 = k_1$,

$$(c_2, k_2) \in \mathfrak{F}(k_1). \tag{7.24}$$

Define $\Delta c_3 = \Delta k_2 / a_{21}$. Then, by (7.23),

$$\Delta c_3 = \frac{\mu^2}{\gamma} \Delta c_1. \tag{7.25}$$

Since $\Delta c_1 < 0$, $\Delta c_3 < 0$. Let $c_3 = c_3^* + \Delta c_3$.

Now, suppose $c_3^* > 0$. Then, there is k_2', $0 < k_2' \leqslant k_2^*$, such that $(c_3^*, k_3^*) \in \mathfrak{K}(k_2')$. Since $(c_3^*, k_3^*) \in \mathfrak{F}(k_2')$ and $\Delta k_2 < 0$, by $c_3^* > 0$ and $k_2' > 0$, it is possible to choose $\Delta c_1 < 0$ sufficiently close to 0 so that $(c_3^* + \Delta c_3, k_3^*) = (c_3, k_3^*) \in \mathfrak{F}(k_2' + \Delta k_2)$. Since $k_2' \leqslant k_2^*$ implies $(c_3^* + \Delta c_3, k^*) = (c_3, k_3^*) \in \mathfrak{F}(k_2^* + \Delta k_2) = \mathfrak{F}(k_2)$, therefore, $(c_1, k_1, c_2, k_2, c_3, k_3^*, c_4^*, \ldots)$ satisfies the constraints of problem (7.3) and gives rise to the value of the objective function $\sum_{t=1}^{\infty} \rho^{t-1} c_t^* + \sum_{t=1}^{3} \rho^{t-1} \Delta c_t$. By (7.23) and (7.25), it holds that

$$\begin{aligned} \sum_{t=1}^{3} \rho^{t-1} \Delta c_t &= \frac{1}{\gamma} [\rho^2 \mu^2 - (\gamma + 1)\rho\mu + \gamma] \Delta c_1 \\ &= \frac{1}{\gamma} (\rho\mu - 1)(\rho\mu - \gamma) \Delta c_t > 0, \end{aligned} \tag{7.26}$$

since $\Delta c_1 < 0$ by definition, and since $\rho\mu - 1 > 0$ and $\rho\mu - \gamma < 0$ by the hypothesis of the lemma. This contradicts the fact that $(c_1^*, k_1^*, \ldots)$ is a solution to problem (7.3).

Lemma 4. *Let* $k_1^* \in H(k)$. *Suppose that* $(k, k_1^*) \in \mathfrak{I}^\circ$, $k_2^* > 0$ *and* $k_3^* > 0$. *If* $1 < \rho\mu < (-1 + \sqrt{(1 + 4\gamma)})/2$, *then* $c_3^* = c_4^* = 0$.

Proof. Keep c_1, Δc_1, k_1, Δk_1, c_2, Δc_2, k_2 and Δk_2 in the proof of Lemma 3. However, we change the definition of Δc_3 to $\Delta c_3 = 0$ and that of $c_3 = c_3^* + \Delta c_3$ accordingly. Recall $\Delta k_2 = a_{21}\mu^2 \Delta c_1/\gamma < 0$. Let $\Delta k_3 = \Delta k_2/a_{22}$ and $k_3 = k_3^* + \Delta k_3$. Note $\Delta k_3 < 0$. Since $(-1 + \sqrt{(1 + 4\gamma)})/2 < \gamma$, the hypothesis of Lemma 3 is satisfied. It holds that $c_3^* = 0$. From these facts, it is possible to choose Δc_1 sufficiently close to 0 so that

$$(c_3, k_3) \in \mathfrak{F}(k_2). \tag{7.27}$$

Define $\Delta c_4 = \Delta k_3/a_{21}$. Then, by $\Delta k_3 = \Delta k_2/a_{22}$ and (7.23),

$$\Delta c_4 = \frac{\mu^3}{\gamma}\Delta c_1. \tag{7.28}$$

Since $\Delta c_1 < 0$, $\Delta c_4 < 0$. Let $c_4 = c_4^* + \Delta c_4$.

Now, suppose $c_4^* > 0$. Then, there is k_3', $0 < k_3' \leqslant k_3^*$, such that $(c_4^*, k_4^*) \in \mathfrak{K}(k_3')$. Since $(c_4^*, k_4^*) \in \mathfrak{F}(k_3')$, by $c_4^* > 0$ and $k_3' > 0$, it is possible to choose $\Delta c_1 < 0$ sufficiently close to 0 so that $(c_4^* + \Delta c_4, k_4^*) \in \mathfrak{F}(k_3' + \Delta k_3)$. Since $k_3' \leqslant k_3^*$ implies $(c_4^* + \Delta c_4, k_4^*) = (c_4, k_4^*) \in \mathfrak{F}(k_3^* + \Delta k_3) = \mathfrak{F}(k_3)$, therefore, $(c_1, k_1, c_2, k_2, c_3, k_3, c_4, k_4^*, c_5^*, \ldots)$ satisfies the constraints of problem (7.3) and gives rise to the value of the objective function $\sum_{t=1}^{\infty} \rho^{t-1} c_t^* + \sum_{t=1}^{4} \rho^{t-1} \Delta c_t$. By $\Delta c_3 = 0$, (7.23) and (7.28), it holds that

$$\begin{aligned}\sum_{t=1}^{4} \rho^{t-1}\Delta c_t &= \frac{1}{\gamma}[\rho^3\mu^3 - (\gamma + 1)\rho\mu + \gamma]\Delta c_1 \\ &= \frac{1}{\gamma}(\rho\mu - 1)(\rho^2\mu^2 + \rho\mu - \gamma)\Delta c_1 > 0,\end{aligned} \tag{7.29}$$

which follows from the hypothesis of the lemma and $\Delta c_1 < 0$. This contradicts the fact that $(c_1^*, k_1^*, \ldots)$ is a solution to problem (7.3).

In order to complete the proof, as is noted above, it suffices to prove $(k, k_1^*) \notin \mathfrak{I}^\circ$ for $k_1^* \in H(k)$. Suppose $(k, k_1^*) \in \mathfrak{I}^\circ$. Then, by Lemma 2, $(c_2^*, k_2^*) \in \mathfrak{L} \cap \mathfrak{F}(k_1^*)$. This holds, by Fact 3, only if $k_1^* \in I''$ and $(k_1^*, k_2^*) \in \mathfrak{D}\backslash\mathfrak{I}^\circ$. Moreover, as Fig. 7.1 indicates, $(k, k_1^*) \in \mathfrak{I}^\circ$ implies

$$k_1^* < \mu/(\gamma + 1). \tag{7.30}$$

Thus, by Lemma 1, $(k_1^*, k_2^*) \in \mathfrak{D} \backslash \mathfrak{I}^\circ$ implies

$$(k_1^*, k_2^*) \in [PS), \tag{7.31}$$

where $[PS)$ is the segment between points P and S including P but not S.

We will first derive a contradiction under Condition A of the theorem. Under Condition A, the slope of line PS is larger than or equal to -1. Since this implies that the second coordinate of point S is larger than $1/(\gamma + 1)$, (7.31) implies $k_2^* > 1/(\gamma + 1)$. Since this guarantees that the hypothesis of Lemma 3 is satisfied, it holds that $c_3^* = 0$. Since this implies $c(k_2^*, k_3^*) = c_3^* = 0$, Fact 2 together with $k_2^* > 1/(\gamma + 1)$ implies that (k_2^*, k_3^*) must lie on half line PR excluding point P. This contradicts Lemma 1.

In order to complete the proof, it suffices to derive a contradiction under Condition B of the theorem. Since $(k_1^*, k_2^*) \in [PS)$, $k_2^* > 0$. Therefore, under Condition B, the hypothesis of Lemma 3 is satisfied. Thus, it follows from Lemma 3 that $c_3^* = 0$. Since, by (7.18), this implies $c(k_2^*, k_3^*) = 0$, by Fact 2 (2), (k_2^*, k_3^*) must lie on the kinked line OPR. Thus, by Lemma 1, $(k_2^*, k_3^*) \in [OP]$.

As noted above, if condition (7.11) is satisfied with strict inequality, the slope of line PQ is not steep enough for the dynamical system OPS to generate period-3 cycles; if, (7.11) is satisfied with equality, the slope, for the first time, becomes steep enough to generate period-3 cycles. These facts together with $(k_1^*, k_2^*) \in [PS)$ and $(k_2^*, k_3^*) \in [OP]$ imply $k_2^* > 1/[\mu(\gamma + 1)]$ and $k_3^* > 1/(\gamma + 1)$, given (7.11). Thus, given Condition B, the hypothesis of Lemma 4 is satisfied. Thus, it holds that $c_4^* = 0$. Since, by (7.18), this implies $c(k_3^*, k_4^*) = 0$. This fact together with $k_3^* > 1/(\gamma + 1)$ implies that (k_3^*, k_4^*) is on half line PR excluding point P, a contradiction to Lemma 1. This completes the proof of Theorem 1.

References

Dorfman, R., Samuelson, P. and R. Solow (1958), Linear Programming and Economic Analysis, McGraw-Hill, New York.

Bellman, R. (1957), Dynamic Programming, Princeton University Press, Princeton.

Bellman, R. and R. Kalaba (1965), Dynamic Programming and Modern Control Theory, Academic Press, New York.

Lasota, A. and J. Yorke (1974), "On the Existence of Invariant Measures for Piecewise Monotonic Transformations," *Transactions of the American Mathematical Society*, 186, 481–488.

Li, T. Y. and J. Yorke (1975), "Period Three Implies Chaos," *American Mathematical Monthly*, 82, 985–992.

Li, T. Y. and J. Yorke (1978), "Ergodic Transformations from an Interval into Itself," *Transactions of the American Mathematical Society*, 235, 183–192.

Chapter 8
On the Least Upper Bound of Discount Factors that are Compatible with Optimal Period-Three Cycles*

Kazuo Nishimura and Makoto Yano**

8.1 Introduction

In this study, we derive, in the standard class of optimal growth models, the least upper bound of discount factors of future utilities for which a cyclical optimal path of period 3 may emerge.[1] On the one hand, Nishimura and Yano (1994) and Nishimura et al. (1994) construct examples in which a cyclical optimal path of period 3 emerges for discount factors around 0.36. On the other hand, Sorger (1992a,b, 1994), demonstrates that if such a path emerges in an optimal growth model of the standard class, the model's discount factor cannot exceed 0.5478. These results imply that the least upper bound of discount factors that can give rise to cyclical optimal paths of period 3 must lie between 0.36 and 0.5478.[2] We demonstrate that the least upper bound is $\widehat{\rho} = \left(3 - \sqrt{5}\right)/2$.

*Journal of Economic Theory 69, 306–333, 1996

**We are grateful to Mukul Majumdar, Tapan Mitra, and Gerhard Sorger for useful conversations and correspondence.

[1]The focus in the early literature on optimal growth is on the convergence of an optimal path to a stationary state. In that literature, it has been proved that an optimal path converges to a stationary state if the utility function is fixed and if the discounting of future utilities is sufficiently weak (see Brock and Scheinkman 1976; Cass and Shell 1976; McKenzie 1983; Scheinkman 1976). In this study, we are concerned with the existence of a utility function that gives rise to an optimal cyclical period-3 path for a given discount factor of future utilities.

[2]In relation to these results, Nishimura and Yano (1995) prove that for values of the discount factor ρ arbitrarily close to 1, ergodically chaotic optimal paths can emerge that are generated by a unimodal, expansive dynamical system. In that result, it is shown that part of the graph of the optimal

K. Nishimura (✉) · M. Yano
Institute of Economic Research, Kyoto University, Japan
e-mail: nishimura@kier.kyoto-u.ac.jp; yano@kier.kyoto-u.ac.jp

J. Stachurski et al. (eds.), *Nonlinear Dynamics in Equilibrium Models*,
DOI 10.1007/978-3-642-22397-6_8, © Elsevier Inc. 1996. Published by Springer-Verlag
Berlin Heidelberg 2012. All Rights Reserved

The importance of finding this least upper bound stems from the following two facts. (1) Given the long-run real interest rate *per annum,* the discount factor is positively correlated to the length of an individual period of the model[3]; (2) the existence of a cyclical path of period 3 is a fundamental criterion for the emergence of complex non-linear dynamics, in particular of cyclical paths of any other periodicity (Li and Yorke 1975; Sarkovskii 1964). With these considerations, in the existing literature, a number of studies have been concerned with the existence of a cyclical optimal path of period 3 and the corresponding discount factor.[4]

The rest of this study is organized as follows. In Sect. 8.2, we will introduce our main result. In Sect. 8.3, we will discuss the basic idea of a proof. In Sect. 8.4, we will provide preliminary results. In Sect. 8.5, we will prove the main result.

8.2 Model

Think of a standard optimal growth model $M = \{u, \rho\}$ in which path $\{k_t\}$ is chosen so as to maximize $\sum_{t=1}^{\infty} \rho^{t-1} u(k_{t-1}, k_t)$ subject to the initial condition $k_0 = k$. If $\{k_t\}$ solves this optimization problem, it is called an *optimal path* from initial stock k. Function $u(x, y)$, which is called a reduced-form utility function, captures the maximum utility level that can be achieved in an individual period when the economy has stock x at the beginning of that period and is to leave stock y at the end of the period. Discount factor ρ satisfies

$$0 < \rho < 1. \tag{8.1}$$

(The model is standard; for a more detailed explanation, for example, see McKenzie 1986.)

In this study, we take the case of a one-dimensional capital stock; i.e., the domain of function u is a subset of $R^2_+ = \{(x, y) \mid x \geq 0, y \geq 0\}$.

---

dynamical system lies on a von Neumann facet containing the stationary state and that any optimal path is confined in a small neighborhood of the facet. In this respect, the result is closely related to the neighborhood turnpike theorem of McKenzie (1983), which implies that any optimal paths converge into a neighborhood of the von Neumann facet. See also the result of Nishimura et al. (1994) which extends Nishimura and Yano (1995) for the case in which the von Neuman facet is trivial.

[3]In a standard optimal growth model, the discounted sum of utilities u_t, $\sum_{t=1}^{\infty} \rho^{t-1} u_t$, is maximized. It is known that, in such a model, the long-run real interest rate per period is $r = \rho^{-1} - 1$. If, therefore, the long-run real interest rate *per annum* is given by i, the length of an individual period of the model, ℓ, is given by

$$r = (1 + i)^{\ell} - 1$$

In other words, it holds that $\rho = 1/(1 + i)^{\ell}$.

[4]See Deneckere and Pelikan (1986), Boldrin and Montrucchio (1986), Boldrin and Deneckere (1990), Nishimura and Yano (1994) and Majumdar and Mitra (1994). These studies are concerned with deterministic economic fluctuations. For broader issues on deterministic and indeterministic fluctuations, see, for example, Shell (1977) and Cass and Shell (1976).

Definition 1. The single-capital optimal growth model is a combination of u and ρ, $M = \{u, \rho\}$, satisfying the following:

(*A-1*) $u : D \to R$ is a continuous and concave function defined on a closed and convex subset $D \subset R_+^2$.
(*A-2*) $u(k, h)$ is strictly increasing in k and strictly decreasing in h.
(*A-3*) $0 < \rho < 1$.

If, in particular, u is strictly concave, model M is called a strictly concave optimal growth model.

We are concerned with cyclical optimal paths of period 3. A path $\{z_t\}$ is a *cyclical path of period 3* if there are ζ_1, ζ_2 and ζ_3, $\zeta_1 < \zeta_2 < \zeta_3$, such that $z_{3\tau} = \zeta_1$, $z_{3\tau+1} = \zeta_2$ and $z_{3\tau+2} = \zeta_3$ for $\tau = 0, 1, \ldots$. If a dynamical system $z_{t+1} = \zeta(z_t)$, $t = 0, 1, 2, \ldots$, generates a cyclical path of period 3, it is often said to be *chaotic in the sense of Li and Yorke*.

In general, as is well-known, the optimal paths can be characterized by a set valued function that relates each stock level at a beginning of a period to the set of optimal stock levels at the end of the period. To state it more precisely, define $V(k)$ as follow.

$$V(k) = \max_{(k_1, k_2, \ldots)} \sum_{t=1}^{\infty} \rho^{t-1} u(k_{t-1}, k_t) \text{ s.t. } k_0 = k. \tag{8.2}$$

Given this definition, it holds that

$$V(k) = \max_{y} [v(k, y) + \rho V(y)] \tag{8.3}$$

Denote by $F(k)$ the set of y that solves the maximization problem on the right-hand side of (8.3). If and only if $\{k_t\}$ is an optimal path solving maximization problem (8.2), it holds that

$$k_t \in F(k_{t-1}) \tag{8.4}$$

for $t = 1, 2, \ldots$ with $k_0 = k$. Because $F(k)$ describes the behavior of optimal paths of problem (8.2) and because it is generally a set-valued function, we call system (8.4) a *generalized optimal dynamical system*. If, in particular, $F(k)$ is a singleton for any k (at which $V(k)$ is well defined), we write $F(k) = \{f(k)\}$ and call $k_t = f(k_{t-1})$ an *optimal dynamical system*.

Our main result is as follows:

Proposition 1. $\widehat{\rho} = (3 - \sqrt{5})/2$ *is the least upper bound of discount factors with which it is possible to construct an optimal growth model $M = \{u, \rho\}$ in which the optimal dynamical system is chaotic in the sense of Li and Yorke.*

8.3 The Basic Idea of the Proof

In the section, we will explain the basic idea of our proof. Think of activities $(x, y) \in D$, $(y, z) \in D$ and $(z, x) \in D$ such that

$$0 \leq x < y < z \tag{8.5}$$

and think of three numbers, a, b and c. In a three dimensional space, called the k-h-v space, take the three points A, B and C as follows:

$$A = (x, y, a), \quad B = (y, z, b) \quad \text{and} \quad C = (z, x, c). \tag{8.6}$$

By using these points, A, B and C, and their projections to the k-h, h-v, v-k planes, construct the polyhedral P, as illustrated in Fig. 8.1. Denote by $v\,(k, h)$ the function the graph of which coincides with polyhedral P. Given (8.1) and (8.5), $M' = \{v, \rho\}$ is an optimal growth model, i.e., satisfies conditions A-1, A-2 and A-3 of Definition 1, if and only if a, b and c satisfy

$$a < c \quad \text{and} \quad b < c. \tag{8.7}$$

For the moment, we will deal with model M', which we call a polyhedral (optimal growth) model. Denote by $G(k)$ the generalized optimal dynamical system of model $M' = \{v, \rho\}$; because the graph of v consists of flats (is not strictly concave), in general, $G(k)$ is not a dynamical system (cannot be described by a function from R_+ to R_+). We say that the generalized optimal dynamical system $G(k)$ of polyhedral model $M' = \{v, \rho\}$ generates an isolated period-3 path if there is (x, y, z) satisfying

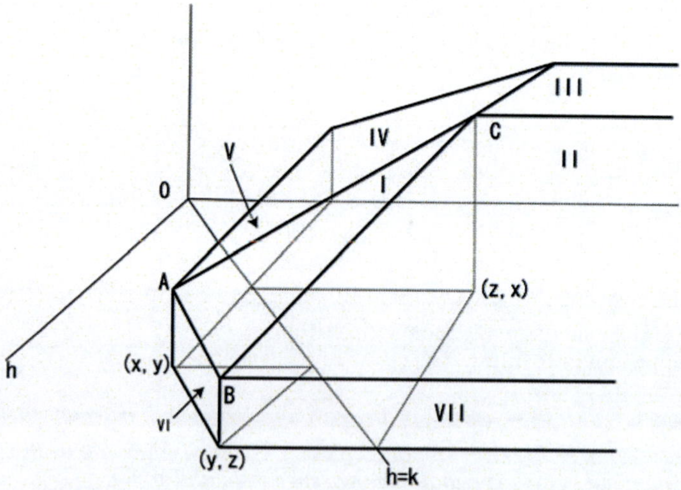

Fig. 8.1 Polyhedral P

$$G(x) = \{y\}, \quad G(y) = \{z\} \quad \text{and} \quad G(z) = \{x\}. \tag{8.8}$$

We will focus on polyhedral models M' satisfying the following condition.

Condition 1 *The generalized optimal dynamical system of polyhedral model M' generates an isolated period-3 path.*

We will first characterize polyhedral P by using x, y, z, a, b and c. Half space S_1 is the region below the plane that goes through points A, B and C (i.e., contains face I of the Fig. 8.1). That is,

$$S_1 = \left\{(k, h, v) \in R^3 \mid d_1 \geq v + H_1 h - K_1 k\right\}, \tag{8.9}$$

where d_1, K_1 and H_1 are positive and given by the following system:

$$\begin{cases} a + H_1 y - K_1 x = d_1 \\ b + H_1 z - K_1 y = d_1 \\ c + H_1 x - K_1 z = d_1. \end{cases} \tag{8.10}$$

This implies

$$\begin{aligned} K_1 &= ((a-b)\,x + (b-c)\,y + (c-a)\,z)\,/\Delta_1 \\ H_1 &= ((b-c)\,x + (c-a)\,y + (a-b)\,z)\,/\Delta_1 \\ \Delta_1 &= \frac{1}{2}(y-x)^2 + \frac{1}{2}(z-y)^2 + \frac{1}{2}(z-x)^2 > 0. \end{aligned} \tag{8.11}$$

Half space S_2 is the region below the plane that goes through points B and C and is perpendicular to the h-v plane (i.e., contains face II of the Fig. 8.1). That is,

$$S_2 = \left\{(k, h, v) \in R^3 \mid d_2 \geq v + H_2 h\right\}, \tag{8.12}$$

where d_2 and H_2 are positive and given by following system:

$$\begin{cases} b + H_2 z = d_2 \\ c + H_2 x = d_2. \end{cases} \tag{8.13}$$

This implies the following:

$$H_2 = (c-b)/\Delta_2, \quad \text{where } \Delta_2 = z - x > 0. \tag{8.14}$$

Half space S_3 is the region below the plane that goes through point C and is perpendicular to the h-v plane and the k-v plane (i.e., contains face III of the Fig. 8.1). That is,

$$S_3 = \left\{(k, h, v) \in R^3 \mid c \geq v\right\}. \tag{8.15}$$

Half space S_4 is the region below the plane that goes through points A and C and is perpendicular to the k-v plane (i.e., contains face IV of the Fig. 8.1). That is,

$$S_4 = \{(k, h, v) \in R^3 \mid d_4 \geq v - K_4 k\}, \tag{8.16}$$

where d_4 and K_4 are positive and given by the following system:

$$\begin{cases} a - K_4 x = d_4 \\ c - K_4 z = d_4 \end{cases}. \tag{8.17}$$

This implies

$$K_4 = (c - a)/\Delta_4, \quad \text{where } \Delta_4 = z - x \tag{8.18}$$

Half space S_5 is the region to the right of the plane that goes through point A and is perpendicular to the k-h plane and the k-v plane (i.e., contains face V of the Fig. 8.1). That is,

$$S_5 = \{(k, h, v) \in R^3 \mid -x \geq -k\}, \tag{8.19}$$

where d_5 and K_5 satisfy

$$-x = d_5 / K_5.$$

Half space S_6 is the region to the northeast of the plane that goes through points A and B and is perpendicular to the k-h plane (i.e., contains face VI of the Fig. 8.1). That is,

$$S_6 = \{(k, h, v) \in R^3 \mid d_6 \geq h - K_6 k\}, \tag{8.20}$$

where d_6, K_6 and H_6 are defined by the following system:

$$\begin{cases} y - K_6 x = d_6 \\ z - K_6 y = d_6 \end{cases}. \tag{8.21}$$

This implies $(y - z) - K_6(x - y) = 0$, and thus

$$K_6 = (z - y)/(y - x). \tag{8.22}$$

Half space S_7 is the region to the north of the plane that goes through point B and is perpendicular to the k-h plane and h-v plane (i.e., contains face VII of the Fig. 8.1). That is,

$$S_7 = \{(k, h, v) \in R^3 \mid z \geq h\}. \tag{8.23}$$

Polyhedral P coincides with the intersection of the above half spaces, $S_{1,\dots,}S_7$ in the non-negative orthant of the k-h-v space. It is said that the price vector $(p, q, 1)$ *supports* polyhedral P (or utility function $v(k, h)$) at $(\widehat{k}, \widehat{h}, v(\widehat{k}, \widehat{h}))$ if for all $(k, h, v) \in P$,

$$v(\widehat{k}, \widehat{h}) + q\widehat{h} - p\widehat{k} \geq v(k, h) + qh - pk. \tag{8.24}$$

Moreover the price vector $(p, q, 1)$ *strongly supports* polyhedral P if (8.24) holds with strict inequality at any point $(k, h, v) \in P$ except $(\widehat{k}, \widehat{h}, v(\widehat{k}, \widehat{h}))$.

Recall points A, B and C are defined by (8.6). We may prove the following lemmas (Lemmas 1–3 are more or less obvious; for a proof of Lemma 4, see McKenzie 1986).

Lemma 1. *Price vector $(p_1, q_{1,} 1)$ strongly supports polyhedral P at point A if and only if there are $t_1 > 0$, $t_4 > 0$, $t_5 > 0$ and $t_6 > 0$ such that $t_1 + t_4 = 1$ and*

$$(p_1, q_1, 1) = t_1(K_1, H_1, 1) + t_4(K_4, 0, 1) + t_5(1, 0, 0) + t_6(K_6, 1, 0). \tag{8.25}$$

Lemma 2. *Price vector $(p_2, q_{2,} 1)$ strongly supports polyhedral P at point B if there are $n_1 > 0$, $n_2 > 0$, $n_6 > 0$ and $n_7 > 0$ such that $n_1 + n_2 = 1$ and*

$$(p_2, q_2, 1) = n_1(K_1, H_1, 1) + n_2(0, H_2, 1) + n_6(K_6, 1, 0) + n_7(0, 1, 0). \tag{8.26}$$

Lemma 3. *Price vector $(p_3, q_{3,} 1)$ strongly supports polyhedral P at point C if and only if there are $m_1 > 0$, $m_2 > 0$, $m_3 > 0$ and $m_4 > 0$ such that $m_1 + m_2 + m_3 + m_4 = 1$ and*

$$(p_3, q_3, 1) = m_1(K_1, H_1, 1) + m_2(0, H_2, 1) + m_3(0, 0, 1) + m_4(K_4, 0, 1). \tag{8.27}$$

Lemma 4. *Condition 1 holds if and only if the following condition is satisfied.*

C.o. *$(p_i, q_i, 1)$, $i = 1, 2, 3$, such that the following conditions are satisfied.*

(i) *$(p_1, q_1, 1)$, $(p_2, q_2, 1)$ and $(p_3, q_3, 1)$ strongly support polyhedral P at points A, B and C, respectively.*

(ii) *$-q_1 + \rho p_2 = 0$, $-q_2 + \rho p_3 = 0$ and $-q_3 + \rho p_1 = 0$.*

This lemma implies that a cyclical optimal path $\{k_t\}$ of period 3 is supported by a cyclical price path $\{p_t\}$ of period 3; that is, there are p_1, p_2 and p_3 such that for any $(k, h) \in D$,

$$v(x, y) + p_2 y - \rho^{-1} p_1 x \geq v(k, h) + p_2 y - \rho^{-1} p_1 k, \tag{8.28}$$

$$v(y, z) + p_3 z - \rho^{-1} p_2 y \geq v(k, h) + p_3 h - \rho^{-1} p_2 k, \tag{8.29}$$

$$v(z, x) + p_1 x - \rho^{-1} p_3 z \geq v(k, h) + p_1 h - \rho^{-1} p_3 k. \tag{8.30}$$

The next Theorem follows from Lemmas 1 through 4.

Theorem 1. *The generalized optimal dynamical system of polyhedral model M' generates an isolated period-3 path (i.e., Condition 1 is satisfied) if and only if the following condition holds:*
C.i. There is $\pi = (m_1, m_2, m_3, m_4, t_1, t_4, t_5, t_6, n_1, n_2, n_6, n_7) \gg 0$ that satisfies the following system[5]*:*

$$\begin{cases} -(H_1 t_1 + t_6) + \rho(K_1 n_1 + K_6 n_6) = 0; \\ -(H_1 n_1 + H_2 n_2 + n_6 + n_7) + \rho(K_1 m_1 + K_4 m_4) = 0; \\ -(H_1 m_1 + H_2 m_2) + \rho(K_1 t_1 + K_4 t_4 + t_5 + K_6 t_6) = 0; \\ m_1 + m_2 + m_3 + m_4 = 1; \\ t_1 + t_4 = 1; \\ n_1 + n_2 = 1. \end{cases} \tag{8.31}$$

Proof. By Lemmas 1–4, C.i is equivalent to C.o of Lemma 4. Thus, the theorem follows from Lemma 4 and Condition 1. □

Our main theorem is as follows (a proof will be given in the next section).

Theorem 2. *If and only if $0 < \rho < \widehat{\rho} = \left(3 - \sqrt{5}\right)/2$, system (8.31) has a strictly positive solution $\pi \gg 0$.*

Given Theorem 2, we may demonstrate the following result.

Proposition 2. *$\widehat{\rho} = \left(3 - \sqrt{5}\right)/2$ is the least upper bound of discount factors with which it is possible to construct a strictly concave optimal growth model $M = \{u, \rho\}$ in which the optimal dynamical system is chaotic in the sense of Li and Yorke.*

Proof. Given Theorem 2, it suffices to demonstrate that if and only if it is possible to construct a polyhedral model $M' = \{v, \rho\}$ satisfying Condition 1, it is possible to construct a strictly concave model $M = \{u, \rho\}$ in which the optimal dynamical system is chaotic in the sense of Li and Yorke. Suppose that model M' satisfies Condition 1. Then, it is possible to construct a strictly concave utility function $u = u(k, h)$ satisfying

$$a = u(x, y), \, b = u(y, z), \, c = u(z, x) \tag{8.32}$$

and that if function u is substituted for v, conditions (8.28), (8.29) and (8.31) hold for any $(k, h) \in D$. This implies that the period-3 path $x, y, z, x, y, z, x, \ldots$ is optimal. Since the solutions to a strictly concave model can be expressed by an optimal dynamical system, the optimal dynamical system of $M = \{u, \rho\}$ is chaotic in the sense of Li and Yorke.

---

[5] $(x_1, \ldots, x_n) \gg 0$ means that $x \in R^n$ and $x_1 > 0, \ldots, x_n > 0$.

Conversely, suppose that the optimal dynamical system of a strictly concave model M is chaotic in the sense of Li and Yorke. Then, it may be proved that there are prices p_1, p_2, and p_3 such that $(\rho^{-1}p_1, p_2)$, $(\rho^{-1}p_2, p_3)$, and $(\rho^{-1}p_3, p_1)$, respectively, strictly support $u(k, h)$ at (x, y), (y, z), and (z, x), respectively; see McKenzie (1986) for a proof. By using (8.32) define a, b and c, and construct polyhedral P and utility function v. Then, $(\rho^{-1}p_1, p_2)$, $(\rho^{-1}p_2, p_3)$, and $(\rho^{-1}p_3, p_1)$ strictly support $v(k, h)$ at (x, y), (y, z), and (z, x), respectively. Thus, by Lemma 4, Condition 1 holds. □

Proof of Proposition 1 In order to prove Proposition 1, we need to extend Proposition 2 to the case including general concave utility functions. As the proof below demonstrates, the possibility of such an extension is straightforward due to the continuity of the model.

8.4 Proof of Theorem 2: Preliminary Steps

We will establish our main results (Propositions 1 and 2) by proving Theorem 2. For this purpose, we use Theorem 1. System (8.31) of Theorem 1, however, involves too may variables. In this section, we will reduce C.i to a more manageable form that does not involve variable π.

Step 1. At the outset, we will erase t_4, n_2 and m_3 from system (8.31) and deal with

$$\begin{cases} -(H_1t_1 + t_6) + \rho(K_1n_1 + K_6n_6) = 0; \\ -((H_1 - H_2)n_1 + H_2 + n_6 + n_7) + \rho(K_1m_1 + K_4m_4) = 0; \\ -(H_1m_1 + H_2m_2) + \rho((K_1 - K_4)t_1 + K_4 + t_5 + K_6t_6) = 0; \\ m_1 + m_2 + m_4 < 1; \\ t_1 < 1; \\ n_1 < 1. \end{cases} \tag{8.33}$$

Sublemma 1. *Condition C.i of Lemma 1 is satisfied if and only if the following condition is satisfied.*

C.ii. There is $\pi' = (m_1, t_1, n_1, t_5, n_7, m_2, m_4, t_6, n_6) \gg 0$ satisfying (8.33).

Before moving to the next step, define $\theta_1 = \rho^3K_1^2(K_4 - K_1) - H_1^2(H_2 - H_1)$. At the next step, we will focus on the case of $\theta_1 > 0$. We will return to the case of $\theta_1 < 0$ at Step 2″ and that of $\theta_1 = 0$ at the end of the next section.

Step 2 (Case $\theta_1 > 0$). Next, we will erase m_1. Solve the second and third conditions of (8.33) for n_1 and t_1, and substitute them into the first condition. By solving the resulting equation for m_1 and using $m_1 > 0$, we obtain

$$(m_1 =)\ \frac{1}{\theta_1}\big[H_1 H_2 (H_2 - H_1) m_2 + \rho^2 (K_4 - K_1)\{(H_2 - H_1) K_6 + K_1\} n_6$$
$$+ \rho^2 K_1 (K_4 - K_1) n_7 - \rho^3 K_1 K_4 (K_4 - K_1) m_4$$
$$- \rho (H_2 - H_1)\{K_4 - K_1 + H_1 K_6\} t_6 - \rho H_1 (H_2 - H_1) t_5$$
$$- \rho H_1 (H_2 - H_1) K_5 + \rho^2 K_1 (K_4 - K_1) H_2\big] > 0. \tag{8.34}$$

By using (8.34), we will erase m_1 from the second and third conditions of (8.33). That is, by $H_2 - H_1 > 0$,[6] the second and third conditions can be transformed as follows.[7]

$$n_7 - \left(\rho^2 K_1/H_1\right) t_5 = A^{n_7} (n_1; m_2, m_4, n_6); \tag{8.35}$$
$$n_7 - \left(\rho^2 K_1/H_1\right) t_5 = B^{n_7} (t_1; m_2, m_4, t_6). \tag{8.36}$$

In order to erase m_1 from the fourth condition of (8.33), again, use (8.34) to obtain

$$(1 - m_1 - m_2 - m_4 =)\ \frac{1}{\theta_1}[-\{H_1 (H_2 - H_1)^2 + \rho^3 K_1^2 (K_4 - K_1)\}m_2$$
$$- \rho^2 (K_4 - K_1)\{(H_2 - H_1) K_6 + K_1\} n_6 - \rho^2 K_1 (K_4 - K_1) n_7$$
$$+ \{H_1^2 (H_2 - H_1) + \rho^3 K_1 (K_4 - K_1)^2\}m_4$$
$$+ \rho (H_2 - H_1)\{(K_4 - K_1) + H_1 K_6\} t_6 + \rho H_1 (H_2 - H_1) t_5$$

---

[6]From (8.10), (8.17), $c - b > 0$ and $z > y > x$

$$H_1 = \frac{c - b - K_1 (z - y)}{z - x} < H_2 = \frac{c - b}{z - x}$$

holds.

[7]$A^{n_7} (n_1; m_2, m_4, n_6)$ and $B^{n_7} (t_1; m_2, m_4, n_6)$ are defined as follows:

$$\begin{aligned} A^{n_7} (n_1; m_2, m_4, n_6) = \tfrac{1}{H_1^2} \big[& -\theta_1 n_1 - \rho K_1 H_1 H_2 m_2 + \rho H_1^2 K_4 m_4 \\ & - \{H_1^2 + \rho^3 K_1 (K_4 - K_1) K_6\} n_6 \\ & + \rho^2 K_1 \{H_1 K_6 + K_4 - K_1\} t_6 \\ & + H_1 (\rho^2 K_1 K_4 - H_1 H_2)\big]; \end{aligned}$$

$$\begin{aligned} B^{n_7} (t_1; m_2, m_4, t_6) = \tfrac{1}{\rho K_1 H_1} \big[& -\rho\theta_1 t_1 - \rho^2 K_1^2 H_2 m_2 + \rho^2 H_1 K_1 K_4 m_4 \\ & - \rho H_1 \{K_1 + (H_2 - H_1) K_6\} n_6 \\ & + \{\rho^3 K_1^2 K_6 + H_1 (H_2 - H_1)\} t_6 \\ & + \rho K_1 (\rho^2 K_1 K_4 - H_1 H_2)\big]. \end{aligned}$$

$$+ \rho H_1 (H_2 - H_1) K_4 - \rho^2 K_1 (K_4 - K_1) H_2 \\ + \rho^3 K_1^2 (K_4 - K_1) - H_1^2 (H_2 - H_1)] > 0. \tag{8.37}$$

In what follows, we will first focus on the case of $\theta_1 > 0$. Note that $K_4 > K_1$ by (8.10) and (8.17).[8] Therefore, by $\theta_1 > 0$, (8.34) and (8.37) can be rewritten as follows[9]:

$$M_1^{n_7}(m_2, m_4, t_6, n_6) < n_7 - \frac{H_1 (H_2 - H_1)}{\rho K_1 (K_4 - K_1)} t_5 < M_3^{n_7}(m_2, m_4, t_6, n_6). \tag{8.38}$$

The next lemma follows from the above construction.

Sublemma 2. *Let $\theta_1 > 0$. Condition C.ii if Sublemma 1 is satisfied if and only if the following condition is satisfied.*

C.iii. There is $\pi'' = (t_1, n_1, t_5, n_7, m_2, m_4, t_6, n_6) \gg 0$ satisfying (8.35) through (8.38) together with $t_1 < 1$ and $n_1 < 1$.

Step 3 (Case $\theta_1 > 0$). Next, we will eliminate t_1 and n_1 from $C.i$. In what follows, we use $A^{n_7}(n_1)$ and $B^{n_7}(t_1)$ as shorthands for $A^{n_7}(n_1; m_2, m_4, n_6)$ and $B^{n_7}(t_1; m_2, m_4, t_6)$ whenever it is not necessary to emphasize that they depend on parameters (m_2, m_4, n_6). Notice that condition (8.38) does not depend on either t_1 or n_1. Condition (8.35) depends on n_1 but not on t_1. Condition (8.36) depends on t_1 but not on n_1. By $\theta_1 > 0$, $A^{n_7}(n_1)$ and $B^{n_7}(t_1)$ are decreasing in n_1 and t_1.

---

[8] From (8.10), (8.17), $c - a > 0$ and $z > y > x$

$$K_1 = \frac{c - a - H_1 (y - x)}{z - x} < K_4 = \frac{c - a}{z - x}$$

holds.

[9] $M_1^{n_7}(m_2, m_4, t_6, n_6)$ and $M_3^{n_7}(m_2, m_4, t_6, n_6)$ are defined as follows:

$$\begin{aligned} &M_1^{n_7}(m_2, m_4, t_6, n_6) \\ &\quad = [-H_1 H_2 (H_2 - H_1) m_2 - \rho^2 (K_4 - K_1) \{(H_2 - H_1) K_6 + K_1\} n_6 \\ &\quad + \rho^3 K_1 K_4 (K_4 - K_1) m_4 + \rho (H_2 - H_1) \{K_4 - K_1 + H_1 K_6\} t_6 \\ &\quad + \rho H_1 (H_2 - H_1) K_4 - \rho^2 K_1 (K_4 - K_1) H_2] / (\rho^2 K_1 (K_4 - K_1)); \end{aligned}$$

$$\begin{aligned} &M_3^{n_7}(m_2, m_4, t_6, n_6) \\ &\quad = [-\{H_1 (H_2 - H_1)^2 + \rho^3 K_1^2 (K_4 - K_1)\} m_2 \\ &\quad - \rho^2 (K_4 - K_1) \{(H_2 - H_1) K_6 + K_1\} n_6 + \{H_1^2 (H_2 - H_1) \\ &\quad + \rho^3 K_1 (K_4 - K_1)^2\} m_4 + \rho (H_2 - H_1) \{K_4 - K_1 + H_1 K_6\} t_6 \\ &\quad + \rho H_1 (H_2 - H_1) K_4 - \rho^2 K_1 (K_4 - K_1) H_2 + \rho^3 K_1^2 (K_4 - K_1) \\ &\quad - H_1^2 (H_2 - H_1)] / (\rho^2 K_1 (K_4 - K_1)). \end{aligned}$$

Moreover, $0 < t_1 < 1$ and $0 < n_1 < 1$. By these facts, conditions (8.35) and (8.36) are equivalent to the following:

$$A^{n_7}(1; m_2, m_4, n_6) < n_7 - (\rho^2 K_1/H_1)t_5 < A^{n_7}(0; m_2, m_4, n_6) \tag{8.39}$$

$$B^{n_7}(1; m_2, m_4, t_6) < n_7 - \left(\rho^2 K_1/H_1\right) t_5 < B^{n_7}(0; m_2, m_4, t_6). \tag{8.40}$$

Sublemma 3. *Let $\theta_1 > 0$. Condition C.iii of Lemma 2 is satisfied if and only if the following condition is satisfied.*

C.iv. There is $\pi''' = (t_5, n_7, m_2, m_4, t_6, n_6) \gg 0$ satisfying (8.38), (8.39) and (8.40).

At the next step, we will eliminate t_5 and n_7 from C.iv. For this purpose, denote by A the region in the t_5-n_7 space defined by condition (8.39). Denote by B that defined by (8.40). In order for condition C.iv of Sublemma 3 to hold (i.e., for π''' to satisfy conditions (8.38)–(8.40)), it is necessary that $A \cap B \neq \phi$. Since the boundaries of regions A and B are parallel to one another, by (8.39) and (8.40), $A \cap B \neq \phi$ if and only if the following conditions are satisfied.

$$A^{n_7}(0) - B^{n_7}(1) > 0; \tag{8.41}$$

$$B^{n_7}(0) - A^{n_7}(1) > 0. \tag{8.42}$$

As explained below, (8.41) and (8.42) are equivalent to

$$-H_1 < t_6 - \rho K_6 n_6 < \rho K_1; \tag{8.43}$$

this follows from $\theta_1 > 0$ and the fact that, by definition, the following holds:

$$A^{n_7}(0) - B^{n_7}(1) = \theta_1(t_6 - \rho K_6 n_6 + H_1)/(\rho K_1 H_1^2); \tag{8.44}$$

$$B^{n_7}(0) - A^{n_7}(1) = -\theta_1(t_6 - \rho K_6 n_6 - \rho K_1)/(\rho K_1 H_1^2). \tag{8.45}$$

Note that the lower boundary of region $A \cap B$, which plays a critical role in the next step, can be characterized as follows:

$$n_7 - (\rho^2 K_1/H_1)t_5 = A^{n_7}(1; m_2, m_4, n_6) \quad \text{if} \quad t_6 - \rho K_6 n_6 \geq \rho K_1 - H_1; \tag{8.46}$$

$$n_7 - (\rho^2 K_1/H_1)t_5 = B^{n_7}(1; m_2, m_4, t_6) \quad \text{if} \quad t_6 - \rho K_6 n_6 < \rho K_1 - H_1. \tag{8.47}$$

Define $\theta_2 = t_6 - \rho K_6 n_6 - (\rho K_1 - H_1)$. At the next step, we will focus on the case of $\theta_2 \geq 0$; we will return to the case of $\theta_2 < 0$ at Step 4′.

Step 4 (Case $\theta_1 > 0$; Subcase $\theta_2 \geq 0$). We now eliminate t_5 and n_7 from condition C.iv. of Sublemma 3. Note that condition (8.38), like (8.39) and (8.40), determines a region between two parallel lines in the t_5-n_7 space. Denote by M the region defined by (8.38). Condition C.iv is equivalent to the condition that $A \cap B$ intersects region M in the strictly positive orthant of the t_5-n_7 space.

See Figs. 8.2 and 8.3. Note the following facts. By $\theta_1 > 0$, the slope of the boundaries of region M, i.e., $H_1(H_2 - H_1)/\rho K_1(K_4 - K_1)$, is smaller than or equal to that of the boundaries of regions A and B, $\rho^2 K_1/H_1$. $A \cap B$ intersects with region M in the strictly positive orthant of the t_5-n_7 space. By $\theta_2 \geq 0$, the lower boundary of $A \cap B$ is given by (8.46). Suppose the vertical intercept of the lower boundary of $A \cap B$ is non-negative ($A^{n_7}(1) \geq 0$). Then, if and only if that intercept lies below the vertical intercept of the upper boundary of M, $A \cap B$ intersects M in the strictly positive orthant of the t_5-n_7 space. Suppose, instead, the vertical intercept of the lower boundary of $A \cap B$ is negative ($A^{n_7}(1) < 0$). Then, if and only if the horizontal intercept of the lower boundary lies to the right of that of the upper boundary of M, $A \cap B$ intersects M.

Denoting by $M_1^{t_5} = M_1^{t_5}(m_2, m_4, t_6, n_6)$ and $M_3^{t_5} = M_3^{t_5}(m_2, m_4, t_6, n_6)$, the horizontal intercepts of the lower and upper boundaries of region M, we may summarize this as follows. Regions $A \cap B$ and M intersect each other in the strictly positive orthant of the t_5-n_7 space if and only if one of the following two cases holds:

$$M_3^{n_7}(m_2, m_4, t_6, n_6) > A^{n_7}(1; m_2, m_4, n_6) \geq 0; \tag{8.48}$$

$$A^{n_7}(1; m_2, m_4, n_6) < 0\,,\; A^{t_5}(1; m_2, m_4, n_6) > M_3^{t_5}(m_2, m_4, t_6, n_6). \tag{8.49}$$

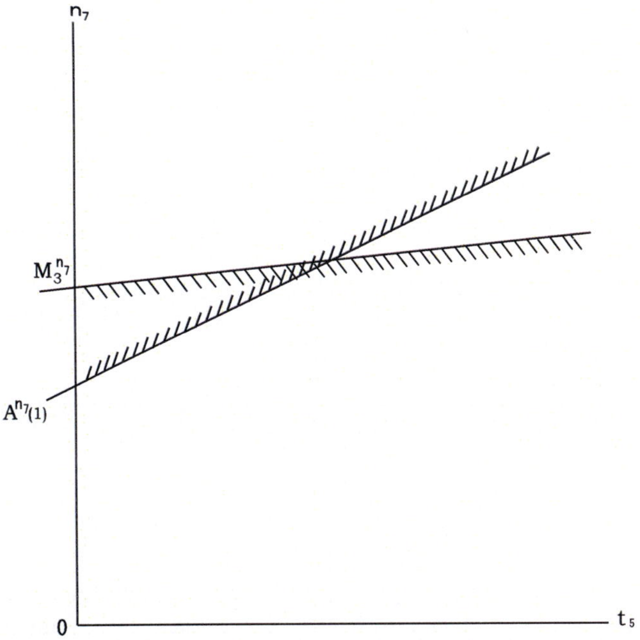

Fig. 8.2 $t_6 - \rho K_6 n_6 \geq 0,\ A^{n_7}(1) \geq 0$

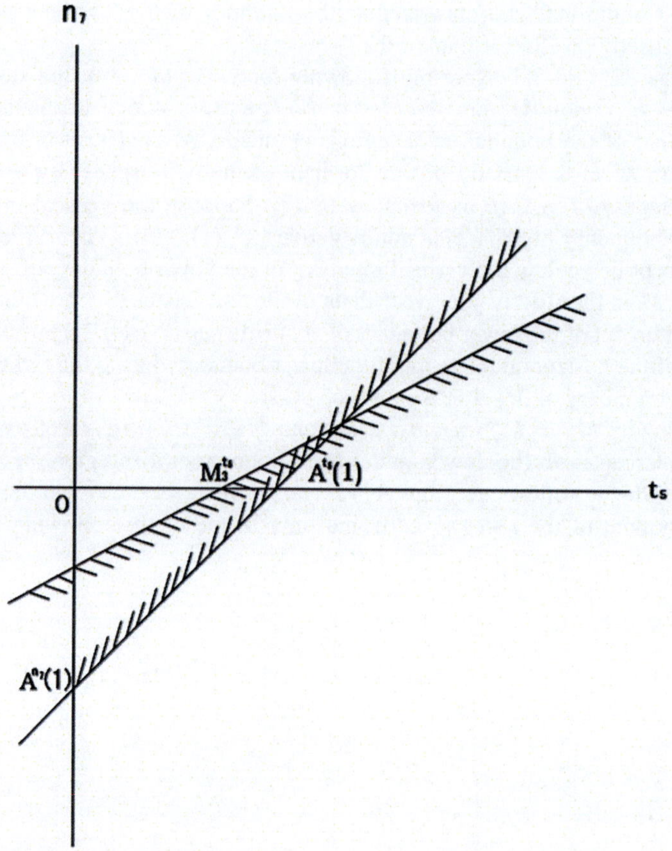

Fig. 8.3 $t_6 - \rho K_6 n_6 > 0,\ A^{n_7}(1) < 0$

Note that by $K_1 > 0$ and $t_6 - \rho K_6 n_6 \geq \rho K_1 - H_1$, (8.43) is equivalent to

$$\rho K_1 - H_1 \leq t_6 - \rho K_6 n_6 < \rho K_1 \tag{8.50}$$

The next lemma follows.

Sublemma 4. *Let* $\theta_1 > 0$ *and* $t_6 - \rho K_6 n_6 \geq \rho K_1 - H_1$. *Condition C.iv of Sublemma 3 is satisfied if and only if the following condition is satisfied.*

C.v. There is $\pi'''' = (m_2, m_4, t_6, n_6) \gg 0$ satisfying (8.50) and either (8.48) or (8.49).

Step 5 (Case $\theta_1 > 0$; Subcase $\theta_2 \geq 0$). Next, we will eliminate m_2 and m_4 from Condition (v). Note that condition (8.50) is independent of m_2 and m_4. We may demonstrate that condition (8.48) is equivalent to the following:

$$
\begin{aligned}
& - \rho K_1 H_1 H_2 m_2 + \rho H_1^2 K_4 m_4 \\
& \quad \geq \{H_1^2 - \rho^3 K_1 (K_4 - K_1) K_6\} n_6 \\
& \qquad - \rho^2 K_1 \{H_1 K_6 + K_4 - K_1\} t_6 \\
& \qquad + \rho^3 K_1^2 (K_4 - K_1) + H_1^3 - \rho^2 K_1 H_1 K_4;
\end{aligned} \tag{8.51}
$$

$$
\begin{aligned}
& - H_1 (H_2 - H_1) m_2 + H_1^2 m_4 \\
& \quad < \rho^2 (K_4 - K_1) K_6 n_6 - \rho (H_1 K_6 + K_4 - K_1) t_6 \\
& \qquad - \rho H_1 K_4 + H_1^2 + \rho^2 K_1 (K_4 - K_1).
\end{aligned} \tag{8.52}
$$

Given $\pi'''''' = (t_6, n_6)$, conditions (8.51) and (8.52) determine a region between two lines in the m_2-m_4 space.

Call this region Γ; see Fig. 8.4. Then, condition (8.48) is satisfied for $(m_2, m_4) \gg 0$ if and only if region Γ intersects the strictly positive orthant of

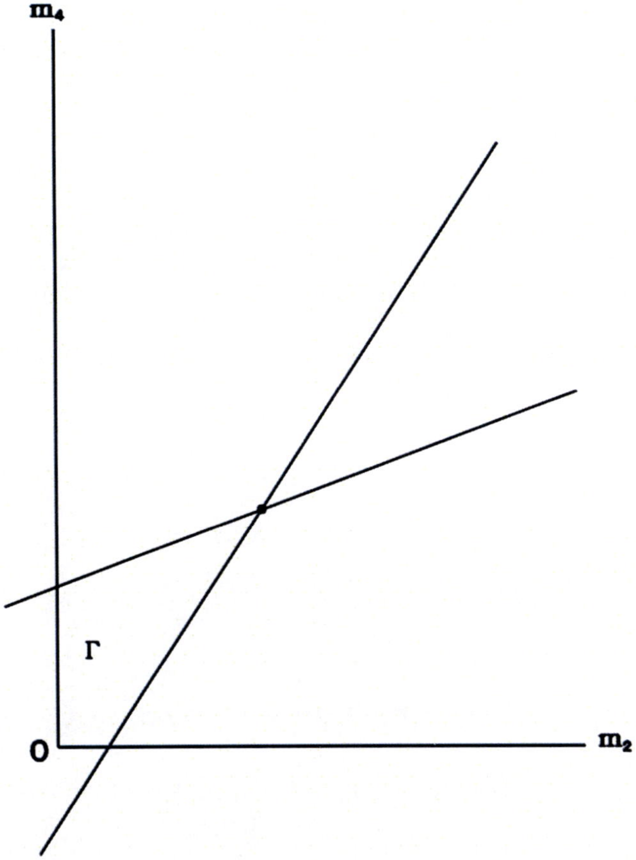

Fig. 8.4

the m_2-m_4 space. Since $K_1H_2 + K_4H_1 - K_4H_2 > 0$,[10] we may demonstrate that the slope of the boundary of (8.51) is larger than that of (8.52) in the m_2-m_4 space. Therefore, region Γ intersects the strictly positive orthant if and only if the boundaries of (8.51) and (8.52) intersect each other in the positive orthant. This condition can be expressed as follows[11]:

$$\begin{aligned} &-\{\rho^3 (K_4 - K_1)^2 K_6 - H_1^2\}n_6 \\ &\quad +\rho^2 (K_4 - K_1)(H_1K_6 + K_4 - K_1)(t_6 - C^{t_6}) < 0; \end{aligned} \tag{8.53}$$

$$\begin{aligned} &-\{\rho^3 K_1 (K_4 - K_1) K_6 - H_1 (H_2 - H_1)\}n_6 \\ &\quad +\rho^2 K_1 (H_1K_6 + K_4 - K_1)(t_6 - D^{t_6}) < 0. \end{aligned} \tag{8.54}$$

Similarly, we may demonstrate that (8.49) is equivalent to the following.

$$\begin{aligned} &-\rho K_1 H_1 H_2 m_2 + \rho H_1^2 K_4 m_4 \\ &\quad < \{H_1^2 + \rho^3 K_1 (K_4 - K_1) K_6\}n_6 \\ &\quad -\rho^2 K_1 \{H_1K_6 + K_4 - K_1\} t_6 \\ &\quad +\rho^3 K_1^2 (K_4 - K_1) + H_1^3 - \rho^2 K_1 H_1 K_4; \end{aligned} \tag{8.55}$$

$$-\rho K_1 m_2 + \rho (K_4 - K_1) m_4 > n_6 - \rho K_1 + H_1. \tag{8.56}$$

In much the same way as (8.51) and (8.52), we may demonstrate that conditions (8.55) and (8.56) as well are equivalent to (8.53) and (8.54).

Sublemma 5. *Let $\theta_1 > 0$ and $t_6 - \rho K_6 n_6 \geq \rho K_1 - H_1$. Condition C.v of Sublemma 4 is satisfied if and only if the following condition is satisfied.*

C.vi. There is $\pi''''' = (t_6, n_6) \gg 0$ satisfying (8.50), (8.53) and (8.54).

---

[10] This may be shown from (8.11), (8.15) and (8.18) and by using

$$\left[(z-y)(y-x)/(z-x)^2\right] < 1.$$

[11] C^{t_6} and D^{t_6} are defined as follows.

$$C^{t_6} = \frac{-\rho^3 K_1 (K_4 - K_1)^2 + \rho^2 H_1 K_4 (K_4 - K_1) - \rho H_1^2 K_4 + H_1^3}{\rho^2 (K_4 - K_1)(H_1K_6 + K_4 - K_1)};$$

$$D^{t_6} = \frac{-\rho^3 K_1^2 (K_4 - K_1) + \rho^2 H_1 K_1 K_4 - \rho H_1 H_2 K_1 + H_1^2 (H_2 - H_1)}{\rho^2 K_1 (H_1K_6 + K_4 - K_1)}.$$

Step 6 (Case $\theta_1 > 0$; Subcase $\theta_2 \geq 0$). This is the last step, in which we eliminate t_6 and n_6 from $C.vi$. Conditions (8.50), (8.53) and (8.54) each determine a region in the n_6-t_6 space. Call those regions Θ, C and D, respectively. Then, it suffices to derive conditions under which these regions have a non-empty intersection in the n_6-t_6 space. It can be demonstrated that, among the slopes of the boundaries of regions Θ, C and D, the slope of the boundaries of region Θ is the largest and that the slope of the boundary of region C is the smallest. Denote by S^C and S^D the slopes of the boundaries of regions C and D. Moreover, denote by Θ^{n_6} and Θ^{t_6} the horizontal and vertical intercepts of the lower boundary of region Θ. For the moment, focus on the case in which $S^C \neq 0$ and $S^D \neq 0$; we will deal with the case of $S^C = 0$ and/or $S^D = 0$, separately.

Then, C.vi of Sublemma 5 is equivalent to the following conditions (see Figs. 8.5 and 8.6):

Condition 2 ***Case I:*** *$\rho K_1 \geq H_1$ and*

$$(I.a)\ C^{t_6} > \Theta^{t_6} \text{ and } D^{t_6} > \Theta^{t_6}.$$

Case II: *$\rho K_1 < H_1$ and one of the following three conditions holds:*

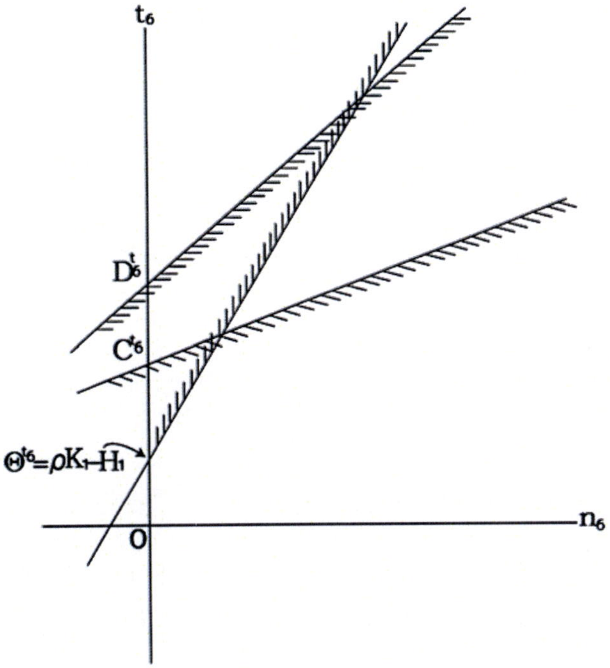

Fig. 8.5 (Ia)$\Theta^{t_6} \geq 0$

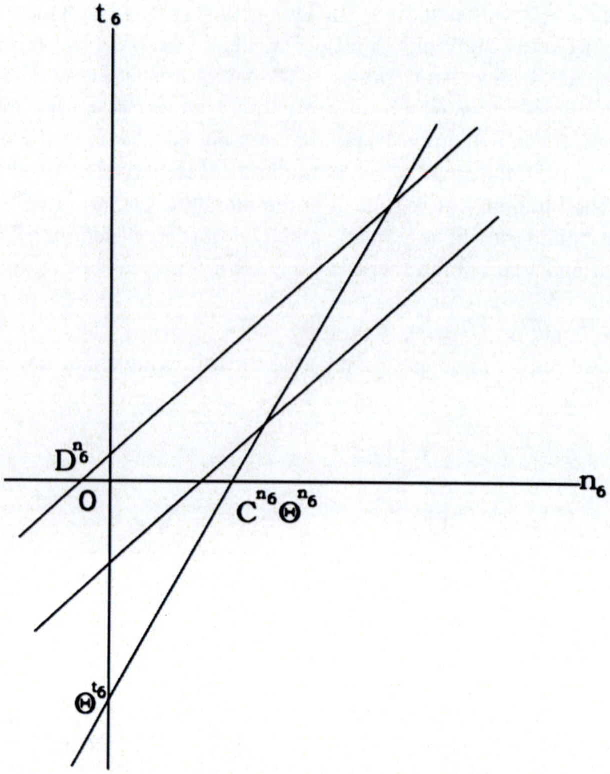

Fig. 8.6 (IIb)$\Theta^{t_6} < 0$

(II.b) $S^C > 0,\ S^D > 0,\ \Theta^{n_6} > C^{n_6},$ *and* $\Theta^{n_6} > D^{n_6};$

(II.c) $S^C > 0,\ S^D < 0,\ C^{n_6} > D^{n_6},$ *and* $\Theta^{n_6} > D^{n_6};$

(II.d) $S^C < 0,\ S^D < 0,\ C^{n_6} > 0,$ *and* $D^{n_6} > 0.$

The next lemma follows from Sublemmas 1 through 5 and the above discussion.

Lemma 5. *Suppose* $S^C \neq 0,\ S^D \neq 0,\ \theta_1 > 0$ *and* $t_6 - \rho K_6 n_6 \geq \rho K_1 - H_1$. *Condition C.i of Theorem 1 is satisfied if and only if Condition 2 is satisfied.*

We now return to the end of Step 3 and focus on the case of $\theta_2 < 0$.

Step 4′ (Case $\theta_1 > 0$; **Subcase** $\theta_2 < 0$**).** In this case, the lower boundary of region $A \cap B$ is given by (8.47). By using this fact, we obtain the counterparts of (8.50), (8.48) and (8.49) as follows.

$$-H_1 < t_6 - \rho K_6 n_6 < \rho K_1 - H_1; \tag{8.57}$$

$$M_3^{n_7}(m_2, m_4, t_6, n_6,) > B^{n_7}(1; m_2, m_4, t_6) \geq 0; \tag{8.58}$$

$$B^{n_7}(1; m_2, m_4, t_6) < 0,\ B^{t_5}(1; m_2, m_4, t_6) > M_3^{t_5}(m_2, m_4, t_6, n_6,). \tag{8.59}$$

The counterpart of Sublemma 4 is as follows:

Sublemma 6. *Let $\theta_1 > 0$ and $t_6 - \rho K_6 n_6 < \rho K_1 - H_1$. Condition C.iv of Sublemma 3 is satisfied if and only if the following condition is satisfied:*

C.v′. There is $\pi'''' = (m_2, m_4, t_6, n_6)$ satisfying (8.57) and either (8.58) or (8.59).

Step 5′ (Case $\theta_1 > 0$; Subcase $\theta_2 < 0$). Following the line of discussion at Step 5, we may obtain the counterparts of (8.55) and (8.56) as follows[12]:

$$\begin{aligned}&\rho H_1 \left[(H_2 - H_1) K_6 + K_1\right] n_6 \\ &\quad + \left[\rho^3 K_1 K_6 (K_4 - K_1) - H_1 (H_2 - H_1)\right](t_6 - E^{t_6}) < 0;\end{aligned} \tag{8.60}$$

$$\begin{aligned}&\rho (H_2 - H_1) \left[(H_2 - H_1) K_6 + K_1\right] n_6 \\ &\quad + \left[\rho^3 K_1^2 K_6 - (H_2 - H_1)^2\right](t_6 - F^{t_6}) < 0.\end{aligned} \tag{8.61}$$

Sublemma 7. *Let $\theta_1 > 0$ and $t_6 - \rho K_6 n_6 < \rho K_1 - H_1$. Condition C.v′ of Sublemma 6 is satisfied if and only if the following condition is satisfied:*

C.vi′. There is $\pi''''' = (t_6, n_6)$ satisfying (8.57), (8.60), and (8.61).

Step 6′(Case $\theta_1 > 0$; Subcase $\theta_2 < 0$). Conditions (8.57), (8.60), and (8.61) each determine a region in the n_6-t_6 space. Call those regions Θ', E and F, respectively. Denote by S^E and S^F the slopes of the boundaries of regions E and F. The equation of the upper boundary of Θ' is the same as that of the lower boundary of Θ. So, the horizontal and the vertical intercepts of the upper boundary of Θ' are, respectively, Θ^{n_6} and Θ^{t_6}. Following the line of discussion at Step 6, we may prove that C.vi′ is equivalent to the following condition, given $S^E \neq 0$ and $S^F \neq 0$.

---

[12] E^{t_6} and F^{t_6} are defined as follows.

$$E^{t_6} = \frac{-\rho^3 K_1^2 (K_4 - K_1) + \rho^2 H_1 K_1 K_4 - \rho H_1 H_2 K_1 + H_1^2 (H_2 - H_1)}{\rho^3 K_1 (K_4 - K_1) K_6 - H_1 (H_2 - H_1)};$$

$$F^{t_6} = \frac{-\rho^3 K_1^3 + \rho^2 H_2 K_1^2 - \rho H_2 (H_2 - H_1) K_1 + H_1 (H_2 - H_1)^2}{\rho^3 K_1^2 K_6 - (H_2 - H_1)^2}.$$

Condition 3 *Case I:* $\rho K_1 \geq H_1$ *and one of the following three conditions:*

$$\text{(I.b) } S^E > 0,\ S^F > 0,\ \Theta^{t_6} > E^{t_6} \text{ and } \Theta^{t_6} > F^{t_6};$$

$$\text{(I.c) } S^E > 0,\ S^F < 0,\ \Theta^{t_6} > E^{t_6} \text{ and } F^{t_6} > E^{t_6};$$

$$\text{(I.d) } S^E < 0,\ S^F > 0,\ E^{t_6} > 0 \text{ and } F^{t_6} > 0.$$

Case II: $\rho K_1 < H_1$ *and*

$$\text{(II.a) } E^{n_6} > \Theta^{n_6} \text{ and } F^{n_6} > \Theta^{n_6}.$$

The counterpart of Lemma 5 is as follows:

Lemma 6. *Suppose* $S^E \neq 0$, $S^F \neq 0$, $\theta_1 > 0$ *and* $t_6 - \rho K_6 n_6 < \rho K_1 - H_1$. *Condition C.i of Theorem 1 is satisfied if and only if Condition 3 is satisfied.*

We now return to the end of Step 1 and focus on the case of $\theta_1 < 0$.

Step 2″/3″ (Case $\theta_1 < 0$**).** Return to the position at Step 2 at which we focused on the case of $\theta_1 > 0$. Instead, we now focus on the case of $\theta_1 < 0$. For the sake of simplicity, here, we now eliminate m_1, t_1 and n_1 by one step, corresponding to Steps 2 and 3. Since $\theta_1 < 0$, the counterparts of (8.35), (8.39) and (8.40) are as follows:

$$M_3^{n_7}(m_2, m_4, t_6, n_6) < n_7 - \frac{H_1 (H_2 - H_1)}{\rho K_1 (K_4 - K_1)} t_5 < M_1^{n_7}(m_2, m_4, t_6, n_6); \quad (8.62)$$

$$A^{n_7}(0; m_2, m_4, n_6) < n_7 - (\rho^2 K_1/H_1) t_5 < A^{n_7}(1; m_2, m_4, n_6); \quad (8.63)$$

$$B^{n_7}(0; m_2, m_4, t_6) < n_7 - (\rho^2 K_1/H_1)\, t_5 < B^{n_7}(1; m_2, m_4, t_6). \quad (8.64)$$

Conditions (8.63) and (8.64) follow since $A^{n_7}(n_1; m_2, m_4, n_6)$ and $B^{n_7}(t_1; m_2, m_4, n_6)$ are increasing in n_1 and t_1, respectively, in the case of $\theta_1 < 0$.

Sublemma 8. *Let* $\theta_1 < 0$. *Condition C.iii of Sublemma 2 is satisfied if and only if the following condition is satisfied.*

C.iv″. There is $\pi''' = (t_5, n_7, m_2, m_4, t_6, n_6)$ satisfying (8.62), (8.63) and (8.64).

At the next step, we will eliminate t_5, n_7, m_2, m_4, t_6 and n_6 by one step. Before moving to this step, as in Step 3, denote by A and B, respectively, the regions defined by (8.63) and (8.64). We may demonstrate that the condition for $A \cap B \neq \phi$ is identical to that derived at Step 4; i.e., $A \cap B \neq \phi$ if and only if (8.43) is satisfied. Recall that at Step 4, we used the lower boundary of $A \cap B$. In the case of $\theta_1 < 0$, instead, we will use the upper boundary of $A \cap B$, which is given as follows:

$$n_7 - (\rho^2 K_1/H_1) t_5 = A^{n_7}(1; m_2, m_4, n_6) \text{ if } t_6 - \rho K_6 n_6 \geq \rho K_1 - H_1; \quad (8.65)$$

$$n_7 - (\rho^2 K_1/H_1) t_5 = B^{n_7}(1; m_2, m_4, t_6) \text{ if } t_6 - \rho K_6 n_6 < \rho K_1 - H_1. \quad (8.66)$$

We will first focus on the case of $\theta_2 \geq 0$, in which (8.65) holds. At Step $4'''/5'''/6'''$, we will return to the case of $\theta_2 < 0$.

Step 4″/5″/6″ (Case $\theta_1 < 0$; Subcase $\theta_2 \geq 0$). Denote by M the region defined by (8.62). In order to eliminate t_5 and n_7, as at Step 4, we derive conditions under which region M intersects region $A \cap B$ in the strictly positive orthant of the t_5-n_7 space.

By $\theta_1 < 0$, the slope of the boundaries of region M, i.e., $H_1\,(H_2 - H_1)\,/\rho K_1\,(K_4 - K_1)$, is larger than or equal to that of the boundaries of regions A and B, $\rho^2 K_1/H_1$. Therefore, by $\theta_2 < 0$, in much the same way as Step 4, we may prove that M intersects region $A \cap B$ in the strictly positive orthant of the t_5-n_7 space if and only if one of the following two conditions is satisfied.

$$A^{n_7}\,(1; m_2, m_4, n_6) \geq 0\,,\; A^{n_7}\,(1; m_2, m_4, n_6) > M_3^{n_7}(m_2, m_4, t_6, n_6) \qquad (8.67)$$

$$A^{n_7}\,(1; m_2, m_4, n_6) < 0\,,\; M_3^{t_5}(m_2, m_4, t_6, n_6) > A^{t_5}\,(1; m_2, m_4, n_6). \qquad (8.68)$$

In order to erase m_2 and m_4, as at Step 5, we may demonstrate that (8.67) is equivalent to (8.51) and (8.52) and that (8.68) is equivalent of (8.55) and (8.56). As is seen at Step 5, these conditions boil down to conditions (8.53) and (8.54). Finally, by $t_6 - \rho K_6 n_6 \geq \rho K_1 - H_1$ and (8.43), (8.50) holds.

This implies that the discussion at Step 6 holds without any change. Therefore, the same result as at Step 6 holds for the case of $\theta_1 < 0$ as well.

Lemma 7. *Suppose $S^C \neq 0$, $S^D \neq 0$, $\theta_1 > 0$ and $t_6 - \rho K_6 n_6 \geq \rho K_1 - H_1$. Condition C.i of Theorem 1 is satisfied if and only if Condition 2 is satisfied.*

Next, we will focus on the case of $\theta_2 < 0$.

Step 4‴/5‴/6‴ (Case $\theta_1 < 0$; Subcase $\theta_2 < 0$). In the case of $t_6 - \rho K_6 n_6 < \rho K_1 - H_1$, we may follow the above discussions to demonstrate that a lemma similar to Lemma 7 holds for the case of $\theta_1 < 0$.

Lemma 8. *Suppose $S^E \neq 0$, $S^F \neq 0$, $\theta_1 < 0$ and $t_6 - \rho K_6 n_6 < \rho K_1 - H_1$. Condition C.i of Theorem 1 is satisfied if and only if Condition 3 is satisfied.*

8.5 The Proof of Theorem 2

We will demonstrate Theorem 2.

8.5.1 *Case of $\theta_1 \neq 0$:*

Lemmas 5 through 8 imply that in the cases of $t_6 - \rho K_6 n_6 \geq \rho K_1 - H_1$ and $t_6 - \rho K_6 n_6 < \rho K_1 - H_1$, respectively, Conditions 2 and 3 can characterize Condition

$C.i$ of Theorem 1 regardless of the sign of $\theta_1 \neq 0$. Therefore, by expressing Conditions 2 and 3 in terms of H_1, H_2, K_1, K_4, K_6 and ρ, we may summarize these lemmas as follows.

Lemma 9. *Let $\theta_1 \neq 0$, $S^C \neq 0$, $S^D \neq 0$, $S^E \neq 0$ and $S^F \neq 0$. Condition $C.i$ of Theorem 1 is satisfied if and only if the following condition is satisfied.*
C.iii. Case I: *$\rho K_1 \geq H_1$ and one of the following conditions is satisfied.*

$$(I.a')\quad \begin{array}{l} \rho^2 (K_4 - K_1) K_6 + \rho (K_4 - K_1) - H_1 < 0, \text{ and} \\ \rho^2 K_1 K_6 + \rho K_1 - (H_2 - H_1) < 0. \end{array}$$

$$(I.b')\quad \begin{array}{l} \rho^3 K_1 (K_4 - K_1) K_6 - H_1 (K_4 - K_1) < 0, \\ \rho^3 K_1^2 K_6 - (H_2 - H_1)^2 < 0, \\ \rho^2 (K_4 - K_1) K_6 + \rho (K_4 - K_1) - H_1 < 0, \text{ and} \\ \rho^2 K_1 K_6 + \rho K_1 - (H_2 - H_1) < 0. \end{array}$$

$$(I.c')\quad \begin{array}{l} \rho^3 K_1 (K_4 - K_1) K_6 - H_1 (H_2 - H_1) < 0, \\ \rho^3 K_1^2 K_6 - (H_2 - H_1)^2 > 0, \\ \rho^2 (K_4 - K_1) K_6 - \rho (K_4 - K_1) - H_1 < 0, \text{ and} \\ \rho^2 K_1 K_6 - \rho (H_2 - H_1) K_6 - (H_2 - H_1) > 0. \end{array}$$

$$(I.d')\quad \begin{array}{l} \rho^3 K_1 (K_4 - K_1) K_6 - H_1 (H_2 - H_1) > 0, \\ \rho^3 K_1^2 K_6 - (H_2 - H_1)^2 > 0, \\ \rho^2 K_1 (K_4 - K_1) - \rho K_1 H_1 + H_1 (H_2 - H_1) < 0, \text{ and} \\ \rho^2 K_1^2 - \rho K_1 (H_2 - H_1) + (H_2 - H_1) > 0. \end{array}$$

Case II: *$\rho K_1 < H_1$ and one of the following conditions is satisfied:*

$$(II.a')\quad \begin{array}{l} \rho^2 (K_4 - K_1) K_6 - \rho H_1 K_6 - H_1 > 0, \text{ and} \\ \rho^2 K_1 K_6 - \rho (H_2 - H_1) K_6 - (H_2 - H_1) > 0. \end{array}$$

$$(II.b')\quad \begin{array}{l} \rho^3 (K_4 - K_1)^2 K_6 - H_1^2 > 0, \\ \rho^3 K_1 (K_4 - K_1) K_6 - H_1 (H_2 - H_1) > 0, \\ \rho^2 (K_4 - K_1) K_6 - \rho H_1 K_6 - H_1 > 0, \text{ and} \\ \rho^2 K_1 K_6 - \rho (H_2 - H_1) K_6 - (H_2 - H_1) > 0. \end{array}$$

$$(II.c')\quad \begin{array}{l} \rho^3 K_1 (K_4 - K_1) K_6 - H_1 (H_2 - H_1) > 0, \\ \rho^3 (K_4 - K_1)^2 K_6 - H_1^2 < 0, \\ \rho^2 K_1 K_6 - \rho (H_2 - H_1) K_6 - (H_2 - H_1) > 0, \text{ and} \\ \rho^2 (K_4 - K_1) K_6 + \rho (K_4 - K_1) - H_1 < 0. \end{array}$$

$$
(II.d')\quad
\begin{array}{l}
\rho^3 K_1 (K_4 - K_1) K_6 - H_1 (H_2 - H_1) < 0, \\
\rho^3 (K_4 - K_1)^2 K_6 - H_1^2 < 0, \\
\rho^2 (K_4 - K_1)^2 - \rho (K_4 - K_1) H_1 + H_1^2 < 0, \textit{ and} \\
\rho^2 K_1 (K_4 - K_1) - \rho K_1 H_1 + H_1 (H_2 - H_1) < 0.
\end{array}
$$

We will rewrite Condition C.vii of Lemma 9 by introducing parameters $\eta = (H_2 - H_1)/K_1$, $\kappa = (K_4 - K_1)/H_1$ and $B = (c-b)/(c-a)$. Then, by (8.11), (8.14) and (8.18), the following relationship holds among the parameters introduced above.

$$
\eta = \frac{H_2 - H_1}{K_1} = \frac{z-y}{z-x}, \quad \kappa = \frac{K_4 - K_1}{H_1} = \frac{y-x}{z-x}, \quad B = \frac{c-b}{c-a}. \tag{8.69}
$$

Moreover, by (8.11), (8.14), (8.18) and (8.22), we can express K_6 and H_1/K_1 as follows:

$$
K_6 = \frac{\eta}{\kappa} \text{ and } \frac{H_1}{K_1} = \frac{\eta - B}{B\kappa - 1}. \tag{8.70}
$$

By (8.5) and (8.7), parameters κ, η and B satisfy the following:

$$
\eta + \kappa = 1; \tag{8.71}
$$

$$
\kappa > 0, \quad \eta > 0, \quad \text{and } B > 0. \tag{8.72}
$$

The next lemma expresses condition C.vii of Lemma 9 in terms of η, κ and B.

Lemma 10. *Let $\theta_1 \neq 0$. Condition C.i of Theorem 1 is satisfied if and only if the following condition is satisfied.*
C.viii. Case I: $(\eta - B)/B\kappa - 1 \leq \rho$ *and one of the following three conditions is satisfied:*

$$(I.a'')\quad \rho^2\eta + \rho\kappa - 1 < 0 \textit{ and } (\eta - \rho)(\kappa - \rho^2) > \rho^3;$$

$$(I.c'')\quad \rho^3 < 1,\ \rho^3 > \eta\kappa,\ \rho^2\eta + \rho\kappa < 1, \textit{ and } \rho\eta + \kappa < \rho^2;$$

$$(I.d'')\quad \rho^3 > 1,\ \rho^3 > \eta\kappa,\ \rho^2\kappa + \eta < \rho, \textit{ and } \rho^2 - \rho\eta + \rho\eta < 0.$$

Case II: $\eta - B/B\kappa - 1 > \rho$ *and one of the following three conditions is satisfied:*

$$(II.a'')\quad (\kappa - \rho^{-1})(\eta - \rho^{-2}) > \rho^{-3}, \textit{ and } \rho\eta + \kappa < \rho^2;$$

$$(II.c'')\quad \rho^3 > 1,\ \rho^3\eta\kappa < 1,\ \rho\eta + \kappa < \rho^2, \textit{ and } \rho^2\eta + \rho\kappa < 1;$$

$$(II.d'')\quad \rho^3 < 1,\ \rho^3\eta\kappa < 1,\ (\rho\kappa)^2 - \rho\kappa + 1 < 0, \textit{ and } \rho^2\kappa + \eta < \rho.$$

Proof. In order to transform Lemma 9 into this lemma, we need to demonstrate that $S^C \neq 0$, $S^D \neq 0$, $S^E \neq 0$, and $S^F \neq 0$. If $S^C = 0$, by (8.53), $\rho^3(K_4 - K_1)^2 K_6 - H_1^2 = 0$. This implies, by the definition of η and κ, $\eta\kappa = \rho^{-3} > 1$, which contradicts (8.71); i.e., $S^C \neq 0$. If $S^D = 0$, by (8.54), $\rho^3 K_1(K_4 - K_1)K_6 - H_1(H_2 - H_1) = 0$. This means $\rho^3 = 1$ by (8.69) and (8.70), which contradicts $\rho < 1$; i.e., $S^D \neq 0$. By (8.60), $S^E = 0$ and $S^F = 0$, respectively, imply $\rho H_1[(H_2 - H_1)K_6 + K_1] = 0$ and $\rho(H_2 - H_1)[(H_2 - H_1)K_6 + K_1] = 0$, both of which contradict $\rho > 0$, $H_2 > H_1$, $K_6 > 0$ and $K_1 > 0$; i.e., $S^E \neq 0$ and $S^F \neq 0$.

Cases (I.b′) and (II.b′) are special cases of (I.a′) and (II.a′), respectively. Therefore, by using (8.69) and (8.70), we may rewrite C.vii of Lemma 9 as C.viii of this lemma. □

We will demonstrate the following lemma.

Lemma 11. *Let $\theta_1 \neq 0$. Condition C.i of Theorem 1 is satisfied if and only if $0 < \rho < \widehat{\rho} = (3 - \sqrt{5})/2$.*

Proof. As the above discussion demonstrates, it suffices to demonstrate that there is (κ, η) that satisfies C.vii if and only if $0 < \rho < \widehat{\rho}$. First, we will demonstrate that all the cases but (I.a″) of C.viii are void. To this end, note the following: The fourth condition of (I.c″) is not compatible with $\eta + \kappa = 1$ for $\rho < 1$; the first condition of (I.d″) contradicts our assumption $\rho < 1$; the second condition of (II.a″) is not compatible with $\eta + \kappa = 1$ for $\rho < 1$; the first condition of (II.b″) contradicts $\rho < 1$; the third condition of (II.d") is never satisfied. Thus, it suffices to focus on case (I.a″).

Given (8.71) and (8.72), $(\eta - B)\,/\,(B\kappa - 1) > 0$ holds if and only if $\kappa < 1/B$ and $\eta < B$. Therefore, $0 < (\eta - B)\,/\,(B\kappa - 1) \leq \rho$ if and only if

$$B > \eta \geq \rho(B\kappa - 1) + B. \tag{8.73}$$

See Fig. 8.7. The intersection of line (8.71) and the lower boundary of the region defined by (8.73) is given by

$$(\kappa(\rho), \eta(\rho)) = ((1 + \rho - B)/(\rho B + 1), (B - \rho(1 - B))/(\rho B + 1)). \tag{8.74}$$

Therefore, given (8.71), (8.73) is satisfied if and only if $\eta(\rho) \leq \eta < B$, which is equivalent to

$$\eta < B \leq \frac{\rho + \eta}{1 + \rho - \rho\eta}. \tag{8.75}$$

In short, conditions on κ, η and B are summarized by (8.71), (8.72) and (8.75). Thus, (I.a″) must be satisfied under conditions (8.71), (8.72) and (8.75). The first condition of (I.a″) is satisfied for any (κ, η) on line $\eta + \kappa = 1$. Note that the region defined by the second condition of (I.a″) intersects line $\eta + \kappa = 1$ if and only if

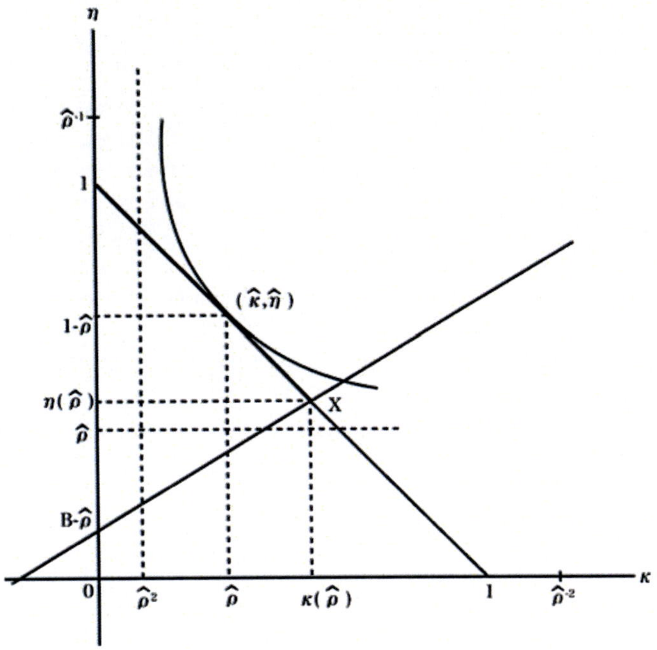

Fig. 8.7 (Ia) with $\hat{\rho} = (3 - \sqrt{5})/2$

$0 < \rho < \widehat{\rho} = \left(3 - \sqrt{5}\right)/2$. In order to prove Lemma 10, therefore, it suffices to demonstrate that we may choose (κ, η) satisfying the second condition of (I.a″) at the same time as (8.71), (8.72) and (8.75) for any ρ satisfying $0 < \rho < \widehat{\rho}$.

Let $0 < \rho < \widehat{\rho}$. Think of the boundary of the region defined by the second condition of (I.a″); i.e.,

$$(\eta - \rho)(\kappa - \rho^2) = \rho^3. \tag{8.76}$$

This boundary intersects $\eta + \kappa = 1$ at two points in the positive orthant of the κ-η space. The value of η at the upper intersection is

$$\eta^* = \frac{1 + \rho - \rho^2 + \sqrt{(1 + \rho - \rho^2)^2 - 4\rho}}{2}.$$

Choose $B = \eta^*$, and choose $\eta < B$ sufficiently close to B. Then, it is easy to demonstrate that condition (8.75) is satisfied. Moreover, choose κ by condition (8.71). Then, the second condition of (I.a″) as well as (8.75), (8.71), and (8.72), is satisfied.

8.5.2 *Proof of Theorem 1*

Lemma 10 implies that in order to complete the proof of Theorem 1, it suffices to demonstrate that in the case of $\theta_1 = 0$, there is no π satisfying condition (8.31) for any ρ, $\widehat{\rho} \leq \rho < 1$. Suppose that $\theta_1 = 0$ and that there is π^* satisfying condition (8.31) for some ρ, $\widehat{\rho} \leq \rho < 1$. Then we may choose parameters x, y, z, a, b and c in such a way that (8.31) is satisfied with π^*. Then, we may demonstrate that it is possible to perturb the value of these parameters in such a way that $\theta_1 \neq 0$ and (8.31) remains to hold with π^*. This contradicts Lemma 10. Theorem 1 is proved.

Note: Recently, Tapan Mitra's article has come to our attention, in which he obtains the same least upper bound by a method different from ours (see Mitra 1996).

References

Boldrin, M. and R. Deneckere (1990), "Sources of Complex Dynamics in Two-Sector Growth Models," *Journal of Economic Dynamics and Control*, 14, 627–653.

Boldrin, M. and L. Montrucchio (1986), "On the Indeterminacy of Capital Accumulation Paths," *Journal of Economic Theory*, 40, 26–39.

Brock, W. and J. Scheinkman (1976), "Global Asymptotic Stability of Optimal Control Systems with Applications to the Theorem of Economic Growth," *Journal of Economic Theory*, 12, 164–190.

Cass, D. and K. Shell (1976), "The Structure and Stability of Competitive Dynamical Systems," *Journal of Economic Theory*, 12, 31–70.

Cass, D. and K. Shell (1983), "Do Sunspots Matter?," *Journal of Political Economy*, 91, 193–227.

Deneckere, R. and S. Pelikan (1986), "Competitive Chaos," *Journal of Economic Theory*, 40, 13–25.

Li, T.-Y. and J. Yorke (1975), "Period Three Implies Chaos," *American Mathematical Monthly*, 82, 985–992.

Majumdar, M. and T. Mitra (1994), "Periodic and Chaotic Programs of Optimal Intertemporal Allocation in an Aggregate Model with Wealth Effects," *Economic Theory*, 4, 649–676.

McKenzie, L. (1983), "Turnpike Theory, Discounted Utility, and the Von Neumann Facet," *Journal of Economic Theory*, 30, 330–352.

McKenzie, L. (1986), Optimal Economic Growth, Turnpike Theorems and Comparative Dynamics, in *Handbook of Mathematical Economics: Vol. III*, ed. by Arrow, K. and M. Intriligator, North-Holland, Amsterdam.

Mitra, T. (1996), "An Exact Discount Factor Restriction for Period Three Cycles in Dynamic Optimization Models," *Journal of Economic Theory*, 69, 281–305.

Montrucchio, L. (1986), "Dynamic Complexity of Optimal Paths and Discount Factors for Strongly Concave Problems," *Journal of Optimization Theory and Applications*, 80, 385–406.

Nishimura, K. and M. Yano (1995), "Non-Linear Dynamics and Chaos in Optimal Growth: An Example," *Econometrica*, 63, 981–1001.

Nishimura, K. and M. Yano (1994), "Optimal Chaos, Non-Linearity and Feasibility Conditions," *Economic Theory*, 4, 689–704.

Nishimura, K., Sorger, G. and M. Yano (1994), "Ergodic Chaos in Optimal Growth Models with Low Discount Rates," *Economic Theory*, 4, 705–717.

Sarkovskii, A. (1964), "Coexistence of Cycles of a Continuous Map of the Line into Itself," *Ukrainian Mathematical Journal*, 16, 61–71.
Scheinkman, J. (1976), "On Optimal Steady States of *n*-Sector Growth Models when Utility is Discounted," *Journal of Economic Theory*, 12, 11–30.
Shell, K. (1977), Monnaie et Allocation Intertemporelle, mimeo, Seminaire d'Econometrie Roy-Malinvaud, Centre National de la Recherche Scientifique, Paris, November 21.
Sorger, G. (1992a), "On the Minimum Rate of Impatience for Complicated Optimal Growth Paths," *Journal of Economic Theory*, 12, 11–30.
Sorger, G. (1992b), Minimum Impatience Theorems for Recursive Economic Models, Springer-Verlag, Berlin.
Sorger, G. (1994), "Period Three Implies Heavy Discounting," *Mathematics of Operations Research*, 19, 1–16.

Part III
Global Dynamics in Optimal Growth Models

Chapter 9
Intertemporal Complementarity and Optimality: A Study of a Two-Dimensional Dynamical System*

Tapan Mitra and Kazuo Nishimura**

9.1 Introduction

The theory of optimal intertemporal allocation has been developed primarily for the case in which the objective function of the planner or representative agent can be written as

$$U(c_1, c_2 \ldots) \equiv \sum_{t=1}^{\infty} \delta^{t-1} w(c_t) \tag{9.1}$$

where c_t represents consumption at date t, w the period felicity function, and $\delta \in (0, 1)$ a discount factor, representing the time preference of the agent.

An objective function like (9.1) leads naturally to the study of dynamic optimization problems of the following "reduced form":

*International Economic Review 46, 93–131, 2005.

**Earlier versions of this article were presented in a conference on nonlinear dynamics in Paris in June, 1999, and in a seminar in Kyoto University in January 2001. We are grateful to Jess Benhabib, Cuong Le Van, Roger Farmer and Takashi Kamihigashi for comments received during these presentations. In particular, the final subsection owes much to an observation made by Takashi Kamihigashi. The current version has benefited from comments by five referees of this journal.

T. Mitra (✉)
Department of Economics, Cornell University, USA
e-mail: tm19@cornell.edu

K. Nishimura
Institute of Economic Research, Kyoto University, Japan
e-mail: nishimura@kier.kyoto-u.ac.jp

J. Stachurski et al. (eds.), *Nonlinear Dynamics in Equilibrium Models*,
DOI 10.1007/978-3-642-22397-6_9,

$$\left.\begin{array}{c}\max \sum_{t=0}^{\infty} \delta^t u(x_t, x_{t+1}) \\ \text{subject to } (x_t, x_{t+1}) \in \Omega \ \text{ for } t \in \{0, 1, 2, \ldots\} \\ x_0 = x\end{array}\right\} \tag{9.2}$$

where $\delta \in (0, 1)$ is the discount factor, X is a compact set (representing the state space), $\Omega \subset X \times X$ is a transition possibility set, $u\colon \Omega \to \mathbb{R}$ is a utility function, and $x \in X$ is the initial state of the system.

The restrictive form of the objective function (9.1) has often been criticized, and alternative forms have been suggested. Since imposing no structure on $U(c_1, c_2, \ldots)$ will yield very little useful information about the nature of optimal programs, the alternative formulations involve some restrictions, of course, and these basically take one of two forms.

First, one can dispense with the *time-additively separable* nature of (9.1), by following Koopmans (1960) and Koopmans et al. (1964), and postulate that there is an *aggregator function*, A, such that

$$U(c_1, c_2, \ldots) = A(c_1, U(c_2, c_3, \ldots)) \tag{9.3}$$

A nice feature of (9.3) is that it preserves the *recursive* nature of the problem inherent in Ramsey-type problems based on (9.1). The restriction is that the *independence of tastes* between periods that was present in (9.1) is also implicit in (9.3). Optimal growth problems with (9.3) as the objective function have been investigated quite extensively, starting with Iwai (1972); a useful reference for this literature is Becker and Boyd (1997).

Second, one can preserve the time-additive separable form, but explicitly model the intertemporal *dependence of tastes* by postulating that the felicity derived by the agent in period t depends on consumption in period t (c_t), but the felicity function itself is (endogenously) determined by past consumption (c_{t-1}). (The fact that "past consumption" is reflected completely in c_{t-1} is a mathematical simplification; consumption in several previous periods can clearly be allowed for at the expense of cumbersome notation and significantly more tedious algebraic manipulations). This formulation leads to the objective function[1]:

$$U(c_1, c_2, \ldots) = \sum_{t=1}^{\infty} \delta^{t-1} w(c_t, c_{t+1}) \tag{9.4}$$

[1] This objective function also arises in a somewhat different class of models, which study economic growth with *altruistic preferences*. For this literature, see, for example, Dasgupta (1974), Kohlberg (1976), Lane and Mitra (1981), and Bernheim and Ray (1987). The focus of this literature is however not on the socially optimal solution, but the intergenerational Nash equilibrium solutions.

Models of optimal growth with intertemporal dependence in tastes, in which the objective function is similar to (9.4), have been examined by several authors.[2] To the best of our knowledge, the specific form (9.4) was first used by Samuelson (1971), to capture the essential features of such intertemporal dependence of tastes.

An objective function like (9.4) leads to the study of dynamic optimization problems of the following "reduced form":

$$\left.\begin{array}{c}\max \sum_{t=0}^{\infty} \delta^t u(x_t, x_{t+1}, x_{t+2}) \\ \text{subject to } (x_t, x_{t+1}, x_{t+2}) \in \Lambda \text{ for } t \in \{0, 1, 2, \ldots\} \\ (x_0, x_1) = (x, y)\end{array}\right\} \quad (9.5)$$

where $\delta \in (0, 1)$ is the discount factor, X is a compact set, $\Omega \subset X \times X$ is a transition possibility set, $\Lambda = \{(x, y, z): (x, y) \in \Omega \text{ and } (y, z) \in \Omega\}$, $u: \Lambda \to \mathbb{R}$ is a utility function, and $(x, y) \in \Omega$ is the initial state of the system.

Note that even under intertemporal dependence in tastes, we have a recursive structure in the dynamic optimization problem (9.5) very much like in (9.2) (and in optimization problems involving (9.3) as the objective function). The difference is that in dealing with a one capital good model (like the standard one- or two-sector models of neoclassical growth theory), the state space is X in problem (9.2), whereas it is a subset of X^2 in problem (9.5). Thus, for problem (9.2), (optimal) value and policy functions are defined on X, and for problem (9.5), these functions are defined on $\Omega \subset X^2$. In terms of examining the dynamic behavior of optimal programs, we are therefore, dealing with a one-dimensional dynamical system for problem (9.2) and a two-dimensional dynamical system for problem (9.5).

The structure of recursive problems like (9.5) are not as well understood as that of (9.2), and we feel that it is worthy of a systematic study. Specifically, one might explore two themes: (1) identifying the conditions under which the results of the traditional Ramsey-type theory are preserved even when the intertemporal independence assumption is relaxed; (2) examining alternative scenarios in which the asymptotic behavior of an optimal program is qualitatively different (from its traditional Ramsey counterpart) because of the presence of intertemporal complementarity. Local analysis of the first theme has been presented by Samuelson (1971) and of the second by Boyer (1978) and others. Our principal interest in this article is in establishing global results on the first theme, and in relating them to the local results, by using the mathematical theory of two-dimensional dynamical systems.[3]

The plan of the article is as follows. After describing the model in Sect. 9.2, we develop the basic properties of the (optimal) value function, V, and the (optimal) policy function, h, in Sect. 9.3. A useful tool for our study is the *ϕ-policy function*, defined on X, by

$$\phi(x) = h(x, x) \; for \;\; x \in X \quad (9.6)$$

[2]The earlier literature on this topic includes, among others, Chakravarty and Manne (1968), and Wan (1970). Heal and Ryder (1973) present a continuous-time model which accommodates a more general dependence structure.

[3]The second theme is explored in detail in Mitra and Nishimura (2001).

It is introduced in Sect. 9.3, and the circumstances under which it satisfies a "single-crossing condition" are examined.

Section 9.4 might be considered as providing the global analytical counterpart to Samuelson's (1971) local analysis of "turnpike behavior" in this model. We show that when the (reduced-form) utility function, u, is *supermodular* on its domain, Λ, then the optimal policy function is monotone increasing in both arguments. This property, together with the "single-crossing condition" on ϕ allows us to establish global asymptotic stability of optimal programs with respect to the (unique) stationary optimal stock, by using an interesting stability result for second-order difference schemes.

In Sect. 9.5, we provide an analysis of the local dynamics of optimal solutions. To this end, we study the fourth order difference equation, which represents the linearized version of the Ramsey–Euler equations near the stationary optimal stock. This equation yields four characteristic roots and we show how two of them are selected by the optimal solution (assuming that the optimal policy function is continuously differentiable in a neighborhood of the stationary optimal stock). The roots selected by the optimal solution provide information about the speed of convergence of nonstationary optimal trajectories to the stationary optimal stock.

The theory linking the derivative of the optimal policy function to the "dominated" characteristic root associated with the Ramsey–Euler equation, for the optimization problem (9.2) is, of course, well-known. To our knowledge, the corresponding theory for problem (9.5) has not been developed in the literature.

In Sect. 9.5.3, the optimal policy function is shown to be continuously differentiable in a neighborhood of the stationary optimal stock, by using the Stable Manifold Theorem.[4] This validates the conclusions which are reached in Sects. 9.5.1 and 9.5.2, by assuming this property.

9.2 Preliminaries

9.2.1 *Model*

Our framework is specified by a *transition possibility set*, Ω, a (reduced form) *utility function*, u, and a *discount factor*, δ. We describe each of these objects in turn.

A *state space* (underlying the transition possibilities) is specified as an interval $X \equiv [0, B]$, where $0 < B < \infty$. The transition possibility set, Ω, is a subset of X^2, satisfying the following assumptions:

(A1) $(0, 0)$ and (B, B) are in Ω; if $(0, y) \in \Omega$ then $y = 0$.
(A2) Ω is closed and convex.

[4]The global differentiability of the optimal policy function for problem (9.2) has been studied by Araujo (1991), Santos (1991) and Montrucchio (1998). The relation of the characteristic roots associated with the optimal policy function to those associated with the Ramsey–Euler equation at the steady state has been studied for problem (9.2) by Araujo and Scheinkman (1977) and Santos (1991).

(A3) If $(x, y) \in \Omega$ and $x \leq x' \leq B$, $0 \leq y' \leq y$, then $(x', y') \in \Omega$.
(A4) There is $(\bar{x}, \bar{y}) \in \Omega$ with $\bar{y} > \bar{x}$.

These assumptions are standard in the literature. Note that (A3) means that the transition possibility set Ω allows *free disposal*, so long as the stock level does not exceed B. Assumption (A4) implies the existence of *expansible* stocks.

Note that for all $x \in [0, B]$, we have $(x, x) \in \Omega$. Associated with Ω is the correspondence $\Psi: X \to X$, given by $\Psi(x) = \{y: (x, y) \in \Omega\}$. Define the set

$$\Lambda = \{(x, y, z): (x, y) \in \Omega \text{ and } (y, z) \in \Omega\}.$$

The utility function, u, is a map from Λ to $\mathbb{R}$. It is assumed to satisfy:

(A5) u is continuous and concave on Λ, and strictly concave in the third argument.
(A6) u is nondecreasing in the first argument, and nonincreasing in the third argument.

In what follows, we will normalize $u(0, 0, 0) = 0$; also, we will denote $\max\limits_{(x,y,z) \in \Lambda} |u(x, y, z)|$ by $\bar{B}$.

The discount factor, δ, reflects how future utilities are evaluated compared to current ones. We assume

(A7) $0 < \delta < 1$.

9.2.2 Programs

The *initial condition* (which should be considered to be historically given) is specified by a *pair* (x, y) in Ω. A *program* (x_t) from (x, y) is a sequence satisfying

$$x_0 = x,\ x_1 = y,\ (x_t, x_{t+1}) \in \Omega \quad \text{for } t \geq 1. \tag{9.7}$$

Thus, in specifying a program, the period 0 and period 1 states are historically given. Choice of future states starts from $t = 2$. Note that for a program (x_t) from $(x, y) \in \Omega$, we have $(x_t, x_{t+1}, x_{t+2}) \in \Lambda$ for $t \geq 0$.

An *optimal program* $(\bar{x}_t)$ from $(x, y) \in \Omega$ is a program from (x, y) satisfying

$$\sum_{t=0}^{\infty} \delta^t u(x_t, x_{t+1}, x_{t+2}) \leq \sum_{t=0}^{\infty} \delta^t u(\bar{x}_t, \bar{x}_{t+1}, \bar{x}_{t+2}) \tag{9.8}$$

for every program (x_t) from (x, y).

Under our assumptions, a standard argument suffices to ensure the existence of an optimal program from every initial condition $(x, y) \in \Omega$. Using Assumptions (A2) and (A5), it can also be shown that this optimal program is unique.

9.2.3 *Value and Policy Functions*

We can define a *value function*, $V : \Omega \to \mathbb{R}$ by

$$V(x, y) = \sum_{t=0}^{\infty} \delta^t u(\bar{x}_t, \bar{x}_{t+1}, \bar{x}_{t+2}) \tag{9.9}$$

where $(\bar{x}_t)$ is the optimal program from (x, y). Then, V is concave and continuous on Ω.

It can be shown that for each $(x, y) \in \Omega$, the Bellman equation

$$V(x, y) = \max_{(y,z)\in\Omega} [u(x, y, z) + \delta V(y, z)] \tag{9.10}$$

holds. Also, V is the unique continuous function on Ω, which solves the functional equation (9.10).

For each $(x, y) \in \Omega$, we denote by $h(x, y)$ the value of z that maximizes $[u(x, y, z) + \delta V(y, z)]$ among all z satisfying $(y, z) \in \Omega$. Then, a program (x_t) from $(x, y) \in \Omega$ is an optimal program from (x, y) if and only if

$$V(x_t, x_{t+1}) = u(x_t, x_{t+1}, x_{t+2}) + \delta V(x_{t+1}, x_{t+2}) \text{ for } t \geq 0. \tag{9.11}$$

This, in turn, holds if and only if

$$x_{t+2} = h(x_t, x_{t+1}) \text{ for } t \geq 0. \tag{9.12}$$

We will call h the (optimal) *policy function*. It can be shown by using standard arguments that h is continuous on Ω.

9.2.4 *Two Examples*

9.2.4.1 Optimal Growth with Intertemporally Dependent Preferences

The example (which follows Samuelson (1971) and Boyer (1978) closely) captures the feature that tastes between periods are intertemporally dependent. Such a model can be described in terms of a *production function*, f, a *welfare function*, w, and a *discount factor*, δ.

Let $X = [0, B]$ be the state space with $0 < B < \infty$. The production function, f, is a function from X to itself which satisfies

(f) $f(0) = 0, f(B) = B$; f is increasing, concave and continuous on X.

The welfare function, w, is a function from X^2 to $\mathbb{R}$, which satisfies

(w) w is continuous and concave on X^2, and strictly concave in the second argument; it is nondecreasing in both arguments.[5]

The discount factor, δ, is as usual assumed to satisfy:

(d) $0 < \delta < 1$.

A program, in this framework, is described by a sequence (k_t, c_t), where k_t denotes the capital stock and c_t the consumption in period t. The initial condition is specified by $(k, c) \geq 0$, where $k + c \leq B$.

Formally, a program (k_t, c_t) from (k, c) is a sequence satisfying

$$\left.\begin{array}{r}(k_1, c_1) = (k, c),\ k_{t+1} = f(k_t) - c_{t+1} \text{ for } t \geq 1 \\ 0 \leq c_{t+1} \leq f(k_t) \text{ for } t \geq 1.\end{array}\right\} \tag{9.13}$$

An optimal program from (k, c) is a program $(\bar{k}_t, \bar{c}_t)$ satisfying

$$\sum_{t=1}^{\infty} \delta^{t-1} w(c_t, c_{t+1}) \leq \sum_{t=1}^{\infty} \delta^{t-1} w(\bar{c}_t, \bar{c}_{t+1}) \tag{9.14}$$

for every program (k_t, c_t) from (k, c).

To reduce the optimality exercise in (9.14) subject to (9.13) to the one in (9.5), we can proceed as follows. First, the transition possibility set, Ω, can be defined as

$$\Omega = \{(x, y) : x \in X,\ 0 \leq y \leq f(x)\}.$$

Second, the reduced form utility function can be defined, for (x, y, z) in Λ as

$$u(x, y, z) = w(f(x) - y,\ f(y) - z).$$

Finally, the initial condition (k, c) in the example translates to the initial condition in the framework of Sect. 9.2.2 as $(x, y) = (f^{-1}(k_1 + c_1), k_1)$. That is, x is the capital stock (in period 0) which produced the output $(k_1 + c_1)$ in period 1, that was split up between consumption (c_1) and capital stock (k_1) in period 1; y is the capital stock in period 1. The choice of consumption decisions, c_t, starts from $t \geq 2$; correspondingly, the state variable, x_t, is determined for $t \geq 2$ by the following equation:

$$x_{t+1} = k_{t+1} = f(k_t) - c_{t+1} \text{ for } t \geq 1. \tag{9.15}$$

It is worth noting that in the model of Samuelson (1971) there is no maximization with respect to c_1. This is because in order for his problem to be well posed, one

[5] Boyer (1978) assumes that w is increasing in both arguments. Samuelson (1971) does not; he assumes instead that $w(c, c)$ is increasing in c. It is this latter assumption that is crucial in proving the uniqueness of a stationary optimal stock in this model, and therefore of our "single-crossing property"; see Sect. 9.3.3.

needs to know both k_0 and c_1 (and, therefore, both x_0 and x_1 in terms of the problem stated in (9.5)) The information about c_1 is needed to *define* the welfare function $w(c_1, c_2)$. That is, the welfare in period 2 depends on the choice of c_2, but the welfare function itself is determined endogenously by past consumption (c_1).[6]

9.2.4.2 Optimal Harvesting of a Renewable Resource with Delayed Recruitment

The theory of management of renewable resources deals with the issue of optimal harvesting of biological populations, such as various species of marine life. For many species, recruitment to the breeding population takes place only after a delay. Clark (1976) has modeled this phenomenon by describing the population dynamics by a delay-difference equation, instead of the standard first-order difference equation that is commonly used in the literature on renewable resources. We describe a simple version of his model where the delay involved is two periods.[7]

The model can be described formally in terms of a *recruitment function*, F, a *return function*, W, a *survival coefficient*, λ, and a *discount factor*, δ.

The recruitment function, F, is a function from $\mathbb{R}_+$ to itself which satisfies

(F) $F(0) = 0$; F is increasing, concave and continuous on X; $\lim_{x\to 0}[F(x)/x] > 1$, $\lim_{x\to\infty}[F(x)/x] = 0$.

The return function, w, is a function from $\mathbb{R}_+$ to $\mathbb{R}$, which satisfies

(W) W is continuous, nondecreasing and strictly concave on $\mathbb{R}_+$.

The survival coefficient, λ, satisfies

(s) $0 < \lambda < 1$.

The discount factor, δ, is as usual assumed to satisfy

(d) $0 < \delta < 1$.

Given (F), there is a unique positive number B, such that $[F(B)/B] = (1 - \lambda)$. Then, defining $f(x) = F(x) + \lambda x$ for all $x \in \mathbb{R}_+$, we see that (1) $f(B) = B$, (2) $B > f(x) > x$ for $x \in (0, B)$, and (3) $B < f(x) < x$ for $x > B$. Thus, it is natural to choose the state space to be $X = [0, B]$.

A program, in this framework, is described by a sequence (k_t, c_t), where k_t denotes the biomass of the adult breeding population and c_t the harvest of this

[6]Of course, variations of problem (9.5) can arise, where the last line of (9.5) would simply say $x_0 = x$. Solving such a problem would involve solving (9.5) and in addition solving for the "correct" x_1. Clearly, an optimal solution of such a problem must solve (9.5), and therefore inherit all the dynamic properties of such a solution, as described in this article.

[7]The modeling of the recruitment delay as two periods in our formulation of the model of renewable resource management is a mathematical simplification; recruitment delays of longer duration can clearly be allowed for. The corresponding theory is somewhat harder to present and analyze.

population in period t. The initial condition is specified by $(k, k') \geq 0$, where $k \leq B$ and $k' \leq B$.

Formally, a program (k_t, c_t) from (k, k') is a sequence satisfying

$$\left.\begin{aligned} (k_0, k_1) = (k, k'), \; k_{t+1} = \lambda k_t + F(k_{t-1}) - c_{t+1} \text{ for } t \geq 1 \\ 0 \leq c_{t+1} \leq \lambda k_t + F(k_{t-1}) \;\text{ for } t \geq 1. \end{aligned}\right\} \tag{9.16}$$

Note that for a program (k_t, c_t) from $(k, k') \leq (B, B)$, we have $(k_t, c_t) \leq (B, B)$ for all $t \geq 2$, and this justifies our choice of the state space as $X = [0, B]$.

An optimal program from (k, k') is a program $(\bar{k}_t, \bar{c}_t)$ satisfying

$$\sum_{t=1}^{\infty} \delta^{t-1} W(c_{t+1}) \leq \sum_{t=1}^{\infty} \delta^{t-1} W(\bar{c}_{t+1}) \tag{9.17}$$

for every program (k_t, c_t) from (k, k').

To explain the population dynamics, the adult breeding population k_0 at time 0 yields a "recruitment" to the population in period 2 (that is, after a delay of two periods) of $F(k_0)$. Part of the adult breeding population k_1 at time 1 does not survive beyond period 1; the remaining part is λk_1. The total available output of the renewable resource at time 2 is, therefore, $F(k_0) + \lambda k_1$. A part of this resource (c_2) is harvested in period 2. The remainder of the resource $(F(k_0) + \lambda k_1 - c_2)$ becomes the adult breeding population k_2 at time 2. This process is then repeated indefinitely.

To reduce the optimality exercise in (9.17) subject to (9.16) to the one in (9.5), we can proceed as follows. First, the set, Λ, can be defined as

$$\Lambda = \{(x, y, z) : x \in X, \; y \in X, 0 \leq z \leq \lambda y + F(x)\}.$$

Second, the reduced form utility function can be defined, for (x, y, z) in Λ as

$$u(x, y, z) = W(\lambda y + F(x) - z).$$

Finally, the initial condition (k, k') in the example translates to the initial condition in the framework of Sect. 9.2.2 as $(x, y) = (k, k')$. The choice of consumption decisions, c_t, starts from $t \geq 2$; correspondingly, the state variable, x_t, is determined for $t \geq 2$ by the following equation:

$$x_{t+1} = k_{t+1} = \lambda k_t + F(k_{t-1}) - c_{t+1} \text{ for } t \geq 1. \tag{9.18}$$

Note that the dynamic optimization problem of the form (9.5) arises in the renewable resource example from the (biological) production side of the model rather than the preference side.

9.3 Basic Properties of Value and Policy Functions

In this section, we examine some basic properties of the value and policy functions. These properties will be useful in conducting the analysis in the following sections.

9.3.1 Value Function

We proceed under the following additional assumption:

(A8) There is $\hat{x}$ in $(0, B)$, such that $(\hat{x}, \hat{x}/\delta, \hat{x}/\delta^2) \in \Lambda$, and $\theta \equiv u(\hat{x}, \hat{x}/\delta, \hat{x}/\delta^2) > u(0,0,0) = 0$.

Assumption (A8) is a *δ-productivity assumption* jointly on (Λ, u, δ). It is analogous to the δ-productivity assumption in the usual reduced-form model, where it is used to establish the existence of a non-trivial stationary optimal stock.

Lemma 1. *Let $N \geq 2$ be a given positive integer. Defining $x = \delta^N \hat{x}$, we have $(x, x/\delta) \in \Omega$, and*

$$V(x, x/\delta) \geq [(N-1)\theta/\hat{x}]x. \tag{9.19}$$

Proof. Since $(\hat{x}, \hat{x}/\delta) \in \Omega$ and $(0,0) \in \Omega$, we have $(\delta^n \hat{x}, \delta^n(\hat{x}/\delta)) \in \Omega$ for $n \geq 1$. Using this observation, the sequence $(x_t) = \big(x, (x/\delta), (x/\delta^2), \ldots, (x/\delta^N), (x/\delta^{N+1}), 0, 0, \ldots\big)$ is a program from $(x, (x/\delta))$. Note that $(x/\delta^N) = \hat{x}$, $(x/\delta^{N+1}) = (\hat{x}/\delta)$, and since $(\hat{x}, \hat{x}/\delta, \hat{x}/\delta^2) \in \Lambda$ by (A8), we have $(\hat{x}, \hat{x}/\delta, 0) \in \Lambda$ by (A3), and $u(\hat{x}, \hat{x}/\delta, 0) \geq u(\hat{x}, \hat{x}/\delta, \hat{x}/\delta^2) > 0$. Also $((\hat{x}/\delta), 0) \in \Omega$ and $(0,0) \in \Omega$ imply that $(\hat{x}/\delta, 0, 0) \in \Lambda$, and $u(\hat{x}/\delta, 0, 0) \geq u(0,0,0)$ (by (A6)) $= 0$. For $0 \leq t \leq N-2$

$$\begin{aligned} u(x_t, x_{t+1}, x_{t+2}) &= u(x/\delta^t, x/\delta^{t+1}, x/\delta^{t+2}) = u(\hat{x}\delta^{N-t}, \hat{x}\delta^{N-t-1}, \hat{x}\delta^{N-t-2}) \\ &\geq \delta^{N-t} u(\hat{x}, \hat{x}/\delta, \hat{x}/\delta^2) + (1 - \delta^{N-t}) u(0,0,0). \end{aligned}$$

Thus, for $0 \leq t \leq N-2$, $\delta^t u(x_t, x_{t+1}, x_{t+2}) \geq \delta^N u(\hat{x}, \hat{x}/\delta, \hat{x}/\delta^2)$, and we have

$$V(x, x/\delta) \geq \sum_{t=0}^{N-2} \delta^t u(x_t, x_{t+1}, x_{t+2}) \geq (N-1)\delta^N \theta = [(N-1)\theta/\hat{x}]x$$

which establishes the Lemma. □

Proposition 1. *The value function, V, satisfies the property*

$$[V(x, x/\delta)/x] \to \infty \text{ as } x \to 0. \tag{9.20}$$

Proof. For $(x, x/\delta) \in \Omega$, and $0 < \lambda < 1$, we have $V(\lambda x, \lambda x/\delta) \geq \lambda V(x, x/\delta) + (1-\lambda)V(0,0) = \lambda V(x, x/\delta)$. Using Lemma 1, and defining the sequence $\{x(N)\}$

by $x(N) = \delta^N \hat{x}$ for $N = 2, 3, \ldots$, we have $[V(x(N), x(N)/\delta)/x(N)] \to \infty$ as $N \to \infty$. Then, (9.20) follows since for $x \in [\delta^{N+1}\hat{x}, \delta^N \hat{x}]$, $V(x, x/\delta)/x \geq [V(\delta^N \hat{x}, \delta^N(\hat{x}/\delta))/\delta^N \hat{x}]$. □

Proposition 2. *The value function, V, satisfies the property*

$$[V(x, x)/x] \to \infty \text{ as } x \to 0. \tag{9.21}$$

Proof. For $0 < x \leq \hat{x}$, we have $(x, x/\delta) \in \Omega$, and $(x, 0) \in \Omega$, so $(\delta x + (1 - \delta)x, \delta(x/\delta) + (1 - \delta) \cdot 0) \in \Omega$; that is $(x, x) \in \Omega$. By concavity of V, we have

$$\begin{aligned} V(x, x) &= V(\delta x + (1 - \delta)x, \delta(x/\delta) + (1 - \delta) \cdot 0) \\ &\geq \delta V(x, x/\delta) + (1 - \delta)V(x, 0) \\ &\geq \delta V(x, x/\delta). \end{aligned}$$

Thus $[V(x, x)/x] \to \infty$ as $x \to 0$ by Proposition 1. □

9.3.2 Policy Function

A useful tool, related to the policy function, is the *ϕ-policy function* defined for $x \in X$ by

$$\phi(x) = h(x, x) \text{ for } x \in X.$$

That is, ϕ gives us the optimal policy when the arguments in h happen to take on identical values.

In the standard reduced-form model, if x_t were constant for two successive periods along an optimal program, the constant value would have to be a stationary optimal stock. Here, given $x_{t-1} = x_t = x$ in X, $\phi(x)$ is not necessarily equal to x; in fact, it will typically be different from x. If $\phi(x) = x$, then x would be a stationary optimal stock in the present framework.

We proceed under the following additional assumption:

(A9) There is $A > 0$, such that for all (x, y, z), (x', y', z') in Λ, $|u(x, y, z) - u(x', y', z')| \leq A \|(x, y, z) - (x', y', z')\|$.

Assumption (A9) is a bounded-steepness assumption on the utility function, and this is ensured by making u Lipschitz-continuous, with Lipschitz constant A. The norm used in (A9) is the sum-norm; that is, $\|(x, y, z)\| = |x| + |y| + |z|$ for (x, y, z) in $\mathbb{R}^3$. (In the usual reduced-form model, a condition like (A9) was introduced by Gale (1967), to establish the existence of shadow-prices, associated with optimal programs).

Proposition 3. *There is $a > 0$ such that for all $x \in (0, a)$, $\phi(x) > x$.*

Proof. Suppose, on the contrary, there is a sequence (x^s), such that $x^s \to 0$ as $s \to \infty$, and $x^s > 0$, $\phi(x^s) \leq x^s$ for all s.

Using Proposition 2, we can find $a_1 > 0$, such that for $x \in (0, a_1)$, we have

$$[V(x,x)/x] > 4A/(1-\delta). \tag{9.22}$$

Since $x^s \to 0$, we can find s large enough for which $0 < x^s < a_1$. Pick such an x^s and call it x. Then $x \in (0, a_1)$ and $\phi(x) \leq x$. Denote $\phi(x)$ by y, and $h(x, y)$ by z.

Since $y \leq x$, and $(y, z) \in \Omega$, we have $(x, z) \in \Omega$, and $(x, y, z) \in \Lambda$,and

$$\begin{aligned} V(x,x) &\geq u(x,x,z) + \delta V(x,z) \\ &\geq u(x,x,z) + \delta V(y,z) \\ &= [u(x,x,z) - u(x,y,z)] + \delta V(y,z) + u(x,y,z) \\ &= [u(x,x,z) - u(x,y,z)] + V(x,y) \\ &\geq V(x,y) - Ax \end{aligned}$$

the final inequality following from (A9). We can now write

$$\begin{aligned} V(x,x) &= u(x,x,y) + \delta V(x,y) \\ &\leq u(x,x,y) + \delta V(x,x) + \delta Ax \end{aligned}$$

so that

$$\begin{aligned} V(x,x) &\leq [u(x,x,y)/(1-\delta)] + \delta Ax/(1-\delta) \\ &\leq [A(2x+y) + \delta Ax]/(1-\delta) \end{aligned}$$

by using (A9) again. Thus, using $y \leq x$,we have $[V(x,x)/x] \leq (3+\delta)A/(1-\delta)$ which contradicts (9.22).

We now introduce an additional assumption for our next result.

(A10) $u(B, B, B) \leq u(0, 0, 0) = 0$.

Assumption (A10) is an expression of the fact that "inaction" can produce at least as much utility as "excessive action". In the context of the standard aggregative growth model, considered in Sect. 9.2.4.1, it translates to the fact that the consumption level associated with the maximum sustainable stock is 0 and so is the consumption associated with the zero stock. Actually the standard aggregative growth model does not model disutility of effort directly. Typically, maintaining high stocks involves considerable effort, which has disutility, and this will reinforce the circumstances under which (A10) will hold.

Proposition 4. *The ϕ-policy function satisfies*

$$\phi(B) < B. \tag{9.23}$$

Proof. Suppose, on the contrary, that $h(B, B) = B$. Then, we have $V(B, B) = u(B, B, B)/(1-\delta) \leq u(0, 0, 0)/(1-\delta) = 0$. Define the sequence (x_t) as follows:

$$(x_0, x_1) = (B, B); \ x_t = 0 \ \text{ for } t \geq 2.$$

Then, (x_t) is a program from (B, B), and we have

$$\sum_{t=0}^{\infty} \delta^t u(x_t, x_{t+1}, x_{t+2}) = u(B, B, 0) + \delta u(B, 0, 0) \geq 0$$

the inequality following from (A3) and (A6). This means that $(B, B, 0, 0, \ldots)$ and $(B, B, B, B, \ldots)$ are both optimal from (B, B). But this contradicts the fact that u is strictly concave in the third argument. This establishes the result. □

Proposition 5. *There is some $x^* \in (0, B)$, such that x^* is a stationary optimal stock; that is,*

$$h(x^*, x^*) = x^*. \tag{9.24}$$

Proof. By Proposition 4, $\phi(B) < B$. By Proposition 3, we can find $x \in (0, B)$, such that $\phi(x) > x$. By continuity of ϕ, there is $x^* \in (0, B)$ such that $\phi(x^*) = x^*$. □

9.3.3 A Single Crossing Condition

In the following sections, we will find it useful to assume that the ϕ-policy function (introduced in Sect. 9.3.2) has the following "single-crossing property":

$$\left.\begin{array}{c}\text{There is } 0 < x^* < B, \text{ such that} \\ \phi(x^*) = x^*; x < \phi(x) \text{ for } 0 < x < x^*; x > \phi(x) \text{ for } x > x^*.\end{array}\right\} \tag{SC}$$

Given Propositions 3 and 4, there is a stationary optimal stock in $(0, B)$, and the single-crossing property holds if there is a unique stationary optimal stock in $(0, B)$. Note that if $x \in (0, B)$ is a stationary optimal stock, then $h(x, x) = x$, and so

$$u_3(x, x, x) + \delta u_2(x, x, x) + \delta^2 u_1(x, x, x) = 0. \tag{9.25}$$

In the example discussed in Sect. 9.2.4.1, with w and f both C^1, $w_1 > 0$ and $w_2 > 0$, denoting $f(x) - x$ by c, condition (9.25) is satisfied only if

$$[w_2(c, c) + \delta w_1(c, c)] = \delta f'(x)[w_2(c, c) + \delta w_1(c, c)].$$

This implies that $\delta f'(x) = 1$. Since f is strictly concave, there can be only one stationary optimal stock in $(0, B)$, and the single-crossing property is verified.

If u is C^2 on Λ, then (9.25) has a unique solution if the function

$$H(x) \equiv u_3(x, x, x) + \delta u_2(x, x, x) + \delta^2 u_1(x, x, x) \tag{9.26}$$

has a negative derivative, wherever it has a zero. This amounts to the condition

$$[\delta^2 u_{11} + \delta u_{22} + u_{33}] + (\delta^2 + 1)u_{13} + \delta(\delta + 1)u_{12} + (\delta + 1)u_{23} < 0 \qquad (9.27)$$

being satisfied at any x (the derivatives being evaluated at (x, x, x)) at which $H(x) = 0$. For $\delta \approx 1$, (9.27) is clearly satisfied if u has a negative-definite Hessian.

9.4 Turnpike Behavior

In this section, we will provide sufficient conditions under which one can establish global asymptotic stability of the stationary optimal stock (turnpike property). This demonstrates that "one can relax the independence assumption somewhat and still derive the usual known results", a point indicated earlier by Samuelson (1971), using local analysis around the turnpike. A crucial role in our global analysis is played by the assumption of *supermodularity* of the utility function[8] in its *three* variables, a concept we define below analogously to the more familiar two-variable case.

9.4.1 *Supermodularity of the Utility Function*

A function $G: \Omega \rightarrow \mathbb{R}$ is *supermodular* if whenever (x, y), $(x', y') \in \Omega$ with $(x', y') \geq (x, y)$, we have

$$G(x, y) + G(x', y') \geq G(x', y) + G(x, y')$$

provided (x', y) and $(x, y') \in \Omega$. If G is C^2 on Ω, then it is well-known that G is supermodular on Ω if and only if $G_{12} \geq 0$ on Ω

In our case, the utility function $u: \Lambda \rightarrow \mathbb{R}$ is a function of three variables, and we may define supermodularity of it as follows. In the C^2 case, we would now like to have all the three cross-partials of u to be non-negative; that is u_{12}, u_{13} and $u_{23} \geq 0$ on Λ. In the general (not necessarily differentiable) case, this translates to the following definition.

The utility function $u: \Lambda \rightarrow \mathbb{R}$ is called *supermodular* on Λ if whenever (x, y, z), $(x', y', z') \in \Lambda$ with $(x', y', z') \geq (x, y, z)$, we have

(i) $u(x, y, z) + u(x', y', z') \geq u(x', y, z) + u(x, y', z')$ provided (x', y, z) and $(x, y', z') \in \Lambda$
(ii) $u(x, y, z) + u(x', y', z') \geq u(x, y', z) + u(x', y, z')$ provided (x, y', z) and $(x', y, z') \in \Lambda$

[8]The supermodularity concept is due to Topkis (1968). A nice exposition of the concept in the two-variable case, and its relation to the nonnegativity of the cross partial derivative is given in Ross (1983). Benhabib and Nishimura (1985) introduced its use in optimal economic dynamics in the two variable case in the form of this derivative condition. More recent comprehensive studies involving the concept of supermodularity can be found in Amir et al. (1991), Amir (1996) and Topkis (1998).

(iii) $u(x, y, z) + u(x', y', z') \geq u(x, y, z') + u(x', y', z)$ provided (x, y, z') and $(x', y', z) \in \Lambda$.

If u is C^2 on Λ with u_{12}, u_{13} and $u_{23} \geq 0$ on Λ, then (i) can be verified as follows $A = [u(x', y', z') - u(x, y', z')] - [u(x', y, z) - u(x, y, z)] = \int_x^{x'} u_1(t, y', z')dt - \int_x^{x'} u_1(t, y, z)dt$. Now $u_1(t, y', z') - u_1(t, y, z) \geq 0$ for all $t \in [x, x']$, since $u_{12} \geq 0$ and $u_{13} \geq 0$, $(y' - y) \geq 0$ and $(z' - z) \geq 0$. Thus, we get $A \geq 0$, establishing (i). Conditions (ii) and (iii) can be verified similarly.

Both of the above definitions are, of course, special cases of the general definition of supermodularity of a function on a lattice, as given by Topkis (1968, 1998).

9.4.2 An Example

In order to understand the restriction imposed by the assumption of supermodularity of u, we consider the assumption in the context of the example discussed in Sect. 9.2.4.1, when w and f are both C^2 on their domains.

We can calculate the first-order partial derivatives of u as follows:

$$u_1(x, y, z) = w_1(f(x) - y,\ f(y) - z)f'(x)$$
$$u_2(x, y, z) = w_2(f(x) - y,\ f(y) - z)f'(y) - w_1(f(x) - y,\ f(y) - z)$$
$$u_3(x, y, z) = -w_2(f(x) - y,\ f(y) - z).$$

Since $w_1 > 0$ and $w_2 > 0$, it follows that $u_1 > 0$ and $u_3 < 0$, as required in (A6).

The second-order cross partial derivatives of u can be calculated as follows:

$$u_{12}(x, y, z) = [w_{12}(f(x) - y,\ f(y) - z)f'(y) - w_{11}(f(x) - y,\ f(y) - z)]f'(x)$$
$$u_{13}(x, y, z) = -f'(x)w_{12}(f(x) - y,\ f(y) - z)$$
$$u_{23}(x, y, z) = w_{12}(f(x) - y,\ f(y) - z) - w_{22}(f(x) - y,\ f(y) - z)f'(y).$$

Thus, in order for u to be supermodular, (i) we need the marginal utility of present consumption to be declining in past consumption ($w_{12} < 0$), sometimes referred to as Edgeworth-Pigou substitutability, and (ii) we need the magnitude of this cross effect ($-w_{12}$) to be "small" relative to the magnitudes of the own effects ($-w_{11}$) and ($-w_{22}$). We state requirement (ii) loosely, since the magnitude of the marginal product of capital is involved beside the second-order derivatives of w. However, the requirement (ii) can be seen most transparently at the steady state, where $\delta f'(x^*) = 1$. There, the requirement of supermodularity translates to the condition that in the symmetric matrix

$$W = \begin{bmatrix} \delta^2(-w_{11}) & \delta(-w_{12}) \\ \delta(-w_{12}) & (-w_{22}) \end{bmatrix}$$

the diagonal terms dominate the off-diagonal terms.

We now provide a specific example of the framework discussed in Sect. 9.2.4.1, to show that all the assumptions made on the reduced-form model can be verified with suitable restrictions on the parameters of the primitive-form. Consider the production function, f, defined by

$$f(x) = px - qx^2 \text{ for } x \in [0, (p-1)/q] \equiv [0, B]$$

where $1 < p < 2$ and $q > 0$, and the welfare function w defined by

$$w(c,d) = ac - bc^2 + \alpha d - \beta d^2 - \theta\, cd \text{ for } (c,d) \in X^2$$

where $a > 0, b > 0, \alpha > 0, \beta > 0$ and $\theta > 0$. Note that at $x = B = (p-1)/q$, we have $[f(x)/x] = p - qx = p - q[(p-1)/q] = 1$. Also, $f'(x) = p - 2qx$ for all $x \in X$, so $f'(0) = p$ and $f'(B) = p - 2q[(p-1)/q] = 2 - p$. Since $1 < p < 2$, we have $f'(0) > 1 > f'(B) > 0$.

To ensure that w is increasing in both components of consumption, we impose the following restrictions:

$$a - (\theta + 2b)B > 0; \quad \alpha - (\theta + 2\beta)B > 0. \tag{R1}$$

These restrictions ensure that $u_1(x,y,z) > 0$ and $u_3(x,y,z) < 0$ on Λ, since $f'(x) > 0$ on X.

Notice that $w_{11} = -2b < 0$ and $w_{22} = -2\beta < 0$, so, to ensure concavity of w, we can assume

$$\theta^2 < 4b\beta. \tag{R2}$$

This ensures that u is concave on Λ, since f is concave on X. Further, since $u_{33}(x,y,z) = w_{22}(f(x) - y,\ f(y) - z) = -2\beta < 0$, u is strictly concave in its third argument.

We have $w_{12} = -\theta < 0$ so that $u_{13} > 0$ on Λ. To ensure that $u_{23} > 0$ on Λ, we assume that $(-w_{22})f'(B) > (-w_{12})$; that is,

$$2\beta(2-p) > \theta. \tag{R3}$$

Finally, to ensure that $u_{12} > 0$ on Λ, we assume that $(-w_{11}) > (-w_{12})f'(0)$; that is,

$$2b > \theta p. \tag{R4}$$

Thus, under the restrictions (R1)–(R4), assumptions (w), (f) are satisfied, and so are Assumptions (A1)–(A6). Further, u is supermodular on Λ.

For specific numerical values of the parameters, ensuring that all the above restrictions are simultaneously satisfied, take $p = (3/2)$, $q = (1/2)$, so that $B = 1$ and $X = [0,1]$. Choosing $b = \beta = 1$, $a = 3$, $\alpha = 5$, and $\theta = (1/2)$, it is easy to check that the restrictions (R1)–(R4) are satisfied.

9.4.3 Monotonicity of the Policy Function

The principal result (Theorem 1) of this section is that if the utility function is supermodular then the (optimal) policy function is monotone nondecreasing in each component.

In the case usually treated, where the reduced form utility function is a function of two variables, if the utility function is supermodular, then the policy function is monotone nondecreasing, and this can be established by ensuring that the value function (a function of a single variable) is monotone nondecreasing. This property of the value function is straightforward, given the free-disposal property of the transition possibility set and the fact that the utility function is monotone nondecreasing in its first argument.

In the present context, the value function is a function of two variables, and we need to show that the value function is *supermodular* in these two variables (Proposition 6), when the utility function is supermodular in its three variables. To obtain the supermodularity of the value function from the supermodularity of the utility function, the natural route suggested is to establish supermodularity for each finite-horizon value function, and then obtain this property for the infinite horizon value function as a limit of the finite-horizon ones. The first part of this two-step procedure (Lemma 2) follows from the general result of Topkis (1968), so we state the result without a proof.

Lemma 2. *Let $G: \Omega \to \mathbb{R}$ be a concave, continuous and supermodular function on Ω. If u is supermodular on Λ, then the function $H: \Omega \to \mathbb{R}$ given by*

$$H(x, y) = \max_{z \epsilon \Psi(y)} [u(x, y, z) + \delta G(y, z)] \tag{P}$$

is well defined, and is a concave, continuous and supermodular function on Ω.

Proposition 6. *If u is supermodular on Λ, then V is supermodular on Ω.*

Proof. Define a sequence of functions, $V^t: \Omega \to \mathbb{R}$ given by

$$V^0(x, y) = u(x, y, 0) \text{ and } V^{t+1}(x, y) = \max_{z \epsilon \Psi(y)} [u(x, y, z) + \delta V^t(y, z)].$$

Then V^0 is a concave, continuous and supermodular function on Ω. Using Lemma 2, V^t is a concave, continuous and supermodular function on Ω for each $t \geq 0$.

Since $|u(x, y, z)| \leq \bar{B}$ on Λ, we have $|V^t(x, y)| \leq \bar{B}/(1-\delta)$ on Ω for all $t \geq 0$. To see this, note that it is clearly true for $t = 0$. Assuming this is true for $t = T \geq 0$, we have

$$\left|V^{T+1}(x, y)\right| \leq \bar{B} + \delta[\bar{B}/(1-\delta)] = \bar{B}/(1-\delta).$$

Thus $|V^t(x, y)| \leq \bar{B}/(1-\delta)$ on Ω for all $t \geq 0$ by induction.

We now proceed to show that $V^{t+1}(x, y) \geq V^t(x, y)$ for $t \geq 0$, for all $(x, y) \in \Omega$. For $t = 0$, we have

$$\begin{aligned} V^1(x, y) &= \max_{z \in \Psi(y)} [u(x, y, z) + \delta V^0(y, z)] \\ &\geq u(x, y, 0) + \delta V^0(y, 0) \\ &= u(x, y, 0) + \delta u(y, 0, 0) \\ &\geq u(x, y, 0) = V^0(x, y) \end{aligned}$$

since u is nondecreasing in its first argument and $u(0, 0, 0) = 0$.

Suppose $V^{t+1}(x, y) \geq V^t(x, y)$ for $t = 0, \ldots, T$ where $T \geq 0$. We now show that the inequality must hold for $t = T + 1$ as well. Let $\bar{z}$ be the solution of the maximization problem

$$\max_{z \in \Psi(y)} [u(x, y, z) + \delta V^T(y, z)]$$

given $(x, y) \in \Omega$. Then, by definition of V^{T+2}, we have

$$\begin{aligned} V^{T+2}(x, y) &\geq u(x, y, \bar{z}) + \delta V^{T+1}(y, \bar{z}) \\ &\geq u(x, y, \bar{z}) + \delta V^T(y, \bar{z}) \\ &= V^{T+1}(x, y). \end{aligned}$$

This completes the induction proof.

For each $(x, y) \in \Omega$, define

$$\bar{V}(x, y) = \lim_{t \to \infty} V^t(x, y)$$

Then $\bar{V}$ is well defined and is a concave, continuous and supermodular function on Ω.

Given $(x, y) \in \Omega$, let z^t be the solution to the maximization problem

$$\max_{z \in \Psi(y)} [u(x, y, z) + \delta V^t(y, z)].$$

Then, we have

$$V^{t+1}(x, y) = u(x, y, z^t) + \delta V^t(y, z^t).$$

The sequence $\{z^t\}$ is bounded, and has a convergent subsequence, converging to some $\bar{z}$; clearly $\bar{z} \in \Psi(y)$. For the subsequence on which z^t converges to $\bar{z}$, taking limits we have

$$\bar{V}(x, y) = u(x, y, \bar{z}) + \delta \bar{V}(y, \bar{z}). \tag{9.28}$$

Also, for all $z \in \Psi(y)$, we have

$$V^{t+1}(x, y) \geq u(x, y, z) + \delta V^t(y, z)$$

and so

$$\bar{V}(x, y) \geq u(x, y, z) + \delta \bar{V}(y, z). \tag{9.29}$$

Using (9.28) and (9.29) we have

$$\bar{V}(x, y) = \max_{z \in \Psi(y)} [u(x, y, z) + \delta \bar{V}(y, z)].$$

Thus, $\bar{V}$ is the value function, V, of problem (9.10), and V is supermodular on Ω.

Theorem 1. *If u is supermodular on Λ, then h is nondecreasing in each component.*

Proof. Let (x, y) and $(x', y') \in \Omega$ with $(x', y') \geq (x, y)$. Define $z = h(x, y)$ and $z' = h(x', y')$. We claim that $z' \geq z$. Suppose, on the contrary, that $z' < z$. We know that

$$V(x, y) = u(x, y, z) + \delta V(y, z)$$
$$V(x', y') = u(x', y', z') + \delta V(y', z').$$

Since $(y, z) \in \Omega$ and $z' < z$, $(y, z') \in \Omega$ and

$$V(x, y) > u(x, y, z') + \delta V(y, z').$$

Since $(y, z) \in \Omega$ and $y' \geq y$, $(y', z) \in \Omega$ and

$$V(x', y') > u(x', y', z) + \delta V(y', z).$$

Thus, we get

$$\begin{aligned} &[u(x, y, z) + u(x', y', z')] + \delta[V(y, z) + V(y', z')] \\ &> [u(x, y, z') + u(x', y', z)] + \delta[V(y, z') + V(y', z)]. \end{aligned} \tag{9.30}$$

Since u is supermodular on Λ, and $(x', y') \geq (x, y)$ and $z > z'$,

$$u(x, y, z') + u(x', y', z) \geq u(x, y, z) + u(x', y', z'). \tag{9.31}$$

Since V is supermodular on Ω, and $y' \geq y$ and $z > z'$,

$$\delta[V(y, z') + V(y', z)] \geq \delta[V(y, z) + V(y', z')]. \tag{9.32}$$

Adding (9.31) and (9.32), we contradict (9.30).

9.4.4 Global Dynamics

In this section we study the global dynamics of the two-dimensional dynamical system, (Ω, Γ) where Γ is a map from Ω to Ω given by

$$\Gamma(x, y) = (y, h(x, y)).$$

For $(x, y) \in \Omega$, we have $h(x, y) \in \Psi(y)$, and so $(y, h(x, y)) \in \Omega$.

We maintain the assumption that u is supermodular on Λ, and so h is nondecreasing in both its arguments. We also maintain the single-crossing condition on ϕ, introduced in Sect. 9.3.

The principal result of this section (Theorem 2) is that if (x_t) is an optimal program from (x, y), where $(x, y) \in \Omega$ and $(x, y) \gg 0$, then x_t converges to x^* as $t \to \infty$, thus exhibiting global asymptotic stability ("turnpike property").

Theorem 2. *Let (x_t) be an optimal program from $(x, y) \in \Omega$ with $(x, y) \gg 0$. Then $\lim_{t\to\infty} x_t = x^*$.*

Proof. Define $m = \min\{x_0, x_1, x^*\}$ and $M = \max\{x_0, x_1, x^*\}$, where x^* is given by the single-crossing condition (SC).

We show that $x_t \geq m$ for all $t \geq 0$. This is clear for $t = 0, 1$. Suppose $x_t \geq m$ for $t = 0, 1, \ldots T$, where $T \geq 1$; then

$$x_{T+1} = h(x_{T-1}, x_T) \geq h(m, m) \geq m. \tag{9.33}$$

The first inequality in (9.33) follows from the monotonicity of h in both arguments, the definition of m, and the fact that x_{T-1} and x_T are at least as large as m. The second inequality follows from the fact that $m \leq x^*$and condition (SC). This establishes by induction that $x_t \geq m$ for $t \geq 0$.

We show that $x_t \leq M$ for all $t \geq 0$. This being clear for $t = 0, 1$, suppose $x_t \leq M$ for $t = 0, 1, \ldots, T$, where $T \geq 1$. Then

$$x_{T+1} = h(x_{T-1}, x_T) \leq h(M, M) \leq M. \tag{9.34}$$

The first inequality in (9.34) follows from the monotonicity of h in both arguments, the definition of M, and the facts that $x_{T-1} \leq M$, $x_T \leq M$ by hypothesis. The second inequality follows from the fact that $M \geq x^*$ and condition (SC). This establishes by induction that $x_t \leq M$ for $t \geq 0$.

We also note that

$$x_{t+1} = h(x_{t-1}, x_t) \leq h(B, B) = \phi(B). \tag{9.35}$$

Define $a = \liminf_{t\to\infty} x_t$. We claim that $a \geq x^*$. Otherwise, if $a < x^*$, then using $a \geq m > 0$, we have $\phi(a) > a$ and so we can find $\varepsilon > 0$ such that $(a-\varepsilon) > 0$ and $\phi(a - \varepsilon) > a + \varepsilon$. By definition of a, we can find N such that for $t \geq N$,

$x_t \geq (a - \varepsilon)$. Thus, for $t \geq N$,

$$x_{t+2} = h(x_t, x_{t+1}) \geq h(a - \varepsilon, a - \varepsilon) = \phi(a - \varepsilon) > a + \varepsilon.$$

But this means that $\liminf_{t\to\infty} x_t \geq a + \varepsilon$ is a contradiction. Thus, we must have $a \geq x^*$.

Define $A = \limsup_{t\to\infty} x_t$. We claim that $A \leq x^*$. Suppose, on the contrary, $A > x^*$. We know that $A \leq \phi(B)$ (by (9.35))$< B$ (by Proposition 4). Using condition (SC) we have $\phi(A) < A$, and so we can find $\varepsilon > 0$ such that $(A+\varepsilon) < B$, and $\phi(A + \varepsilon) < (A - \varepsilon)$. By definition of A, we can find N such that for $t \geq N$, $x_t \leq (A + \varepsilon)$. Thus for $t \geq N$,

$$x_{t+2} = h(x_t, x_{t+1}) \leq h(A + \varepsilon, A + \varepsilon) = \phi(A + \varepsilon) < A - \varepsilon.$$

But this means that $\limsup_{t\to\infty} x_t \leq A - \varepsilon$, a contradiction. Thus, we must have $A \leq x^*$.

Since $A \geq a$, we have

$$x^* \geq A \geq a \geq x^* \tag{9.36}$$

which proves that $a = A = x^*$, and so (x_t) converges and $\lim_{t\to\infty} x_t = x^*$.

Remark 1. The style of proof is similar to that used in Hautus and Bolis (1979), but since the domain of definition of h and ϕ are different in our framework from theirs, we cannot appeal directly to their result.

9.4.5 Remarks on Models of Habit Formation

The literature on habit formation[9] studies optimization problems of the type described in Sect. 9.2.4.1. The model of Boyer (1978) on habit formation, where utility is assumed to be increasing both in current and in past consumption, can be treated as a special case of the model we described in Sect. 9.2.4.1.[10] However, in many models of habit formation, utility is assumed to be increasing in present consumption, but *decreasing* in past consumption. The idea is that a high consumption in the past means that a person gets used to a higher standard, and therefore this has a negative effect on her evaluation of current consumption.

---

[9]See Boyer (1978), Abel (1990) and Deaton (1992) and the references cited by them for the main contributions to this literature.

[10]He does not look at the corresponding reduced-form model, and does not assume conditions on the primitive form which would ensure the supermodularity of the reduced-form utility function. Thus, in his model, unlike ours, it is "possible to experience cycles in consumption, investment, capital and the interest rate" (Boyer 1978, p. 594).

Assumption (w) in the example described in Sect. 9.2.4.1 (and more generally assumption (A6) of the reduced-form model described in Sect. 9.2.1) rules out such environments of habit formation. However, the methods and results of our article continue to be applicable to *some* of these environments. We elaborate on this remark by presenting an example of habit formation where utility is assumed to be increasing in present consumption and decreasing in past consumption, and yet the main monotonicity and global asymptotic stability results of our article continue to hold.

The framework of the example is similar to the one described in Sect. 9.2.4.1, with a *production function*, f, satisfying assumption (f), a *discount factor*, δ, satisfying assumption (d); however, the *welfare function*, w, is a function from X^2 to $\mathbb{R}$, which satisfies

(w') w is continuous and concave on X^2, and strictly concave in the second argument; it is nondecreasing in the second argument and nonincreasing in the first argument.

We now provide a specific example of this framework, which is a variation of the example discussed in Sect. 9.4.2.. Consider the production function, f, defined by

$$f(x) = px - qx^2 \text{ for } x \in [0, (p-1)/q] \equiv [0, B] = X.$$

Here $p = (3/2)$ and $q = (1/2)$, so that $B = 1$.

The welfare function, w, is defined by

$$w(c,d) = A[d/(1+d)] - ac^b \text{ for } (c,d) \in X^2$$

where $a \in (0,1)$, $b \geq 2$, and $A > 4ab$.

The discount factor, δ, is chosen to be in $(0.8, 1)$.

For any $x \in (0,1)$, we have $f(x) > x$, and so the stationary program $(x, x, x, \ldots)$ is feasible. Denoting $f(x) - x$ by c, we note that $c \in (0,1)$ is the constant consumption along this program. Given the form of w, we have $w(c,c) > 0$.

Note that w is C^2 on X^2; the partial derivatives of w can be calculated as follows:

$$\begin{aligned}
w_1(c,d) &= -abc^{b-1} < 0 \\
w_2(c,d) &= A/(1+d)^2 > 0 \\
w_{11}(c,d) &= -b(b-1)ac^{b-2} < 0 \\
w_{12}(c,d) &= 0 \\
w_{22}(c,d) &= -2A/(1+d)^3 < 0.
\end{aligned}$$

Thus, w is clearly increasing in d and decreasing in c on X^2. Further, it is strictly concave in (c,d) on X^2. It can be checked that for all $c \in (0,1)$, we have $w_1(c,c)$

$+w_2(c,c) > 0$, as required in the study of Samuelson (1971) and the habit formation model of Sundaresan (1989).

The properties (A1)–(A4) of the corresponding reduced form model (Λ, u, δ) can be verified quite easily. To verify other properties, we can compute the relevant first- and second-order partial derivatives of u as in Sect. 9.4.2. Since $w_1 < 0$ and $w_2 > 0$, it follows that $u_1 < 0$ and $u_3 < 0$, while $u_2 > 0$. Note that (A6) is clearly violated. By definition of u, it is clearly concave (given concavity of w and f) and continuous on Λ, so that (A5) is satisfied.

Since $w_{12} = 0$ and $w_{11} < 0$, $w_{22} < 0$, we have $u_{13} = 0$ and $u_{12} > 0$, $u_{23} > 0$. Thus, u is supermodular on Λ.[11]

Assumption (A7) is clearly satisfied by definition of δ. To verify (A8), define $\hat{x} = (\delta/4)$. Then $(\hat{x}/\delta) = (1/4)$ and $(\hat{x}/\delta^2) = (1/4\delta)$. Note that $f(\hat{x}/\delta) = f(1/4) = (11/32) > (10/32) > (1/4\delta) = (\hat{x}/\delta^2)$; it follows that $f(\hat{x}) \geq \delta f(\hat{x}/\delta) > (\hat{x}/\delta)$. Thus, $(\hat{x}, (\hat{x}/\delta), (\hat{x}/\delta^2)) \in \Lambda$. Note that for $x \in (0, \hat{x}/\delta)$, defining $g(x) = f(x) - (x/\delta)$, we have $g'(x) = (3/2) - x - (1/\delta) > (3/2) - (1/4) - (5/4) = 0$. Thus, we have $f(\hat{x}/\delta) - (\hat{x}/\delta^2) > f(\hat{x}) - (\hat{x}/\delta)$, and so using $w_2 > 0$, we have

$$u(\hat{x}, (\hat{x}/\delta), (\hat{x}/\delta^2)) \geq w(f(\hat{x}) - (\hat{x}/\delta), f(\hat{x}) - (\hat{x}/\delta)) > 0$$

the last inequality following from the fact that $w(c,c) > 0$ for all c in $(0,1)$. Thus, (A8) is verified. Assumption (A9) is clearly satisfied since both f and w have bounded steepness. Finally, Assumption (A10) is satisfied, since $u(0,0,0) = 0 = u(B,B,B)$. To summarize, in this example, all the assumptions except (A6) are satisfied, further, u is supermodular on Λ.

Since the analysis in our article relies at various points on the use of assumption (A6), our methods are not directly applicable to this example of habit formation. However, slight modifications of our methods can be used to verify that (1) the policy function satisfies the single-crossing property; (2) the value function is supermodular on its domain, and the policy function is nondecreasing in each of its arguments. Thus, Theorem 2, which uses only these properties of the model, continues to be valid in this example of habit formation. The details of this verification can be found in Mitra and Nishimura (2003).

We should add that there are clearly models of habit formation which *cannot* be analyzed in terms of the methods used in our article, and the results of our article *do not* apply to those frameworks. For example, one might follow Abel (1990) and consider a particular specification of the habit-formation model where

$$w(c_t, c_{t+1}) = (1/(1-\alpha))(c_{t+1}/c_t)^{(1-\alpha)} \quad \text{where } \alpha \in (0,1).$$

[11] In order to keep our example simple, we have used a form for $w(c,d)$ for which $w_{12} = 0$. However, this is not essential to the example. One can allow for functions $w(c,d)$ in which $w_{12} < 0$, and the cross effect is small relative to the direct effects of w_{11} and w_{22}, (as explained in Sect. 9.4.2) and still preserve the main results of this example.

In this case, not only is $w_1 < 0$ and $w_2 > 0$, so that the corresponding reduced-form model violates (A6), but w itself is *not* a concave function of (c_t, c_{t+1}), so that the corresponding reduced-form model also violates (A5). The dynamic optimization problem in (9.5) is then one involving a nonconcave objective function. This takes one beyond the scope of environments that can be handled with the methods used in our article.

9.5 Local Dynamics

In this section, we provide an analysis of the local dynamics of optimal solutions near a stationary optimal stock. To this end, we study (in Sect. 9.5.1) the behavior of the optimal policy function (assuming that it is continuously differentiable in a neighborhood of the stationary optimal stock) and obtain restrictions on the two characteristic roots associated with the linearized version of it near the stationary optimal stock. Next, we show (in Sect. 9.5.2) that each of these characteristic roots must also be a characteristic root of the linearized version of the Ramsey–Euler equation near the stationary optimal stock. We then examine the fourth order difference equation, which represents the linearized version of the Ramsey–Euler equations near the stationary optimal stock, and we show which two of them are selected by the optimal solution. The roots selected by the optimal solution provide information about the speed of convergence of nonstationary optimal trajectories to the stationary optimal stock. The assumption of supermodularity of the utility function is not used in the above analysis.

In Sect. 9.5.3, the optimal policy function is shown to be continuously differentiable in a neighborhood of the stationary optimal stock, by using the Stable Manifold Theorem. This provides a rigorous basis for the analysis carried out in Sects. 9.5.1 and 9.5.2.[12]

9.5.1 Characteristic Roots Associated with the Optimal Policy Function

We proceed with our local analysis of the optimal policy function by making strong smoothness assumptions.

We assume that there is $\varepsilon > 0$ such that the utility function is C^2 in a neighborhood $N \equiv Q^3$ of (x^*, x^*, x^*), (where $Q = (x^* - \varepsilon, x^* + \varepsilon)$) with $u_1 > 0$,

[12]Our proof of the continuous differentiability of the optimal policy function near the stationary optimal stock involves using the result as well as the method of proof of Theorem 2. Since the latter result was proved by us under the assumption of supermodularity of the utility function, we are not able to totally dispense with the supermodularity assumption in Sect. 9.5.

$u_3 < 0$ and $u_{13} > 0$, and a negative-definite Hessian on N. Further, we assume that there is a neighborhood M' of (x^*, x^*) on which V is C^2 and h is C^1.[13] Clearly, we can choose a smaller neighborhood M of M' such that for all (x, y) in M, $(x, y, h(x, y))$ is in N and $(y, h(x, y))$ is in M'.

In terms of the example of Sect. 9.2.4, the restriction $u_{13} > 0$ is satisfied if w is C^2 with $w_{12} < 0$ (and f is C^1, with $f' > 0$). This restriction is quite important. It implies that the policy function is monotone increasing in the first argument on M.

Proposition 7. *The policy function, h, satisfies $h_1(x, y) > 0$ for $(x, y) \in M$.*

Proof. Let $(x, y) \in M$. Then $h(x, y)$ solves the maximization problem:

$$\max_{(y,z)\in\Omega} [u(x, y, z) + \delta V(y, z)].$$

Since $(y, h(x, y)) \in M'$ and $(x, y, h(x, y))$ is in N,

$$u_3(x, y, h(x, y)) + \delta V_2(y, h(x, y)) = 0. \tag{9.37}$$

This is an identity in $(x, y) \in M$, and so, differentiating with respect to x,

$$u_{31}(x, y, h(x, y)) + u_{33}(x, y, h(x, y))h_1(x, y) + \delta V_{22}(y, h(x, y))h_1(x, y) = 0.$$

We have $V_{22} \leq 0$ (by concavity of V), and $u_{33} < 0$ (since the Hessian of u is negative definite); thus $u_{33}(x, y, h(x, y)) + \delta V_{22}(y, h(x, y)) < 0$, and so $h_1(x, y) > 0$.

If $x^* > 0$ is the unique positive stationary optimal stock, another useful property of the optimal policy function may be obtained, namely, $h_1(x^*, x^*) + h_2(x^*, x^*) \leq 1$. Recall that the circumstances under which there is a unique positive stationary optimal stock were discussed in connection with the single-crossing condition in Sect. 9.3.3.

Proposition 8. *Suppose x^* is the unique positive stationary optimal stock. Then*

$$h_1(x^*, x^*) + h_2(x^*, x^*) \leq 1.$$

Proof. Since $[h(x, x) - x] \geq 0$ for $0 \leq x \leq x^*$, and $[h(x^*, x^*) - x^*] = 0$, we must have $[h(x, x) - x]$ minimized at $x = x^*$ among all $x \in [0, x^*]$. Thus,

$$h_1(x^*, x^*) + h_2(x^*, x^*) - 1 \leq 0$$

which establishes the result.

[13]The circumstances under which these smoothness assumptions hold are given in Sect. 9.5.3.

Given the non-linear difference equation

$$x_{t+2} = h(x_t, x_{t+1})$$

the linear difference equation associated with it (near the stationary optimal stock, x^*) is given by

$$a_{t+2} = qa_t + pa_{t+1} \tag{9.38}$$

where q denotes $h_1(x^*, x^*)$ and p denotes $h_2(x^*, x^*)$, and a_t is to be interpreted as $(x_t - x^*)$ for $t \geq 0$.

The characteristic equation associated with the Eq. 9.38 is

$$\lambda^2 = q + p\lambda. \tag{9.39}$$

Denoting by λ_1 and λ_2 the roots of (9.39), we observe that

$$\left.\begin{aligned} \lambda_1 + \lambda_2 &= p \\ \text{and } \lambda_1\lambda_2 &= -q. \end{aligned}\right\} \tag{9.40}$$

These are explicitly given by the formula

$$\lambda = [p \pm \sqrt{p^2 + 4q}]/2. \tag{9.41}$$

Under our assumptions we have the information that

$$q > 0,\ p \geq 0,\ p + q \leq 1. \tag{9.42}$$

Since $q > 0$, we can use (9.40) to infer that the roots λ_1, λ_2 are real and they are of opposite signs. Without loss of generality, let us denote the positive root by λ_1 and the negative root by λ_2.

Using (9.40), (9.42), we have

$$1 \geq p + q = \lambda_1 + \lambda_2 - \lambda_1\lambda_2 = \lambda_1 + (1 - \lambda_1)\lambda_2$$

so that

$$(1 - \lambda_1) \geq (1 - \lambda_1)\lambda_2. \tag{9.43}$$

Now, if $\lambda_1 > 1$, then we would get $(1 - \lambda_1) < 0$, and $(1 - \lambda_1)\lambda_2 > 0$ (since $\lambda_2 < 0$) contradicting (9.43). Thus, we can conclude that

$$0 < \lambda_1 \leq 1. \tag{9.44}$$

Now, using (9.40), we have $(-\lambda_2) = \lambda_1 - p \leq \lambda_1 \leq 1$. Thus, neither characteristic root can exceed 1 in absolute value.

9.5.2 *Characteristic Roots Associated with the Ramsey–Euler Equation*

Consider the Ramsey–Euler equation:

$$u_3(x_t, x_{t+1}, x_{t+2}) + \delta u_2(x_{t+1}, x_{t+2}, x_{t+3}) + \delta^2 u_1(x_{t+2}, x_{t+3}, x_{t+4}) = 0. \quad (9.45)$$

In particular, of course, $x_{t+s} = x^*$ for $s = 0, 1, 2, 3, 4$ satisfies (9.45):

$$u_3(x^*, x^*, x^*) + \delta u_2(x^*, x^*, x^*) + \delta^2 u_1(x^*, x^*, x^*) = 0. \quad (9.46)$$

If we use the Mean-Value theorem around (x^*, x^*, x^*) to evaluate the difference between the left-hand sides of (9.45) and (9.46), but ignore the second-order terms (so that one obtains a "first-order" or "linear" approximation to the difference) we get (dropping the point of evaluation (x^*, x^*, x^*) to ease the writing) the expression

$$\delta^2 u_{13}\varepsilon_{t+4} + (\delta^2 u_{12} + \delta u_{23})\varepsilon_{t+3} + (\delta^2 u_{11} + \delta u_{22} + u_{33})\varepsilon_{t+2} + (\delta u_{21} + u_{32})\varepsilon_{t+1} + u_{31}\varepsilon_t.$$

If we substitute β^{t+s} for $\varepsilon_{t+s}(s = 0, 1, 2, 3, 4)$, and equate the resulting expression to zero, we get the characteristic equation associated with the Ramsey–Euler equation (9.45):

$$\delta^2 u_{13}\beta^4 + (\delta^2 u_{12} + \delta u_{23})\beta^3 + (\delta^2 u_{11} + \delta u_{22} + u_{33})\beta^2 + (\delta u_{21} + u_{32})\beta + u_{31} = 0. \quad (9.47)$$

The idea is that the roots of this characteristic equation will reflect local behavior around the stationary optimal stock, x^*, of solutions to Ramsey–Euler equations.

We now show that the characteristic roots associated with the optimal policy function, which we analyzed in Sect. 9.5.1, must be solutions to the characteristic equation (9.47). By continuity of the optimal policy function, we can choose a neighborhood M of (x^*, x^*) such that for all (x, y) in M, $(y, h(x, y))$, $(h(x, y), h(y, h(x, y)))$ and $(h(y, h(x, y)), h(h(x, y), h(y, h(x, y))))$ are in M', and $(x, y, h(x, y)), (y, h(x, y), h(y, h(x, y)))$ and $(h(x, y), h(y, h(x, y)), h(h(x, y), h(y, h(x, y))))$ are in N. Thus, the Ramsey–Euler equation yields the following *identity* in (x, y):

$$\begin{aligned} W(x, y) &= u_3(x, y, h(x, y)) \\ &\quad + \delta u_2(y, h(x, y), h(y, h(x, y))) \\ &\quad + \delta^2 u_1(h(x, y), h(y, h(x, y)), h(h(x, y), h(y, h(x, y)))) \\ &= 0. \end{aligned} \quad (9.48)$$

If we differentiate W with respect to x and evaluate the derivatives of u at (x^*, x^*, x^*), and the derivatives of h at (x^*, x^*), then the derivative $\partial W(x^*, x^*)/\partial x$

must be equal to zero. We can write the derivative (after dropping the points of evaluation (x^*, x^*, x^*) and (x^*, x^*) to ease the writing) as

$$\begin{aligned}\partial W(x^*, x^*)/\partial x &= u_{31} + u_{33}h_1 + \delta[u_{22}h_1 + u_{23}h_2h_1] \\ &\quad + \delta^2[u_{11}h_1 + u_{12}h_2h_1 + u_{13}(h_1)^2 + u_{13}(h_2)^2h_1] \\ &= u_{31} + [u_{33} + \delta u_{22} + \delta^2 u_{11}]h_1 \\ &\quad + \delta[u_{23} + \delta u_{12}]h_1h_2 + \delta^2 u_{13}[(h_1)^2 + (h_2)^2h_1].\end{aligned}$$

Denote $[u_{33} + \delta u_{22} + \delta^2 u_{11}]$ by $\hat{C}$, and $[u_{23} + \delta u_{12}]$ by $\hat{D}$. Then, we have

$$\partial W(x^*, x^*)/\partial x = u_{31} + \hat{C}h_1 + \delta\hat{D}h_1h_2 + \delta^2 u_{13}[(h_1)^2 + (h_2)^2h_1] = 0. \quad (9.49)$$

Similarly, if we differentiate W with respect to y, and evaluate the derivatives of u at (x^*, x^*, x^*), and the derivatives of h at (x^*, x^*), then the derivative $\partial W(x^*, x^*)/\partial y$ must be equal to zero. We can write the derivative (after dropping the points of evaluation (x^*, x^*, x^*) and (x^*, x^*) to ease the writing) as

$$\begin{aligned}\partial W(x^*, x^*)/\partial y &= u_{32} + u_{33}h_2 + \delta[u_{21} + u_{22}h_2 + u_{23}h_1 + u_{23}(h_2)^2] \\ &\quad + \delta^2[u_{11}h_2 + u_{12}h_1 + u_{12}(h_2)^2 + 2u_{13}h_1h_2 + u_{13}(h_2)^3] \\ &= u_{32} + \delta u_{21} + \hat{C}h_2 + \delta\hat{D}h_1 + 2\delta^2 u_{13}h_1h_2 + \delta\hat{D}(h_2)^2 + \delta^2 u_{13}(h_2)^3.\end{aligned}$$

Rearranging terms yields the derivative:

$$\begin{aligned}\partial W(x^*, x^*)/\partial y &= (u_{32} + \delta u_{21}) + \hat{C}h_2 \\ &\quad + \delta\hat{D}[h_1 + (h_2)^2] + \delta^2 u_{13}[2h_1 + (h_2)^2]h_2 = 0. \quad (9.50)\end{aligned}$$

We recall from Sect. 9.5.1 that if λ is a characteristic root associated with the optimal policy function, then $(\lambda)^2 = h_2\lambda + h_1$. Using this information in (9.49) and (9.50), we get:

$$\begin{aligned}\partial W(x^*, x^*)/\partial x + \lambda\partial W(x^*, x^*)/\partial y &= u_{31} + [u_{32} + \delta u_{21}]\lambda + \hat{C}(h_1 + h_2\lambda) \\ &\quad + \delta\hat{D}[h_1\lambda + (h_2)^2\lambda + h_1h_2] \\ &\quad + \delta^2 u_{13}[(h_1)^2 + (h_2)^2h_1 + 2h_1h_2\lambda + (h_2)^3\lambda] \\ &= u_{31} + [u_{32} + \delta u_{21}]\lambda + \hat{C}\lambda^2 + \delta\hat{D}[h_1\lambda + h_2\lambda^2] \\ &\quad + \delta^2 u_{13}[h_1(h_1 + h_2\lambda) + h_1h_2\lambda \\ &\quad + (h_2)^2[h_1 + h_2\lambda]] \\ &= u_{31} + [u_{32} + \delta u_{21}]\lambda + \hat{C}\lambda^2 + \delta\hat{D}\lambda^3\end{aligned}$$

$$+\delta^2 u_{13}[h_1\lambda^2 + h_1 h_2 \lambda + (h_2)^2\lambda^2]$$
$$= u_{31} + [u_{32} + \delta u_{21}]\lambda + \hat{C}\lambda^2 + \delta\hat{D}\lambda^3$$
$$+\delta^2 u_{13}[h_1\lambda^2 + h_2\lambda^3]$$
$$= u_{31} + [u_{32} + \delta u_{21}]\lambda + \hat{C}\lambda^2 + \delta\hat{D}\lambda^3 + \delta^2 u_{13}\lambda^4$$
$$= 0.$$

This completes the verification of our claim.

We now show how the characteristic roots associated with the optimal policy function (analyzed in Sect. 9.5.1) can be found by calculating the characteristic roots of (9.47).

Notice that $\beta = 0$ is *not* a solution to (9.47) since $u_{13} \neq 0$. We can, therefore, use the transformed variable

$$\mu = \delta\beta + (1/\beta)$$

to examine the roots of (9.47). Using this transformation, (9.47) becomes

$$u_{13}\mu^2 + (\delta u_{12} + u_{23})\mu + [\delta^2 u_{11} + \delta u_{22} + u_{33} - 2\delta u_{13}] = 0. \tag{9.51}$$

Let us define $\mathbb{G}: \mathbb{R} \to \mathbb{R}$ by

$$\mathbb{G}(\mu) = u_{13}\mu^2 + (\delta u_{12} + u_{23})\mu + [\delta^2 u_{11} + \delta u_{22} + u_{33} - 2\delta u_{13}]. \tag{9.52}$$

Since the Hessian of u is negative definite, we have $u_{11} < 0$, $u_{22} < 0$, $u_{33} < 0$, and since $u_{13} > 0$, we have

$$[\delta^2 u_{11} + \delta u_{22} + u_{33} - 2\delta u_{13}]/u_{13} < 0.$$

Denoting the roots of (9.51), which is a quadratic in μ, by μ_1 and μ_2, we note that

$$\mu_1\mu_2 < 0 \tag{9.53}$$

so these roots are necessarily real. We denote the positive root by μ_1 and the negative root by μ_2.

Given $\mu_i (i = 1, 2)$, we can obtain the corresponding roots of β by solving the quadratic

$$\delta\beta + (1/\beta) = \mu_i. \tag{9.54}$$

We denote the roots of (9.54) corresponding to μ_1 by β_1 and β_2 (with $|\beta_1| = \min(|\beta_1|, |\beta_2|)$) and the roots of (9.54) corresponding to μ_2 by β_3 and β_4 (with $|\beta_3| = \min(|\beta_3|, |\beta_4|)$).

Define the function $\mathbb{F}: \mathbb{R}^2 \to \mathbb{R}$ by

$$\mathbb{F}(\beta; m) = \delta\beta^2 - m\beta + 1. \tag{9.55}$$

Then β_1 and β_2 are the roots of $\mathbb{F}(\beta;\mu_1) = 0$, and β_3 and β_4 are the roots of $\mathbb{F}(\beta;\mu_2) = 0$.

Using our analysis in Sect. 9.5.1, we can show that the roots β_1 and β_2 are real, and

$$0 < \beta_1 \leq 1 < \beta_2. \tag{9.56}$$

To see this, recall that λ_1 and λ_2 are solutions of (9.39). These are real and of opposite signs. Thus, examining (9.55), it is clear that λ_1 and λ_2 must correspond to different μ_i. This means that β_1 and β_2 are real, and so are β_3 and β_4.

Now, note that since β_1 and β_2 solve the equation

$$\delta\beta^2 - \mu_1\beta + 1 = 0 \tag{9.57}$$

and $\mu_1 > 0$, we have $\beta_1\beta_2 = (1/\delta) > 0$ and $(\beta_1 + \beta_2) = (\mu_1/\delta) > 0$. Thus, β_1 and β_2 are both positive.

Since β_3 and β_4 are roots of the equation

$$\delta\beta^2 - \mu_2\beta + 1 = 0 \tag{9.58}$$

we have $\beta_3\beta_4 = (1/\delta) > 0$ and $\beta_3 + \beta_4 = (\mu_2/\delta) < 0$. Thus β_3 and β_4 are of the same sign, and they must both be negative.

It follows from the above analysis that λ_1 must be one of the roots β_1 and β_2, and λ_2 must be one of the roots β_3 and β_4. Further, since $\lambda_1 \leq 1$ and $\beta_1\beta_2 = (1/\delta) > 1$, $\lambda_1 = \beta_1$ and $\beta_2 \geq (1/\delta)$. This establishes (9.56).

Similarly, we can show that

$$0 > \beta_3 \geq -1 > \beta_4 \tag{9.59}$$

Since $(-\lambda_2) \leq 1$ and $\beta_3\beta_4 = (1/\delta) > 1$, $(-\lambda_2) = (-\beta_3)$ and $(-\beta_4) \geq (1/\delta)$. This establishes (9.59).

9.5.3 *Differentiability of the Optimal Policy Function*

In Sect. 9.5.1, we assumed that the optimal policy function was continuously differentiable in a neighborhood of the steady state, x^*. We used this to obtain the characteristic roots associated with the optimal policy function, and to relate them (in Sect. 9.5.2) to the characteristic roots associated with the Ramsey–Euler equation. To complete our analysis, we need to show that the optimal policy function is indeed continuously differentiable in a neighborhood of the steady state, x^*. We do this by applying the Stable Manifold Theorem.[14]

[14]The use of stable manifold theory to optimal growth was pioneered by Scheinkman (1976). Since then, it has figured prominently in the theoretical work of Araujo and Scheinkman (1977) and Santos (1991), and in numerous applications of this theory to dynamic macroeconomic models.

We have seen in Sect. 9.5.2 that the characteristic roots $(\beta_1, \beta_2, \beta_3, \beta_4)$ associated with Eq. 9.47 satisfy the restrictions

$$\beta_4 < -1 \leq \beta_3 < 0 < \beta_1 \leq 1 < \beta_2. \tag{9.60}$$

We *assume* now that the generic case in (9.60) holds; that is, the weak inequalities in (9.60) are replaced by strict inequalities

$$\beta_4 < -1 < \beta_3 < 0 < \beta_1 < 1 < \beta_2. \tag{9.61}$$

We wish to analyze the behavior of the Ramsey–Euler dynamical system near the steady state, x^*. To this end, we define

$$F(v, w, x, y, z) = u_3(v, w, x) + \delta u_2(w, x, y) + \delta^2 u_1(x, y, z)$$

in a neighborhood $N' \equiv Q^5$ of $(x^*, x^*, x^*, x^*, x^*)$ (where $Q = (x^* - \varepsilon, x^* + \varepsilon)$ and ε is as given in Sect. 9.5.1). Then, F is C^1 on N'. We note that

$$D_5 F(x^*, x^*, x^*, x^*, x^*) = \delta^2 u_{13}(x^*, x^*, x^*) \neq 0$$

and so we can apply the implicit function theorem[15] to obtain an open set $\tilde{U}$ containing (x^*, x^*, x^*, x^*), and an open set V containing x^*, and a unique function $\Phi: \tilde{U} \to V$, such that

$$u_3(v, w, x) + \delta u_2(w, x, y) + \delta^2 u_1(x, y, \Phi(v, w, x, y)) = 0 \text{ for all } (v, w, x, y) \in \tilde{U} \tag{9.62}$$

and

$$\Phi(x^*, x^*, x^*, x^*) = x^*. \tag{9.63}$$

Further, Φ is C^1 on $\tilde{U}$. Clearly, we can pick an open set $\hat{U} \subset \tilde{U}$, with $\hat{U}$ containing (x^*, x^*, x^*, x^*), such that $\Phi(\hat{U}) \subset Q$.

Define the set $U' = \{(v', w', x', y') \in \mathbb{R}^4: (v', w', x', y') = (v - x^*, w - x^*, x - x^*, y - x^*)$ for some $(v, w, x, y) \in \hat{U}\}$. Thus, U' is a neighborhood of $(0, 0, 0, 0)$, a translation of the set $\hat{U}$ by subtraction of the point (x^*, x^*, x^*, x^*) from each point $(v, w, x, y) \in \hat{U}$. Now, define $G: U' \to \mathbb{R}^4$ as follows:

$$\left.\begin{aligned} G^1(X_1, X_2, X_3, X_4) &= X_2 \\ G^2(X_1, X_2, X_3, X_4) &= X_3 \\ G^3(X_1, X_2, X_3, X_4) &= X_4 \\ G^4(X_1, X_2, X_3, X_4) &= \Phi(x^* + X_1, x^* + X_2, x^* + X_3, x^* + X_4) - x^*. \end{aligned}\right\} \tag{9.64}$$

Note that $G(0, 0, 0, 0) = (0, 0, 0, 0)$, using (9.63).

---

[15] See, for example, Rosenlicht (1986), pp. 205–209.

The Ramsey–Euler dynamics near the steady state is governed by (9.62). This gives rise to the (four-dimensional) dynamical system

$$X_{t+1} = G(X_t). \tag{9.65}$$

In order to apply the standard form of the Stable Manifold Theorem, however, we need to transform the variables appearing in this dynamical system.

To this end, we proceed as follows. Given G, we can calculate the Jacobian matrix of G at $(0, 0, 0, 0)$ as

$$J_G(0) = \begin{bmatrix} 0 & 1 & 0 & 0 \\ 0 & 0 & 1 & 0 \\ 0 & 0 & 0 & 1 \\ \Phi_1(x^*, x^*, x^*, x^*) & \Phi_2(x^*, x^*, x^*, x^*) & \Phi_3(x^*, x^*, x^*, x^*) & \Phi_4(x^*, x^*, x^*, x^*) \end{bmatrix}. \tag{9.66}$$

The entries in the last row of $J_G(0)$ can be related to the second-order derivatives of u at (x^*, x^*, x^*). Differentiating (9.62) with respect to v, w, x, y and evaluating the relevant derivatives at $(v, w, x, y) = (x^*, x^*, x^*, x^*)$, we obtain

$$\begin{aligned}
&u_{31}(x^*, x^*, x^*) + \delta^2 u_{13}(x^*, x^*, x^*)\Phi_1(x^*, x^*, x^*, x^*) = 0 \\
&u_{32}(x^*, x^*, x^*) + \delta u_{21}(x^*, x^*, x^*) + \delta^2 u_{13}(x^*, x^*, x^*)\Phi_2(x^*, x^*, x^*, x^*) = 0 \\
&u_{33}(x^*, x^*, x^*) + \delta u_{22}(x^*, x^*, x^*) + \delta^2 u_{11}(x^*, x^*, x^*) \\
&\qquad + \delta^2 u_{13}(x^*, x^*, x^*)\Phi_3(x^*, x^*, x^*, x^*) = 0 \\
&\delta u_{23}(x^*, x^*, x^*) + \delta^2 u_{12}(x^*, x^*, x^*) + \delta^2 u_{13}(x^*, x^*, x^*)\Phi_4(x^*, x^*, x^*, x^*) = 0.
\end{aligned} \tag{9.67}$$

These equations yield

$$\left.\begin{aligned}
\Phi_1(x^*, x^*, x^*, x^*) &= -(1/\delta^2) \\
\Phi_2(x^*, x^*, x^*, x^*) &= -\frac{[u_{32}(x^*, x^*, x^*) + \delta u_{21}(x^*, x^*, x^*)]}{\delta^2 u_{13}(x^*, x^*, x^*)} \\
\Phi_3(x^*, x^*, x^*, x^*) &= -\frac{[u_{33}(x^*, x^*, x^*) + \delta u_{22}(x^*, x^*, x^*) + \delta^2 u_{11}(x^*, x^*, x^*)]}{\delta^2 u_{13}(x^*, x^*, x^*)} \\
\Phi_4(x^*, x^*, x^*, x^*) &= -\frac{[\delta u_{23}(x^*, x^*, x^*) + \delta^2 u_{12}(x^*, x^*, x^*)]}{\delta^2 u_{13}(x^*, x^*, x^*)}.
\end{aligned}\right\} \tag{9.68}$$

Define the Vandermonde matrix

$$P = \begin{bmatrix} 1 & 1 & 1 & 1 \\ \beta_1 & \beta_3 & \beta_2 & \beta_4 \\ \beta_1^2 & \beta_3^2 & \beta_2^2 & \beta_4^2 \\ \beta_1^3 & \beta_3^3 & \beta_2^3 & \beta_4^3 \end{bmatrix}. \tag{9.69}$$

Note that the unusual order in the Vandermonde matrix is to be explained by the fact that the characteristic roots β_1 and β_3 are less than one in absolute value, while β_2 and β_4 are greater than one in absolute value. (This order becomes important in the application of the Stable Manifold Theorem below). Define the diagonal matrix of characteristic values

$$\mathbb{B} = \begin{bmatrix} \beta_1 & 0 & 0 & 0 \\ 0 & \beta_3 & 0 & 0 \\ 0 & 0 & \beta_2 & 0 \\ 0 & 0 & 0 & \beta_4 \end{bmatrix}. \tag{9.70}$$

Now, denoting by $\mathbb{A}$ the Jacobian matrix $J_G(0)$, we can verify (using (9.66), (9.68) and (9.47)) that

$$\mathbb{A}P = P\mathbb{B} = \begin{bmatrix} \beta_1 & \beta_3 & \beta_2 & \beta_4 \\ \beta_1^2 & \beta_3^2 & \beta_2^2 & \beta_4^2 \\ \beta_1^3 & \beta_3^3 & \beta_2^3 & \beta_4^3 \\ \beta_1^4 & \beta_3^4 & \beta_2^4 & \beta_4^4 \end{bmatrix}. \tag{9.71}$$

This means that $(\beta_1, \beta_3, \beta_2, \beta_4)$ are the characteristic roots of $\mathbb{A}$, with the column vectors of P constituting a set of characteristic vectors of $\mathbb{A}$, corresponding to these characteristic roots. The Vandermonde matrix is known to be nonsingular,[16] so we get the spectral decomposition

$$P^{-1}\mathbb{A}P = \mathbb{B}. \tag{9.72}$$

Returning now to our dynamical system (9.65), we rewrite it as

$$X_{t+1} = \mathbb{A}X_t + [G(X_t) - \mathbb{A}X_t]. \tag{9.73}$$

Multiplying through in (9.73) by P^{-1}, we obtain

$$P^{-1}X_{t+1} = (P^{-1}\mathbb{A}P)P^{-1}X_t + [P^{-1}G(PP^{-1}X_t) - (P^{-1}\mathbb{A}P)P^{-1}X_t]. \tag{9.74}$$

Thus, using (9.72), and defining new variables $Y = P^{-1}X$, we get

$$Y_{t+1} = \mathbb{B}Y_t + [P^{-1}G(PY_t) - \mathbb{B}Y_t]. \tag{9.75}$$

Denote by U the set $\{Y : Y = P^{-1}X \text{ for some } X \in U'\}$, and define $g : U \to \mathbb{R}^4$ as follows:

$$g(Y) = P^{-1}G(PY) - \mathbb{B}Y. \tag{9.76}$$

Note that by (9.64), we have

$$g(0,0,0,0) = (0,0,0,0). \tag{9.77}$$

[16] Several methods are known for computing the inverse of a Vandermonde matrix. For one such approach, see Parker (1964).

Also, we obtain by differentiating (9.76) and evaluating the derivatives at (0, 0, 0, 0)

$$J_g(0) = P^{-1}J_G(P0)P - \mathbb{B} = P^{-1}J_G(0)P - \mathbb{B} = P^{-1}\mathbb{A}P - \mathbb{B} = 0. \quad (9.78)$$

Thus, the dynamical system (9.75) can now be written as

$$Y_{t+1} = \mathbb{B}Y_t + g(Y_t) \quad (9.79)$$

with $g(0) = 0$ and $J_g(0) = 0$.

The Stable Manifold Theorem can be applied to the dynamical system (9.79). We give below the particular statement of it (due to Irwin 1970) that is directly applicable.[17]

Stable Manifold Theorem for a Fixed Point (Irwin):
Let $E = E_1 \times E_2$ be a Banach Space and let $T_1 : E_1 \to E_1$ and $T_2 : E_2 \to E_2$ be isomorphisms with $\max\{\|T_1\|, \|T_2^{-1}\|\} < 1$. Let U be an open neighborhood of 0 in E and let $g : U \to E$ be a C^r map ($r \geq 1$) with $g(0) = 0$ and $Dg(0) = 0$. Let $f = T_1 \times T_2 + g$. Then, there exist open balls C and D centered at 0 in E_1 and E_2, respectively, and a unique map $H : C \to D$ such that $f(graph(H)) \subset graph(H)$. The map H is C^r on the open ball C and $DH(0) = 0$. Further, for all $z \in C \times D$, $f^n(z) \to 0$ as $n \to \infty$ if and only if $z \in graph(H)$.

To apply the theorem, we define the maps $T_1 : \mathbb{R}^2 \to \mathbb{R}^2$ and $T_2 : \mathbb{R}^2 \to \mathbb{R}^2$ as follows:

$$T_1(z) = \begin{bmatrix} \beta_1 & 0 \\ 0 & \beta_3 \end{bmatrix} \begin{bmatrix} z_1 \\ z_2 \end{bmatrix}; \; T_2(z) = \begin{bmatrix} \beta_2 & 0 \\ 0 & \beta_4 \end{bmatrix} \begin{bmatrix} z_1 \\ z_2 \end{bmatrix}.$$

Note that

$$T_2^{-1}(z') = \begin{bmatrix} (1/\beta_2) & 0 \\ 0 & (1/\beta_4) \end{bmatrix} \begin{bmatrix} z_1' \\ z_2' \end{bmatrix}$$

so that, using (9.61), we have $\|T_1\| < 1$ and $\|T_2^{-1}\| < 1$. Applying the theorem in our context (with $r = 1$) we get the C^1 function H with the properties stated above. We wish to conclude from this that the policy function, h, is C^1 in a neighborhood of (x^*, x^*).

First, we note that $H(0, 0) = (0, 0)$. To see this, we check that $f(0, 0, 0, 0) = g(0, 0, 0, 0) = (0, 0, 0, 0)$ by (9.77), so that $f^n(0, 0, 0, 0) = (0, 0, 0, 0)$, and so by the Stable Manifold Theorem, $(0, 0, 0, 0) \in graph(H)$. That is, $H(0, 0) = (0, 0)$.

Next, we define a function, $K : \mathbb{R}^2 \times \mathbb{R}^2 \times C \to \mathbb{R}^4$ as follows:

$$K(a, b, z) = P^{-1}(a, b) - (z, H(z)). \quad (9.80)$$

[17] A good exposition of Irwin's result can be found in Franks (1979).

Clearly, K is C^1 on its domain, and $K(0,0,0,0,0,0) = (0,0,0,0)$, since $H(0,0) = (0,0)$. Further, the matrix $(D_j K^i(0,0,0,0,0,0))$, where $i = 1,2,3,4$ and $j = 3,4,5,6$ can be checked to be nonsingular. To see this, denote P^{-1} by R, and write R as follows:

$$R = \begin{bmatrix} R_{11} & R_{12} \\ R_{21} & R_{22} \end{bmatrix}$$

where each R_{ij} (with $i = 1,2$; $j = 1,2$) a 2×2 matrix. Then, we have

$$(D_j K^i(0,0,0,0,0,0)) = \begin{bmatrix} R_{12} & -I \\ R_{22} & 0 \end{bmatrix}$$

where I is the 2×2 identity matrix, and 0 is the 2×2 null matrix. Thus, the matrix $(D_j K^i(0,0,0,0,0,0))$ is nonsingular if and only if R_{22} is nonsingular. To verify that R_{22} is nonsingular, we write (by definition of R)

$$\begin{bmatrix} R_{11} & R_{12} \\ R_{21} & R_{22} \end{bmatrix} \begin{bmatrix} P_{11} & P_{12} \\ P_{21} & P_{22} \end{bmatrix} = \begin{bmatrix} I & 0 \\ 0 & I \end{bmatrix}$$

where each P_{ij} (with $i = 1,2$; $j = 1,2$) is a 2×2 sub-matrix of P. This yields the equations

$$\left.\begin{aligned} R_{21}P_{11} + R_{22}P_{21} &= 0 \\ R_{21}P_{12} + R_{22}P_{22} &= I. \end{aligned}\right\} \tag{9.81}$$

Clearly, P_{11} is nonsingular, since $\det(P_{11}) = \beta_3 - \beta_1 < 0$ (by (9.61)). Thus, $R_{21} = -R_{22}P_{21}P_{11}^{-1}$ (from the first equation of (9.81)) and using this in the second equation of (9.81), we obtain $R_{22}[P_{22} - P_{21}P_{11}^{-1}P_{12}] = I$. This establishes that R_{22} is nonsingular.

We can now use the implicit function theorem to obtain an open set $E' \subset \mathbb{R}^2$ containing $(0,0)$, an open set $C' \subset C$ containing $(0,0)$, and an open set $E'' \subset \mathbb{R}^2$ containing $(0,0)$, and unique functions $L^1 \colon E' \to E''$ and $L^2 \colon E' \to C'$, such that

$$K(a, L^1(a), L^2(a)) = 0 \text{ for all } a \in E' \tag{9.82}$$

and

$$L^1(0,0) = (0,0); \;\; L^2(0,0) = (0,0). \tag{9.83}$$

Further, L^1 and L^2 are C^1 on E'. Using the definition of K, we have from (9.82)

$$P^{-1}(a, L^1(a)) = (L^2(a), H(L^2(a))) \text{ for all } a \in E'. \tag{9.84}$$

Now, we look at the optimal policy function, h. Pick $0 < \varepsilon' < \varepsilon$ (where ε is given as in Sect. 9.5.1) so that $(-\varepsilon', \varepsilon')^4 \subset U'$ (where U' is given as in (9.64)), and $P^{-1}z \in C \times D$ for all $z \in (-\varepsilon', \varepsilon')^4$. Denote $(-\varepsilon', \varepsilon')$ by S.

Pick any $(z_1, z_2) \in S^2$. Define $(x_1, x_2) = (x^*, x^*) + (z_1, z_2)$. Then the sequence $\{x_t\}$ satisfying $x_{t+2} = h(x_t, x_{t+1})$ for $t \geq 1$ is well defined and $x_t \to x^*$ as $t \to \infty$.

Thus, the sequence $\{z_t\}$ satisfying $z_t = x_t - x^*$ for $t \geq 1$ is well defined and $z_t \to 0$ as $t \to \infty$. Further, since $(z_1, z_2) \in S^2$, we have $z_t \in S$ for all $t \geq 1$ (by the proof of Theorem 2). Then, we have

$$(z_t, z_{t+1}, z_{t+2}, z_{t+3}) \in U' \text{ for } t \geq 1 \tag{9.85}$$

and

$$P^{-1}(z_t, z_{t+1}, z_{t+2}, z_{t+3}) \in C \times D \text{ for } t \geq 1. \tag{9.86}$$

Using (9.64) and (9.85), we can write for $t \geq 1$,

$$f(P^{-1}(z_t, z_{t+1}, z_{t+2}, z_{t+3})) = P^{-1}(z_{t+1}, z_{t+2}, z_{t+3}, \Phi(x^* + z_t, x^* + z_{t+1}, x^* + z_{t+2}, x^* + z_{t+3}) - x^*). \tag{9.87}$$

Using (9.48), we have for $t \geq 1$,

$$u_3(x^* + z_t, x^* + z_{t+1}, x^* + z_{t+2}) + \delta u_2(x^* + z_{t+1}, x^* + z_{t+2}, x^* + z_{t+3}) + \delta^2 u_1(x^* + z_{t+2}, x^* + z_{t+3}, x^* + z_{t+4}) = 0. \tag{9.88}$$

Using (9.62) and (9.85), we have for $t \geq 1$,

$$u_3(x^* + z_t, x^* + z_{t+1}, x^* + z_{t+2}) + \delta u_2(x^* + z_{t+1}, x^* + z_{t+2}, x^* + z_{t+3}) + \delta^2 u_1(x^* + z_{t+2}, x^* + z_{t+3}, \Phi(x^* + z_t, x^* + z_{t+1}, x^* + z_{t+2}, x^* + z_{t+3})) = 0. \tag{9.89}$$

Note that by (9.85), $\Phi(x^* + z_t, x^* + z_{t+1}, x^* + z_{t+2}, x^* + z_{t+3}) \in Q$. Since $u_{13} > 0$ on Q^3, (9.88) and (9.89) yield (by an application of the Mean Value Theorem):

$$\Phi(x^* + z_t, x^* + z_{t+1}, x^* + z_{t+2}, x^* + z_{t+3}) = x^* + z_{t+4}. \tag{9.90}$$

Using (9.90) in (9.87), we obtain

$$f(P^{-1}(z_t, z_{t+1}, z_{t+2}, z_{t+3})) = P^{-1}(z_{t+1}, z_{t+2}, z_{t+3}, z_{t+4}) \text{ for } t \geq 1. \tag{9.91}$$

We can infer from (9.91) that

$$f^n(P^{-1}(z_1, z_2, z_3, z_4)) = P^{-1}(z_{n+1}, z_{n+2}, z_{n+3}, z_{n+4}) \text{ for } n \geq 1. \tag{9.92}$$

Since the right hand-side of (9.92) converges to $(0, 0, 0, 0)$ as $n \to \infty$, we must have $f^n(P^{-1}(z_1, z_2, z_3, z_4)) \to (0, 0, 0, 0)$ as $n \to \infty$. By the Stable Manifold Theorem, then, we must have

$$P^{-1}(z_1, z_2, z_3, z_4) \in graph(H). \tag{9.93}$$

Define a function $\psi\colon S^2 \to \mathbb{R}^2$ by

$$\psi(z_1, z_2) = (h(x^*+z_1, x^*+z_2)-x^*, h(x^*+z_2, h(x^*+z_1, x^*+z_2))-x^*) \text{for all } z\in S^2.$$

Then $\psi(0,0) = (0,0)$ and (9.93) shows that, given any $z = (z_1, z_2) \in S^2$, we must have $P^{-1}(z, \psi(z)) \in graph(H)$. Thus, given any $z \in S^2$, there is $z' \in C$, such that

$$P^{-1}(z, \psi(z)) = (z', H(z')).$$

Clearly, such a z' must be unique. Thus, there is a function, $\mathbb{K}: S^2 \to C$ such that

$$P^{-1}(z, \psi(z)) = (\mathbb{K}(z), H(\mathbb{K}(z))) \text{ for all } z \in S^2. \tag{9.94}$$

Note that since $\psi(0,0) = (0,0)$, (9.94) implies that $\mathbb{K}(0,0) = (0,0)$. Defining $S' = S^2 \cap E'$, we have from (9.94),

$$P^{-1}(z, \psi(z)) = (\mathbb{K}(z), H(\mathbb{K}(z))) \text{ for all } z \in S'. \tag{9.95}$$

On the other hand, from (9.84), we have:

$$P^{-1}(z, L^1(z)) = (L^2(z), H(L^2(z))) \text{ for all } z \in S'. \tag{9.96}$$

Since L^1 and L^2 are the unique functions satisfying (9.96) and (9.83), and since $\psi(0,0) = (0,0)$ and $\mathbb{K}(0,0) = (0,0)$, we must have $\psi = L^1$ and $\mathbb{K} = L^2$ on S'. Since L^1 is C^1 on S', we can conclude that ψ is C^1 on S'. Using the definition of ψ , it follows that the optimal policy function, h, is C^1 on S'.

9.6 Concluding Remarks

The purpose of the article was to complete the program sketched in the contribution of Samuelson (1971), by providing both a complete local and global analysis of the model under which the standard results of the Ramsey model continue to hold even with dependence of tastes between periods. We approached the problem by trying to identify structures in the *reduced form* of the model under which this would be true. The reduced form model involves a utility function which depends on the values of the state variable (capital stock) at *three* successive dates (instead of the usual two). We showed that supermodularity of the reduced form utility function (in the three variables), and a single-crossing condition provides such a structure. The methods used indicate that our results should generalize to situations in which the reduced form utility function depends on the values of the state variable in more than three periods (which correspond to situations in which utility function used to evaluate current consumption depends on several periods of past consumption).

We examined the implication of this structure for the model of Samuelson (1971) with intertemporal dependence of tastes. Our analysis indicated the conditions on the primitive form of the model under which the assumptions on the corresponding reduced form are met. It also indicated plausible scenarios in which the stated assumptions on the reduced-form *would not* be satisfied. Thus, identifying these assumptions provides a good handle on the richer dynamics that this model can generate when these assumptions fail; exploration of this topic is undertaken in Mitra and Nishimura (2001).

Application of our methods to models of habit formation is a natural direction of enquiry. The model of Boyer (1978) on habit formation, where utility is assumed to be increasing both in current and in past consumption, can be accommodated in our framework. Other frameworks of habit formation, where utility is increasing in present consumption, but decreasing in past consumption, are ruled out by the basic assumptions of our model. However, we indicated with an example that our methods and results are valid in *some* of these frameworks of habit formation as well. Models of habit formation leading to nonconcave utility functions cannot be directly addressed by the methods of this article, and constitute a potentially interesting area of future research.

References

Abel, A. B. (1990), "Asset Prices under Habit Formation and Catching up with the Joneses," *American Economic Review Papers and Proceedings*, 80, 38–42.

Amir, R. (1996), "Sensitivity Analysis of Multisector Optimal Economic Dynamics," *Journal of Mathematical Economics*, 25, 123–141.

Amir, R., Mirman, L. J. and W. R. Perkins (1991), "One-Sector Nonclassical Optimal Growth: Optimality Conditions and Comparative Dynamics," *International Economic Review*, 32, 625–792.

Araujo, A. (1991), "The Once but not Twice Differentiability of the Policy Functions," *Econometrica*, 59, 1381–1393.

Araujo, A. and J. A. Scheinkman (1977), "Smoothness, Comparative Dynamics, and the Turnpike Property," *Econometrica*, 45, 601-620.

Becker, R. and J. Boyd (1997), Capital Theory, Equilibrium Analysis, and Recursive Utility, Basil Blackwell: Oxford.

Benhabib, J. and K. Nishimura (1985), "Competitive Equilibrium Cycles," *Journal of Economic Theory*, 35, 284–306.

Bernheim, B. D. and D. Ray (1987), "Economic Growth with Intergenerational Altruism," *Review of Economic Studies*, 54, 227–243.

Boyer, M. (1978), "A Habit Forming Optimal Growth Model," *International Economic Review*, 19, 585–609.

Clark, C. W. (1976), "A Delayed-Recruitment Model of Population Dynamics, with an Application to Baleen Whale Populations," *Journal of Mathematical Biology*, 3, 381–391.

Chakravarty, S. and A. S. Manne (1968), "Optimal Growth when the Instantaneous Utility Function Depends upon the Rate of Change of Consumption," *American Economic Review*, 58, 1351–1354.

Dasgupta, P. (1974), "On Some Problems Arising from Professor Rawls' Conception of Distributive Justice," *Theory and Decision*, 4, 325–344.

Deaton, A. (1992), Consumption, Oxford University Press.

Franks, J. (1979), Manifolds of C^r Mappings and Applications to Differentiable Dynamical Systems, in *Studies in Analysis*, ed. by Rota, G.-C., New York: Academic Press, 271–290.
Gale, D. (1967), "On Optimal Development in a Multi-Sector Economy," *Review of Economic Studies*, 34, 1–18.
Hautus, M. L. J. and T. S. Bolis (1979), "A Recursive Real Sequence," Ǎmerican Mathematical Monthly, 86, 865–866.
Heal, G. M. and H. E. Ryder (1973), "Optimal Growth with Intertemporally Dependent Preferences," *Review of Economic Studies*, 40, 1–31.
Irwin, M. C. (1970), "On the Stable Manifold Theorem," *Bulletin of the London Mathematical Society*, 2, 196–198.
Iwai, K. (1972), "Optimal Economic Growth and Stationary Ordinal Utility: A Fisherian Approach," *Journal of Economic Theory*, 5, 121–151.
Kohlberg, E. (1976), "A Model of Economic Growth with Altrusim Between Generations," *Journal of Economic Theory*, 13, 1–13.
Koopmans, T. C. (1960), "Stationary Ordinal Utility and Impatience," *Econometrica*, 28, 287–309.
Koopmans, T. C., Diamond, P. A. and R. E. Williamson (1964), "Stationary Ordinal Utility and Time Perspective," *Econometrica*, 32, 82–100.
Lane, J. and Mitra, T. (1981), "On Nash Equilibrium Programs of Capital Accumulation under Altruistic Preferences," *International Economic Review*, 22, 309–331.
Mitra, T. and K. Nishimura (2001), Cyclical and Chaotic Optimal Paths in a Model with Intertemporal Complementarity, mimeo, Cornell University.
Mitra, T. and K. Nishimura (2003), Simple Dynamics in a Model of Habit Formation, mimeo, Cornell University.
Montrucchio, L. (1998), "Thompson Metric, Contraction Property and Differentiability of Policy Functions," *Journal of Economic Behavior and Organization*, 33, 449–466.
Parker, F. D. (1964), "Inverses of Vandermonde Matrices," *American Mathematical Monthly*, 71, 410–411.
Rosenlicht, M. (1986), Introduction to Analysis, Dover: New York.
Ross, S. M. (1983), Introduction to Stochastic Dynamic Programming, Academic Press: New York.
Samuelson, P. A. (1971), "Turnpike Theorems even though Tastes are Intertemporally Dependent," *Western Economic Journal*, 9, 21–26.
Santos, M. S. (1991), "Smoothness of the Policy Function in Discrete-Time Economic Models," *Econometrica*, 59, 1365–1382.
Scheinkman, J. A. (1976), "On Optimal Steady States of n-Sector Growth Models when Utility is Discounted," *Journal of Economic Theory*, 12, 11–20.
Sundaresan, S. M. (1989), "Intertemporally Dependent Preferences and the Volatility of Consumption and Wealth," *Review of Financial Studies*, 2, 73–89.
Topkis, D. (1968), "Minimizing a Submodular Function on a Lattice," *Operations Research*, 26, 305–321.
Topkis, D. (1998), Supermodularity and Complementarity, Princeton University Press.
Wan, H. Y. (1970), "Optimal Savings Programs under Intertemporally Dependent Preferences," *International Economic Review*, 11, 521–547.

Part IV
Dynamic Models with Non-concave Technologies

Chapter 10
A Complete Characterization of Optimal Growth Paths in an Aggregated Model with a Non-Concave Production Function*

W. Davis Dechert and Kazuo Nishimura**

10.1 Introduction

The convexity of technology has played a crucial role in economic analyses of optimal one-sector growth problems. For example, two of the key results on the traditional model of Ramsey (1928) that have relied on the convexity of the technology are that optimal intertemporal growth involves moving monotonically towards a unique steady state (as in Cass 1965; Koopmans 1965), and that the value function is a concave differentiable function of the initial capital stock (as in Benveniste and Scheinkman 1979). Moreover, convexity is a convenient assumption in that it guarantees that the sequence of optimal stocks is uniquely determined and that the first-order conditions (i.e., the Euler equation and the transversality condition) are sufficient as well as necessary for optimality (as in Weitzman 1973).

Clark (1971) initiated an analysis, subsequently completed by Majumdar and Mitra (1980), for a problem that was the equivalent of an optimal growth model

***Journal of Economic Theory 31**, 332–354, 1983

**We wish to thank Professor W. A. Brock for calling our attention to the topic discussed in this paper. We thank Professors W. A. Brock, David Cass and especially Tapan Mitra for many helpful conversations and comments about the problem. We have also benefitted greatly from the comments of the referee and the assistance of Mr. Kenji Yamamoto in preparing this draft. An earlier version of the paper was presented at seminars at the University of Southern California and the California Institute of Technology. Thanks are due to the participants of those seminars, too.

D. Dechert (✉)
Department of Economics, University of Houston, USA
e-mail: wdechert@gmail.com

K. Nishimura
Institute of Economic Research, Kyoto University, Japan
e-mail: nishimura@kier.kyoto-u.ac.jp

J. Stachurski et al. (eds.), *Nonlinear Dynamics in Equilibrium Models*,
DOI 10.1007/978-3-642-22397-6_10, © Elsevier B.V. 2006. Published by
Springer-Verlag Berlin Heidelberg 2012. All Rights Reserved

with a linear utility function and a non-convex technology. Clark was interested in determining the optimal rate of harvest of a renewable resource. He analyzed the case where the rate of growth and reproduction is a convex function of the size of the resource stock for small values of the stock and a concave function for large ones. What Clark speculated and Majumdar and Mitra proved is that if the interest rate is greater than the marginal productivity of the stock at the origin but less than the maximum average productivity, there must exist a critical level of the stock below which the optimal harvest is to deplete the stock, causing the resource to become extinct. Only if the size of the stock is above this critical level does the optimal harvest allow for an accumulation of the stock to a steady-state level.

Majumdar and Mitra (1982) extended the work of their earlier paper to the growth problem where the utility function is strictly concave and the technology has the same non-concave properties as before. In that paper the authors proved that for low interest rates the sequence of optimal stocks converges to the steady state, while for high interest rates they converge to the origin, causing the economy's extinction. As a result the authors demonstrated that it is not the initial level of capital stock that determines whether the economy moves to a steady state or to extinction, but rather it is whether the interest rate is in the low range or the high range. They were able to achieve these results by using only two properties of optimal programs: that the optimal paths are both Euler programs and efficient programs. However, the authors were not able to derive definitive results for the difficult case when interest rate is in the intermediate range.

In this paper we use the Principle of Optimality in addition to the Euler equation in order to provide a characterization of optimal one-sector growth for all ranges of interest rates when the technology is not convex. As a result we are able to analyze the behavior of optimal programs when interest rates are in the intermediate range. We depart from the previous literature by using the Principle of Optimality to prove our key result that the sequence of optimal stocks is necessarily monotonic. In earlier works, due to the assumption that the technology is convex, the Euler equation was sufficient to prove this monotonicity property. However, it appears that it is not sufficient in the case that the technology is not convex. In this paper we show that the Euler equation in conjunction with the Principle of Optimality is sufficient to prove the monotonicity property for non-concave technologies. However, the proof of our theorem on the monotonicity of optimal programs applies to *both* convex and non-convex technologies, and as such provides an alternative proof for the convex technology case as well. While the power of the Principle of Optimality in analyzing dynamic economics is well known, our proofs in this paper demonstrate its usefulness when applied to dynamic problems with non-convexities.

Furthermore, in Sect. 10.5 of this paper we also use the Principle of Optimality to analyze the differentiability properties of the value function as well as the uniqueness of the optimal paths. For the convex technology case, Benveniste and Scheinkman (1979) showed that the value function is everywhere differentiable as well as concave. What we show here is that when the technology is not convex the value function is not necessarily differentiable everywhere. However, we demonstrate that it does have right- and left-hand derivatives everywhere, and that the set of discontinuities of its derivative is at most countable. As to the uniqueness of optimal paths, it is well known that when the technology is convex the optimal path is uniquely determined by the initial level of capital stock. In contrast, we prove that when the technology is not convex the set of points at which the value function is differentiable *exactly* coincides with the set of initial capital stocks for which the optimal path is unique. Thus we are able to show that the set of initial stocks for which the optimal paths are not determined uniquely is at most a countable set.

10.2 Model

As in the standard Ramsey-Cass-Koopmans one-sector growth model we assume that the preferences of the economic agents in each time period can be summarized by a strictly concave utility function of consumption in each period. The per period utility of consumption, $u(c)$, satisfies:

(U1) u is twice continuously differentiable on $(0, \infty)$, and if u is bounded from below then u is continuous (from the right) at 0;
(U2) $u' > 0$, $u'' < 0$ on $(0, \infty)$ and $\lim_{c \downarrow 0} u'(c) = +\infty$.

The time rate of preference of the economic agents is given by a positive discount factor, ρ, less than unity, and the intertemporal objective is to maximize the present discounted value of the utility of consumption in each period:

$$\sum_{t=0}^{\infty} \rho^t u(c_t). \tag{10.1}$$

The productivity of the economy can be characterized by a non-negative function of capital, and the economy's output is divided between current consumption and next period's capital stock:

$$c_t + k_{t+1} = f(k_t), \quad t = 0, 1, \dots .$$

In this paper we are analyzing the case that the production process exhibits increasing returns to scale for small values of the stock and decreasing returns for large values. Moreover we shall assume that there is a maximum sustainable level of the stock which implies that all paths of capital accumulation are necessarily bounded. This means that the one-sector production function satisfies:

(P1) f is twice continuously differentiable on $[0, \infty)$, with $f' > 0$ and $f(0) = 0$;
(P2) There exists an inflection point, $k_I > 0$ such that $f''(k) \lesseqgtr 0$ if $k \gtreqless k_I$;
(P3) There exists a point $k_{max} > k_I$ such that $f(k_{max}) = k_{max}$ and $f(k) < k$ if $k > k_{max}$.

Condition (P2) is where we depart from the classical growth model in that the production function is *convex* on the interval $[0, k_I]$ while on $[k_I, \infty)$ it is *concave*. Notice that when we view this production function as the relationship between output and a (single) variable input, the average product curve has the tradition

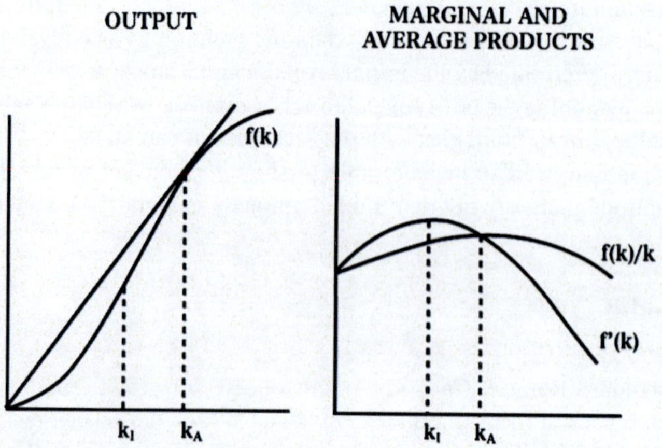

Fig. 10.1

upside down U-shape, with the maximum average product occurring at some level of capital stock, call it k_A. Furthermore, the marginal product is greater (less) than average product if the stock is less (greater) than k_A. This implies that the region of increasing returns to scale is $[0, k_A]$, and $[k_A, \infty)$ is the region of decreasing returns to scale. It is easy to check that $k_I < k_A \leqslant k_{max}$ (Fig. 10.1).

We shall refer to a *path* as a sequence of stocks $\{k_t\}$ that satisfies the feasibility condition of

$$0 \leqslant k_{t+1} \leqslant f(k_t), \quad t = 0, 1, ...$$

and an *interior path* as a path for which

$$c_t > 0, \quad t = 0, 1, ...$$

or equivalently, $k_t > 0$. An *optimal path* from k_0 is any path that maximizes the objective function (10.1).

It is straightforward to show the existence of optimal paths by the following two observations: the set of paths starting at k_0 is compact in the product topology, and the functional $\sum_t \rho^t u(f(k_t) - k_{t+1})$ is coordinatewise continuous on the set of paths. The details of this type of argument on spaces of sequences can be found in Gale (1967), Brock (1970) and Majumdar (1975).

An interior path will be called an *Euler path* if it satisfies the discrete time Euler equation:

$$u'(c_{t-1}) = \rho u'(c_t) f'(k_t), \quad t = 1, 2, ... \tag{10.2}$$

Because we are assuming that the marginal utility of consumption becomes unbounded as the level of consumption diminishes to zero, it is the case that every optimal path (from a positive initial stock, $k_0 > 0$) is an Euler path. The details of this argument (which parallel the case when f is concave) can be found in Majumdar and Mitra (1982, Lemma 5.1(i)).

10.3 Properties of Optimal Paths

As in the traditional one-sector growth model we define the value function by

$$V(k_0) = \max \sum_{t=0}^{\infty} \rho^t u(f(k_t) - k_{t+1})$$

subject to

$$k_0 > 0$$
$$0 \leqslant k_{t+1} \leqslant f(k_t)$$

which (by virtue of the non-satiation of utility) is a strictly increasing function of the initial level of capital stock. The main result of this section is that any optimal path is a monotonic sequence.

Throughout this paper we shall call the origin a *steady state* as well as any solution to $f'(k) = 1/\rho$. Notice that there may be no solutions to this equation, one solution, or two solutions. In this latter case we shall use the notation of k^* (k_*) for the steady state in the concave (convex) region of the function f, and we shall call it the upper (lower) steady state.

Theorem 1. *Let* $\{k_t\}$ *and* $\{k'_t\}$ *be optimal paths starting from* k_0 *and* k'_0, *respectively. If* $k_0 < k'_0$ *then* $k_1 < k'_1$.

Proof. First, assume that $k_1 > k'_1$. By the Principle of Optimality,

$$\begin{aligned} V(k_0) &= u(f(k_0) - k_1) + \rho V(k_1) \\ &\geqslant u(f(k_0) - k'_1) + \rho V(k'_1) \end{aligned}$$

since $k'_1 < k_1 \leqslant f(k_0)$. Similarly

$$\begin{aligned} V(k'_0) &= u(f(k'_0) - k'_1) + \rho V(k'_1) \\ &\geqslant u(f(k'_0) - k_1) + \rho V(k_1). \end{aligned}$$

By combining the two inequalities we get that

$$u(f(k_0) - k_1) + u(f(k'_0) - k'_1) \geqslant u(f(k_0) - k'_1) + u(f(k'_0) - k_1). \tag{10.3}$$

For simplicity, define

$$c_0 = f(k_0) - k_1, \quad c'_0 = f(k'_0) - k'_1$$
$$\overline{c}_0 = f(k_0) - k'_1, \quad \overline{c}'_0 = f(k'_0) - k_1$$

and note that $c_0 + c'_0 = \overline{c}_0 + \overline{c}'_0$, and that $c_0 < \overline{c}_0 < c'_0$, $c_0 < \overline{c}'_0 < c'_0$. Thus there exists $0 < \theta < 1$ such that

$$\overline{c}_0 = \theta c_0 + (1-\theta)c_0'$$
$$\overline{c}_0' = (1-\theta)c_0 + \theta c_0'$$

and so by the strict concavity of u,

$$u(\overline{c}_0) > \theta u(c_0) + (1-\theta)u(c_0')$$
$$u(\overline{c}_0') > (1-\theta)u(c_0) + \theta u(c_0')$$

and upon adding these two inequalities we get a contradiction to inequality (10.3).

Now suppose that $k_1' = k_1$. Then both paths $\{k_0, k_1', k_2', ...\}$ and $\{k_0', k_1', k_2', ...\}$ are optimal as well as interior paths, and so both are Euler paths which satisfy

$$u'(\overline{c}_0) = \rho u'(c_1')f'(k_1')$$
$$u'(c_0') = \rho u'(c_1')f'(k_1')$$

where $c_1' = f(k_1') - k_2'$, and c_0', $\overline{c}_0$ are as above. Thus $\overline{c}_0 = c_0'$, which implies that $k_0 = k_0'$. Hence for $k_0 < k_0'$ it is the case that $k_1 < k_1'$. □

Notice that the conclusion of this theorem can be weakened to: $k_0 \leq k_0'$ implies that $k_1 \leq k_1'$.

Corollary 1. *Let $\{k_t\}$ be an optimal path starting from k_0. Then either $k_t \leqslant k_{t+1}$ for all t, or $k_t \geqslant k_{t+1}$ for all t.*

Proof. Suppose $k_1 \geqslant k_0$. By the principle of optimality $\{k_2, k_3, ...\}$ is an optimal path starting from k_1. By Theorem 1, $k_2 \geqslant k_1$, and by induction on t, $k_{t+1} \geqslant k_t$ for all t. For the case that $k_1 \leqslant k_0$ the proof is the same. □

Assumption (P3) implies that all feasible paths are bounded sequences, which along with the monotonicity of optimal paths imply that the latter form convergent sequences.

Theorem 2. *Every optimal path converges to a steady state.*

Proof. Denote a solution of $f(x) = x$ and $f'(x) < 1$ by $\overline{K}$, $f(x) = x$, $f'(x) > 1$ and $x > 0$ by $\underline{K}$. $\overline{K}$ always exists by assumption (P3) but $\underline{K}$ may or may not exist. We note that for $k = 0$, $\underline{K}$, and $\overline{K}$, $f(k) - k$ gives rise to zero consumption. It is clear that staying at $\underline{K}$ or $\overline{K}$ is not optimal behavior since it yields $u(0)/(1-\rho)$ and can be dominated by the path $k_{t+1} = \frac{1}{2}f(k_t)$. (Any feasible path with positive consumption will do.)

Now suppose that $\{k_t\}$ converges to some $\overline{k} > 0$. It is clear $\overline{k} \leqslant \overline{K}$. Then by the Euler equation

$$\begin{aligned} u'(f(\overline{k}) - \overline{k}) &= \lim_{t\to\infty} u'(f(k_{t-1}) - k_t) \qquad (10.4) \\ &= \lim_{t\to\infty} \rho u'(f(k_t) - k_{t+1})f'(k_t) \\ &= \rho u'(f(\overline{k}) - \overline{k})f'(\overline{k}). \end{aligned}$$

First consider the case that $\overline{k} \neq \underline{K}$ and $\overline{k} \neq \overline{K}$. Since $0 < \overline{k} < \overline{K}$, $f(\overline{k}) - \overline{k} > 0$ and hence $\rho f'(\overline{k}) = 1$ follows from (10.4). This means that $\overline{k}$ is a steady state.

Next consider the case that $\overline{k} = \underline{K}$. If $k_0 < \underline{K}$, then $\{k_t\}$ is a decreasing sequence. Hence $k_0 \geqslant \underline{K}$ must be the case. We have already dealt with the case that $k_0 = \underline{K}$, so suppose $k_0 > \underline{K}$. As k_t converges to $\underline{K}$, $f(k_t) - k_{t+1}$ converges to zero and so we can choose $\varepsilon > 0$ so that $u(\varepsilon) < (1-\rho)u(f(\underline{K}))$ and T so that $f(k_t) - k_{t+1} < \varepsilon$ for all $t > T$. Choose $\delta > 0$ so that $u(\varepsilon) = (1-\rho)u(f(\underline{K}) - \delta)$. Let an alternative path be defined by

$$k'_t = \begin{cases} k_t & t \leqslant T \\ \delta & t = T+1 \\ \frac{1}{2} f(k'_{t-1}) & t \geqslant T+2. \end{cases}$$

Then since $u(f(k_t) - k_{t+1}) < (1-\rho)u(f(\underline{K}) - \delta)$ for $t \geqslant T+1$ we have

$$\begin{aligned} \sum_{t=0}^{\infty} \rho^t u(f(k_t) - k_{t+1}) &= \sum_{t=0}^{T} \rho^t u(f(k_t) - k_{t+1}) + \sum_{t=T+1}^{\infty} \rho^t u(f(k_t) - k_{t+1}) \\ &< \sum_{t=0}^{T} \rho^t u(f(k_t) - k_{t+1}) + \rho^{T+1} u(f(\underline{K}) - \delta) \\ &< \sum_{t=0}^{\infty} \rho^t u(f(k'_t) - k'_{t+1}). \end{aligned}$$

Therefore an optimal path cannot converge to $\underline{K}$. If an optimal path converges to $\overline{K}$, we can apply the same argument which also leads to a contradiction. □

For an initial stock k_0 with $f(k_0) > k_0$ it is feasible to consume at the stationary rate of $c_t = f(k_0) - k_0$ for all t, and this implies that $u(f(k_0) - k_0) \leqslant (1-\rho)V(k_0)$. By a straightforward application of Jensen's Inequality we can show the stronger result that the present value of consumption at the stationary rate does not exceed the present discounted value of consumption along an optimal path.

Lemma 1. *An optimal path starting from k_0 satisfies*

$$f(k_0) - k_0 \leqslant \sum_{t=0}^{\infty} (1-\rho)\rho^t c_t$$

with strict inequality if $k_t \neq k_0$ for some t.

Proof. Denote the right-hand side of the inequality by $\overline{c}$. By Jensen's Inequality,

$$\sum_{t=0}^{\infty} (1-\rho)\rho^t u(c_t) \leqslant u(\overline{c})$$

with strict inequality if $c_t \neq \overline{c}$ for some t. Hence we have that

$$\sum_{t=0}^{\infty} \rho^t u(c_t) \leqslant \sum_{t=0}^{\infty} \rho^t u(\overline{c})$$

and if $f(k_0) > k_0$, the stationary rate of consumption $f(k_0) - k_0$ would be better than the given optimal path, a contradiction. □

These results are the principal elements of our analysis by which we will show that optimal paths converge to either the origin or the upper steady state. Throughout this section we neither stated nor assumed that there is a unique optimal path.

10.4 The Role of Discounting and the Critical Level of Capital Stock

It is useful to consider three separate cases which depend on the discount factor: mild discounting for which $\rho f'(0) \geqslant 1$; intermediate discounting where $\rho f'(0) < 1 \leqslant \rho \max[f(k)/k]$; and strong discounting which has for all positive stocks, $\rho f(k) < k$. Majumdar and Mitra (1982, Theorem 5.5) have shown that for mild discounting all optimal paths converge to the upper steady state k^*, and in those cases of strong discounting for which $\rho f'(k_I) < 1$ all optimal paths converge to the origin. We extend these results by showing that for all strong discounting if k^* is not an optimal steady state then all optimal paths converge to the origin.

It is for the intermediate discounting case that Clark's conjecture for linear utility functions is also correct for the concave ones as well. When the discount factor is in this range of values, there is a minimum level of the capital stock necessary for it to be optimal to accumulate capital; otherwise it is optimal to deplete the stock.

Mild Discounting: $\rho f'(0) \geqslant 1$

While Majumdar and Mitra (1982, Theorem 5.5) solved the problem for this case, we present here a proof of their proposition based on the monotonicity property of optimal paths.

Theorem 3. *If $\rho f'(0) \geqslant 1$ then all optimal paths starting from $k_0 > 0$ converge to k^*.*

Proof. An optimal path $\{k_t\}$ starting from a positive level of capital stock $0 < k_0 < k^*$ satisfies the Euler equation $u'(c_{t-1}) = \rho f'(k_t) u'(c_t)$ and since $\rho f'(k_t) \geqslant 1$ it follows that $c_t \geqslant c_{t-1}$. No Euler path that converges to the origin has this property, and so by Theorem 2, $\{k_t\}$ must converge to k^*. A similar argument shows that if $\{k_t\}$ is optimal from $k_0 \geqslant k^*$ it can not be the case that $k_t < k^*$ for any t. □

Fig. 10.2

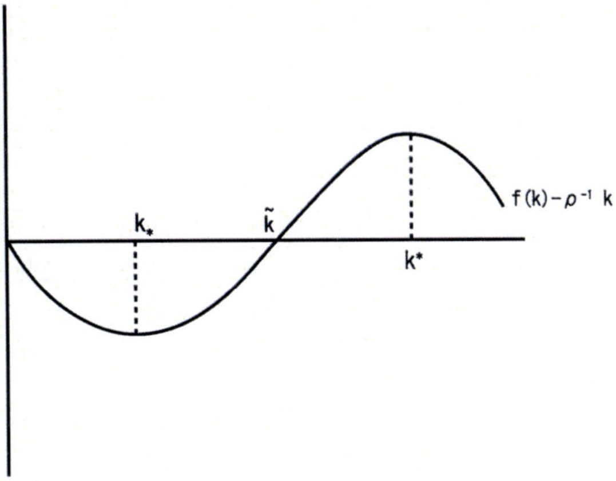

Intermediate Discounting: $\rho f'(0) < 1 \leqslant \rho \max[f(k)/k]$

In this section we show the existence of a unique critical level of the capital stock such that for all initial stocks *below* this level, all optimal paths converge to the origin, while for initial stocks *above* this level, all optimal paths converge to the upper steady state. Figure 10.2 depicts the case of intermediate discounting, along with several reference values of capital stock which we shall use throughout this section.

We consider each of steady states in turn. First we show that the upper steady state is a locally stable optimal steady state:

Lemma 2. *Define* $\widetilde{k}$ *(as shown in Fig. 10.2) by* $f(\widetilde{k}) = \rho^{-1}\widetilde{k}$ *and* $f'(\widetilde{k}) \geqslant \rho^{-1}$. *If* $k_0 \geqslant \widetilde{k}$ *then all optimal paths from* k_0 *converge to* k^*.

Proof. That k^* is an optimal steady state follows from the fact that if $\{k_t\}$ is any feasible path starting from k^* which is not stationary then

$$f(k_t) - \rho^{-1}k_t \leqslant f(k^*) - \rho^{-1}k^*$$

for all t, and with strict inequality holding for some t. (See Fig. 10.2) This implies that

$$\sum_{t=0}^{\infty} \rho^t [f(k_t) - \rho^{-1}k_t] < \frac{f(k^*) - \rho^{-1}k^*}{1-\rho}. \tag{10.5}$$

However, for any k_0

$$\sum_{t=0}^{\infty} \rho^t [f(k_t) - \rho^{-1}k_t] = \sum_{t=0}^{\infty} \rho^t [f(k_t) - k_{t+1}] - \rho^{-1}k_0 \tag{10.6}$$

and therefore the inequality in (10.5) implies that

$$\sum_{t=0}^{\infty}\rho^t[f(k_t)-k_{t+1}] < \frac{f(k^*)-k^*}{1-\rho}.$$

By Lemma 1, this means that the only optimal path starting at k^* is the stationary path of $k_t = k^*$ for all t.

Now if $k_0 > k^*$ and $\{k_t\}$ is the optimal path starting from k_0 then Theorem 1 shows that $k_t > k^*$ for all t and Theorem 2 implies that $k_t \to k^*$. If $\widetilde{k} \leqslant k_0 < k^*$ and if the optimal path converges either to the origin or to k_* then

$$f(k_t) - \rho^{-1}k_t < f(k_0) - \rho^{-1}k_0 \tag{10.7}$$

for all t. By a similar argument to the first part of this proof we are led to the conclusion that

$$\sum_{t=0}^{\infty}\rho^t[f(k_t)-k_{t+1}] < \frac{f(k_0)-k_0}{1-\rho}$$

which is a contradiction to Lemma 1. □

While Lemma 2 only shows that the optimal path converges to k^* as long as $k_0 \geqslant \widetilde{k}$, it is clear from the proof that the conclusion also holds for $k_0 < \widetilde{k}$, provided that k_0 is sufficiently close to $\widetilde{k}$. This follows from the fact that an optimal program starting from a k_0 just less than $\widetilde{k}$ would spend too many periods in the region where the inequality in (10.7) would hold (if it converges to the origin) and this again leads to a contradiction of Lemma 1.

The lower steady state exhibits a certain pathological behavior in that no optimal path that does not start there can ever converge to it; while it is possible that it might be an optimal steady state. This can occur if the utility function has a sharp bend at a level of consumption of $c_* = f(k_*) - k_*$.

Lemma 3. *No optimal path starting from $k_0 \neq k_*$ converge to k_*.*

Proof. By Lemma 2 we need only consider optimal paths starting from $k_0 < \widetilde{k}$. If the optimal path, $\{k_t\}$, converges to k_* then by Theorem 1 this convergence is monotone and so

$$f(k_t) - \rho^{-1}k_t < f(k_0) - \rho^{-1}k_0, \quad t = 1, 2, \ldots$$

and as the proof of Lemma 2 shows, this leads to a contradiction of Lemma 1. □

A further application of the monotonicity property of Theorem 1 leads to a corollary of Lemma 3: if k_* is an optimal steady state then it is the (unique) critical level of capital stock that we are looking for. This follows from the following argument: suppose an optimal path starting from $k_0 > k_*$ converges to the origin (Lemma 3 shows that it can't converge to k_*) and so it crosses over k_*. However, there's an optimal path starting from k_* which remains at k_*, and this is a violation of the monotonicity property shown in Theorem 1. For some pairs of utility and production functions the lower steady state cannot be an optimal steady state.

Lemma 4. *If* $(\rho^{-1}-1)u''(c_*)+\rho u'(c_*)f''(k_*)>0$ *then* k_* *is not an optimal steady state.*

Proof. Define $\phi(\varepsilon)=u(c_*-\varepsilon)+(\rho/(1-\rho))u(f(k_*+\varepsilon)-k_*-\varepsilon)$ for $0\leqslant k_*+\varepsilon\leqslant f(k_*)$, and suppose that k_* is an optimal steady state, in which case $\phi(0)=V(k_*)$. However,

$$\begin{aligned}\phi'(0)&=-u'(c_*)+\frac{\rho}{1-\rho}u'(c_*)(f'(k_*)-1)\\&=-u'(c_*)+\frac{\rho}{1-\rho}u'(c_*)(\rho^{-1}-1)\\&=-u'(c_*)+u'(c_*)=0\end{aligned}$$

and

$$\begin{aligned}\phi''(0)&=u''(c_*)+\frac{\rho}{1-\rho}\left[u''(c_*)(f'(k_*)-1)^2+u'(c_*)f''(k_*)\right]\\&=u''(c_*)+\frac{\rho}{1-\rho}\left[u''(c_*)(\rho^{-1}-1)^2+u'(c_*)f''(k_*)\right]\\&=\frac{1}{1-\rho}\left[(\rho^{-1}-1)u''(c_*)+\rho u'(c_*)f''(k_*)\right]>0\end{aligned}$$

and since $\phi''(0)>0$, there is an $\varepsilon\neq 0$ for which $\phi(\varepsilon)>\phi(0)$, and hence k_* cannot be an optimal steady state, for otherwise $V(k_*)\geqslant\phi(\varepsilon)$ for any ε. □

In a trivial way the origin is always an optimal steady state; however, so long as $f'(0)<\rho^{-1}$ it is also locally stable.

Lemma 5. *There is a* $0<k_0<k_*$ *such that all optimal paths starting from* k_0 *converge to the origin.*

Proof. Notice that if $f'(0)<1$ then for sufficiently small k_0, $f(k_0)<k_0$. This implies that *all* feasible paths from k_0 necessarily converge to 0 and hence the lemma holds. For the case that $f'(0)\geqslant 1$, suppose to the contrary that for all $0<k_0<k_*$ at least one optimal path converges to k^*, and let $\{k_t\}$ be an optimal path from k_0 such that $k_0<k_1<k_*$. Our first step is to show that there is a minimum level of consumption, ξ, such that for any $k_0<k_0'<k_1$, and any optimal path $\{k_t'\}$ from k_0', it is necessarily the case that $f(k_1')-k_1'\geqslant\xi$. Let $c_t=f(k_t)-k_{t+1}$ and $c_t'=f(k_t')-k_{t+1}'$. Since all optimal paths are Euler paths,

$$\frac{u'(c_0)}{u'(c_t)}=\prod_{s=1}^{t}\rho f'(k_s),$$

and since $\lim_{t\to\infty} c_t = c^* = f(k^*) - k^*$, we have that $u'(c_0) = u'(c^*)\Pi_{s=1}^{\infty}\rho f'(k_s)$. Similarly, $u'(c_0') = u'(c^*)\Pi_{s=1}^{\infty}\rho f'(k_s')$, and hence

$$\frac{u'(c_0)}{u'(c_0')} = \prod_{s=1}^{\infty} \frac{f'(k_s)}{f'(k_s')}.$$

Let k_I be the inflection point where $f''(k_I) = 0$. Then for some T, $k_T \leqslant k_I < k_{T+1}$, and by the monotonicity property, $k_0 < k_0' < k_1 < k_1' < \cdots < k_T < k_T'$ and we may assume that $k_T' \leqslant k_I$ (the case that $k_I < k_T' < k_{T+1}$ can be handled in a similar way). Then

$$\frac{u'(c_0)}{u'(c_0')} = \frac{f'(k_1)}{f'(k_1')} \cdot \frac{f'(k_2)}{f'(k_2')} \cdots \frac{f'(k_T)}{f'(k_T')} \cdot \prod_{s=T+1}^{\infty} \frac{f'(k_s)}{f'(k_s')}.$$

Since for all $s \geqslant T+1$, $k_I < k_s < k_s'$, $f'(k_s) > f'(k_s')$ and so

$$\prod_{s=T+1}^{\infty} \frac{f'(k_s)}{f'(k_s')} > 1.$$

Also for all $s \leqslant T$, $k_s < k_s' \leqslant k_I$ and so $f'(k_s) > f'(k_{s-1}')$. Therefore

$$\frac{u'(c_0)}{u'(c_0')} > \frac{f'(k_I)}{f'(k_T')} \geqslant \frac{f'(k_1)}{f'(k_I)}$$

or

$$u'(c_0') \leqslant \frac{u'(c_0) f'(k_I)}{f'(k_1)}.$$

If we let $u'(\xi) = u'(c_0) f'(k_I)/f'(k_1)$ then the above inequality implies that $c_0' \geqslant \xi$ for any $k_0 < k_0' < k_1$ and $c_0' = f(k_0') - k_1'$. Let $k_0'' < k_0$ be any positive stock with $f(k_0'') - k_0'' < \xi$, and let $\{k_t''\}$ be the sequence of stocks along an optimal path. By assumption, k_t'' converges to k^*, and so for some t_0, $k_{t_0-1}'' < k_0 < k_{t_0}''$, and therefore by the monotonicity property $k_{t_0}'' < k_1$. By our argument given above, any optimal path starting at k_{t_0}'' has consumption $f(k_{t_0}'') - k_{t_0+1}'' \geqslant \xi$. However, the optimal path $\{k_t''\}$ is an Euler path with $\rho f'(k_t'') < 1$, and from the Euler equation it follows that $u'(c_{t-1}'') < u'(c_t'')$ and hence the consumption $c_t'' = f(k_t'') - k_{t+1}''$ satisfies

$$c_0'' > c_1'' > \cdots > c_{t_0}''.$$

But by construction

$$c_0'' = f(k_0'') - k_1'' < f(k_0'') - k_0'' \leqslant \xi$$

and therefore $c''_{t_0} < \xi$, which is a contradiction. Therefore, there is some initial stock $k''_0 > 0$ for which all optimal paths must converge to the origin. □

With these results we can now derive our main conclusion about the critical level of capital stock in this intermediate discounting case;

Theorem 4. *There is a* $0 < k_c < \widetilde{k}$ *so that every optimal path starting from* $k_0 > k_c$ *converges to* k^* *and every optimal path starting from* $k_0 < k_c$ *converges to the origin.*

Proof. Let $\underline{k}_c$ be the supremum of the set of initial stocks such that all optimal paths that begin at any point of the set converge to 0; and similarly, let $\overline{k}_c$ be the infimum of the set of initial stocks such that all optimal paths that begin in the set converge to k^*. It is consequence of Theorem 1 and Lemmas 2 and 5 that $0 < \underline{k}_c \leqslant \overline{k}_c < k^*$. Now suppose $\underline{k}_c < \overline{k}_c$, and pick any k_0 and k'_0 so that $\underline{k}_c < k_0 < k'_0 < \overline{k}_c$. By definition of $\underline{k}_c$ and $\overline{k}_c$, there is an optimal path from k_0 that converges to 0 and also one that converges to k^*. Call these optimal paths $\{\widetilde{k}_t\}$ and $\{\widehat{k}_t\}$, respectively, and similarly for k'_0 let $\{\widetilde{k}'_t\}$ and $\{\widehat{k}'_t\}$ be the optimal paths. Then for some t, $\widehat{k}_t > \widetilde{k}'_t$, which is a violation of Theorem 1 since $k_0 < k'_0$. Therefore $\underline{k}_c = \overline{k}_c$. Furthermore, by Lemma 5, $k_c > 0$, and by Lemma 2, $k_c < \widetilde{k}$. □

As can be seen in Fig. 10.2, as the discount factor ρ increases, the value of $\widetilde{k}$ decreases. Nevertheless, the critical value of the capital stock remains positive, until the mild discounting condition of $\rho f'(0) \geqslant 1$ is reached.

In order to further characterize the critical level of capital stock, we show the following continuity property of optimal paths.

Lemma 6. *If* $k_c \neq k_*$, *then there is an optimal path starting from* k_c *and converging to* k^*, *and an optimal path starting from* k_c *and converging to* 0.

Proof. Let $k^n_0 > k_c$ converge monotonically to k_c and let $\{k^n_t\}$ any optimal path from k^n_0. Also let $\{x_t\}$ be any feasible path from k_c and $\{x^n_t\}$ be any feasible path from k^n_0 such that $x^n_t \to_{n\to\infty} x_t$ for all t. By the compactness of the set of feasible capital stocks there is a subsequence of $\{k^n_t\}$ that converges coordinatewise, so we may assume that $k^n_t \to_{n\to\infty} k_t$ for some sequence $\{k_t\}$. By the optimality of $\{k^n_t\}$ and the feasibility of $\{x^n_t\}$,

$$\sum_{t=0}^{\infty} \rho^t u(f(k^n_t) - k^n_{t+1}) > \sum_{t=0}^{\infty} \rho^t u(f(x^n_t) - x^n_{t+1}).$$

Since both sums are uniformly convergent we have in the limit

$$\sum_{t=0}^{\infty} \rho^t u(f(k_t) - k_{t+1}) \geqslant \sum_{t=0}^{\infty} \rho^t u(f(x_t) - x_{t+1}).$$

Since this holds for any arbitrary feasible path $\{x_t\}$, it is the case that the path $\{k_t\}$ is an optimal path starting from $k_0 = k_c$. By definition of k_c and by choice of $k_0^n > k_c$ we have that $k_c < k_0^n < k_1^n$ for all n, and hence the limit $k_c \leqslant k_1$. Therefore it cannot be the case that *all* optimal paths from k_c converge to 0, and by Theorem 2 and Lemma 3, there must be an optimal path starting from k_c which converges to k^*. A similar argument with a sequence $k_0^n < k_c$ converging to k_c shows that there must also be an optimal path which converges to 0. □

In the proof of this lemma we used the fact that $k_c \neq k_*$ when we assumed that an optimal path starting from k_c converged either to 0 or to k^*. When k_* is an optimal steady state (and so $k_c = k_*$) the above proof fails, since as we have shown the *only* optimal path starting from k_* is the one that remains at k_*.

Strong Discounting: $\rho \max[f(k)/k] < 1$

When the discount factor is very small, there are two cases to consider: the extreme case that even the inflection point is not a steady state, $\rho f'(k_I) < 1$; and the more moderate case of $1 \leqslant \rho f'(k_I)$.

In the extreme case there are no positive steady states, and so by Theorem 2 all optimal paths converge to the origin. In the more moderate case, the behavior of the optimal path depends on whether or not k^* is an optimal steady state.

Theorem 5. *If* $\rho \max[f(k)/k] < 1 < \rho f'(k_I)$*then*

(i) If k^* *is not an optimal steady state, all optimal paths converge to the origin;*
(ii) If k^* *is an optimal steady state then there is a unique* $0 < k_c \leqslant k^*$ *such that all optimal paths starting from* $k_0 < k_c$ *converge to the origin, and all optimal paths starting from* $k_0 > k_c$ *converge to* k^*.

Proof. (i) Suppose the optimal path $\{k_t\}$ starting from k_0 converges to k^*. Define $k_0^n = k_n$ for $n = 1, 2, ...,$ and note that starting from each k_0^n there is an optimal path $\{k_t^n\}$ that converges to k^*. Furthermore $k_t^n \to_{n\to\infty} k^*$ for each $t = 0, 1, 2, ...$. Hence the continuity argument of Lemma 6 can be applied to show that the stationary path of $\{k^*, k^*, ...\}$ is optimal starting from k^*.

(ii) If $\{k_t\}$ is an optimal path starting from $k_0 \geqslant k^*$ then the monotonicity property of Theorem 1 shows that $k_t < k^*$ can never occur.

The argument used in Lemma 3 shows that if $k_0 > 0$ is sufficiently small so that $f(k_0) - \rho^{-1}k_0 > f(k^*) - \rho^{-1}k^*$ (see Fig. 10.3), then it cannot be the case that $\{k_t\}$ converges to k^*, and hence by Theorem 2 it must converge to the origin. The uniqueness of the critical level k_c then follows from the same argument as in Theorem 3. □

Whether or not k^* is an optimal steady state in this strong discounting case depends on the actual utility and production functions. Majumdar and Mitra (1982) have given an example where it is not. We present here an example where it is optimal, and moreover the critical level of capital stock is strictly less than the upper steady state.

Fig. 10.3

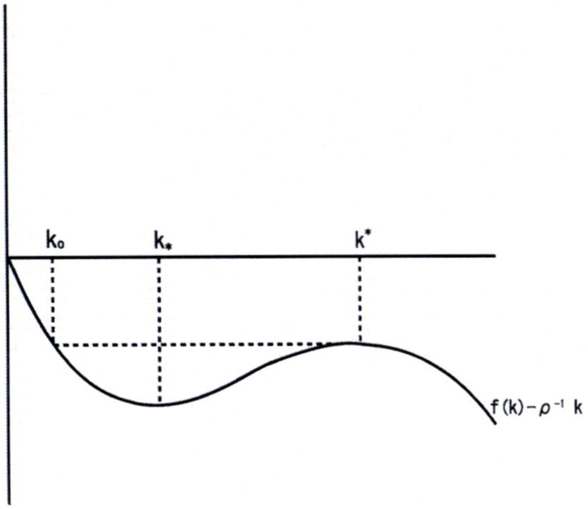

Example 1. When $f'(k_I) > \rho^{-1} > f(k^*)/k^*$ there is a $\overline{k} < k^*$ so that $f(\overline{k}) = f(k^*)$, as in Fig. 10.3.

Let $\{k_t\}$ be any sequence with $k_0 = k^*$ and which converges monotonically to the origin. Then there is a $T \geqslant 1$ for which $k_T \leqslant \overline{k} < k_{T-1}$, and

$$\sum_{t=0}^{\infty} \rho^t [u(c_t) - u(c^*)] = \sum_{t=0}^{T-1} \rho^t [u(c_t) - u(c^*)] + \sum_{t=T}^{\infty} \rho^t [u(c_t) - u(c^*)] \quad (10.8)$$

$$< u'(c^*) \sum_{t=0}^{T-1} \rho^t (c_t - c^*) + \frac{\rho^T}{1-\rho} [u(f(\overline{k})) - u(c^*)].$$

But

$$\sum_{t=0}^{T-1} \rho^t (c_t - c^*)$$

$$= \begin{cases} k^* - k_1 \text{ if } T = 1 \\ \sum_{t=1}^{T-1} \rho^t [f(k_t) - f(k^*) - \rho^{-1}(k_t - k^*)] + \rho^{T-1}(k^* - k_T) \text{ if } T \geqslant 2 \end{cases}$$

and since $\overline{k} < k_t$ for $t = 1, ..., T-1$

$$f(k_t) - f(k^*) < \rho^{-1}(k_t - k^*)$$

Therefore,

$$\sum_{t=0}^{T-1} \rho^t (c_t - c^*) \leqslant \rho^{T-1}(k^* - k_T) \leqslant \rho^{T-1} k^*$$

and in (10.8) we get

$$\sum_{t=0}^{\infty} \rho^t [u(c_t) - u(c^*)] < \rho^{T-1} u'(c^*) k^* + \frac{\rho^T}{1-\rho} [u(f(\overline{k})) - u(c^*)].$$

Thus for any strictly concave utility function for which

$$u'(c^*) k^* \leqslant \frac{\rho}{1-\rho} [u(c^*) - u(f(\overline{k}))] \tag{10.9}$$

it will be the case that

$$\sum_{t=0}^{\infty} \rho^t [u(c_t) - u(c^*)] < 0 \tag{10.10}$$

along *any* path $\{k_t\}$ which converges monotonically to the origin. For such a utility function, k^* would be an optimal steady state for values of ρ and $\overline{k}$ which satisfy inequality (10.9). Furthermore, it is clear from this argument and the strict inequality (10.10) that if $k_0 < k^*$ is sufficiently close to k^*, then it will not be optimal to converge to the origin. Hence for such a utility function the critical level of the capital stock is strictly below the upper steady state.

10.5 The Differentiability of the Value and the Uniqueness of Optimal Paths

In this section we derive several results about the value function, the most important of which is that the set of points at which the value function is differentiable coincides with the set of starting points from which the optimal paths are uniquely determined. Furthermore, we show that the set of points at which the value is not differentiable is a sparse set and that the value function is sufficiently well behaved that it has both right-and left-hand derivatives everywhere.

Theorem 6. *V has right-and left-hand derivatives on* $(0, \infty)$.

Proof. Let $k_0 > 0$, $\underline{k}_0^n \uparrow k_0$, and let $\{\underline{k}_t^n\}$ be optimal from $\underline{k}_0^n$. By the continuity property shown in Lemma 6 we can assume that $\underline{k}_t^n \to \underline{k}_t$ for each t and that $\{\underline{k}_t\}$ is optimal from k_0. Furthermore, we can assume that $\underline{k}_0^n$ is sufficiently close to k_0 so that $\underline{k}_1 \leqslant f(\underline{k}_0^n)$ for all n. By the Principle of Optimality

$$V(k_0) = u(f(k_0) - \underline{k}_1) + \rho V(\underline{k}_1)$$

$$V(\underline{k}_0^n) \geqslant u(f(\underline{k}_0^n) - \underline{k}_1) + \rho V(\underline{k}_1)$$

which leads to

$$V(k_0) - V(\underline{k}_0^n) \leqslant u(f(k_0) - \underline{k}_1) - u(f(\underline{k}_0^n) - \underline{k}_1)$$
$$\leqslant u'(f(\underline{k}_0^n) - \underline{k}_1)[f(k_0) - f(\underline{k}_0^n)]$$

and so

$$\frac{V(\underline{k}_0^n) - V(k_0)}{\underline{k}_0^n - k_0} \leqslant u'(f(\underline{k}_0^n) - \underline{k}_1)\frac{f(\underline{k}_0^n) - f(k_0)}{\underline{k}_0^n - k_0}. \tag{10.11}$$

Similarly,

$$V(k_0) \geqslant u(f(k_0) - \underline{k}_1^n) + \rho V(\underline{k}_1^n)$$
$$V(\underline{k}_0^n) = u(f(\underline{k}_0^n) - \underline{k}_1^n) + \rho V(\underline{k}_1^n)$$

which leads to

$$\frac{V(\underline{k}_0^n) - V(k_0)}{\underline{k}_0^n - k_0} \geqslant u'(f(k_0) - \underline{k}_1^n)\frac{f(\underline{k}_0^n) - f(k_0)}{\underline{k}_0^n - k_0}. \tag{10.12}$$

The right-hand sides of both (10.11) and (10.12) have $u'(f(k_0)-\underline{k}_1)f'(k_0)$ as limits, and therefore

$$V'(k_0^-) = u'(f(k_0) - \underline{k}_1)f'(k_0). \tag{10.13}$$

Following the same technique, if $\overline{k}_0^n \downarrow k_0$, and $\{\overline{k}_t^n\}$ is an optimal path from $\overline{k}_0^n$, then we can assume that $\overline{k}_t^n \to k_t$ for each t and that $\{\overline{k}_t\}$ is optimal from k_0. We get that

$$V'(k_0^+) = u'(f(k_0) - \overline{k}_1)f'(k_0) \tag{10.14}$$

□

Corollary 2. *$V'(k_0^-) \leqslant V'(k_0^+)$ for all $k_0 > 0$.*

Proof. Since $\underline{k}_1 \leqslant \overline{k}_1$ the result follows from formulae (10.13) and (10.14). □

Corollary 3. *If k_* is an optimal steady state (and hence $k_c = k_*$) then $V'(k_*) = \rho^{-1}u'(c_*)$ and the optimal path (which remains at k_*) is unique.*

Proof. Let $\underline{k}_0^n \uparrow k_*$ and $\bar{k}_0^n \downarrow k_*$. Then from

$$V(k_*) = u(f(k_*) - k_*) + \rho V(k_*) \geqslant u(f(k_*) - \underline{k}_0^n) + \rho V(\underline{k}_0^n)$$

we get that

$$\rho\frac{V(\underline{k}_0^n) - V(k_*)}{\underline{k}_0^n - k_*} \geqslant u'(f(k_*) - \underline{k}_0^n)$$

and so $V'(k_*^-) \geqslant \rho^{-1}u'(f(k_*) - k_*)$. Similarly, from

$$V(k_*) = u(f(k_*) - k_*) + \rho V(k_*) \geqslant u(f(k_*) - \overline{k}_0^n) + \rho V(\overline{k}_0^n)$$

we get that $V'(k_*^+) \leqslant \rho^{-1}u'(f(k_*) - k_*)$. Therefore by Corollary 2, $V'(k_*) = \rho^{-1}u'(c_*)$. Finally, if there were an optimal path $\{k_t\}$ starting from k_* with $k_1 < k_*$, then by the Theorem, $V'(k_*^-) \geqslant \rho^{-1}u'(f(k_*) - k_t)$, a contradiction to the first part of this corollary. □

Corollary 4. *If $k_c \neq k_*$ then $V'(k_c^-) < V'(k_c^+)$.*

Proof. If $\underline{k}_0^n \uparrow k_c$ and $\bar{k}_0^n \downarrow k_c$, then optimal paths $\{\underline{k}_t^n\}$ and $\{\overline{k}_t^n\}$ have limits $\{\underline{k}_t\}$ and $\{\overline{k}_t\}$ with $\underline{k}_t \to 0$ and $\overline{k}_t \to k^*$, and thus $\underline{k}_1 < k_c < \overline{k}_1$. The rest follows by Corollary 2. □

While Theorem 6 and its Corollaries give a fairly complete characterization of the derivative of the value function, we can sharpen this result by showing that the set of discontinuities of the derivative is a countable set. First note that once started, all optimal paths are unique:

Lemma 7. *Let $\{k_t\}$ be any optimal path starting from any k_0. Then the unique optimal path starting from k_t is $\{k_{t+s}, s \geqslant 1\}$ for every $t \geqslant 1$.*

Proof. Suppose from k_1 there are two optimal paths, say, $\{k'_{1+s}\}$ and $\{k''_{1+s}\}$. By the Principle of Optimality, both $k_0, k_1, k'_2, k'_3, \ldots$ and $k_0, k_1, k''_2, k''_3, \ldots$ must be optimal starting from k_0, and so the Euler equation must hold along both paths:

$$u'(f(k_0) - k_1) = \rho f'(k_1)u'(f(k_1) - k'_2)$$
$$u'(f(k_0) - k_1) = \rho f'(k_1)u'(f(k_1) - k''_2)$$

which can only occur if $k'_2 = k''_2$. Since $\{k_2, k_3, \ldots\}$ must also be optimal from k_1 it follows that $k_2 = k'_2 = k''_2$. The rest follows by induction on t. □

Corollary 5. *Let $\{k_t\}$ any optimal path starting from $k_0 > 0$. Then V is differentiable at each k_t, for $t \geqslant 1$.*

Proof. Since the optimal path from each k_t is unique, the result follows from Theorem 6. □

To show that the set of discontinuities of the derivative of V forms a countable set, let $a_n \to 0$, $b_n \to \infty$ with $0 < a_n < b_n$ and define

$$D_n = \{k \in [a_n, b_n] | V'(k^-) < V'(k^+)\}.$$

Lemma 8. *$D = \cup_{n=1}^{\infty} D_n$ is a countable set.*

Proof. It is clear that V is a continuous increasing function on $(0, \infty)$, and since $V'(k^-) \leq V'(k^+)$ for all $k \in (0, \infty)$,

$$V(b_n) - V(a_n) \geqslant \sum_{k \in D_n} [V'(k^+) < V'(k^-)]$$

and therefore D_n is countable.[1] Since D is a countable union of countable sets it is also countable. □

An implication of Theorem 6 is that if the derivative of the value function exists at k_0 then the optimal path from k_0 is unique since there is a unique k_1 that solves

$$u'(f(k_0) - k_1) = \frac{V'(k_0)}{f'(k_0)}.$$

Furthermore, if there are two (or more) optimal paths from k_0, say, $\{\underline{k}_t\}$ and $\{\overline{k}_t\}$ with $\underline{k}_1 < \overline{k}_1$, then the value function is not differentiable at k_0 since

$$V'(k_0^-) \leqslant u'(f(k_0) - \underline{k}_1) f'(k_0) < u'(f(k_0) - \overline{k}_1) f'(k_0) \leqslant V'(k_0^+).$$

Thus the set of discontinuities of the derivative of V precisely coincides with the set of initial stocks for which the optimal paths are not uniquely determined, and Lemma 8 shows that this set is at most countable.

10.6 The Necessity of the Transversality Condition

Weitzman (1973) showed that the necessary and sufficient conditions for a path $k_1, k_2, \ldots$ to be optimal starting from k_0 are

(i) $\{k_t\}$ satisfies the Euler equation (cf. Definition(2))
(ii) $\lim_{t\to\infty} \rho^t u'(f(k_{t-1}) - k_t) k_t = 0$.

Condition (ii) is referred to as the transversality condition, and it plays the same role in the infinite horizon case that a boundary condition at the terminal time plays in a finite horizon model. In both the finite and infinite horizon models the optimal path must satisfy the Euler equation, along with the initial condition that the path starts at k_0. Since the Euler equation is a second-order difference equation, there are many solutions to it that start at k_0, not all of which are optimal. The optimal solutions also satisfy a boundary condition in the finite horizon case, or condition (ii) in the infinite horizon case. Weitzman proved this for a concave production function.

As we mentioned in Sect. 10.2, optimal paths are Euler paths for the convex-concave production function we consider in this paper. It is also the case that an optimal path satisfies the transversality condition,[2] which we show.

[1]For any $g : R \to [0, \infty)$ and $S \subset R$,

$$\sum_{k\in S} g(k) = \sup_{F\subset S} \{\sum_{k\in F} g(k) |\ \text{F is finite}\}.$$

[2]This necessity of the transversality condition was conjectured by David Cass in a private conversation.

Theorem 7. *If $f'(0) \geqslant 1$, $\lim_{t\to\infty} \rho^t u'(f(k_{t-1}) - k_t)k_t = 0$ along an optimal path.*

Proof. We need only consider the cases that $k_t \to k^*$ or $k_t \to 0$. If $k_t \to k^*$ then $f(k_{t-1}) - k_t \to f(k^*) - k^* > 0$ and since $\rho^t \to 0$ the conclusion follows. If $k_t \to 0$ then, as in the proof of Lemma 5,

$$u'(c_{t-1}) = \frac{u'(c_0)}{\Pi_{s=1}^{t-1} \rho f'(k_s)}$$

so

$$\rho^t u'(f(k_{t-1}) - k_t)k_t = \frac{\rho u'(c_0)k_t}{\Pi_{s=1}^{t-1} f'(k_s)}$$

along a path for which $k_s \to 0$, there exists a T so that $f'(k_s) \geqslant 1$ if $s \geqslant T$. Thus for $t > T$,

$$\prod_{s=1}^{t-1} f'(k_s) \geqslant \prod_{s=1}^{T} f'(k_s) > 0$$

and since $k_t \to 0$, the conclusion follows. □

To see that the converse of this theorem is false, consider the stationary path that starts and remains at the lower steady state k_*. It is an Euler path and satisfies the transversality condition. However, if the utility and production functions are such that k_* is *not* an optimal steady state, then this stationary path is not optimal. Hence the transversality condition (along with the Euler equation) is not sufficient for optimality.

10.7 A Final Remark

Clark (1971) referred to the critical level of stock as the minimum safe standard for conservation of the species. In Clark's model of a profit maximizing fishery (which corresponds to a linear utility of consumption growth model) it turns out that the optimal program is a most rapid approach path: either deplete the stock completely in one period, or else go as rapidly as possible to the upper steady state depending on whether the initial stock is below or above the critical value. By applying the results of our paper to the fishery model, we have shown that even for a risk averse, centrally managed renewable resource (i.e., where the per period utility is a strictly concave function of per capita harvest) there is a critical level of the stock below which it is optimal to cause the species to become extinct. However, the distinction between the linear and strictly concave cases is that when the utility function is strictly concave (and $u'(0^+) = +\infty$) the optimal paths approach the steady states *asymptotically*. Thus for the model as we have analyzed it, *actual* extinction would never occur even though depleting the stock is optimal.

References

Benveniste, L. M. and J. A. Scheinkman (1979), "On the Differentiability of the Value Function in Dynamic Models of Economies," *Econometrica*, 47, 727–732.

Brock, W. A. (1970), "On Existence of Weakly Maximal Programs in a Multi Sector Economy," *The Review of Economic Studies*, 37, 275–280.

Cass, D. (1965), "Optimum Growth in an Aggregative Model of Capital Accumulation," *The Review of Economic Studies*, 32, 233–240.

Clark, C. W. (1971), "Economically Optimal Policies for the Utilization of Biologically Renewable Resources," *Mathematical Biosciences*, 17, 245–268.

Gale, D. (1967), "On Optimal Development of a Multi Sector Economy," *Review of Economic Studies*, 34, 1–18.

Koopmans, T. C. (1965), On the Concept of Optimal Economic Growth, in *The Econometric Approach to Development Planning*, Pontificiae Academice Scientiarum Scriptum Varia, Noth-Holland, Amsterdam.

Majumdar, M. (1975), "Some Remarks on Optimal Growth with Intertemporally Dependent Preferences in the Neoclassical Model," *Review of Economic Studies*, 42, 147–153.

Majumdar, M. and T. Mitra (1980), On Optimal Exploitation of a Renewable Resource in a Non-Convex Environment and the Minimum Safe Standard of Conservation, Working Paper No. 223, Cornell University.

Majumdar, M. and T. Mitra (1982), "Intertemporal Allocation with a Non-Convex Technology: The Aggregative Framework," *Journal of Economic Theory*, 27, 101–136.

Ramsey, F. (1928), "A Mathematical Theory of Savings," *The Economic Journal*, 38, 543–559.

Weitzman, M. L. (1973), "Duality Theory for Infinite Horizon Convex Models," *Management Science*, 19, 783–789.

Part V
Stochastic Optimal Growth Models

Chapter 11
Stochastic Optimal Growth with Nonconvexities*

Kazuo Nishimura, Ryszard Rudnicki, and John Stachurski**

11.1 Introduction

The stochastic optimal growth model (Brock and Mirman 1972) is a foundation stone of modern macroeconomic and econometric research. To accommodate the data, however, economists are often forced to go beyond the convex production technology used in these original studies. Nonconvexities lead to technical difficulties which applied researchers would rather not confront. Value functions are in general no longer smooth, optimal policies contain jumps, and the Euler equation may not hold. This reality precludes the use of many standard tools. Further, convergence of state variables to a stationary equilibrium is no longer assured. The latter is a starting point of much applied analysis (see, e.g., Kydland and Prescott 1982; Long and Plosser 1983) and fundamental to the rational expectations hypothesis (Lucas 1986).

*Journal of Mathematical Economics **42**, 74–96, 2006

**The authors thank Takashi Honda, Takashi Kamihigashi, Kevin Reffett and an anonymous referee for many helpful comments, and the grant-in-aid for twenty-first Century COE Research, Australian Research Council Grant DP0557625, and the State Committee for Scientific Research (Poland) grant no. 2P03A 031 25 for financial support.

K. Nishimura (✉)
Institute of Economic Research, Kyoto University, Japan
e-mail: nishimura@kier.kyoto-u.ac.jp

R. Rudnicki
Department of Biomathematics, Institute of Mathematics, University of Silesia, Poland
e-mail: rudnicki@us.edu.pl

J. Stachurski
Research School of Economics, Australian National University, Australia
e-mail: john.stachurski@anu.edu.au

J. Stachurski et al. (eds.), *Nonlinear Dynamics in Equilibrium Models*,
DOI 10.1007/978-3-642-22397-6_11,

Although nonconvexities are technically challenging, the richer dynamics that they provide help to replicate key time series. For example, nonconvexities often lead to the kind of regime-switching behavior found in aggregate income data (e.g., Prescott 2003), or the growth miracles and growth disasters in cross-country income panels. Also, nonconvexities can arise directly from micro-level modeling, taking the form of fixed costs, threshold effects, ecological properties of natural resource systems, economies of scale and scope, network and agglomeration effects, and so on.

The objective of this paper is to investigate in depth the fundamental properties of stochastic nonconvex one-sector models and the series they generate using assumptions which facilitate integration with empirical research.[1]

Previously, in the deterministic case, optimal growth models with nonconvex technology were studied in continuous time by Skiba (1978). In discrete time, Majumdar and Mitra (1982) examined efficiency of intertemporal allocations. Dechert and Nishimura (1983) studied the standard discounted model with convex/concave technology, and characterized the dynamics of the model for every value of the discount factor. More recently, Amir et al. (1991) used lattice programming techniques to study solutions of the Bellman equation and associated comparative dynamics. Kamihigashi and Roy (2007) study nonconvex optimal growth without differentiability or even continuity.

In the stochastic case, a rigorous early treatment of optimal growth with nonconvex technology is given in Majumdar et al. (1989). Amir (1997) studies optimal growth in economies that have some degree of convexity. Using martingale arguments, Joshi (1997) analyzes the classical turnpike properties when technology is nonstationary. Schenk-Hoppé (2005) considers dynamic stability of stochastic overlapping generations models with S-shaped production function. Mitra and Roy (2006) study nonconvex renewable resource exploitation and stability of the resource stock.

The above papers assume that the shock which perturbs activity in each period has compact support. We extend their analysis by assuming instead that the shock is multiplicative, and its distribution has a density—which may in general have bounded or unbounded support. This formulation is relatively standard in quantitative applications. It provides considerable structure, which can be exploited when investigating optimality and dynamics.

Without convexity many standard results pertaining to the optimal policy and the value function can fail. In this study the density representation of the shock is used to prove interiority of the optimal policy and Ramsey–Euler type results. Based on these findings and some additional assumptions, we then obtain a fundamental dichotomy: every economy is either globally stable in a strong sense to be made precise, or globally collapsing to the origin. This result (a version of the Foguel Alternative) simplifies greatly the range of possible dynamics. We connect the two

[1] We consider only optimal dynamics. There are many studies of nonoptimal competitive dynamics in nonconvex environments. See for example Mirman et al. (2005).

possibilities to the discount rate, and also provide conditions to determine which outcome prevails for specific parameterizations.[2]

One caveat is that we consider only one-sector models. Multi-sector models are common in applications, but their dynamics are vastly more complex. Thus it is an important open question whether or not the results on dynamics presented in this paper can be extended to multi-sector models.

Section 11.2 introduces the model. Section 11.3 discusses optimization and properties of the optimal policies. Section 11.4 considers the dynamics of the processes generated by these policies (i.e., the optimal paths). All of the proofs are given in Sect. 11.5 and the Appendix.

11.2 Outline of the Model

Let $\mathbb{R}_+ := [0, \infty)$ and let $\mathcal{B}$ be the Borel subsets of $\mathbb{R}_+$. At the start of each period t a representative agent receives current income $y_t \in \mathbb{R}_+$ and allocates it between current consumption c_t and savings. On current consumption c the agent receives instantaneous utility $u(c)$. Savings determines the stock k_t of available capital, where $0 \leq k_t + c_t \leq y_t$. Production then takes place, delivering at the start of the next period output

$$y_{t+1} = f(k_t)\varepsilon_t, \tag{11.1}$$

which is net of depreciation. Here ε_t is a shock taking values in $\mathbb{R}_+$.

The productivity shocks $(\varepsilon_t)_{t=0}^{\infty}$ form an independent and identically distributed sequence on probability space $(\Omega, \mathcal{F}, \mathbf{P})$. When the time t savings decision is made shocks $\varepsilon_0, \ldots, \varepsilon_{t-1}$ are observable. The distribution of ε_t is represented by density φ.[3] We let $\mathbb{E}[\varepsilon_t] = \int z\varphi(z)dz = 1$. The bold symbol $\boldsymbol{\varphi}$ is used to denote the probability measure on $\mathbb{R}_+$ corresponding to the density φ, so that $\boldsymbol{\varphi}(dz)$ and $\varphi(z)dz$ have the same meaning.

The agent seeks to maximize the expectation of a discounted sum of utilities. Future utility is discounted according to $\rho \in (0, 1)$.

Assumption 1 *The function u is strictly increasing, strictly concave, and continuously differentiable on* $(0, \infty)$. *It satisfies (U1)* $\lim_{c \to 0} u'(c) = \infty$; *and (U2) u is bounded with* $u(0) = 0$.

The condition (U1) is needed to obtain the Ramsey–Euler equation. Strict concavity is critical to the proof of monotonicity of the optimal policy, on which all subsequent results depend. Note that if u is required to be bounded, then assuming $u(0) = 0$ sacrifices no additional generality.

---

[2]For further discussion of dynamics, including a specific condition on the primitives that ensures global stability, see Nishimura and Stachurski (2005).

[3]That is, $\mathbf{P}[\varepsilon_t^{-1}(B)] = \int_B \varphi(z)dz$ for all $B \in \mathcal{B}$. Here and in what follows, by a density is meant a nonnegative $\mathcal{B}$-measurable function on $\mathbb{R}_+$ that integrates to unity.

Assumption 2 *The production function f is strictly increasing and continuously differentiable on $(0, \infty)$. In addition, (F1) $f(0) = 0$; (F2) $\lim_{k\to\infty} f'(k) = 0$, and (F3) $\lim_{k\to 0} f'(k) > 1$.*

Condition (F2) is the usual decreasing returns assumption. Actually for the proofs we require only that f is majorized by an affine function with slope less than one. This is implied by (F2), as can be readily verified from the Fundamental Theorem of Calculus.

An economy is defined by the collection (u, f, φ, ρ), for which Assumptions 1 and 2 are always taken to hold.

By a control policy is meant a function $\pi: \mathbb{R}_+ \ni y \mapsto k \in \mathbb{R}_+$ associating current income to current savings. The policy is called feasible if it is $\mathcal{B}$-measurable and $0 \leq \pi(y) \leq y$ for all y. An initial condition and a feasible policy complete the dynamics of the model (11.1), determining a stochastic process $(y_t)_{t\geq 0}$ on $(\Omega, \mathcal{F}, \mathbf{P})$, where $y_{t+1} = f(\pi(y_t))\varepsilon_t$ for all $t \geq 0$.

Investment behavior is determined by the solution to the problem

$$\sup_{\pi} \mathbb{E}\left[\sum_{t=0}^{\infty} \rho^t u(y_t - \pi(y_t))\right], \tag{11.2}$$

where $\mathbb{E}$ denotes integration over Ω with respect to $\mathbf{P}$, an initial condition y_0 is given, and the supremum is over the set of all feasible policies. By (U2) the functional inside the integral is bounded independent of π, and the supremum always exists. A policy is called optimal if it is feasible and attains the supremum (11.2).

11.3 Optimization

In this section we solve the optimization problem by dynamic programming, and characterize the properties of the value function and control policy. To begin, define as usual the value function V by setting $V(y)$ as the real number given in (11.2) when $y = y_0$ is the initial condition. Let $b\mathcal{B}$ be the space of real bounded $\mathcal{B}$-measurable functions on $\mathbb{R}_+$. Define also the Bellman operator T mapping $b\mathcal{B}$ into itself by

$$(Tv)(y) = \sup_{0\leq k\leq y} \left\{u(y-k) + \rho \int v[f(k)z]\varphi(dz)\right\}. \tag{11.3}$$

It is well-known that T is a uniform contraction on $b\mathcal{B}$ in the sense of Banach, and that the value function V is the unique fixed point of T in $b\mathcal{B}$.

Lemma 1. *For any economy (u, f, φ, ρ), the value function V is continuous, bounded and strictly increasing. An optimal policy π exists. Moreover, if π is optimal, then*

$$V(y) = u(y - \pi(y)) + \rho \int V[f(\pi(y))z]\varphi(dz), \quad \forall y \in \mathbb{R}_+.$$

The proof does not differ from the neoclassical case (Stokey et al. 1989, see for example) and is omitted.

As a matter of notation, define

$$\Sigma(y) := \arg\max_{0 \leq k \leq y} \left\{ u(y-k) + \rho \int V[f(k)z]\varphi(dz) \right\},$$

so that $y \mapsto \Sigma(y)$ is the optimal correspondence, and π is an optimal policy if and only if it is a $\mathcal{B}$-measurable selection from Σ.

11.3.1 Monotonicity

Monotone policy rules play an important role in economics, particularly with regards to the characterization of equilibria. That monotonicity of the optimal investment function holds in deterministic one-sector nonconvex growth environments was established by Dechert and Nishimura (1983) and is now well-known. Indeed, monotone controls are a feature of many very general environments. See, for example, Mirman et al. (2005) and Kamihigashi and Roy (2007). Our model is no exception:

Lemma 2. *Let an economy (u, f, φ, ρ) be given, and let π be a feasible policy. If π is optimal, then it is nondecreasing on $\mathbb{R}_+$.*

Put differently, one cannot construct a measurable selection from the optimal correspondence Σ that is not nondecreasing. (On the other hand, in nonconvex models consumption is *not* generally monotone with income.)

One supposes that as ρ decreases the propensity to save will fall. The following result was established for the stochastic neoclassical case in Danthine and Donaldson (1981, Theorem 5.1), and in the nonconvex, deterministic case by Amir et al. (1991) using lattice programming.

Lemma 3. *The optimal policy is nondecreasing in the discount factor ρ, in the sense that if (u, f, φ, ρ_0) and (u, f, φ, ρ_1) are two economies, and if π_0 (resp. π_1) is optimal for the former (resp. latter), then $\rho_1 \geq \rho_0$ implies $\pi_1 \geq \pi_0$ pointwise on $\mathbb{R}_+$.*

In fact we can say more:

Lemma 4. *For u, f and φ given, let (ρ_n) be a sequence of discount factors in $(0, 1)$, and for each n let π_n be a corresponding optimal policy. If $\rho_n \downarrow 0$, then $\pi_n \downarrow 0$ pointwise, and the convergence is uniform on compact sets.*

11.3.2 Derivative Characterization of the Policy

Optimal behavior in growth models is usually characterized by the Ramsey–Euler equation. In stochastic models, where sequential arguments are unavailable, the obvious path to the Ramsey–Euler equation is via differentiability of the value function and a well-known envelope condition (Mirman and Zilcha 1975, Lemma 1). See also Blume et al. (1982), who demonstrated differentiability of the optimal *policy* under convexity and absolute continuity of the shock by way of the Implicit Function Theorem. Amir (1997) considered a weaker convexity requirement.

Without any convexity there may be jumps in the optimal policy, which in turn affect the smoothness of the value function. The validity of the Ramsey–Euler characterization is not clear. However, Dechert and Nishimura (1983, Theorem 6, Lemma 8) showed that in their model the value function has left and right derivatives at every point, and that these agree off an at most countable set.[4] These results were extended to the stochastic case by Majumdar et al. (1989). In addition to the above results concerning the value function, they were able to show that the Ramsey–Euler equations holds *everywhere*, irrespective of jumps in the optimal policy. We extend their analysis, starting from the essential idea of Blume et al. (1982), but without convexity or compact state. From this we prove interiority of the policy and the Ramsey–Euler equation for standard econometric shocks.

Assumption 3 *The shock ε_t is such that (S1) the density φ is continuously differentiable on $(0, \infty)$; and (S2) the integral $\int z\,|\varphi'(z)|\,dz$ is finite.*

The set of densities satisfying (S1) and (S2) is norm-dense in the set of all densities when the latter are viewed as a subset of $L_1(\mathbb{R}_+)$. They also hold for many standard econometric shocks on $\mathbb{R}_+$, such as the lognormal distribution. With these assumptions in hand we can establish the following without convexity or bounded shocks.

Proposition 1. *Let (u, f, φ, ρ) satisfy Assumptions 1–3.*

1. *If policy π is optimal, then it is interior. That is, $0 < \pi(y) < y$ for all $y \in (0, \infty)$.*
2. *The value function V has right and left derivatives V'_- and V'_+ everywhere on $(0, \infty)$.*
3. *If policy π is optimal, then it satisfies $V'_-(y) \leq u'(y - \pi(y)) \leq V'_+(y)$ for all $y \in (0, \infty)$.*
4. *The functions V'_- and V'_+ disagree on an at most countable subset of $\mathbb{R}_+$.*

In the stochastic nonconvex case, Part 1 of Proposition 1 was proved by Majumdar et al. (1989, Theorem 4). Their proof requires that the shock has compact support bounded away from zero, and there exists an $a > 0$ such that $f(k)\varepsilon > k$ with probability one whenever $k \in (0, a)$. Part 2 was proved in the deterministic case by Dechert and Nishimura, as was Part 4 (1983, Theorem 6 and Lemma 8).[5]

[4]The intuition is that nondifferentiability of the value function coincides pointwise with jumps in optimal investment. But by Lemma 2, the only optimal jumps are increases. To each jump, then, can be associated a distinct rational, which precludes uncountability.

[5]On Part 2 see also Askri and Le Van (1998, Proposition 3.2) and Mirman et al. (2005).

Part 3 is due in the stochastic neoclassical case to Mirman and Zilcha (1975, Lemma 1), and the proof for the nonconvex case is the same.[6]

Corollary 1. *For a given economy* (u, f, φ, ρ), *any two optimal policies are equal almost everywhere.*

Proof. Immediate from Parts 3 and 4. □

It will turn out that under the maintained assumptions, differences on null sets do not really concern us (see Lemma 5). So we can in some sense talk about *the* optimal policy (when a.e.-equivalent policies are identified).

One of our main results is that under Assumptions 1–3 the Ramsey–Euler equation can still be established.

Proposition 2. *Let Assumptions 1–3 hold. If* π *is optimal for* (u, f, φ, ρ), *then for all* $y > 0$,

$$u'(y - \pi(y)) = \rho \int u'[f(\pi(y))z - \pi(f(\pi(y))z)]f'(\pi(y))z\boldsymbol{\varphi}(dz).$$

Using Proposition 2 we can strengthen the monotonicity result for the optimal policy (Lemma 2). The proof is straightforward and is omitted.

Corollary 2. *For a given economy* (u, f, φ, ρ), *every optimal policy is strictly increasing.*

11.4 Dynamics

Next we discuss the dynamics of the optimal process $(y_t)_{t\geq 0}$. For the nonconvex deterministic case a detailed characterization of dynamics was given by Dechert and Nishimura (1983). Not surprisingly, for some parameter values multiple equilibria obtain. On the other hand, for the convex stochastic growth model, Mirman (1970) and Brock and Mirman (1972) proved that the sequence of marginal distributions for the process converge to a unique limit independent of the initial condition. Subsequently this problem has been treated by many authors.[7]

Our main contribution in this paper is to show that many convex and nonconvex optimal processes satisfy a fundamental dichotomy: either they are globally stable, or they are globally collapsing to the origin, independent of the initial condition. This result reduces considerably the possible range of asymptotic outcomes.

---

[6]Note that if V is concave on some open interval, then the subdifferentials exist everywhere on that interval, and $V'_+ \leq V'_-$. If follows from Part 3 of the Proposition, then, that concavity immediately gives differentiability, and, moreover, $V'(y) = u'(y - \pi(y))$. See Mirman and Zilcha (1975, Lemma 1).

[7]See Nishimura and Stachurski (2005) and references.

For example, path dependence never holds. More importantly, global stability can now be established by showing only that an economy does not collapse to the origin.

To begin, let $\mathcal{P}$ be the set of probability measures on $(\mathbb{R}_+, \mathcal{B})$. Let $E := (u, f, \varphi, \rho)$ be given. For a fixed policy π and initial condition y_0, we consider the evolution of the income process $(y_t)_{t \geq 0}$ satisfying $y_{t+1} = f(\pi(y_t))\varepsilon_t$, and the corresponding sequence of marginal distributions $(\boldsymbol{\psi}_t)_{t \geq 0} \subset \mathcal{P}$.[8] Evidently the process is Markovian, with y_t independent of ε_t. From this independence it follows that for any bounded Borel function $h \colon \mathbb{R}_+ \to \mathbb{R}$ we have

$$\mathbb{E}h(y_{t+1}) = \mathbb{E}h[f(\pi(y_t))\varepsilon_t] = \int \int h[f(\pi(y))z]\varphi(dz)\psi_t(dy).$$

Specializing to the case $h = \mathbb{1}_B$ and using $y_t \sim \psi_t$ gives the recursion

$$\psi_{t+1}(B) = \int \left[\int \mathbb{1}_B[f(\pi(y))z]\varphi(dz) \right] \psi_t(dy). \tag{11.4}$$

When π is optimal for E, the sequence of marginal distributions (ψ_t) defined inductively by (11.4) is called an optimal path for (E, π). Every initial condition $y_0 \sim \psi_0$ defines a (unique) optimal path. If initial income is zero the dynamics require no additional investigation. Henceforth, by an *initial condition* is meant a distribution $\boldsymbol{\psi}_0 \in \mathcal{P}$ for y_0 which puts no mass on $\{0\}$. This convention makes the results neater, and is maintained throughout the proofs without further comment.

When studying convergence of probabilities two topologies are commonly used. One is the so-called weak topology, under which distribution functions converge if and only if they converge pointwise at all continuity points. The other is the norm topology, or strong topology, generated by the total variation norm. Under the latter, the distance between μ and ν in $\mathcal{P}$ is $\sup_{B \in \mathcal{B}} |\mu(B) - \nu(B)|$.

Definition 1. Let an economy $E := (u, f, \varphi, \rho)$ be given, and let π be an optimal policy for E. A (nontrivial, stochastic) steady state for (E, π) is a measure $\boldsymbol{\psi}^* \in \mathcal{P}$, such that $\boldsymbol{\psi}^*(\{0\}) = 0$ and

$$\int \left[\int \mathbb{1}_B[f(\pi(y))z]\varphi(dz) \right] \psi^*(dy) = \psi^*(B), \quad \forall B \in \mathcal{B}. \tag{11.5}$$

The policy π is called globally stable if for (E, π) there is a unique steady state $\psi^* \in \mathcal{P}$, and the (E, π)-optimal path (ψ_t) satisfies $\psi_t \to \psi^*$ in the norm topology as $t \to \infty$ for all initial conditions ψ_0.

It is clear from (11.4) to (11.5) that if $y_t \sim \psi^*$, then $y_{t+k} \sim \psi^*$ for all $k \geq 0$. Note also that the stability condition in Definition 1 is particularly strong. It implies

[8] As before, $(y_t)_{t \geq 0}$ is a stochastic process on $(\Omega, \mathcal{F}, \mathbf{P})$. By the marginal distribution $\boldsymbol{\psi}_t \in \mathcal{P}$ of y_t is meant its distribution on $\mathbb{R}_+$ in the usual sense. Precisely, $\boldsymbol{\psi}_t := \mathbf{P} \circ y_t^{-1}$, the image measure induced on $(\mathbb{R}_+, \mathcal{B})$ by y_t.

many standard stability conditions for Markov processes, such as recurrence, and also convergence of the marginal distributions in the weak topology.[9]

Instability of stochastic growth models has been studied less than stability. There are various notions which capture instability; we borrow a relatively strong one from the Markov process literature referred to as sweeping.[10]

Definition 2. Let an economy $E := (u, f, \varphi, \rho)$ be given, and let π be an optimal policy. Let $\mathcal{B}_0 \subset \mathcal{B}$. In general, the Markov process generated by (E, π) is called sweeping with respect to $\mathcal{B}_0$ if each optimal path (ψ_t) satisfies $\lim_{t\to\infty} \psi_t(A) = 0$ for all $A \in \mathcal{B}_0$ and all initial conditions ψ_0. We say that (E, π) is globally collapsing to the origin if it is sweeping with respect to the collection of intervals $\mathcal{B}_0 := \{[a, \infty) : a > 0\}$.

Nonconvex technology introduces the possibility that many optimal policies exist for the one economy. For these models it has been shown (Dechert and Nishimura 1983, Lemma 6) that different optimal trajectories can have very different dynamics, even from the same initial condition. Indeed, there may be two optimal policies π and π' for E such that the optimal path from ψ_0 generated by (E, π) sustains a nontrivial long run equilibrium, whereas that generated by (E, π') leads to economic collapse. For our stochastic model this is not possible:

Lemma 5. *Let an economy $E := (u, f, \varphi, \rho)$ be given. If (E, π) is globally stable for some optimal π, then (E, π') is globally stable for every optimal policy π'. Similarly, if (E, π) is globally collapsing to the origin, then so is (E, π') for every optimal policy π'.*

As a result we may simply say that E is globally stable or globally collapsing, without specifying the particular optimal policy π. Next, we introduce a new assumption as a preliminary to our main dynamics result.

Assumption 4 *The density φ of the productivity shock is strictly positive (Lebesgue almost) every where on $\mathbb{R}_+$.*

Many standard shocks on $\mathbb{R}_+$ have this property. An example is the lognormal distribution. The following result indicates that when this assumption holds there is a fundamental dichotomy for the dynamic behavior of the economy. The proof is based on the Foguel Alternative for Markov chains (Foguel 1969; Rudnicki 1995). Monotonicity and interiority of the optimal policy also play key roles.

Proposition 3. *Let an economy $E := (u, f, \varphi, \rho)$ be given. If in addition to Assumptions 1–3, Assumption 4 also holds, then there are only two possibilities. Either*

1. *E is globally stable, or*
2. *E is globally collapsing to the origin.*

---

[9] In the present case it also implies uniform convergence of distribution functions, which is the criterion of Brock and Mirman (1972). See Dudley (2002, p. 389).

[10] See, for example, Lasota and Mackey (1994, Sect. 5.9).

Remark 1. Assumption 4 can be weakened at the cost of more complicated proofs. See Rudnicki (1995, Lemma 3 and Theorem 2).

It follows that multiple long run equilibria are never observed, regardless of nonconvexities in production technology. Instead long run outcomes are completely determined by the structure of the model, and historical conditions are asymptotically irrelevant. However, the steady state distribution may well be multi-modal, concentrated on areas that are locally attracting on average.

We have seen that a decrease in ρ is associated with lower savings and investment, which in turn should increase the likelihood of collapse to the origin. Conversely, higher ρ should increase the likelihood that the economy is stable. Indeed,

Lemma 6. *For economies $E_0 := (u, f, \varphi, \rho_0)$ and $E_1 := (u, f, \varphi, \rho_1)$ with $\rho_0 \leq \rho_1$, the following implications hold.*

1. *If E_1 is globally collapsing to the origin, then so is E_0.*
2. *If E_0 is globally asymptotically stable, then so is E_1.*

Combining the above results we can deduce that the dynamic behavior of the stochastic optimal growth model has only three possible types. Precisely,

Proposition 4. *For u, f, and φ given, either*

1. *(u, f, φ, ρ) is globally stable for all $\rho \in (0, 1)$,*
2. *(u, f, φ, ρ) is globally collapsing for all $\rho \in (0, 1)$, or*
3. *there is a $\hat{\rho} \in (0, 1)$ such that (u, f, φ, ρ) is globally stable for all $\rho > \hat{\rho}$, and globally collapsing for all $\rho < \hat{\rho}$.*

Under the current hypotheses one cannot rule out either of the first two possibilities. For example, Kamihigashi (2003) shows that very general one-sector growth models converge almost surely to zero when $f'(0) < \infty$ and shocks are sufficiently volatile. Determining which of the above three possibilities holds is far from trivial. However, we now show that one need only consider behavior of the model in the neighborhood of the origin.

Assumption 5 *The shock satisfies $\mathbb{E}\,|\ln \varepsilon| = \int |\ln z|\,\varphi(dz) < \infty$.*

Proposition 5. *Let $E := (u, f, \varphi, \rho)$ be given, and let π be an optimal policy. Suppose that Assumptions 1–5 hold. Define*

$$p := \limsup_{y \to 0} \frac{f(\pi(y))}{y}, \quad q := \liminf_{y \to 0} \frac{f(\pi(y))}{y}.$$

1. *If $p < \exp(-\mathbb{E} \ln \varepsilon)$, then E is globally collapsing to the origin.*
2. *If $q > \exp(-\mathbb{E} \ln \varepsilon)$, then E is globally stable.*

Also, in the light of Lemma 4, one might suspect that even in the situation where an economy is globally stable for every ρ, the stationary distribution will become more and more concentrated around the origin when $\rho \downarrow 0$. In this connection,

Proposition 6. *Let u, f and φ be given. Suppose that (u, f, φ, ρ) is globally stable for all $\rho \in (0, 1)$. If $\rho_n \to 0$, then $\psi_n^* \to \delta_0$ in the weak topology, where ψ_n^* is the stationary distribution corresponding to ρ_n, and δ_0 is the probability measure concentrated at zero.*

Remark 2. As δ_0, and ψ_n^* are mutually singular, norm convergence is impossible.

11.5 Proofs

In the proofs, $L_1(X)$ refers as usual to all integrable Borel functions on given space X, and $C^n(X)$ is the n times continuously differentiable functions.

11.5.1 Monotonicity

The proof of monotonicity of the optimal policy is as follows.

Proof. [Proof of Lemma 2] Let π be optimal, and take any nonnegative $y \leq y'$. If $y = y'$ then monotonicity is trivial. Suppose the inequality is strict. By way of contradiction, suppose that $\pi(y) > \pi(y')$. Define $c := y - \pi(y)$, $c' = y' - \pi(y')$, and $\widehat{c} := \pi(y) - \pi(y') > 0$. Note first that

$$c' - \widehat{c} = y' - \pi(y) > y - \pi(y) = c \geq 0. \tag{11.6}$$

Also, since $c + \widehat{c} + \pi(y) = y$, we have

$$u(c) + \rho \int V[f(\pi(y))z]\varphi(dz) \geq u(c + \widehat{c}) + \rho \int V[f(\pi(y'))z]\varphi(dz),$$

and since $c' - \widehat{c} + \pi(y) = y'$,

$$u(c') + \rho \int V[f(\pi(y'))z]\varphi(dz) \geq u(c' - \widehat{c}) + \rho \int V[f(\pi(y))z]\varphi(dz).$$

$$\therefore u(c') - u(c' - \widehat{c}) \geq u(c + \widehat{c}) - u(c).$$

As $c' - \widehat{c} > c$ by (11.6), this contradicts the strict concavity of u. □

Proof. [Proof of Lemma 3] Pick any $y \geq 0$. Let $k_0 := \pi_0(y)$ and $k_1 := \pi_1(y)$. By definition,

$$u(y - k_0) + \rho_0 \int V[f(k_0)z]\varphi(dz) \geq u(y - k_1) + \rho_0 \int V[f(k_1)z]\varphi(dz)$$

and

$$u(y-k_1)+\rho_1\int V[f(k_1)z]\varphi(dz)\geq u(y-k_0)+\rho_1\int V[f(k_0)z]\varphi(dz).$$

Multiplying the first inequality by ρ_1 and the second by ρ_0 and adding gives

$$\rho_1 u(y-k_0)+\rho_0 u(y-k_1)\geq \rho_1 u(y-k_1)+\rho_0 u(y-k_0).$$
$$\therefore (\rho_1-\rho_0)(u(y-k_0)-u(y-k_1))\geq 0.$$
$$\therefore \rho_1\geq\rho_2 \Longrightarrow u(y-k_0)-u(y-k_1)\geq 0 \Longrightarrow k_1\geq k_0.$$

□

Proof. [Proof of Lemma 4] Since u is concave, for any $y>0$ and any $k\leq y$,

$$u(y-k)\leq u(y)-u'(y)k. \tag{11.7}$$

Also, since $u(y)\leq M<\infty$ for all y,

$$V(y):=\sup_{\pi}\mathbb{E}\left[\sum_{t=0}^{\infty}\rho^t u(y_t-\pi(y_t))\right]\leq\frac{1}{1-\rho}M. \tag{11.8}$$

Since $\pi(y)=0$ is feasible,

$$u(y-\pi(y))+\rho\int V(f(\pi(y))z)\varphi(dz)\geq u(y)+\rho\int V(f(0)z)\varphi(dz)=u(y).$$

$$\therefore u(y)-u(y-\pi(y))\leq\rho\int V(f(\pi(y))z)\varphi(dz)\leq\frac{\rho}{1-\rho}M.$$

Using the bound (11.7) gives us

$$u'(y)\pi(y)\leq\frac{\rho}{1-\rho}M,\quad \forall y>0.$$

$$\therefore \pi(y)\leq\frac{\rho}{1-\rho}\frac{M}{u'(y)}:=b(y;\rho).$$

The function $y\to b(y;\rho)$ is continuous and converges pointwise to zero as $\rho\to 0$. The statement follows (uniform convergence on compact sets is by Dini's Theorem).

□

11.5.2 The Ramsey–Euler Equation

Next Propositions 1 and 2 are established. We use the following lemma, which can be thought of as a kind of convolution argument designed to verify precisely the conditions necessary for the Ramsey–Euler equation to hold. The proof is rather long, and is relegated to the Appendix.

Lemma 7. *Let g and h be nonnegative real functions on $\mathbb{R}$. Define*

$$\mu(r) := \int_{-\infty}^{\infty} h(x+r)g(x)\,dx. \tag{11.9}$$

Consider the following conditions:

1. $g \in L_1(\mathbb{R}) \cap C^1(\mathbb{R})$, $g' \in L_1(\mathbb{R})$
2. *h is bounded*
3. *h is nondecreasing*
4. *h is absolutely continuous on compact intervals*
5. *h' is bounded on compact subsets of $\mathbb{R}$,*
   where h' is defined as the derivative of h when it exists and zero elsewhere.

If (1) and (2) hold, then $\mu \in C^1(\mathbb{R})$, and

$$\mu'(r) = -\int_{-\infty}^{\infty} h(x+r)g'(x)\,dx. \tag{11.10}$$

If, in addition, (3)–(5) hold, then μ' also has the representation

$$\mu'(r) = \int_{-\infty}^{\infty} h'(x+r)g(x)\,dx. \tag{11.11}$$

Remark 3. Note that higher order derivatives are immediate if g has high order derivatives that are all integrable. In the first part of the proof, where differentiability and the representation $\mu'(r) = -\int h(x+r)g'(x)dx$ are established we do not use nonnegativity of g—it may be any real function. So now suppose that g is twice differentiable, and that $g'' \in L_1(\mathbb{R})$. Then by applying the same result, this time using g' for g, differentiability of μ' is verified.

To prove Proposition 1, the following preliminary observation is important.

Lemma 8. *Assume the hypotheses of Proposition 1, and let V be the value function. The map $k \mapsto \int V[f(k)z]\boldsymbol{\varphi}(dz)$ is continuously differentiable on the interior of $\mathbb{R}_+$.*

Proof. By a simple change of variable,

$$\int_{0}^{\infty} V[f(k)z]\varphi(z)dz = \int_{-\infty}^{\infty} V[\exp(\ln f(k) + x)]\varphi(e^x)e^x dx.$$

Let $h(x) := V[\exp(x)]$, $g(x) := \varphi(e^x)e^x$, and let μ be defined as in (11.9). Then $\int V[f(k)z]\varphi(z)dz = \mu[\ln f(k)]$. Regarding μ, conditions (1) and (2) of Lemma 7 are satisfied by (U2), (S1) and (S2). Hence $\int V[f(k)z]\varphi(z)dz$ is continuously differentiable as claimed. □

Now let us consider the interiority result.

Proof. [Proof of Proposition 1, Part 1] Pick any $y > 0$. Consider first the claim that $\pi(y) \neq 0$. Suppose instead that $0 \in \sum(y)$, so that

$$V(y) = u(y) - \rho \int V[f(0)z]\varphi(dz) = u(y), \tag{11.12}$$

where we have used $u(0) = 0$ in (U2). Define also

$$V_\xi := u(y - \xi) + \rho \int V[f(\xi)z]\varphi(dz), \tag{11.13}$$

where ξ is a positive number less than y. By (F3), there exists a $\delta > 0$ such that $f(\xi) > \xi$ whenever $\xi < \delta$. Therefore,

$$V_\xi \geq u(y - \xi) + \rho \int V(\xi z)\varphi(dz), \quad \forall \xi < \delta. \tag{11.14}$$

In addition, $V \geq u$ every where on $\mathbb{R}_+$. Using this bound along with (11.12) and (11.14) gives

$$0 \leq \frac{V(y) - V_\xi}{\xi} \leq \frac{u(y) - u(y - \xi)}{\xi} - \rho \int \frac{u(\xi z)}{\xi}\varphi(dz), \quad \forall \xi < \delta. \tag{11.15}$$

Take a sequence $\xi_n \downarrow 0$. If $H_n(z) = u(\xi_n z)/\xi_n$, then $H_n \geq 0$ on $\mathbb{R}_+$ and $H_{n+1}(z) \geq H_n(z)$ for all z and all n. Moreover $\lim_{n\to\infty} H_n = \infty$ almost everywhere. By the Monotone Convergence Theorem, then

$$\lim_{n\to\infty} \int \frac{u(\xi_n z)}{\xi_n}\varphi(dz) = \int \infty\varphi(dz) = \infty,$$

which induces a contradiction in (11.15).

Now consider the claim that $\pi(y) \neq y$. Let

$$v(k) := u(y - k) + w(k), \ \ w(k) := \rho \int V[f(k)z]\varphi(dz), \quad k \in [0, y].$$

If $y \in \sum(y)$, then for all positive ε,

$$0 \leq \frac{v(y) - v(y - \varepsilon)}{\varepsilon} = -\frac{u(\varepsilon)}{\varepsilon} + \frac{w(y) - w(y - \varepsilon)}{\varepsilon}. \tag{11.16}$$

Since $w(k)$ is differentiable at y (Lemma 8), the second term on the right-hand side converges to a finite number as $\varepsilon \downarrow 0$. In this case clearly there will be a contradiction of inequality (11.16). This completes the proof that $y \notin \sum(y)$. □

Proof. [Proof of Proposition 1, Part 2] Regarding the existence of left and right derivative, pick any $y > 0$, any $\xi_n \downarrow 0$, $\xi_n > 0$, and any optimal policy π. By monotonicity, $\pi(y + \xi_n)$ converges to some limit k_+, and the value k_+ is independent of the choice of sequence (ξ_n). Moreover, upper hemi-continuity of π implies that k_+ is maximal at y. It follows from this and interiority of optimal policies that $0 < k_+ < y$ and

$$V(y) = u(y - k_+) + \rho \int V[f(k_+)z]\varphi(dz).$$

Also, for all $n \in N$,

$$\begin{aligned} V(y + \xi_n) &= u(y + \xi_n - \pi(y + \xi_n)) + \rho \int V[f(\pi(y + \xi_n))z]\varphi(dz) \\ &\geq u(y - k_+ + \xi_n) + \rho \int V[f(k_+)z]\varphi(dz). \\ &\therefore u(y - k_+ + \xi_n) - u(y - k_+) \leq V(y + \xi_n) - V(y), \quad \forall n \in \mathbb{N}. \end{aligned}$$

On the other hand, since $\pi(y + \xi_n) \downarrow k_+ < y$, there exists an $N \in \mathbb{N}$ such that

$$V(y) \geq u(y - \pi(y + \xi_n)) + \rho \int V[f(\pi(y + \xi_n))z]\varphi(dz), \quad \forall n \geq N.$$

$$\therefore V(y + \xi_n) - V(y) \leq u(y + \xi_n - \pi(y + \xi_n)) - u(y - \pi(y + \xi_n)), \quad \forall n \geq N.$$

$$\therefore V(y + \xi_n) - V(y) \leq u'(y - \pi(y + \xi_n))\xi_n, \quad \forall n \geq N,$$

where the last inequality is by concavity of u. In summary, then

$$u(y - k_+ + \xi_n) - u(y - k_+) \leq V(y + \xi_n) - V(y) \leq u'(y - \pi(y + \xi_n))\xi_n$$

for all n sufficiently large. Dividing through by $\xi_n > 0$ and taking limits gives $V'_+(y) = u'(y - k_+)$, which is of course finite by $k_+ < y$.[11]

Now consider the analogous argument for V'_-. Let y, (ξ_n) and π be as above. Again, as π is monotone, $\pi(y - \xi_n) \uparrow k_-$, where k_- is independent of the precise sequence (ξ_n), maximal at y and satisfies $0 < k_- < y$. Since $k_- > 0$, the sequence $\pi(y - \xi_n)$ will be positive for large enough n and we can assume this is so for all n. By maximality,

$$V(y) = u(y - k_-) + \rho \int V[f(k_-)z]\varphi(dz).$$

[11] We are using continuity of u', which is guaranteed by Assumption 1.

Also, since $k_- < y$, there exists an $N \in \mathbb{N}$ with $k_- \leq y - \xi_n$ for all $n \geq N$. Hence, $\forall n \geq N$,

$$V(y - \xi_n) = u(y - \xi_n - \pi(y - \xi_n)) + \rho \int V[f(\pi(y - \xi_n))z]\varphi(dz)$$
$$\geq u(y - k_- - \xi_n) + \rho \int V[f(k_-)z]\varphi(dz).$$
$$\therefore u(y - k_- - \xi_n) - u(y - k_-) \leq V(y - \xi_n) - V(y), \quad \forall n \geq N.$$

One the other hand, since $0 < \pi(y - \xi_n) \uparrow k_- < y$,

$$V(y) \geq u(y - \pi(y - \xi_n)) + \rho \int V[f(\pi(y - \xi_n))z]\varphi(dz), \quad \forall n \in \mathbb{N}.$$
$$\therefore V(y - \xi_n) - V(y) \leq u(y - \xi_n - \pi(y - \xi_n)) - u(y - \pi(y - \xi_n)), \quad \forall n \in \mathbb{N}.$$
$$\therefore V(y - \xi_n) - V(y) \leq -u'(y - \pi(y - \xi_n))\xi_n, \quad \forall n \in \mathbb{N},$$

where again the last inequality is by concavity of u. Putting the inequalities together gives

$$u(y - k_- - \xi_n) - u(y - k_-) \leq V(y - \xi_n) - V(y) \leq u'(y - \pi(y - \xi_n))(-\xi_n)$$

for all n sufficiently large. Dividing through by $-\xi_n$ and taking limits gives $V'_-(y) = u'(y - k_-)$.

Proof. [Proof of Proposition 1, Part 3] The proof is identical to that given in Mirman and Zilcha (1975, Lemma 1). □

Proof. [Proof of Proposition 1, Part 4] The proof is essentially the same as that of Majumdar et al. (1989, Lemma 4). Briefly, it is clear from the proof of Part 2 of Proposition 1 that $V'_-(y)$ and $V'_+(y)$ will agree whenever $\sum(y)$ is a singleton. If y_1 and y_2 are any two distinct points where $\sum$ is multi-valued, then $\sum(y_1)$ and $\sum(y_2)$ can intersect at most one point, otherwise we can construct a non-monotone optimal policy, contradicting Lemma 2. It follows that for each y where $\sum(y)$ is multi-valued, $\sum(y)$ can be allocated a unique rational number. □

Next we come to the proof of the Ramsey–Euler equation. We need the following lemma, which was first proved (under different assumptions) by Majumdar et al. (1989, Lemma 2A).

Lemma 9. *For every compact $K \subset (0, \infty)$, $\inf\{y - \pi(y) : y \in K\}$ is strictly positive.*

Proof. Suppose to the contrary that on some compact set $K \subset (0, \infty)$, there exists for each n a y_n with $\pi(y_n) > y_n - 1/n$. By compactness (y_n) has a convergent subsequence, and without loss of generality we assume that the whole sequence converges to $y^* \in K$. The bounded sequence $\pi(y_n)$ itself has a convergent subsequence $\pi(y_{n(i)}) \to k^*$ as $i \to \infty$. Since the subsequence $(y_{n(i)}$ converges to y^* too, k^* is optimal at y^* by upper hemicontinuity. But then $y^* - \frac{1}{n(i)} \le k^* \le y^*$ for all $i \in \mathbb{N}$. This contradicts the interiority of the optimal policy, which has already been established. □

The next lemma is fundamental to our results.

Lemma 10. *Define V' to be the derivative of V when it exists and zero elsewhere. For all $k > 0$,*

$$\frac{d}{dk}\int V[f(k)z]\varphi(z)dz = \int V'(f(k)z)f'(k)z\varphi(z)dz.$$

Proof. We change variables to shift the problem to the real line. Our objective is to apply Lemma 7. Let $w(k) := \int V[f(k)z]\varphi(z)dz$. As before, we can use a change of variable to obtain

$$w(k) = \int_{-\infty}^{\infty} V(f(k)e^x)\varphi(e^x)e^x dx = \int_{-\infty}^{\infty} h(x + \ln f(k))g(x)dx,$$

where $g(x) := \varphi(e^x)e^x$ and $h(x) := V(e^x)$. All of the hypotheses of Lemma 7 are satisfied.[12] Therefore, using the representation (11.11),

$$w'(k) = \frac{f'(x)}{f(x)}\int_{-\infty}^{\infty} h'(x + \ln f(k))g(x)dx = f'(k)\int_{-\infty}^{\infty} V'(e^x f(k))e^x g(x)dx.$$

Changing variables again gives the desired result:

$$w'(k) = \int_0^{\infty} V'(f(k)z)f'(k)g(\ln z)dz = \int_0^{\infty} V'(f(k)z)f'(k)z\varphi(z)dz.$$ □

Now the proof of the Ramsey–Euler equation can be completed.

Proof. [Proof of Proposition 2] Evidently $\pi(y)$ solves

$$u'(y-k) - \rho\frac{d}{dk}\int V[f(k)z]\varphi(z)dz = 0.$$

The result now follows from Lemma 10, given that $V'(y) = u'(y - \pi(y))$ Lebesgue almost everywhere. □

[12]In particular, h' is bounded on compact sets, because $h'(x) = V'(e^x)e^x$, and $V'(y) = u'(y - \pi(y))$ when it exists (i.e., when the function V' is not set to zero). The latter is bounded on compact sets by Lemma 9. Also, V is absolutely continuous because countinuous functions of bounded variation (provided by monotonicity here) fail to be absolutely continuous only if they have infinite derivative on an uncountable set (Saks 1937, p. 128). This is impossible by Proposition 1, Part 4.

11.5.3 Dynamics

In the following discussion let an optimal policy π be given. We simplify notation by defining the map S by $S(y) := f(\pi(y))$. The most important properties of S are that S is nondecreasing and $S(y) = 0$ implies $y = 0$ (see Lemma 2 and Proposition 1, Part 1). Also, let $\mathcal{D}$ be all $\boldsymbol{\psi} \in \mathcal{P}$ that are absolutely continuous with respect to Lebesgue measure.

Define the Markov operator $\mathbf{M}: \mathcal{P} \ni \boldsymbol{\psi} \to \mathbf{M}\boldsymbol{\psi} \in \mathcal{P}$ corresponding to π by

$$(\mathbf{M}\boldsymbol{\psi})(B) = \int \int \mathbb{1}_B[S(y)z]\boldsymbol{\varphi}(dz)\boldsymbol{\psi}(dy). \tag{11.17}$$

It is immediate from (11.4) that the sequence of marginal distributions $(\boldsymbol{\psi}_t)$ for income satisfies $\boldsymbol{\psi}_{t+1} = \mathbf{M}\boldsymbol{\psi}_t$ for all $t \geq 0$.

We note the following facts, which are easy to verify. First, if $\boldsymbol{\psi}(\{0\}) = 0$, then $\mathbf{M}\boldsymbol{\psi} \in \mathcal{D}$. It follows immediately that $\mathbf{M}(\mathcal{D}) \subset \mathcal{D}$, and that $(\boldsymbol{\psi}_t)$, the sequence of marginal distributions for income, satisfies $\boldsymbol{\psi}_t \in \mathcal{D}$ for all $t \geq 1$. Also, if $\boldsymbol{\psi} \in \mathcal{D}$, then the simple change of variable $y' = S(y)z$ gives

$$(\mathbf{M}\boldsymbol{\psi})(B) = \int \int_B k(y, y')dy'\boldsymbol{\psi}(dy), \tag{11.18}$$

where dy' is of course integration with respect to Lebesgue measure, and

$$k(y, y') := \varphi\left(\frac{y'}{S(y)}\right)\frac{1}{S(y)}. \tag{11.19}$$

It is immediate from Definition 1 that a steady state is a fixed point of the Markov operator in $\mathcal{P}$ which puts zero mass on $\{0\}$. Since such distributions are mapped into $\mathcal{D}$ by $\mathbf{M}$, when a steady state $\boldsymbol{\psi}^*$ exists it must be in $\mathcal{D}$.

The next lemma is just elementary manipulation of the definitions.

Lemma 11. *Let π be a fixed optimal policy, and let $\mathbf{M}$ be the corresponding Markov operator.*

1. *The economy is globally stable in the sense of Definition 1 if and only if there is a unique $\boldsymbol{\psi}^* \in \mathcal{D}$ with $\mathbf{M}\boldsymbol{\psi}^* = \boldsymbol{\psi}^*$ and $\mathbf{M}^t\boldsymbol{\psi} \to \boldsymbol{\psi}^*$ in norm as $t \to \infty$ for every $\boldsymbol{\psi} \in \mathcal{D}$.*
2. *The economy is globally collapsing to the origin in the sense of Definition 2 if and only if $\mathbf{M}^t\boldsymbol{\psi}([a, \infty)) \to 0$ for every $\boldsymbol{\psi} \in \mathcal{D}$ and every $a > 0$.*

Proof. [Proof of Lemma 5] By Corollary 1, any pair of optimal policies is equal almost everywhere. Inspection of (11.18) and (11.19) indicates that they will have identical Markov operators on $\mathcal{D}$, in the sense that if $\mathbf{M}$ corresponds to one optimal policy and **M'** to another, then $\mathbf{M}\psi = \mathbf{M}'\psi$ for all $\psi \in \mathcal{D}$. The rest of the proof of Part 1 follows immediately from Lemma 11. The proof of Part 2 is similar. □

Proof. [Proof of Proposition 3] Let $\mathbf{M}$ be the Markov operator corresponding to π, and let k be as in (11.19). Consider the following two conditions:

1. $\mathbf{M}\psi$ dominates the Lebesgue measure ($\mathbf{M}\psi$-null sets are Lebesgue null) for all $\psi \in \mathcal{D}$.
2. $\forall \widehat{y} > 0, \exists \varepsilon > 0$ and $\eta \geq 0$ with $\int \eta(x)dx > 0$ and

$$k(y, y') \geq \eta(y')1_{(\widehat{y}-\varepsilon, \widehat{y}+\varepsilon)}(y), \qquad \forall y, y'.$$

Here by Rudnicki (1995, Theorem 2 and Corollary 3), (1) and (2) imply the Foguel Alternative; in particular that either $\mathbf{M}$ has a unique fixed point $\psi^* \in \mathcal{D}$ and $\mathbf{M}^t\psi \to \psi^*$ in norm for all $\psi \in \mathcal{D}$, or alternatively $\mathbf{M}$ is sweeping with respect to the compact sets, so that $\lim_{t\to\infty} \mathbf{M}^t\psi([a,b]) = 0$ for any $\psi \in \mathcal{D}$ and any $0 < a < b < \infty$. In the light of Lemma 11, then, to prove Proposition 3 it is sufficient to check (1), (2) and, in addition,

$$\lim_{b\to\infty} \limsup_{t\to\infty} \int \mathbf{M}^t\psi([b,\infty)) = 0, \qquad \forall \psi \in \mathcal{D}, \tag{11.20}$$

where (11.20) demonstrates that sweeping occurs not just with respect to any interval $[a,b], a > 0$, but in fact to any interval $[a,\infty)$.

Condition (1) is immediate from the assumption that φ is everywhere positive, in light of (11.18) and (11.19). Regarding condition (2), pick any $\widehat{y} > 0$ and any ε such that $\widehat{y} - \varepsilon > 0$. Also let $0 > \gamma_0 > \gamma_1 < \infty$. Define

$$\delta_0 := \frac{\gamma_0}{S(\widehat{y}+\varepsilon)}, \quad \delta_1 := \frac{\gamma_1}{S(\widehat{y}-\varepsilon)}.$$

Note that $\inf_{z\in[\delta_0,\delta_1]} \varphi(z) > 0$ by (S1) and strict positivity. Set

$$r := \frac{\inf_{z\in[\delta_0,\delta_1]} \varphi(z)}{S(\widehat{y}+\varepsilon)}, \quad \eta = r1_{[\gamma_0,\gamma_1]}.$$

Then η has the required properties.

Regarding (11.20), from (F2) there exists an $\alpha \in (0,1)$ and $m < \infty$ such that $S(y) \leq \alpha y + m$ for all $y \in \mathbb{R}_+$. Then

$$y_{t+1} \leq (\alpha y_{t+m})\varepsilon_t. \tag{11.21}$$

Since y_t and ε_t are independent and $\mathbb{E}\varepsilon = 1$ we have

$$\mathbb{E}y_{t+1} \leq \alpha \mathbb{E}y_t + m. \tag{11.22}$$

Using an induction argument gives

$$\mathbb{E}y_t \leq \alpha^t \mathbb{E}y_0 + (1 + \alpha + \cdots + \alpha^{t-1})m \leq \alpha^t \mathbb{E}y_0 + \frac{m}{1-\alpha}. \tag{11.23}$$

Suppose that $\mathbb{E} y_0 < \infty$. Then from (11.23) it follows that

$$\limsup_{t \to \infty} \mathbb{E} y_t \leq \frac{m}{1-\alpha}. \tag{11.24}$$

By the Chebychev inequality, $\mathbf{M}^t \psi([b, \infty)) \leq \mathbb{E} y_t b^{-1}$. From (11.24) it then follows that (11.20) holds for all ψ with $\mathbb{E} y_0 := \int y \psi(dy) < \infty$. This set (all densities with finite first moments) is normdense in $\mathcal{D}$, and $\mathbf{M}$ is an L_1 contraction on $\mathcal{D}$. Together, these facts imply that condition(11.20) in fact holds for every $\psi \in \mathcal{D}$ (Lasota and Mackey, 1994, p. 126). □

Proof. [Proof of Lemma 6] Regarding Part 1, let π_0 (resp. π_1) be an optimal policy for E_0 (resp. E_1), let $\mathbf{M}_0$ and $\mathbf{M}_1$ he the corresponding Markov operators and let $(y_t^0)_{t \geq 0}$ and $(y_t^1)_{t \geq 0}$ be the respective income processes. By Lemmas 5 and 11 it is sufficient to show that for any $\psi \in \mathcal{D}$ and any $a > 0$ we have

$$\lim_{t \to \infty} \mathbf{M}_0^t \psi([a, \infty)) = 0. \tag{11.25}$$

From Lemma 3 we have $\pi_1 \geq \pi_0$ pointwise on $\mathbb{R}_+$, so it is clear (by induction) that

$$y_t^1 \geq y_t^0 \text{ pointwise on } \Omega \text{ for any } t.$$
$$\therefore \{y_t^0 \geq a\} \subset \{y_t^1 \geq a\}.$$
$$\therefore \mathbf{M}_0^t \psi([a, \infty)) = \mathbf{P}\{y_t^0 \geq a\} \leq \mathbf{P}\{y_t^1 \geq a\} = \mathbf{M}_1^t \psi([a, \infty)).$$

By Lemma 11 and the hypothesis, the right hand side converges to zero as $t \to \infty$, which proves (11.25). □

Proof. [Proof of Proposition 5] For this proof we set $x_t = \ln y_t$, and define $\alpha := \mathbb{E} \ln \varepsilon$, $\eta := \ln \varepsilon - \alpha$ and $T : R \ni x \to \ln f(\pi(e^x)) + \alpha$, so that $x_{t+1} = T(x_t) + \eta_t$, where $\mathbb{E}\eta_t = 0$.

(Part 1) By the condition, $\limsup_{x \to -\infty}(T(x) - x) < 0$, implying the existence of an $m \in \mathbb{R}$ and $a > 0$ such that $T(x) \leq x - 2a$, for all $x \leq m$.

$$\therefore x_{t+1} \leq x_t + \eta_t - 2a, \quad \forall x_t \leq m.$$

Let $\widehat{x}_t := x_t - m$ and $\widehat{\eta}_t := \eta_t - a$. Then

$$\widehat{x}_{t+1} \leq \widehat{x}_t + \widehat{\eta}_t - a, \quad \forall \widehat{x}_t \leq 0. \tag{11.26}$$

Define $\Omega_0 := \{\omega \in \Omega : \sup_{T \geq 0} \sum_{t=0}^{T} \widehat{\eta}_t(\omega) \leq 0\}$. Since $\mathbb{E}\widehat{\eta}_t = -a < 0$, it follows that $\mathbf{P}(\Omega_0) > 0$ (Borovkov 1998, Chapter 11). From (11.26) we have

$$\widehat{x}_t \leq \widehat{x}_0 + \widehat{\eta}_0 + \cdots + \widehat{\eta}_{t-1} - ta \text{ for } \omega \in \Omega_0,$$

so if $\mathbf{P}\{\widehat{x}_0 \leq 0\} = 1$, then $\mathbf{P}\{x_t \leq -at\} \geq \mathbf{P}(\Omega_0) > 0$ for all t. Since $\{\widehat{x}_t \leq -at\} = \{y_t \leq e^{m-at}\}$, we have shown the existence of an initial condition $y_0(\mathbf{P}\{\widehat{x}_0 \leq 0\} = 1$ if y_0 is chosen s.t. $\mathbf{P}\{y_0 \leq e^m\} = 1)$ with the property

$$\liminf_{t\to\infty} \mathbf{P}\{y_t \leq c\} = \liminf_{t\to\infty} \psi_t([0,c]) \geq \mathbf{P}(\Omega_0) > 0.$$

But then ψ_t cannot converge in norm to any $\psi^* \in \mathcal{D}$. (If $\psi_t \to \psi^* \in \mathcal{D}$ then $\psi_t([0,c]) \to \psi^*([0,c])$, so choosing $c > 0$ such that $\psi^*([0,c]) < \mathbf{P}(\Omega_0)$ leads to a contradiction.) Therefore, the economy is not globally stable, and it follows from Proposition 3 that it must be collapsing to the origin.

(Part 2) By the condition, $\liminf_{x\to-\infty}(T(x) - x) > 0$, there is an $m \in \mathbb{R}$ and $a > 0$ such that $T(x) \geq x + a$ whenever $x \leq m$. Let $\widehat{x} := x - m$ and $\widehat{\eta} := \eta + a$. Then $\widehat{x}_{t+1} \geq \widehat{x}_t + \widehat{\eta}_t$ whenever $\widehat{x} \leq 0$. Also, since T is nondecreasing, $\widehat{x} \geq 0$ implies $T(x) \geq m + a$. Therefore, $\widehat{x}_t \geq 0 \Longrightarrow \widehat{x}_{t+1} \geq +\widehat{\eta}_t$.

$$\therefore \widehat{x}_{t+1} \geq -\widehat{x}_t^- + \widehat{\eta}_t \geq -(-\widehat{x}_t^- + \widehat{\eta}_t)^-, \tag{11.27}$$

where we are using the standard notation $x^- := -\min(0,x)$ and $x^+ := \max(0,x)$.

Assume to the contrary that the economy is not globally stable, in which case it must be sweeping from the sets $[a,\infty)$, all $a > 0$, so that for each $c \in \mathbb{R}$ we have

$$\lim_{t\to\infty} \mathbf{P}\{\widehat{x}_t \leq c\} = 1. \tag{11.28}$$

Let us introduce now the process (z_t) defined by $z_0 := -\widehat{x}_0^-$, $z_{t+1} := -(z_t + \widehat{\eta}_t)^-$. By (11.27) we have $z_t \leq \widehat{x}_t$ for all t. Since $\widehat{\eta}_0$ is $\mathbf{P}$-integrable, there is an $L > 0$ such that $\mathbb{E}(\widehat{\eta}_0 - L)^+ < a/3$. Let y_0 be chosen so that $\widehat{x}_0$ is also integrable. Then $\mathbb{E}\,|z_0| < \infty$, and in fact $\mathbb{E}\,|z_t| < \infty$ for all t. From (11.28) and $z_t \leq \widehat{x}_t$ we have

$$\lim_{t\to\infty} \mathbf{P}\{z_t \leq -L\} = 1.$$

Choose t_0 so that $\mathbf{P}\{z_t \leq -L\} < a/(3L)$ when $t \geq t_0$. Since $z_t \leq 0$, then, $t \geq t_0$ implies $\mathbb{E}(z_t + L)^+ < a/3$. Therefore,

$$\begin{aligned}\mathbb{E}z_{t+1} = -\mathbb{E}(z_t + \widehat{\eta}_t)^- &= \mathbb{E}(z_t + \widehat{\eta}_t) - \mathbb{E}(z_t + \widehat{\eta}_t)^+ \geq \mathbb{E}z_t + \mathbb{E}\widehat{\eta}_t - \mathbb{E}(z_t + L)^+ \\ -\mathbb{E}(\widehat{\eta}_t - L)^+ &> \mathbb{E}z_t + \frac{a}{3},\end{aligned}$$

which contradicts $z_t \leq 0$ for all t. □

Proof. [Proof of Proposition 6] By the Portmanteau Theorem ((Shiryaev, 1996, Theorem III, 1.1)), $\psi_n^* \to \delta_0$ weakly if and only if

$$\liminf_{n\to\infty} \psi_n^*(G) \geq \delta_0(G) \quad \text{for every open set } G \subset \mathbb{R}_+.$$

Here by "open" we refer to the relative topology on $\mathbb{R}_+$. Evidently the above condition is equivalent to $\liminf_{n\to\infty} \psi_n^*(G) = 1$ for all open G containing 0, which in turn is equivalent to

$$\lim_{n\to\infty} \psi_n^*([a,\infty)) = 0, \quad \forall a > 0.$$

Take (π_n) to be any sequence of optimal policies corresponding to $\rho_n \to 0$. Let (y_t^n) be the Markov chain generated by π_n and fixed initial distribution $y_0 \sim \psi_0$ (i.e., $y_{t+1}^n = f(\pi_n(y_t^n))\varepsilon_t$). Here $y_0 = y_0^n$ is chosen so that $\mathbb{E}\, y_0 < \infty$.

Consider the probability that y_t^n exceeds a. For each real R we have

$$\mathbf{P}\{y_t^n \geq a\} = \mathbf{P}\{y_t^n \geq a\} \cap \{y_{t-1}^n \leq R\}) + \mathbf{P}(\{y_t^n \geq a\} \cap \{y_{t-1}^n > R\}). \quad (11.29)$$

Consider the second term. We claim that

$$\forall r > 0, \exists R \in \mathbb{R} \text{ s.t. } \sup_{n\in\mathbb{N}} \sup_{t\geq 0} \mathbf{P}\{y_t^n > R\} < r. \quad (11.30)$$

To see this, fix $r > 0$, and pick any $n \in \mathbb{N}$. Define a sequence (ξ_t) of random variables on $\{\Omega, F, \mathbf{P}\}$ by $\xi_0 = y_0$, $\xi_{t+1} = \{\alpha\xi_t + \beta)\varepsilon_t$, where $y \mapsto \alpha y + \beta$ is an affine function dominating f on $\mathbb{R}_+$ and satisfying $\alpha < 1$ (see the comment after Assumption 2). From the definition of y_t^n, the fact that $\pi_n(y) \leq y$ and $f(y) \leq \alpha y + \beta$, it is clear that $y_t^n \leq \xi_t$ pointwise on Ω for all t, and hence

$$\forall R \in \mathbb{R}, \quad \{y_t^n > R\} \subset \{\xi_t > R\}.$$
$$\therefore \mathbf{P}\{y_t^n > R\} \leq \mathbf{P}\{\xi_t > R\}, \quad \forall t \geq 0. \quad (11.31)$$

Since ξ_t and ε_t are independent, $\mathbb{E}\xi_{t+1} = \alpha\mathbb{E}\xi_t + \beta$. It follows that

$$\mathbb{E}\xi_t \leq \alpha^t \mathbb{E}\xi_0 + \frac{\beta}{1-\alpha} \leq \mathbb{E}\xi_0 + \frac{\beta}{1-\alpha}$$

for all t. Since $\mathbb{E}\xi_0 = \mathbb{E}y_0 < \infty$ we see that $\mathbb{E}\xi_t \leq C$ for all t, where C is a finite constant. By the Chebychev inequality, then

$$\mathbf{P}\{\xi_t > R\} \leq \frac{\mathbb{E}\xi_t}{R} \leq \frac{C}{R}, \quad \forall t \geq 0. \quad (11.32)$$

Combining (11.31) and (11.32) gives $\mathbf{P}\{y_t^n > R\} < C/R$ for all t and n. Since R is arbitrary the claim (11.30) is established.

Our objective was to bound the second term in (11.29). So fix $r > 0$. By (11.30) we can choose R so large that

$$\mathbf{P}\{y_t^n \geq a\} = \mathbf{P}(\{y_t^n \geq a\} \cap \{y_{t-1}^n \leq R\}) + \frac{r}{2} \quad (11.33)$$

for all t and all n. It remains to bound the first term. Let $(\psi_t^n) \subset \mathcal{P}$ be the sequence of marginal distributions associated with (y_t^n). From the well-known expression for the finite dimensional distribution of Markov chains on measurable rectangles (e.g., Shiryaev 1996, Theorem II. 9.2) we have

$$\mathbf{P}(\{y_t^n \geq a\} \cap \{y_{t-1}^n \leq R\}) = \int_0^R \int_a^\infty \varphi\left(\frac{y'}{f(\pi_n(y))}\right) \frac{1}{f(\pi_n(y))} dy' \psi_{t-1}(dy).$$

A change of variable gives

$$\int_a^\infty \varphi\left(\frac{y'}{f(\pi_n(y))}\right) \frac{1}{f(\pi_n(y))} dy' = \varphi([a/f(\pi_n(y)), \infty)).$$

From the proof of Lemma 4, we know that π_n is dominated by an increasing function b_n which converges pointwise to zero. Therefore, $f \circ \pi_n$ is dominated by $f \circ b_n$, again an increasing function, which must by continuity of f converge pointwise and hence uniformly to zero on $[0, R]$. Combining this with the fact that $a > 0$ and φ is a finite measure, there is an $N \in \mathbb{N}$ such that $n \geq N$ implies

$$\varphi\left(\left[\frac{a}{f}(\pi_n(y)), \infty\right)\right) < \frac{r}{2}, \qquad \forall y \in [0, R].$$

But then

$$\mathbf{P}(\{y_t^n \geq a\} \cap \{y_{t-1}^n \leq R\}) \leq \int_0^R \frac{r}{2} \psi_{t-1}(dy) \leq \frac{r}{2}.$$

Using this inequality together with (11.29) and (11.33), we conclude that for all $r > 0$ there is an $N \in \mathbb{N}$ such that $n \geq N$ and $t \geq 0$ implies $\mathbf{P}\{y_t^n \geq a\} = \psi_t^n([a, \infty)) < r$. Since $\psi_t^n \to \psi_n^*$ in norm it follows that $\psi_t^n([a, \infty)) \longrightarrow \psi_n^*([a, \infty))$ in $\mathbb{R}$ as $t \to \infty$, so that $\psi_n^*([a, \infty)) \leq r$ is also true. That is, $\lim_{n \to \infty} \psi_n^*([a, \infty)) = 0$, as was to be proved. □

Appendix

First we need the following lemma regarding continuity of translations in L_1, which is well-known.

Lemma A.1 *Let g be in $L_1(\mathbb{R})$. If $\tau(t) := \|g(x - t) - g(x)\|$, then τ is bounded on $\mathbb{R}$, and $\tau(t) \to 0$ as $t \to 0$.*

Now define the real number $\mu'(r)$ to be $-\int h(x + r) g'(x)\, dx$, which is clearly finite. By the Fundamental Theorem of Calculus,

$$\mu(r+t)-\mu(r)-\mu'(r)t = \int h(x+r)(g(x-t)-g(x)+g'(x)t)\,dx$$
$$= -t\int h(x+r)\int_0^1 (g'(x-ut)-g'(x))\,du\,dx.$$

Taking absolute values, using (ii) and Fubini's theorem,

$$\left|\frac{\mu(r+t)-\mu(r)}{t}-\mu'(r)\right| \le M\int_0^1\int |g'(x-ut)-g'(x)|\,dx\,du \tag{A.1}$$

for some M. By Lemma A.1, $\int |g'(x-ut)-g'(x)|\,dx$ is uniformly bounded in u and converges to zero as $t\to 0$ for each $u\in[0,1]$. By Lebesgue's Dominated Convergence Theorem the term on the right hand side of (A.1) then goes to zero and

$$\mu'(r) = -\int h(x+r)g'(x)\,dx.$$

Regarding continuity of the derivative, we have

$$|\mu'(r+t)-\mu'(r)| \le \int h(x)|g'(x-r-t)-g'(x-r)|\,dx$$
$$\le M\int |g'(x-t)-g'(x)|\,dx.$$

Continuity now follows from Lemma A.1.

Next we argue that under (iii)–(v),

$$\mu'(r) = \int h'(x+r)g(x)\,dx \tag{A.2}$$

is also valid. To begin, define $\mu_h'(r)$ to be the right hand side of (A.2). This number exists in $\mathbb{R}$, because

$$h'(x+r) = \liminf_{t\downarrow 0}\frac{h(x+r+t)-h(x+r)}{t}$$

almost everywhere by either (iii) or (iv), and hence

$$\mu_h'(r) = \int \liminf_{t\downarrow 0}\frac{h(x+r+t)-h(x+r)}{t}g(x)dx$$
$$\le \liminf_{t\downarrow 0}\int \frac{h(x+r+t)-h(x+r)}{t}g(x)dx = \mu'(r).$$

Here the inequality follows from the assumption that h is increasing, which gives nonnegativity of the difference quotient, and Fatou's Lemma.

By (iv) the Fundamental Theorem of Calculus applies to h, and

$$\begin{aligned}\mu(r+t)-\mu(r)-\mu_h'(r)t &= \int (h(x+t)-h(x)-h'(x)t)g(x-r)\,dx \\ &= t\int\int_0^1 (h'(x+ut)-h'(x))g(x-r)\,dx\,du.\end{aligned}$$

Some simple manipulation gives

$$\mu_h'(r)=\mu'(r)-\lim_{t\to 0}\int\int_0^1 (h'(x+ut)-h'(x))g(x-r)\,dx\,du.$$

Thus it is sufficient to now show that

$$\lim_{t\to 0}\int_0^1\int |h'(x+ut)-h'(x)|g(x-r)\,dx\,du=0.$$

The inner integral is bounded independent of u, because it is less than

$$\int h'(x+ut)g(x-r)\,dx+\int h'(x)g(x-r)\,dx \le \mu'(r+ut)+\mu'(r),$$

which is bounded for $u\in[0,1]$ by continuity of μ'. Thus by Lebesgue's Dominated Convergence Theorem we need only prove that

$$\lim_{t\to 0}\int |h'(x+ut)-h'(x)|g(x-r)\,dx=0.$$

Adding and subtracting appropriately, this integral is seen to be less than

$$\begin{aligned}&\int |h'(x+ut)g(x-r+ut)-h'(x)g(x-r)|dx \\ &\quad+\int |h'(x+ut)g(x-r)-h'(x+ut)g(x-r+ut)|\,dx. \qquad \text{(A.3)}\end{aligned}$$

Consider the first integral in the sum. By Lemma A.1, we can choose a $\delta_0>0$ such that $|t|\le\delta_0$ implies

$$\int |h'(x+ut)g(x-r+ut)-h'(x)g(x-r)|dx<\frac{\varepsilon}{3}.$$

The second integral in the sum can be written as

$$\int_{|x|\le R} |h'(x+ut)g(x-r) - h'(x+ut)g(x-r+ut)|\,dx$$
$$+ \int_{|x|\ge R} |h'(x+ut)g(x-r) - h'(x+ut)g(x-r+ut)|\,dx.$$

By the usual property of L_1 functions, we can choose R such that the integral over $|x| \ge R$ is less than $\varepsilon/3$ for all t with $|t| \le \delta_0$.

To summarize the results so far, we have $|t| \le \delta_0$ implies

$$\int |h'(x+ut) - h'(x)|g(x-r)\,dx$$
$$< \frac{2\varepsilon}{3} + \int_{|x|\le R} |h'(x+ut)g(x-r) - h'(x+ut)g(x-r+ut)|\,dx.$$

Finally, since h' is bounded on compact sets,

$$h'(x+ut) \le M, \quad \forall x, t \text{ with } |x| \le R,\ |t| \le \delta_0.$$

Therefore $|t| \le \delta_0$ implies

$$\int |h'(x+ut) - h'(x)|g(x-r)\,dx$$
$$< \frac{2\varepsilon}{3} + M \int |g(x-r) - g(x-r+ut)|\,dx.$$

By Lemma A.1 there is a $\delta_1 > 0$ such that

$$M \int |g(x-r) - g(x-r+ut)|\,dx < \frac{\varepsilon}{3}$$

whenever $|t| < \delta_1$. Now setting $\delta := \delta_0 \wedge \delta_1$ gives

$$|t| < \delta \implies \int |h'(x+ut) - h'(x)|g(x-r)\,dx < \varepsilon$$

as required.

References

Amir, R. (1997), "A new look at optimal growth under uncertainty," *Journal of Economic Dynamics and Control*, 22, 67–86.

Amir, R., Mirman, L. J. and W. R. Perkins (1991), "One-sector nonclassical optimal growth: optimality conditions and comparative dynamics," *International Economic Review*, 32, 625–644.

Askri, K. and C. Le Van (1998), "Differentiability of the value function of nonclassical optimal growth models," *Journal of Optimization Theory and Applications*, 97, 591–604.

Blume, L., Easley, D. and M. O'Hara (1982), "Characterization of optimal plans for stochastic dynamic economies," *Journal of Economic Theory*, 28, 221–234.

Borovkov, A. A. (1998), Probability Theory, Gordon and Breach Scientific Publishers, The Netherlands.

Brock, W. A. and L. Mirman (1972), "Optimal economic growth and uncertainty: the discounted case," *Journal of Economic Theory*, 4, 479–513.

Danthine, J-P. and J. B. Donaldson (1981), "Stochastic properties of fast vs. slow growth economies," *Econometrica*, 49, 1007–1033.

Dechert, W. D. and K. Nishimura (1983), "A complete characterization of optimal growth paths in an aggregated model with non-Concave production function," *Journal of Economic Theory*, 31, 332–354.

Dudley, R. M. (2002), Real Analysis and Probability, Cambridge Studies in *Advanced Mathematics* No. 74, Cambridge.

Foguel, S. R. (1969), The Ergodic Theory of Markov Processes, Van Nostrand Reinhold, New York.

Joshi, S. (1997), "Turnpike Theorems in Nonconvex Nonstationary Environments," *International Economic Review*, 38, 225–248.

Kamihigashi, T. (2003), Almost Sure Convergence to Zero in Stochastic Growth Models, mimeo, RIEB, Kobe University.

Kamihigashi, T. and S. Roy (2007), "A Non-smooth, Non-convex Model of Economic Growth," *Journal of Economic Theory*, 132, 435-460.

Kydland, F. and E. C. Prescott (1982), "Time to Build and Aggregate Fluctuations," *Econometrica*, 50, 1345–1370.

Lasota, A. and M. C. Mackey (1994), Chaos, Fractals and Noise: Stochastic Aspects of Dynamics, 2nd edition, Springer-Verlag, New York.

Long, J. B. and C. I. Plosser (1983), "Real Business Cycles," *Journal of Political Economy*, 91, 39–69.

Lucas, R. E. (1986), "Adaptive Behavior and Economic Theory," *The Journal of Business*, 59, 385–399.

Majumdar, M. and T. Mitra (1982), "Intertemporal Allocation with Nonconvex Technology," *Journal of Economic Theory*, 27, 101–136.

Majumdar, M., Mitra, T. and Y. Nyarko (1989), Dynamic Optimization under Uncertainty: Non-convex Feasible Set in *Joan Robinson and Modern Economic Theory*, ed. by Feiwel, G., R., Macmillan Press, New York.

Mirman, L. J. (1970), Two Essays on Uncertainty and Economics, Ph.D. Thesis, University of Rochester.

Mirman, L. J., Morand, O. F. and K. Reffett (2005), A Qualitative Approach to Markovian Equilibrium in Infinite Horizon Economies with Capital, manuscript.

Mirman, L. J. and I. Zilcha (1975), "On Optimal Growth under Uncertainty," *Journal of Economic Theory*, 11, 329–339.

Mitra, T. and S. Roy (2006), "Optimal Exploitation of Resources under Uncertainty and the Extinction of Species," *Economic Theory*, 28, 1-23.

Nishimura, K. and J. Stachurski (2005), "Stability of Stochastic Optimal Growth Models: A New Approach," *Journal of Economic Theory*, 122, 100–118.

Prescott, E. C. (2003), "Non-convexities in Quantitative General Equilibrium Studies of Business Cycles," *Staff Report*, No. 312, Federal Reserve Bank of Minneapolis.

Rudnicki, R. (1995), "On Asymptotic Stability and Sweeping for Markov Operators," *Bulletin of the Polish Academy of Science: Mathematics*, 43, 245–262.

Saks, S. (1937), Theory of the Integral, Monografie Matematyezne, Warsaw.

Schenk-Hoppé, K. R. (2005), "Poverty Traps and Business Cycles in a Stochastic Overlapping Generations Economy with S-shaped Law of Motion," *Journal of Macroeconomics*, 27, 275-288.

Shiryaev, A. N. (1996), Probability, Springer-Verlag, New York.

Skiba, A. K. (1978), "Optimal Growth with a Convex-Concave Production Function," *Econometrica*, 46, 527–539.

Stokey, N. L., Lucas, R.E., and E. C. Prescott (1989), Recursive Methods in Economic Dynamics, Harvard University Press, Massachusetts.

Chapter 12
Stability of Stochastic Optimal Growth Models: A New Approach*

Kazuo Nishimura and John Stachurski**

12.1 Introduction

Many economic models are now explicitly dynamic and stochastic. Their state variables evolve in line with the decisions and actions of individual economic agents. These decisions are identified in turn by imposing rationality. Depending on technology, market structure, time discount rates and other primitives, rational behavior may lead either to stability or to instability.[1]

A classic study of stability in dynamic stochastic models is Brock and Mirman (1972). They show that for many convex one-sector growth models, the optimizing behavior of agents implies convergence for the sequence of per-capita income distributions to a unique, nondegenerate limiting distribution, or stochastic steady state. Put differently, the optimal process for the state variable is ergodic.

Their study laid the foundations for a vast and growing literature, spanning economic development, public finance, fiscal policy, environmental and resource economics, monetary policy and asset pricing. For much of this research the

* **Journal of Economic Theory 122**, 100–118, 2005

**The authors thank Takashi Kamihigashi, Yasusada Murata and Kevin Reffett for helpful comments. Financial support provided by a Grant-in-Aid for the 21st Century C.O.E. program in Japan is gratefully acknowledged.

[1] Rationality itself has no intrinsic stability implications—any sufficiently smooth function can be rationalized as the solution to a discounted dynamic program (see Boldrin and Montrucchio 1986).

K. Nishimura (✉)
Institute of Economic Research, Kyoto University, Japan
e-mail: nishimura@kier.kyoto-u.ac.jp

J. Stachurski
Research School of Economics, Australian National University, Australia
e-mail: john.stachurski@anu.edu.au

J. Stachurski et al. (eds.), *Nonlinear Dynamics in Equilibrium Models*,
DOI 10.1007/978-3-642-22397-6_12,

existence of a unique nontrivial long-run equilibrium distribution is fundamental (implicitly, often explicitly).

In this paper we illustrate a new method for determining whether optimal accumulation problems are stable. It combines the Euler equation of the optimal program with the Foster–Lyapunov theory of Markov chains. In fact the Euler equation is identified directly with a Lyapunov equation, connecting stochastic optimal growth with strong ergodicity results from Foster–Lyapunov theory.

The inherent simplicity of this technique allows us to eliminate several conditions required for ergodicity in earlier studies. For example, technology need not be convex, productivity shocks need not be bounded, and marginal product of capital at zero need not be infinite: We give a precise condition on the latter in terms of the shock distribution and the agent's discount factor.

The problem of establishing a general stability result for *nonconvex* stochastic optimal growth has resisted fundamental innovation for many years. This is not because convexity is somehow intrinsic to stable investment behavior. Rather it is due to the difficulty of untangling the implications of the Euler equation without the extra structure that convexity provides. For it appeared that only when these implications had been recovered could work begin on finding appropriate Lyapunov functions (or applying some other standard dynamics machinery). In this paper, the Euler equation *is* the Lyapunov equation. We find that under standard productivity shock distributions local nonconvexities do not alter Brock and Mirman's essential conclusion.

Research on Lyapunov techniques for studying Markov dynamics is still very active. For a recent survey see the (excellent) monograph of Meyn and Tweedie (1993). We exploit in particular the powerful theory of V-uniform ergodicity (see Meyn and Tweedie 1993, Chap. 16). By linking our Euler equation method with this theory, we prove that in general optimal stochastic growth models are not only ergodic but *geometrically* ergodic. That is, for any given starting point, the distance between the current distribution and the limiting distribution decreases at a geometric rate.

Geometric ergodicity has numerous theoretical and empirical applications. As an example of the former, the *rate* at which stochastically growing economies tend to their steady state is a central component of the "convergence" debate; of the latter, geometric convergence is required by Duffie and Singleton (1993) for consistency of the Simulated Moments Estimator. Geometric ergodicity also has applications to numerical procedures: When computing ergodic distributions, rates of convergence can determine bounds on algorithm run-times for a prescribed level of accuracy.

Finally, we use V-uniform ergodicity to prove that under standard econometric assumptions on the noise process the series for the state variable also satisfies both the Law of Large Numbers (LLN) and Central Limit Theorem (CLT). The former states that sample means of the optimal process converge asymptotically to their long-run expected value. The latter associates asymptotic distributions to estimators, from which confidence intervals and hypothesis tests are constructed.

It is shown that the number of moments of the optimal income process for which the LLN and CLT apply depend on the number of finite moments possessed by the productivity shock. For example, when the shock is lognormal—and all moments are finite—we have the remarkable conclusion that the LLN and the CLT hold for all moments of the income process.

12.1.1 Existing Literature

The problem considered in this paper is one of deducing stability of the state variable process from the model *primitives* and the restrictions imposed by optimizing behavior. Key references include Brock and Mirman (1972), Mirman and Zilcha (1975), Donaldson and Mehra (1983), Stokey et al. (1989) and Hopenhayn and Prescott (1992).

These studies contain many important contributions to the theory of stochastic growth. On the precise problem considered in this paper, however, we obtain a stronger form of stability (geometric ergodicity) without many of their assumptions, such as convex technology, bounded shocks and infinite marginal product of capital at zero.[2]

In the deterministic case, stability of optimal nonconvex planning problems was studied by Skiba (1978), Majumdar and Mitra (1982), and Dechert and Nishimura (1983), among others. More general studies of stochastic optimal growth under nonconvexities include Majumdar et al. (1989) and Nishimura et al. (2003).

In earlier research, LLN and CLT results for stochastic optimal growth models were usually proved using restrictions on the support of the productivity shock (cf., e.g., Stokey et al. 1989). This is because when Markov processes are ergodic and have compact state space they are typically geometrically ergodic, and geometric ergodicity is in turn associated with LLN and CLT. LLN and CLT results for some stochastic growth models without bounded shocks are given in Stachurski (2003), but the assumptions on technology are too strict for the general Brock–Mirman problem. Evstigneev and Flåm (1997) and Amir and Evstigneev (2000) studied CLT related properties of competitive equilibrium economies.

There is of course much general research in economics on ergodicity and stability of Markov chains which does not deal directly with the problem considered here (i.e., inferring stability from model primitives and optimizing behavior). Classic studies include Mirman (1970) and Futia (1982). A recent survey is Bhattacharya and Majumdar (2003).[3] They consider chains that satisfy a splitting condition, as well as those that are contracting on average. This paper advocates an alternative approach using Lyapunov functions.

---

[2]Previously, Stachurski (2002) relaxed the assumption that productivity shocks have compact support. Mitra and Roy (2004) studied stability and instability of stochastic optimal planning problems with bounded shocks where technology may be nonconvex and marginal productivity at zero is not generally infinite. Although they do not consider ergodicity, they do give sufficient conditions on the primitives to avoid collapse of the process to the origin.

[3]See also related papers in this symposium, edited by the same authors.

The remainder of the paper is structured as follows. Section 12.2 formulates the problem. Section 12.3 states our results. Section 12.4 outlines the method of proof. Main proofs are given in Sect. 12.5. The proofs of some lemmas are deferred to an appendix.

12.2 Model

The economy produces a single good, which can either be consumed or invested. When technology is convex we can and do assume the existence of a single social planner, who implements a state-contingent savings policy to maximize the discounted sum of expected utilities. When technology is not convex we study the planning problem (rather than competitive equilibria), again solved by a single agent.

At the start of time t, the agent observes income y_t, which is then divided between savings and consumption. Savings is added one-for-one to the existing capital stock. For simplicity we assume that depreciation is total: current savings and the capital stock k_t are identified. Labor is supplied inelastically; we normalize the total quantity to one.

After the time t investment decision is made a shock ε_t is drawn by nature and revealed to the agent. Production then takes place, yielding at the start of next period output

$$y_{t+1} = f(k_t)\,\varepsilon_t.$$

The sequence $(\varepsilon_t)_{t=0}^{\infty}$ is independent; f describes technology.

The production technology is smooth and as usual the derivative is eventually zero—the standard Inada assumption. On the other hand, it may not be convex:

Assumption 1 *The function* $f : R_+ \to R_+$ *is strictly increasing, continuously differentiable and satisfies* $f(0) = 0$, $\liminf_{k\downarrow 0} f'(k) > 1$ *and* $\limsup_{k\uparrow\infty} f'(k) = 0$.

A standard condition for stability in *stochastic* models is that $f'(0) = \infty$. We will be developing a tighter condition. In the meantime, however, one needs sufficient productivity at low investment to be sure that the optimal policy will be interior, and that the Euler equation does in fact hold. This is the purpose of requiring $\liminf_{k\downarrow 0} f'(k) > 1$.

To formalize uncertainty, let each random variable ε_t be defined on a fixed probability space $(\Omega, \mathfrak{F}, \mathbb{P})$, where Ω is the set of outcomes, $\mathfrak{F}$ is the set of events, and $\mathbb{P}$ is a probability. The symbol $\mathbb{E}$ will denote integration with respect to $\mathbb{P}$.

Assumption 2 *The shock* ε *is distributed according to* ψ, *a density on* R_+. *The density* ψ *is continuous and strictly positive on the interior of its domain. In addition, the moments* $E\,(\varepsilon^p)$ *and* $E(1/\varepsilon)$ *are both finite for some* $p \geqslant 1$.

For example, the class of lognormal distributions satisfies Assumption 2 for every $p \in \mathbb{N}$. By definition, $\mathbb{P}\{a \leqslant \varepsilon_t \leqslant b\} = \int_a^b \psi(dz)$ for all a, b and t, where here and in all of what follows $\psi(dz)$ represents $\psi(z)dz$.

In earlier studies it was commonly assumed that the shock ε only took values in a closed interval $[a, b] \subset (0, \infty)$. In this case $\mathbb{E}(\varepsilon^p)$ and $\mathbb{E}(1/\varepsilon)$ are automatically finite. For unbounded shocks the last two restrictions can be interpreted as bounds on the size of the right- and left-hand tails of ψ, respectively. Without such bounds the stability of the economy is jeopardized.

The assumption that the shock distribution has a density representation which is continuous and everywhere positive is needed to complete the proof of geometric ergodicity. It is not related to the basic idea of the paper, and without it alternative stability results will be available. However, global stability may fail—for example when technology is nonconvex and the shock is degenerate (see Dechert and Nishimura 1983).

The larger p can be taken in Assumption 2, the tighter the conclusions of the paper will be. For example, we prove that the Law of Large Numbers holds for all moments of the optimal process up to order p, and the Central Limit Theorem holds for all moments up to order q, where $q \leqslant p/2$.

A feasible savings policy is a (Borel) function π from $\mathbb{R}_+$ to itself such that $0 \leqslant \pi(y) \leqslant y$ for all y. The set of all feasible policies will be denoted by Π. Corresponding to each $\pi \in \Pi$ there is a consumption policy $c^\pi(y) := y - \pi(y)$.

Every $\pi \in \Pi$ defines a Markov process on $(\Omega, \mathfrak{F}, \mathbb{P})$ for income via

$$y_{t+1} = f(\pi(y_t))\,\varepsilon_t. \tag{12.1}$$

The problem for the agent is to choose a policy which solves

$$\max_{\pi \in \Pi} \mathbb{E}\left[\sum_{t=0}^{\infty} \rho^t u(c^\pi(y_t))\right], \tag{12.2}$$

where, for given π, the sequence $(y_t)_{t=0}^{\infty}$ is determined by (12.1). The number $\rho \in (0, 1)$ is the discount factor, and u is the period utility function.

Assumption 3 *The function* $u : R_+ \to R_+$ *is strictly increasing, bounded, strictly concave, continuously differentiable, and* $\lim_{c \downarrow 0} u'(c) = \infty$.

12.2.1 Optimal Policies

A policy π is called optimal if it is an element of Π and attains the maximum in (12.2). We are particularly interested in situations where optimal policies exist, are interior and satisfy the Euler equation. There are essentially two routes to the Euler equation. One is via convexity of the feasible set (Mirman and Zilcha 1975). The second is to use smoothness of the productivity shock distribution (see Blume et al. 1982; Nishimura et al. 2003). Thus, Assumption 4:

Assumption 4 *Either (i) the production function* f *is concave; or (ii) the density* ψ *of the shock is continuously differentiable on* $(0, \infty)$*, and the integral* $\int z\,|\psi'(z)|\,dz$ *is finite.*

The class of lognormal shocks satisfies (ii) of Assumption 4.

Theorem 1. *Under Assumptions 1–3 there is at least one optimal policy for (12.2), and every optimal policy is nondecreasing in income. If in addition Assumption 4 holds, then every optimal policy π is interior, and satisfies the Euler equation*

$$u' \circ c^{\pi}(y) = \rho \int u' \circ c^{\pi}[f(\pi(y))z] f'(\pi(y)) z \psi(dz), \quad \forall\, y \in (0, \infty).$$

Here $u' \circ c^{\pi}$ is of course the composition of u' and c^{π}, so that $u' \circ c^{\pi}(y)$ is the marginal utility of consumption when income equals y. By interiority is meant that $0 < \pi(y) < y$ for all $y \in (0, \infty)$.

Proof. For f concave the proof of Theorem 1 is well known (see Mirman and Zilcha 1975; Stokey et al. 1989). For the other case (i.e., (ii) of Assumption 4), see Nishimura et al. (2003) (Lemmas 3.1 and 3.2, Propositions 3.1 and 3.2).[4]

12.2.2 Dynamics

Once an initial condition for income is specified, each optimal policy completely defines the process $(y_t)_{t=0}^{\infty}$ for income via the recursion (12.1). Formally, $(y_t)_{t=0}^{\infty}$ is a Markov process on $(\Omega, \mathfrak{F}, \mathbb{P})$. It is simplest in what follows to take the state space for the optimal process $(y_t)_{t=0}^{\infty}$ to be $(0, \infty)$ rather than $\mathbb{R}_+$. After all, each optimal policy is interior, and since the shock is distributed according to a density, $(y_t)_{t=0}^{\infty}$ remains in $(0, \infty)$ with probability one provided that $y_0 > 0$. We can always assume that $y_0 > 0$, as dynamics from $y_0 = 0$ are trivial.

Since each y_t is a random variable taking values in $(0, \infty)$ it has a distribution on the same.[5] A characteristic of Markov chains is that the sequence of distributions corresponding to the sequence of state variables satisfies a fundamental recursion, now to be described.[6]

To begin, let π be a fixed optimal policy, and let $\Gamma(y, \cdot)$ be the distribution for y_{t+1} given that $y_t = y$. Recall that if X is a real-valued random variable with density f_X, if $a > 0$, and if $Y := a \cdot X$, then Y has density $f_Y(x) = f_X(x/a)(1/a)$. From this expression, the strict positivity of $f(\pi(y))$ for each y in the state space (Theorem 1) and (12.1) it follows that $\Gamma(y, \cdot)$ is a density given by

$$\Gamma(y, y') = \psi\left(\frac{y'}{f(\pi(y))}\right) \frac{1}{f(\pi(y))}. \tag{12.3}$$

---

[4]The assumption in Nishimura et al. (2003) that the shock has mean one is not used in the proofs of these four results, and hence is omitted.

[5]Precisely, the distribution of y_t is the image measure $\mathbb{P} \circ y_t^{-1}$.

[6]More discussion of what follows can be found in Stokey et al. (1989), Stachurski (2002) and Bhattacharya and Majumdar (2003).

Now let $\mathfrak{D}$ be the collection of densities on $(0, \infty)$.[7] Suppose for the moment that the initial condition y_0 is a random variable, with distribution equal to $\varphi_0 \in \mathfrak{D}$. It then follows that the distribution of y_t is a density φ_t for all t, and the sequence $(\varphi_t)_{t=0}^{\infty}$ satisfies

$$\varphi_{t+1}(y') = \int \Gamma(y, y')\varphi_t(y)dy \tag{12.4}$$

for all $t \geqslant 0$. The intuition is that $\Gamma(y, y')$ is the probability of moving from income y to income y' in one period; in which case (12.4) simply states that the probability of being at y' next period is the probability of moving to y' via y, summed across all y, weighted by the probability that current income is equal to y.

It is perhaps more natural to regard y_0 as a single point, rather than a random variable with a density. In this case, provided $y_0 > 0$, one can take φ_0 to be the degenerate probability at y_0, and set $\varphi_1(\cdot) = \Gamma(y_0, \cdot) \in \mathfrak{D}$. The remaining sequence of densities is then defined recursively by (12.4). Let us agree to write $\varphi_t^{y_0}$ for the t-th element so defined.

A $\varphi^* \in \mathfrak{D}$ is called stationary for the optimal process (12.1) if it satisfies

$$\varphi^*(y') = \int \Gamma(y, y')\varphi^*(y)dy \quad \forall\, y' \in (0, \infty). \tag{12.5}$$

It is clear from (12.4) and (12.5) that if y_t has distribution φ^*, then so does y_{t+n} for all $n \in \mathbb{N}$. A density satisfying (12.5) is also called a stochastic steady state. At such a long-run equilibrium the *probabilities* are stationary over time, even though the state variable is not.[8]

Finally, we define ergodicity, which is the fundamental result of Brock and Mirman (1972). Ergodicity of the optimal process (12.1) means that there is a unique $\varphi^* \in \mathfrak{D}$ satisfying (12.5), and, moreover,

$$\|\varphi_t^y - \varphi^*\| \to 0 \quad \text{as} \quad t \to \infty \quad \text{for all } y > 0,$$

where $\|\cdot\|$ is the L_1 norm.[9] Geometric ergodicity essentially means that the process is ergodic and, in addition, $\|\varphi_t^y - \varphi^*\| = O(\alpha^t)$ for some $\alpha < 1$. A more precise definition will be given in the statement of results.

12.3 Results

Our main stability results are now presented. All of Assumptions 1–4 are maintained without further comment.

---

[7]To be precise, $\mathfrak{D}$ is the set of nonnegative Borel functions on $(0, \infty)$ that integrate to one.

[8]Why have we not defined stationary *distributions*, which are more general than stationary densities? The answer is that for our model all stationary distributions will in fact be densities. A proof is available from the authors.

[9]That is, $\|\varphi_t - \varphi^*\| = \int |\varphi_t - \varphi^*|$. Some studies use a weaker topology.

For the remainder of the paper, let $\pi \in \Pi$ be a fixed optimal savings policy. As before, c^π is the corresponding consumption policy. Define

$$V(y) := \sqrt{u' \circ c^\pi(y)} + y^p + 1, \quad p \text{ as in Assumption 2.} \tag{12.6}$$

In the proofs V will play the role of Lyapunov function.

We now state our main result. For an outline of the proof see Sect. 12.4. A full proof is given in Sect. 12.5.

Theorem 2. *Let V be the function defined in (12.6), and let $f'(0) := \underline{\lim}_{k\downarrow 0} f'(k)$. If the inequality*

$$f'(0) > \frac{\mathbb{E}(1/\varepsilon)}{\rho} \tag{12.7}$$

holds, then the process $(y_t)_{t=0}^\infty$ defined by (12.1) is geometrically ergodic. Precisely, $(y_t)_{t=0}^\infty$ has a unique stationary distribution φ^*, and, moreover, there is a constant $\alpha < 1$ and an $R < \infty$ such that

$$\|\varphi_t^y - \varphi^*\| \leqslant \alpha^t R\, V(y) \quad \forall\, y > 0 \;\; \forall\, t \geqslant 0.$$

In (12.7) the term $\mathbb{E}(1/\varepsilon) = \int (1/z)\psi(dz)$ will be large when unfavorable shocks are likely. Note that if the shock is degenerate and puts all probability mass on one, then the stability condition reduces to $f'(0) > 1/\rho$, which is the usual deterministic condition for stability.[10]

Now let h be a real function on the state space, and define $S_n(h) := \sum_{t=1}^n h \circ y_t$. The next result is to some extent a corollary of Theorem 2.

Theorem 3. *Let (12.7) hold, and let φ^* be the unique stationary distribution for the optimal process $(y_t)_{t=0}^\infty$. If $|h| \leqslant V$, then the Law of Large Numbers holds for h, i.e.,*

$$\mathbb{E}_{\varphi^*}(h) := \int h\, d\varphi^* < \infty \;\; \text{and} \;\; \lim_{n\to\infty} \frac{S_n(h)}{n} = \mathbb{E}_{\varphi^*}(h) \;\; \mathbb{P}\text{-a.s.} \tag{12.8}$$

If in addition $h^2 \leqslant V$, then the Central Limit Theorem also holds for h. Precisely, there is a constant $\sigma^2 \in R_+$ such that

$$\frac{S_n(h - \mathbb{E}_{\varphi^*}(h))}{\sqrt{n}} \xrightarrow{d} N(0, \sigma^2). \tag{12.9}$$

In the statement of the theorem the symbol $\xrightarrow{d}$ means convergence in distribution. If $\sigma^2 = 0$ then the right-hand side of (12.9) is interpreted as the probability measure concentrated on zero. Also, $(h - \mathbb{E}_{\varphi^*}(h))(y_t) := h(y_t) - \mathbb{E}_{\varphi^*}(h)$.

[10] Kamihigashi (2003) provides a partial converse to Theorem 2. He shows that $y_t \to 0$ holds $\mathbb{P}$-a.s., whenever $f'(0) < \exp(-\mathbb{E} \ln \varepsilon)$.

These results are in a rather convenient form. In particular, since $x^p \leqslant V(x)$ we see that all moments of the income process up to order p satisfy the LLN, and all moments up to order q satisfy the CLT, where q is the largest integer such that $2q \leqslant p$.

12.4 Outline of Techniques

In this section, we describe the main ideas used in the proof (The details are in Sect. 12.5). First, the Foster–Lyapunov approach to stability of Markov chains is outlined. As in deterministic dynamical systems, Lyapunov methods often provide the most readily applicable techniques for stability proofs.[11]

In essence, Lyapunov functions are constructed to assign large values to the "edges" of the state space, and small values to the center. If one can show using the law of motion for the system that the value the Lyapunov function assigns to the next-period state is expected to contract relative to the value assigned to the current state, then it must be that the state is moving away from the edges of the state space and towards the center. This behavior is associated with stability.

However, the choice of appropriate Lyapunov function and proof of the above-mentioned contraction property is rarely trivial. The main task, then, is to find an appropriate function and establish the contraction condition. We suggest using marginal utility of consumption as a Lyapunov function, and the Euler equation as the contraction condition. In short, the idea is that for low levels of income marginal returns to investment are likely to be large, which via the Euler equation requires a high marginal willingness to substitute future for current consumption. In other words, expected marginal utility of consumption (the value of the Lyapunov function) for the next period state is small relative to that of the current state. This provides the contraction property.

Actually it is more accurate to say that marginal utility forms a component of our Lyapunov function. It is large at the "edge" of the state space near the origin, but not near the other edge, at plus infinity. Hence contraction with respect to marginal utility only means that the state does not get too small. So another component to the Lyapunov function (which is large at plus infinity) will have to be spliced on. The mechanics are detailed below.

We begin with a general discussion of Foster–Lyapunov methods. The usual way to implement the notion that a potential Lyapunov function V takes small values in the center of the state space and is large towards the edges is to require that all its sublevel sets be precompact. Here precompactness means having compact closure, and sublevel sets of V are sets of the form $\{x : V(x) \leqslant a\}$ for real-valued a.

[11]In the case of Markov chains they are very general too—existence of Lyapunov functions satisfying contraction conditions often characterize stability properties.

Definition 1. A Lyapunov function on a topological space S is a nonnegative real function on S with the property that all sublevel sets are precompact.

Lemma 1. *If* $S = (0, \infty)$*, then* $V: S \to R_+$ *is a Lyapunov function if and only if* $\lim_{x \downarrow 0} V(x) = \lim_{x \uparrow \infty} V(x) = \infty$.[12]

The proof is straightforward and we omit it. From Lemma 1 and the fact that $u' \circ c^\pi \geqslant u'$ it is easy to see that V defined in (12.6) is a Lyapunov function.

Suppose now that for some Lyapunov function V on $(0, \infty)$ and constants $\lambda < 1$ and $b < \infty$ we have

$$\int V[f(\pi(y))z]\psi(dz) \leqslant \lambda V(y) + b \quad \forall y \in (0, \infty). \tag{12.10}$$

This in essence says that when the value assigned by V to the state is large (i.e., when current income y is either close to zero or very large), the value assigned to the next period state is expected to be less than the current value (loosely speaking, income moves back towards the center of the state space).[13]

To gain some understanding of the implications of (12.10), recall that (by the Markov property)

$$\mathbb{E}[V(y_{t+1}) \mid y_t] = \int V[f(\pi(y_t))z]\psi(dz) \quad \mathbb{P}\text{-a.s.}$$

$$\therefore \quad \mathbb{E}[V(y_{t+1}) \mid y_t] \leqslant \lambda V(y_t) + b \quad \mathbb{P}\text{-a.s.}$$

Taking expectations of both sides and using the law of iterated expectations gives $\mathbb{E}V(y_{t+1}) \leqslant \lambda \cdot \mathbb{E}V(y_t) + b$. Extrapolating this inequality from time zero (where y_0 is a given constant) forward to time t and using the fact that $\lambda < 1$ gives the bound

$$\mathbb{E}V(y_t) \leqslant V(y_0) + \frac{b}{1-\lambda} =: M \quad \forall t \in \mathbb{N}. \tag{12.11}$$

$$\therefore \quad \mathbb{P}\{V(y_t) > n\} \leqslant \frac{M}{n} \quad \forall t, n \in \mathbb{N} \qquad (\because \text{ Chebychev's ineq.})$$

Thus, for each $n \in \mathbb{N}$ there is a compact $K_n \subset (0, \infty)$ containing $\{x : V(x) \leqslant n\}$ with

$$\int_{K_n^c} \varphi_t^{y_0}(y)dy = \mathbb{P}\{y_t \notin K_n\} \leqslant \frac{M}{n} \quad \forall t \in \mathbb{N},$$

[12] We stress that on $(0, \infty)$ topological concepts such as precompactness always refer to the *relative* Euclidean topology.

[13] To see this, write (12.10) as $\int V[f(\pi(y))z]\psi(dz)/V(y) \leq \lambda + b/V(y)$. Since $\lambda < 1$ and $b < \infty$ clearly $\int V[f(\pi(y))z]\psi(dz) < V(y)$ for sufficiently large $V(y)$.

where $K_n^c := K_n \setminus (0, \infty)$.[14] It is now evident that

$$\forall \varepsilon > 0, \ \exists \text{ a compact } K \subset (0, \infty) \text{ s.t. } \sup_{t \in \mathbb{N}} \int_{K^c} \varphi_t^{y_0}(y) dy < \varepsilon.$$

This statement says precisely that (for any given starting point y_0) the sequence of distributions $(\varphi_t^{y_0})_{t=0}^{\infty}$ generated by (12.4) is *tight*. Tightness is very closely linked with many stability properties. But, even without invoking these results one can see directly that apart from an arbitrarily small ε, all probability mass for the sequence $(y_t)_{t=0}^{\infty}$ stays on a compact subset of $(0, \infty)$, in which case it cannot be escaping to zero or plus infinity.

One way to construct a Lyapunov function satisfying (12.10) is to take a pair of real functions w_1 and w_2 with the properties $\lim_{x \downarrow 0} w_1(x) = \infty$, $\lim_{x \uparrow \infty} w_2(x) = \infty$, and

$$\int w_i[f(\pi(y))z]\psi(dz) \leqslant \lambda_i w_i(y) + b_i \quad \forall y \in (0, \infty), \quad i = 1, 2 \qquad (12.12)$$

for some $\lambda_1, \lambda_2, b_1, b_2$ with $\lambda_i < 1$ and $b_i < \infty$. In this case it is easy to show that

Proposition 1. *If* $V := w_1 + w_2 + 1$, *where* w_1 *and* w_2 *are two nonnegative real functions on* $(0, \infty)$ *satisfying (12.12), and* $\lim_{x \downarrow 0} w_1(x) = \lim_{x \uparrow \infty} w_2(x) = \infty$, *then* V *is a Lyapunov function on* $(0, \infty)$ *satisfying (12.10) for* $\lambda := \max\{\lambda_1, \lambda_2\}$ *and* $b := b_1 + b_2 + 1$.[15]

Using Proposition 1 we will show that V defined in (12.6) is a Lyapunov function satisfying (12.10) for some $\lambda < 1$ and $b < \infty$. The advantage of constructing V from two components w_1 and w_2 is that we can treat the two problems of diverging to zero (economic collapse) and diverging to infinity (unbounded growth) separately. Obtaining a contraction with respect to w_1 prevents the former (because w_1 is large at zero), while obtaining one for w_2 prevents the latter (because w_2 is large at infinity).

When diminishing returns are present, eliminating the possibility of unbounded growth is straightforward. The reason is that unbounded growth cannot occur for *any* feasible savings policy—even one that invests all output. There is no need to deal with the subtleties of optimization and the Euler equation.

Eliminating the possibility of collapse—finding a suitable function w_1 satisfying a contraction in the form of (12.12)—is considerably more difficult. The question is whether or not economies invest sufficiently to sustain a nontrivial long-run equilibrium. This depends on preferences (particularly rates of time discount), returns to investment, and the distribution of the productivity shock.

[14] Such a compact set exists by the definition of V.

[15] The reason for adding 1 to $w_1 + w_2$ in the definition of V will become clear in the proof of Proposition 4.

Our method for finding a suitable w_1 satisfying (12.12) is to use the Euler equation directly. This leads to the most important result of the paper (the rest is something of a mopping up operation):

Proposition 2. *Let* $w_1 := \sqrt{u' \circ c^\pi}$*. If the inequality (12.7) is satisfied, then there exists a* $\lambda_1 < 1$ *and a* $b_1 < \infty$ *such that*

$$\int w_1[f(\pi(y))z]\psi(dz) \leqslant \lambda_1 w_1(y) + b_1 \quad \forall y \in (0, \infty). \tag{12.13}$$

Proof. We will make use of the fact that if g and h are positive real functions on $(0, \infty)$, then, by the Cauchy-Schwartz inequality,

$$\int (gh)^{1/2} d\psi \leqslant \left(\int g d\psi \cdot \int h d\psi \right)^{1/2}. \tag{12.14}$$

Recall the Euler equation:

$$u' \circ c^\pi(y) = \varrho f'(\pi(y)) \int u' \circ c^\pi[f(\pi(y))z] z\psi(dz). \tag{12.15}$$

Set $w_1 := \sqrt{u' \circ c^\pi}$, as in the statement of the proposition. From (12.14) we have

$$\begin{aligned} \int w_1[f(\pi(y))z]\psi(dz) &= \int [u' \circ c^\pi(f(\pi(y))z) z(1/z)]^{1/2} \psi(dz) \\ &\leqslant \left(\int u' \circ c^\pi[f(\pi(y))z] z\psi(dz) \cdot \int (1/z)\psi(dz) \right)^{1/2}. \end{aligned}$$

Combining this with (12.15) now gives

$$\int w_1[f(\pi(y))z]\psi(dz) \leqslant \left[\frac{\mathbb{E}(1/\varepsilon)}{\rho f'(\pi(y))} \right]^{1/2} w_1(y).$$

From (12.7) one can deduce the existence of a $\delta > 0$ and a $\lambda_1 \in (0, 1)$ such that

$$\left[\frac{\mathbb{E}(1/\varepsilon)}{\rho f'(\pi(y))} \right]^{1/2} < \lambda_1 < 1 \quad \forall y < \delta.$$

$$\therefore \int w_1[f(\pi(y))z]\psi(dz) \leqslant \lambda_1 w_1(y) \quad \forall y < \delta. \tag{12.16}$$

We also need the following technical bound, the proof of which is given in the appendix.

Lemma 2. *Given δ in (12.16), there exists a $b_1 < \infty$ such that*

$$\int w_1[f(\pi(y))z]\psi(dz) \leqslant b_1 \quad \forall y \geqslant \delta. \tag{12.17}$$

From (12.16) and (12.17) the inequality (12.13) is immediate.

We now establish the complementary result for w_2.

Proposition 3. *Let p be as in Assumption 2. For w_2 defined by $w_2(y) = y^p$, there exists a $\lambda_2 < 1$ and a $b_2 < \infty$ such that*

$$\int w_2[f(\pi(y))z]\psi(dz) \leqslant \lambda_2 w_2(y) + b_2 \quad \forall y \in (0, \infty). \tag{12.18}$$

Proof. First, choose $\gamma \in (0, 1)$ so that $\gamma^p \mathbb{E}(\varepsilon^p) < 1$. For such a γ it follows from the assumption $\overline{\lim}_{k\to\infty} f'(k) = 0$ that we can find a $d < \infty$ such that whenever $y > d$, we have $f(y) \leqslant \gamma \cdot y$.[16] For all $y \in (0, d]$ we have $f(\pi(y)) \leqslant f(y) \leqslant f(d)$, and hence

$$\int [f(\pi(y))z]^p \psi(dz) \leqslant f(d)^p \mathbb{E}(\varepsilon^p) \quad \forall y \leqslant d.$$

One the other hand, $y > d$ implies $f(\pi(y)) \leqslant f(y) \leqslant \gamma y$, so

$$\int [f(\pi(y))z]^p \psi(dz) \leqslant \gamma^p \mathbb{E}(\varepsilon^p) y^p \quad \forall y \leqslant d.$$

Setting $\lambda_2 := \gamma^p \mathbb{E}(\varepsilon^p)$ and $b_2 = f(d)^p \mathbb{E}(\varepsilon^p)$ gives (12.18).

Summarizing Propositions 1–2,

Proposition 4. *The function V defined in (12.6) is a Lyapunov function on $(0, \infty)$, and under the hypotheses of Theorems 2 and 3 the contraction condition (12.10) holds for some $\lambda < 1$ and $b < \infty$.*

12.5 Proofs

The complete proofs of Theorems 2 and 3 are now given. They center on establishing that the optimal process is V-uniformly ergodic for V specified by (12.6), where V-uniform ergodicity is defined in Meyn and Tweedie (1993, Chap. 16). Essentially

[16] To see this, choose $\bar{y}$ such that $y \geqslant \bar{y}$ implies $f'(y) \leqslant \gamma/2$. By the Fundamental Theorem of Calculus, when $y \geqslant \bar{y}$ we have $f(y) \leqslant f(\bar{y}) + (y - \bar{y})(\gamma/2)$. This function in turn is dominated by γy for y sufficiently large—larger than $\hat{y}$, say. Now set $d := \max\{\bar{y},\ \hat{y}\}$.

this requires geometric drift towards a subset of the state space which satisfies a certain minorization condition.

The following definitions are necessary. Let $\mathfrak{B}$ be the Borel sets on $(0, \infty)$, let $\mathfrak{M}$ be the finite measures on $((0, \infty), \mathfrak{B})$, and let $\mathfrak{P}$ be all $\nu \in \mathfrak{M}$ with $\nu(0, \infty) = 1$. For $B \in \mathfrak{B}$ let 1_B denote the indicator function of B. Proofs of lemmas are given in the appendix.

Definition 2. Let $\mu \in P$. The optimal process (12.1) is called μ-irreducible if

$$\mathbb{P}\{y_t \in B \textit{ for some } t \in \mathbb{N}\} > 0, \quad \forall\, y_0 > 0,\ B \in \mathfrak{B} \textit{ with } \mu(B) > 0.$$

In other words, $(y_t)_{t=0}^{\infty}$ visits every set of positive μ-measure from every starting point.

Definition 3. A set $C \in B$ is called a C-set for the optimal process (12.1) if there is a nontrivial $\nu \in M$ such that

$$y \in C \implies \left\{ \int_B \Gamma(y, y')dy' \geq \nu(B), \quad \forall\, B \in \mathfrak{B} \right\}, \tag{12.19}$$

where Γ is the stochastic kernel defined in (12.3).

Definition 4. The optimal process is called strongly aperiodic if (12.19) holds for some $C \in B$ and nontrivial $\nu \in M$ (i.e., Γ has a C-set), and, moreover, $\nu(C) > 0$.

Definition 5. Let V be as in (12.6). The optimal process $(y_t)_{t=0}^{\infty}$ is called V-uniformly ergodic (see Meyn and Tweedie 1993, Chap. 16) if

$$\sup_{y>0} \left\{ \frac{\|\varphi_t^y - \varphi^*\|}{V(y)} \right\} \to 0 \text{ as } t \to \infty.$$

Almost all the results derived in this paper follow from

Proposition 5. *Under the hypotheses of Theorems 2 and 3, the optimal process* $(y_t)_{t=0}^{\infty}$ *is* V*-uniformly ergodic.*

Proof. By Meyn and Tweedie (1993) (Theorem 16.1.2), The optimal process will be V-uniformly ergodic whenever it is μ-irreducible for some $\mu \in \mathfrak{P}$, strongly aperiodic, and there is a C-set $C \in \mathfrak{B}$, an $\alpha < 1$, a $\beta < \infty$ and a real function $V : (0, \infty) \to (1, \infty)$ such that

$$\int V[f(\pi(y))z]\psi(dz) \leq \alpha V(y) + \beta 1_C(y) \quad \forall\, y \in (0, \infty). \tag{12.20}$$

(Actually the theorem only requires aperiodicity rather than strong aperiodicity, and that C be "petite", which is a generalization of the usual notion of a C-set.)

The following three lemmas are all proved in the appendix.

Lemma 3. *The optimal process is μ-irreducible for every $\mu \in P$ which is absolutely continuous with respect to Lebesgue measure.*

Lemma 4. *Every compact subset of the state space $(0, \infty)$ is a C-set.*

Lemma 5. *The optimal process is strongly aperiodic.*

As result of Lemmas 3–5, to establish Proposition 4 it is sufficient to show that there is an $\alpha < 1$, a $\beta < \infty$ and a *compact* set C such that (12.20) holds for V defined in (12.6).

Let λ and b be as in Proposition 3. Set $\beta := b$. Take any α such that $\lambda < \alpha < 1$. By the Lyapunov property, we can choose a compact set $C \subset (0, \infty)$ such that $V(y) \geqslant \beta/(\alpha - \lambda)$ whenever $y \notin C$. For $y \in C$ the bound (12.20) is trivial, because by Proposition 3

$$\int V[f(\pi(y))z]\psi(dz) \leqslant \lambda V(y) + b \leqslant \alpha V(y) + \beta 1_C(y).$$

For $y \notin C$ the bound (12.20) also holds, because

$$\frac{\int V[f(\pi(y))z]\psi(dz)}{V(y)} \leqslant \lambda + \frac{b}{V(y)} \leqslant \alpha.$$

Finally, $V \geqslant 1$ by construction. This completes the proof of Proposition 4. ■

Proof of Theorem 1 Immediate from Proposition 4 and Meyn and Tweedie, 1993 (Theorem 16.0.1, Part (ii)). ■

Proof of Theorem 2 LLN Result: By Meyn and Tweedie (1993) (Theorem 17.0.1, Part (i)), the LLN holds for h provided that $(y_t)_{t=0}^{\infty}$ is positive Harris and $\int |h| \, d\varphi^* < \infty$. By positive Harris is meant that $(y_t)_{t=0}^{\infty}$ has an invariant distribution and is Harris recurrent. For a definition of Harris recurrence see Meyn and Tweedie (1993) (Chap. 9). For our purposes we need only note that by the same reference, Proposition 9.1.8, Harris recurrence holds when the sublevel sets of V are all C-set, and, in addition, that there is a C-set $C \in \mathfrak{B}$ such that

$$\int V[f(\pi(y))z]\psi(dz) \leqslant V(y) \quad \forall y \notin C. \tag{12.21}$$

We have already shown that sublevel sets of V are C-sets in Lemma 4 (recall that sublevel sets of V are precompact, and that measurable subsets of C-sets are C-sets).

Therefore $(y_t)_{t=0}^{\infty}$ is positive Harris, and it remains only to show that $\int |h| \, d\varphi^* := \int |h(y)| \, \varphi^*(y) dy < \infty$. Since $|h| \leqslant V$, it is sufficient to show that $\int V d\varphi^*$ is finite.

To see that this is the case, pick any initial condition y_0. For $n \in \mathbb{N}$, let $K_n := 1_{[1/n,n]}$. By (12.11) there is an $M < \infty$ satisfying $\int V d\varphi_t^{y_0} \leqslant M$ for all t, and hence $\int K_n V d\varphi_t^{y_0} \leqslant M$ for all t and n. ∎

Lemma 6. *The function V is bounded on compact sets.*

From this lemma, we see that $K_n V$ is bounded, so (since L_1 convergence implies weak convergence) taking the limit with respect to t gives $\int K_n V d\varphi^* \leqslant M$ for all n. Now taking limits with respect to n and using Monotone Convergence gives $\int V d\varphi^* < \infty$.

CLT Result: Immediate from Meyn and Tweedie (1993) (Theorem 17.0.1, Parts (ii)–(iv)) and Proposition 4. The constant σ^2 is given by

$$\sigma^2 = \lim_{n\to\infty} \frac{1}{n} \mathbb{E}_{\varphi^*}[S_n(h - \mathbb{E}_{\varphi^*}(h))^2].$$

12.6 Appendix

Proof of Lemma 1 If part (i) of Assumption 4 holds (f is concave), then c^π is increasing (see Mirman and Zilcha 1975). We already have the bound

$$\int w_1 [f(\pi(y))z]\, \psi(dz) \leqslant \left(\int u' \circ c^\pi [f(\pi(y))z]\, z\psi(dz) \cdot \int (1/z)\psi(dz) \right)^{1/2}$$

for all y. Since $u' \circ c^\pi$ is decreasing, we can set

$$b_1 := \left(\int u' \circ c^\pi [f(\pi(\delta))z]\, z\psi(dz) \cdot \int (1/z)\psi(dz) \right)^{1/2}.$$

Suppose on the other hand that (ii) of Assumption 4 holds. Let $v : (0,\infty) \to \mathbb{R}$ be the value function for the dynamic programming problem, i.e,

$$v(y) = \mathbb{E}\left[\sum_{t=0}^{\infty} \rho^t u(c^\pi(y_t)) \right] \quad \text{when} \quad y_0 = y.$$

Let $r > 0$. By a change of variable we van write

$$\int_0^\infty v(rz)\psi(z)dz = \int_{-\infty}^{\infty} v\,[\exp(\ln r + x)]\, \psi(e^x)e^x dx.$$

Define

$$\mu(r) := \int_{-\infty}^{\infty} h(x+r)g(x)dx,$$

where $h(y) = v[\exp(y)]$ and $g(y) = \psi(e^y)e^y$, so that when μ is differentiable we have

$$\frac{d}{dr}\int_0^\infty v(rz)\psi(dz) = \frac{1}{r}\mu'(\ln r).$$

By Nishimura et al. (2003) (Lemma 3), μ is continuously differentiable, and

$$\mu'(\ln r) = -\int_{-\infty}^\infty h(x+\ln r)g'(x)dx = \int_{-\infty}^\infty h'(x+\ln r)g(x)dx, \qquad \text{(A.1)}$$

where h' is defined as the derivative of h where it exists and zero elsewhere. Reversing the change of variables gives

$$\int_{-\infty}^\infty h'(x+\ln r)g(x)dx = r\int_{-\infty}^\infty v'(rz)z\psi(dz). \qquad \text{(A.2)}$$

Combining (A.1) and (A.2) gives

$$\int_{-\infty}^\infty v'(rz)z\psi(dz) = \frac{1}{r}\mu'(\ln r) = -\frac{1}{r}\int_{-\infty}^\infty h(x+\ln r)g'(x)dx.$$

Therefore, using the fact that v is bounded by a constant M, say (because u is bounded: Assumption 3),

$$\begin{aligned}
\int_0^\infty v'(rz)z\psi(dz) &\leqslant \frac{1}{r}\int_{-\infty}^\infty \left|h(x+\ln r)g'(x)\right| dx \\
&= \frac{1}{r}\int_{-\infty}^\infty v(\exp(x+\ln r))\left|\psi'(e^x)e^{2x} + \psi(e^x)e^x\right| dx \\
&\leqslant \frac{M}{r}\int_{-\infty}^\infty \left|\psi'(e^x)e^{2x} + \psi(e^x)e^x\right| dx \\
&\leqslant \frac{M}{r}\int_0^\infty \left[\left|\psi'(z)z + \psi(z)\right|\right] dz.
\end{aligned}$$

In light of Assumption 4, part (ii), then, there is a constant N such that

$$\int_0^\infty v'(f(\pi(y))z)z\psi(dz) \leqslant \frac{N}{f(\pi(y))}.$$

Given δ as in the statement of the lemma, and using the fact that $u' \circ c^\pi$ is equal to v' almost everywhere (see Nishimura et al. (2003), Proposition 2, parts 3 and 4), we have the bound

$$\int_0^\infty u' \circ c^\pi[f(\pi(y))z]z\psi(dz) \leqslant b_0 := \frac{N}{f(\pi(\delta))} \quad \forall y \geqslant \delta.$$

But then, applying Cauchy-Schwartz again,

$$\int_0^{\infty} w_1\,[f(\pi(y))z]\psi(dz) \leqslant b_1 := [b_0\mathbb{E}(1/\varepsilon)]^{1/2} \quad \forall y \geqslant \delta.$$

■

Proof of Lemma 2 Let $q \in \mathfrak{D}$. Take any $B \in \mathfrak{B}$ with positive q-measure and any $y_0 \in (0,\infty)$. It is easy to check that the set $[f(\pi(y_0))]^{-1} \cdot B$ has positive Lebesgue measure. We have $\mathbb{P}\{y_1 \in B\} = \int_{\{z: f(\pi(y_0))z \in B\}} \psi(dz) = \int_{[f(\pi(y_0))]^{-1}\cdot B} \psi(dz)$. The latter is strictly positive by the strict positivity of ψ (Assumption 2). ■

Proof of Lemma 3 Measurable subsets of C-sets are easily seen to be C-sets. Evidently then it is sufficient to establish that the interval $C_k := [1/k, k]$ is a C-set for every $k \in \mathbb{N}$. Pick and $k \in \mathbb{N}$. By interiority and monotonicity of the optimal policy (Theorem 1) we have

$$0 < f(\pi(1/k)) \leqslant f(\pi(k)) < \infty \quad \forall y \in C_k.$$

Since ψ is continuous and strictly positive it follows that

$$\inf_{C_k \times C_k} \Gamma(y, y') = \inf_{C_k \times C_k} \psi\left(\frac{y'}{f(\pi(y))}\right)\frac{1}{f(\pi(y))} =: r > 0.$$

Now letting ν be the measure defined by $\nu(B) = r \cdot \int_B 1_{C_k}(x)dx$ gives (12.19) for all $y \in C_k$. ■

Proof of Lemma 4 Clearly $\nu(C_k) > 0$ holds for the set C_k and measure ν given in the proof of Lemma 4. ■

Proof of Lemma 5 If Assumption 4 part (i) holds (f concave), then $u' \circ c$ is continuous (see Mirman and Zilcha 1975) and the result is clear. Suppose instead that Assumption 4 part (ii) holds. In that case $u' \circ c$ is again bounded on compacts, this time by Nishimura et al. (2003) (Lemma 5).

References

Amir, R. and I. V. Evstigneev (2000), "A Functional Central Limit Theorem for Equilibrium Paths of Economic Dynamics," *Journal of Mathematical Economics*, 33, 81–99.

Bhattacharya, R. and M. Majumdar (2003), "Dynamical Systems Subject to Random Shocks: An Introduction," *Economic Theory*, 23, 1–13.

Blume, L., Easley, D. and M. O'Hara (1982), "Characterization of Optimal Plans for Stochastic Dynamic Economies," *Journal of Economic Theory*, 28, 221–234.

Boldrin, M. and L. Montrucchio (1986), "On the Indeterminacy of Capital Accumulation Paths," *Journal of Economic Theory*, 40, 26–39.

Brock, W. A. and L. Mirman (1972), "Optimal Economic Growth and Uncertainty: The Discounted Case," *Journal of Economic Theory*, 4, 479–513.

Dechert, W. D. and K. Nishimura (1983), "A Complete Characterization of Optimal Growth Paths in an Aggregated Model with Non-Concave Production Function," *Journal of Economic Theory*, 31, 332–354.

Donaldson, J. B. and R. Mehra (1983), "Stochastic Growth with Correlated Production Shocks," *Journal of Economic Theory*, 29, 282–312.

Duffie, D. and K. J. Singleton (1993), "Simulated Moments Estimation of Markov Models of Asset Prices," *Econometrica*, 61, 929–952.

Evstigneev, I. V. and S. D. Flåm (1997), The Turnpike Property and the Central Limit Theorem in Stochastic Models of Economic Dynamics, in *Statistics and Control of Stochastic Processes*, ed. by Kabanov, Y. M., Rozovskii, B. L. and A. N. Shiryaev, World Scientific, London, 63–101.

Futia, C. A. (1982), "Invariant Distributions and the Limiting Behavior of Markovian Economic Models," *Econometrica*, 50, 377–408.

Hopenhayn, H. A. and E. C. Prescott (1992), "Stochastic Monotonicity and Stationary Distributions for Dynamic Economies," *Econometrica*, 60, 1387–1406.

Kamihigashi, T. (2003), Almost Sure Convergence to Zero in Stochastic Growth Models, mimeo, RIEB, Kobe University.

Majumdar, M. and T. Mitra (1982), "Intertemporal Allocation with Nonconvex Technology," *Journal of Economic Theory*, 27, 101–136.

Majumdar, M., Mitra, T. and Y. Nyarko (1989), Dynamic Optimization Under Uncertainty: Non-convex feasible set, in *Joan Robinson and Modern Economic Theory*, ed. by Feiwel, G. R., MacMillan Press, New York.

Meyn, S. P. and R. L. Tweedie (1993), Markov Chains and Stochastic Stability, Springer-Verlag: London.

Mirman, L. J. (1970), "Two Essays on Uncertainty and Economics," Ph.D. Thesis, University of Rochester.

Mirman, L. J. and I. Zilcha (1975), "On Optimal Growth under Uncertainty," *Journal of Economic Theory*, 11, 329–339.

Mitra, T. and S. Roy (2004), Optimal Exploitation of Resources under Uncertainty and the Extinction of Species, mimeo.

Nishimura, K., Rudnicki, R. and J. Stachurski (2003), Stochastic Growth with Nonconvexities: The Optimal Case, mimeo, Kyoto Institute of Economic Research.

Skiba, A. K. (1978), "Optimal Growth with a Convex-Concave Production Function," *Econometrica*, 46, 527–539.

Stachurski, J. (2002), "Stochastic Optimal Growth with Unbounded Shock," *Journal of Economic Theory*, 106, 40–65.

Stachurski, J. (2003), "Stochastic Growth: Asymptotic Distributions," *Economic Theory*, 21, 913–919.

Stokey, N. L., Lucas, R. E. and E. C. Prescott (1989), Recursive Methods in Economic Dynamics, Harvard University Press, Massachusetts.

Part VI
Indeterminacy in Exogenous Growth Models

Chapter 13
Indeterminacy and Sunspots with Constant Returns*

Jess Benhabib and Kazuo Nishimura**

13.1 Introduction

Recently there has been a renewed interest in the possibility of indeterminacy and sunspots, or alternatively put, in the existence of a continuum of equilibria that arises in dynamic economies with some market imperfections.[1] Much of the research in this area has been concerned with the empirical plausibility of indeterminacy in markets with external effects or with monopolistic competition and which exhibit some degree of increasing returns. While the early results on indeterminacy relied on relatively large increasing returns and high markups, more recently Benhabib and Farmer (1996a) showed that indeterminacy can also occur in two-sector

***Journal of Economic Theory* 81,** 58–96, 1998

**We thank Jordi Gali, Stephanie Schmitt-Grohé, Sharon Harrison, Takashi Kamihigashi, Levent Kockesen, and Roberto Perli for very useful discussions and comments. Technical support from the C.V. Starr Center for Applied Economics at New York University is gratefully acknowledged.

[1]A long but incomplete list of the recent literature includes Beaudry and Devereux (1993), Benhabib and Farmer (1994, 1996a), Benhabib and Perli (1994), Benhabib et al. (1994), Boldrin and Rustichini (1994), Chatterjee and Cooper (1989), Christiano and Harrison (1996), Farmer and Guo (1994, 1995), Gali (1994), Perli (1994), Rotemberg and Woodford (1992), Schmitt-Grohé (1997), Weder (1996) and Xie (1994).

J. Benhabib (✉)
Department of Economics, New York University, USA
e-mail: jess.benhabib@nyu.edu

K. Nishimura
Institute of Economic Research, Kyoto University, Japan
e-mail: nishimura@kier.kyoto-u.ac.jp

J. Stachurski et al. (eds.), *Nonlinear Dynamics in Equilibrium Models*,
DOI 10.1007/978-3-642-22397-6_13,

models with small sector-specific external effects and very mild increasing returns.[2] Nevertheless, a number of empirical researchers, refining the earlier findings of Hall (1988, 1990) on disaggregated US data, have concluded that returns to scale seem to be roughly constant, if not decreasing.[3] While one can argue whether the degree of increasing returns required for indeterminacy in Benhabib and Farmer (1996a) falls within the standard errors of these recent empirical estimates, one may also ask whether increasing returns are at all needed for indeterminacy to arise in a plausible manner. The purpose of this paper is to give a negative answer to this question, and to show how indeterminacy can occur in a standard growth model with constant social returns, decreasing private returns, small or negligible external effects, and standard parameter values that are typically used in the literature on business cycles. Furthermore we will shows that it is possible to realistically calibrate such a model and to obtain a reasonably good match to the moments of aggregate US data.

Indeterminacy or multiple equilibria emerges in dynamic models with small market distortions as a type of coordination problem. Roughly speaking, what is needed for indeterminacy is a mechanism such that, starting from an arbitrary equilibrium, if all agents were to simultaneously increase their investment in an asset, the rate of return on the asset would tend to increase, and in turn set off relative price changes that would drive the economy back towards a stationary equilibrium. One such simple mechanism in one-sector models is increasing returns, typically sustained in a market context via external effects or monopolistic competition. In a multisector model, however, the rates of return and marginal products depend not only on stocks of assets, but also on the composition of output across sectors. Increasing the production and the stock of a capital asset, say due to an increase in its price, may well increase its rate of return. It is possible therefore to have constant aggregate returns in all sectors at the social level, and to still obtain indeterminacy if there are minor or even negligible external effects in some of the sectors. A more detailed intuition for indeterminacy is given at the end of Sect. 13.2 in the case of a simple two-sector model.

Constant social returns coupled with small external effects implies that some sectors must have a small degree of decreasing returns at the private level. This is in contrast to models of indeterminacy with social, increasing but private, constant returns to scale. An implication of decreasing private returns is of course positive profits. In the parameterized examples given in the sections below, these profits will be quite small because the size of external effects, and therefore the degree of decreasing returns needed for indeterminacy will also be small. Nevertheless positive profits would invite entry, and unless the number of firms are fixed, a fixed cost of entry must be assumed to determine the number of firms along the

[2]Since Benhabib and Farmer (1996a) postulate constant returns at the private level, we can measure increasing returns as the sum of all Cobb-Douglas coefficients minus one. Indeterminacy then, for standard parametrization, requires increasing returns of about 0.07.

[3]See for example Basu and Fernald (1994a,b). Burnside et al. (1995), or Burnside (1996), among others.

equilibrium. Such a market structure would then exhibit increasing private marginal costs but constant social marginal costs, which is in line with current empirical work on this subject (see footnote 3, above). It seems therefore that models of indeterminacy based on market imperfections which drive a wedge between private and social returns must have some form of increasing returns, no matter how small, either in variable costs (as in some of the earlier models of indeterminacy), or through a type of fixed cost that prevents entry in the face of positive profits (see also Gali 1994; Gali and Zilibotti 1995). The point is that while some small wedge between private and social returns is necessary for indeterminacy, this in no way requires decreasing marginal costs or increasing marginal returns in production.

For reasons also given at the end of Sect. 13.2, indeterminacy can arise in a constant returns two-sector economy only if the utility of consumption is close to linear. In order to calibrate the model with standard parameters for production and preferences we need a three sector model. Section 13.3 presents such a model in a continuous time framework. In Sect. 13.3.2 we show that this model easily gives rise to indeterminacy with standard parametrizations for utility functions, labor supply elasticities, discount and depreciation rates, and factor shares. Much of the derivations are relegated to Appendix I.

In Sect. 13.4 we present the stochastic, discrete-time version of our model and we calibrate it. We construct some simple sunspot equilibria and show that we can easily find standard parametrizations of our Cobb-Douglas technology and preferences to reasonably match the various moments of US data. The full derivations for this case are given in Appendix II.

13.2 The Two-Sector Model

13.2.1 Basic Structure

We model an economy having an infinitely-lived representative agent with instantaneous utility given by

$$U(c) = (1-\sigma)^{-1} c^{(1-\sigma)} - (1+v)^{-1} L^{(1+v)} \qquad \sigma, v \geqslant 0$$

where c is consumption, L is labor supply, v^{-1} is the labor supply elasticity and σ is the intertemporal elasticity of substitution in consumption. For simplicity of exposition we will start with a two-sector rather than an n-sector Cobb-Douglas production technology with consumption goods c, and investment goods x. The agent's optimization problem will be given by

$$\max \int_0^\infty \left(U(q_c L_c^{\alpha_0} K_c^{\alpha_1} \overline{L_c^{a_0} K_c^{a_1}}) - (1+v)^{-1} L^{(1+v)} \right) e^{-(r-g)t} dt \tag{13.1}$$

with respect to K_c, L_c, K_x, L_x and subject to

$$x = q_x L_x^{\beta_0} K_x^{\beta_1} \overline{L_x^{b_0} K_x^{b_1}}, \tag{13.2}$$

$$c = q_c L_c^{\alpha_0} K_c^{\alpha_1} \overline{L_c^{a_0} K_c^{a_1}},$$

$$\frac{dk}{dt} = x - gk, \tag{13.3}$$

$$K_x + K_c = k, \quad L_x + L_c = L, \tag{13.4}$$

with initial stock of k given. The components of the production functions, $\overline{L_c^{b_0} K_x^{b_1}}$ for x, and $\overline{L_c^{a_0} K_c^{a_1}}$ for c, represent output effects that are external, and are viewed as functions of time by the agent.

We can write the Hamiltonian as follows:

$$H = U(q_c L_c^{\alpha_0} K_c^{\alpha_1} \overline{L_c^{a_0} K_c^{a_1}}) - (1+v)^{-1} L^{(1+v)} + \overline{p}(q_x L_x^{\beta_0} K_x^{\beta_1} \overline{L_x^{b_0} K_x^{b_1}} - gk)$$
$$+ \overline{w}_0(L - L_x - L_c) + \overline{w}(k - K_x - K_c).$$

Here $\overline{p}$, $\overline{w}_0$, and $\overline{w}$ are the Lagrange multipliers which will represent the utility price of the capital good x, the rental rates of capital goods,and the wage rate of labor, all in terms of the price of the consumption good c. The first-order conditions with respect to K_c, L_c, K_x, L_x yield the following

$$\overline{w}_0 = U' \alpha_0 q_c L_c^{\alpha_0 + a_0 - 1} K_c^{\alpha_1 + a_1} = \overline{p} \beta_0 q_x L_x^{\beta_0 + b_0 - 1} K_x^{\beta_1 + b_1},$$
$$\overline{w} = U' \alpha_1 q_c L_c^{\alpha_0 + a_0} K_c^{\alpha_1 + a_1 - 1} = \overline{p} \beta_1 q_x L_x^{\beta_0 + b_0} K_x^{\beta_1 + b_1 - 1}.$$

If we define

$$\overline{w}_0 = U' w_0, \quad \overline{w} = U' w, \quad \overline{p} = U' p,$$

then the first-order conditions become

$$w_0 = \alpha_0 q_c L_c^{\alpha_0 + a_0 - 1} K_c^{\alpha_1 + a_1} = p \beta_0 q_x L_x^{\beta_0 + b_0 - 1} K_x^{\beta_1 + b_1}, \tag{13.5}$$
$$w = \alpha_1 q_c L_c^{\alpha_0 + a_0} K_c^{\alpha_1 + a_1 - 1} = p \beta_1 q_x L_x^{\beta_0 + b_0} K_x^{\beta_1 + b_1 - 1}. \tag{13.6}$$

The first-order conditions with respect to L, after combining with the others, give the labor market equilibrium condition:

$$c^{(1-\sigma)} \alpha_0 L_c^{-1} = L^v. \tag{13.7}$$

If we assume constant returns at the social level, we have

$$\alpha_0 + \alpha_1 + a_0 + a_1 = \beta_0 + \beta_1 + b_0 + b_1 = 1.$$

The equations of motion for the system are given by

$$\left(\frac{dk}{dt}\right) = x - gk, \tag{13.8}$$

$$\left(\frac{d(U'p)}{dt}\right) = U'(c)(rp - w), \tag{13.9}$$

Then (13.9) can be written as

$$\begin{aligned}
\frac{dp}{dt} &= rp - w(p,k) - p\frac{U''(c)\left[\frac{dc}{\partial p}\frac{dp}{dt} + \frac{\partial c}{\partial k}\frac{dk}{dt}\right]}{U'(c)} \qquad (13.10)\\
&= \left[1 + p\left(\frac{U''(c)}{U'(c)}\right)\left(\frac{\partial c}{\partial p}\right)\right]^{-1}\left[rp - w(p,k) - p\frac{U''(c)\left[\frac{\partial c}{\partial k}\frac{dk}{dt}\right]}{U'(c)}\right]\\
&= \left[1 - \sigma\left(\frac{p}{c}\right)\left(\frac{\partial c}{\partial p}\right)\right]^{-1} \cdot \left[rp - w(p,k) + \sigma\left(\frac{p}{c}\right)\left(\frac{\partial c}{\partial k}\right)(x(p,k) - gk)\right],
\end{aligned}$$

where $\sigma = (-U''(c)c/U'(c))$. With logarithmic utility of consumption, we have, of course, $\sigma = 1$. The first-order conditions given by (13.5)–(13.7), and the equations of motion given by (13.8) and (13.10) completely describe the system.

13.2.2 Two-Sector Dynamics

The Jacobian matrix $[J]$ for the differential equations (13.8) and (13.10) is given by

$$[J] = \begin{bmatrix} \frac{\partial x}{\partial k} - g & \frac{\partial x}{\partial p} \\ \sigma E^{-1}\left(\frac{p}{c}\right)\left(\frac{\partial c}{\partial k}\right)\left(\frac{\partial x}{\partial k} - g\right) & E^{-1}\left[\left(-\frac{\partial w}{\partial p} + r\right) + \sigma\left(\frac{p}{c}\right)\left(\frac{\partial c}{\partial k}\right)\left(\frac{\partial x}{\partial p}\right)\right] \end{bmatrix}, \tag{13.11}$$

where $E = [1 - \sigma(p/c)(\partial c/\partial p)]$. Note that E can be written as one minus the product of two elasticities:

$$E = (1 - \sigma\varepsilon_{cp}) \tag{13.12}$$

where

$$\varepsilon_{cp} \equiv \left(\frac{\widehat{c}}{\widehat{p}}\right) = \left(\frac{p}{c}\frac{\partial c}{\partial p}\right).$$

If we multiply the first row of $[J]$ by $-\sigma E^{-1}p$ and add it to the second, we get a matrix with an unchanged determinant. We have

$$DET\,[J] = \left(\frac{\partial x}{\partial k} - g\right)\left(-\frac{\partial w}{\partial p} + r\right)[1 - \sigma\varepsilon_{cp}]^{-1}.$$

If the utility of consumption is linear, then $\sigma = 0$, and it is easy to see that the roots of the matrix $[J]$ become $((\partial x/\partial k) - g)$ and $((-\partial w/\partial p) + r)$. In Appendix I we show, from (13.82), that

$$\left(\frac{\partial x}{\partial k}\right) = \frac{r\alpha_0\left(1+\left(\frac{L}{L_c}\right)\left(\frac{\sigma}{v}\right)\right)}{\left(\beta_1(1-a_0-a_1)-\alpha_1(1-b_0-b_1)+\beta_1\alpha_0\left(\frac{\sigma}{v}\right)\frac{L}{L_c}\right)}$$
$$= \frac{r\alpha_0\left(1+\left(\frac{L}{L_c}\right)\left(\frac{\sigma}{v}\right)\right)}{\left(\beta_1\alpha_0-\alpha_1\beta_0+\beta_1\alpha_0\left(\frac{\sigma}{v}\right)\frac{L}{L_c}\right)}.$$

If $\sigma = 0$, we have

$$\left(\frac{\partial x}{\partial k} - g\right) = \frac{r\alpha_0}{(\beta_1\alpha_0 - \alpha_1\beta_0)} - g.$$

Similarly, from (13.59) in Appendix I, we have

$$\left(-\frac{\partial w}{\partial p} + r\right) = r\left(-\frac{\alpha_0 + a_0}{\alpha_0 + a_0 - \beta_0 - b_0} + 1\right)$$
$$= \left(\frac{r(\beta_0 + b_0)}{\beta_0 + b_0 - \alpha_0 - a_0}\right)$$
$$= \left(\frac{r(\beta_0 + b_0)}{(\alpha_1 + a_1)(\beta_0 + b_0) - (\alpha_0 + a_0)(\beta_1 + b_1)}\right).$$

The last step above follows from multiplying $(\alpha_0 + a_0)$ in the denominator by $(\beta_0 + \beta_1 + b_0 + b_1)$, which under constant returns equals one, similarly multiplying $(\beta_0 + b_0)$ by $(\alpha_0 + \alpha_1 + a_1 + a_0)$, and cancelling to simplify the denominator. It is easily shown that comparing the ratios of Cobb-Douglas exponents of the production function amounts to comparing factor intensities, since the ratios of exponents determine input ratios. These ratios can be defined either with or without the external effects entering the exponents. We may therefore say that the capital good is labor intensive *from the private perspective* if $(\beta_1\alpha_0 - \alpha_1\beta_0 < 0)$, but that it is capital intensive *from the social perspective* if $((\alpha_1 + a_1)(\beta_0 + b_0) - (\alpha_0 + a_0)(\beta_1 + b_1) < 0)$. The expressions above allow us to state the following simple result:

Proposition 1. *In the two-sector model with $\sigma = 0$, if the capital good is labor intensive from the private perspective, but capital intensive from the social perspective, that is if $(\beta_1\alpha_0 - \alpha_1\beta_0 < 0)$ but $(\alpha_1 + a_1)(\beta_0 + b_0) - (\alpha_0 + a_0)(\beta_1 + b_1) < 0$, then the steady state is indeterminate.*

A simple example illustrates the possibility of indeterminacy in the two sector model, for $\sigma = 0$, and $r > 0$, $g \geqslant 0$, and only a small externality of the capital good in the production of the consumption good. Let

$$\beta_0 = 0.34; \quad b_0 = 0.00; \quad \beta_1 = 0.66; \quad b_1 = 0.0;$$
$$\alpha_0 = 0.30; \quad a_0 = 0.05; \quad \alpha_1 = 0.65; \quad a_1 = 0.0.$$

Then we have

$$\beta_1\alpha_0 - \alpha_1\beta_0 < 0,$$
$$(\alpha_1 + a_1)(\beta_0 + b_0) - (\alpha_0 + a_0)(\beta_1 + b_1) < 0,$$

and therefore both roots of $[J]$ are negative. Note also that without some external effects both of the above conditions cannot hold simultaneously. It is clear nevertheless that examples satisfying the above conditions for indeterminacy can be constructed with arbitrarily small external effects.

To establish the intuition behind this result we note the following. Without external effects, the sign of $(\partial w/\partial p)$ depends on the sign of $(\alpha_0\beta_1 - \alpha_1\beta_0)$, which represents the factor intensity difference between the two goods. This dependence on factor intensities is in fact nothing but an expression of the Stolper-Samuelson theorem. Similarly, without external effects, the sign of term $(\partial x/\partial k)$ also depends on the sign of $(\alpha_0\beta_1 - \alpha_1\beta_0)$, and reflects the Rybczynski theorem. We note from the Rybczynski theorem that this effect of stocks on outputs will, at constant prices, be more than proportional, and since at a steady state $x = gk$, it will be strong enough to overwhelm the term g.[4] It should be clear then that without external effects we have $(\partial w/\partial p) = (\partial x/\partial k)$, so that the roots of $[J]$ will be of opposite sign. The example of indeterminacy works above precisely because, through external effects, it destroys the duality between the Stolper-Samuelson and Rybczynski effects. Since input coefficients are determined by factor prices, a change in aggregate inputs with prices fixed requires an adjustment of output levels to maintain full employment. The adjustment must reflect the structure of the input coefficient matrix, as implied by the Rybczynski Theorem. When there are no externalities the same is true, via Shepard's Lemma, for the effect of input prices on outputs, and this reflects the Stolper-Samuelson Theorem. However, with market distortions, true costs are not being minimized, and Shepard's Lemma no longer holds, breaking the reciprocal relation between the Rybczynski and Stolper-Samuelson effects.[5] We will make use of this point to provide a heuristic explanation of our indeterminacy result.

[4]In the same way, the Stolper-Samuelson theorem implies that $-(\partial w_x/\partial p)$ will overwhelm r, since at a steady state $w_x = rp$, but the expression above for $((-\partial w_x/\partial p) + r)$ already incorporates the steady-state relationship.

[5]We should note that other distortions which interfere with true cost minimization are also likely to give rise to similar results.

To understand the intuition for this indeterminacy result consider first a simple one-sector model. Starting from an arbitrary equilibrium, consider another one with a higher rate of investment. A higher investment rate results in higher stocks and, if there are no increasing returns, in a lower marginal return to capital. The only way that this can be an equilibrium is if the other component of the return, the price (or shadow price) appreciation of capital, offsets the decline in the marginal product and justifies the increased holding of such stocks. This appreciating relative price induces a higher production of the capital good, that is a higher rate of investment. The result is a further decline in the marginal product of capital, which then requires an even higher price appreciation to justify the holding of the higher stocks. Transversality conditions rule out such an equilibrium. If there are increasing returns, however, incorporated into the model through some market imperfections, the higher stock levels increase rather than decrease the marginal product of capital, and this higher return justifies the holding of the higher stocks without requiring explosive price appreciations and violating transversality conditions. Such increasing returns to capital are generally introduced indirectly. In Benhabib and Perli (1994) increasing returns to capital are the result of changes induced in labor supply due to the reallocation of production in favor of investment and capital accumulation. In Gali (1994), Rotemberg and Woodford (1992), or Schmitt-Grohé (1997), increasing returns are the result of countercyclical markups.

In a two-sector model another mechanism leading to indeterminacy becomes operational. The return to capital now depends on the composition of output as well as the level of the stock. Let us first consider the case without external effects. Take the case where the capital good is capital intensive, and, again starting from an equilibrium, consider an increase in the rate of investment above the level of its initial equilibrium, induced by an instantaneous increase in the relative price of the investment good. An increase in the stock of capital at constant prices would, from the Rybczynski theorem, lead to a more than proportional rise in its output. From the Stolper-Samuelson theorem, the initial price rise leads to an increase in rate of return of capital given by w, and to maintain the equality of the overall return to capital and the discount rate, the price of the investment good must decline. However this is not enough to check the Rybczynski effect: The increasing capital stock leads to further expansions of investment output despite the retreat of prices towards the steady state levels, and investment output becomes explosive.[6]

To get indeterminacy without relying on increasing returns, there must be a mechanism to nullify the duality between the Rybczynski and Stolper-Samuelson theorems. This is precisely what happens in the two-sector model above in the presence of external effects, and is illustrated by Proposition 1. When the investment good is labor intensive from the private perspective, an increase in the capital stock decreases its output at constant prices through the Rybczynski effect. This checks

[6]The argument for the case in which the capital good is labor intensive is similarly based on the Rybczynski and Stolper-Samuelson theorems, but in this case departing from the initial equilibrium trajectory leads to explosive prices instead of outputs.

the output side. The Stolper-Samuelson theorem, however, operates through the "social" factor intensities, and the investment good is capital intensive from the social perspective. The initial rise in its price causes an increase in one of the components of its return, w, and requires a price decline to maintain the overall return to capital equal to the discount rate. This offsets the initial rise in the relative price of the investment good and prices also reverse direction toward the steady state. Therefore, in the two-sector model, indeterminacy requires the destruction of the duality between the Rybczynski and Stolper-Samuelson effects through the introduction of market imperfections.[7]

Why then do we have to resort to a three-sector model to generate examples of indeterminacy that are empirically plausible? The problem in the two-sector model arises because, when we consider constructing an alternative equilibrium with a higher investment rate, we must initially curtail consumption. If there is some curvature on the utility function, the desire to smooth consumption over time can overwhelm the effects described above (A formal demonstration of this, in terms of the roots of $[J]$, is tedious, but is available from the authors on request). When a third nonconsumption good is introduced however, indeterminacy can arise from compositional changes in outputs, without severely affecting the output of the consumption. Therefore, with a third sector it becomes possible to construct examples of indeterminacy with $\sigma \geqslant 1$, whereas in the two-sector model indeterminacy seems to hold for values of σ in a narrow range above 0.[8] Therefore, for a realistic parametrization and calibration of indeterminacy, we must turn to a three-sector model.

Before focusing on the three-sector model, it may be useful to briefly compare our results to the other two-sector models in the literature. The model of Benhabib and Farmer (1996a) uses a two-sector model with sector specific externalities, but the production functions in the two sectors are identical so that compositional changes in production can affect returns only because of increasing returns in the form of sector specific external effects. The model of Gali (1994) combines a setup of monopolistic competition with variable markups. Output is divided into a consumption and an investment good, and the composition of this division affects average markups and profits because the monopolistic competitors face demand curves that have different slopes for the consumption and the investment goods.

[7] A generalization of the results to a multisector framework will follow from the roots of $(\partial x/\partial k)$ and $(\partial w/\partial p)$, which now are matrices, and which are equal to each other if there are no external effects. It can be shown that only $(\partial w/\partial p)$ depends on external effects. Given any $r > g$, it is then possible to construct robust families of Cobb-Douglas technologies giving rise to indeterminacy for arbitrarily small external effects.

[8] With nonlinear utility we will have indeterminacy in the two-sector model if the trace of $[J]$ is negative and its determinant is positive. It can in fact be shown that the trace will be negative under the assumptions of the proposition above because $(\partial c/\partial k)(\partial x/\partial p)$ will be positive. For a positive determinant we must also assume that the term E in (13.12) is positive. However ε_{cp} appearing in E is endogenous. It can be computed as a function of parameters, and it is possible to produce examples of indeterminate steady states for positive but small values of σ.

The magnitude of average markups required, however, is large (see Schmitt-Grohé 1997). Gali's model is related to a model of Rotemberg and Woodford (1992), which is also analyzed by Schmitt-Grohé (1997). The Rotemberg-Woodford model has a variable markup that depends on aggregate economic activity, rather than a composition effect as in Gali (1994).

13.3 The Three-Sector Model

13.3.1 The Basic Structure

We again model an economy having an infinitely-lived representative agent with instantaneous utility given by

$$U(c) = (1-\sigma)^{-1} c^{(1-\sigma)} - (1+v)^{-1} L^{(1+v)} \qquad \sigma, v \geqslant 0$$

where c is consumption, L is labor supply, v^{-1} is the labor supply elasticity and σ is the intertemporal elasticity of substitution in consumption. For simplicity of exposition we construct a three-sector rather than an n-sector Cobb-Douglas production technology with a consumption good c, and two investment goods, x and y. The agent's optimization problem is given by

$$\max \int_0^{\infty} (U(q_c L_c^{\alpha_0} K_{xc}^{\alpha_1} K_{yc}^{\alpha_2} \overline{L_c^{a_0} K_{xc}^{a_1} K_{yc}^{a_2}}) - (1+v)^{-1} L^{(1+v)}) e^{-(r-g)t} dt \quad (13.13)$$

with respect to K_{xc}, K_{yc}, L_c, L_y, K_{xy}, K_{xx}, K_{yx}, K_{yy}, and subject to

$$x = q_x L_x^{\beta_0} K_{xx}^{\beta_1} K_{yx}^{\beta_2} \overline{L_x^{b_0} K_{xx}^{b_1} K_{yx}^{b_2}} \quad (13.14)$$

$$y = q_y L_y^{\gamma_0} K_{xy}^{\gamma_1} K_{yy}^{\gamma_2} \overline{L_y^{c_0} K_{xy}^{c_1} K_{yy}^{c_2}} \quad (13.15)$$

$$c = q_c L_c^{\alpha_0} K_{xc}^{\alpha_1} K_{yc}^{\alpha_2} \overline{L_c^{a_0} K_{xc}^{a_1} K_{yc}^{a_2}} \quad (13.16)$$

$$\frac{dk_x}{dt} = x - gk_x,$$

$$\frac{dk_y}{dt} = y - gk_y,$$

$$K_{xx} + K_{xy} + K_{xc} = k_x;$$

$$K_{yx} + K_{yy} + K_{yc} = k_y; \quad (13.17)$$

$$L_x + L_y + L_c = L,$$

with the initial stocks of k_x and k_y given. The components of the production functions, $\overline{L_x^{b_0} K_{xx}^{b_1} K_{yx}^{b_2}}$ for x, $\overline{L_y^{c_0} K_{xy}^{c_1} K_{yy}^{c_2}}$ for y, and $\overline{L_c^{a_0} K_{xc}^{a_1} K_{yc}^{a_2}}$ for c, represent output effects that are external and are viewed as functions of time by the agent.

We can write the Hamiltonian as follows:

$$
\begin{aligned}
H = {} & U(q_c L_c^{\alpha_0} K_{xc}^{\alpha_1} K_{yc}^{\alpha_2} \overline{L_c^{a_0} K_{xc}^{a_1} K_{yc}^{a_2}}) - (1+v)^{-1} L^{(1+v)} \\
& + \overline{p}_x (q_x L_x^{\beta_0} K_{xx}^{\beta_1} K_{yx}^{\beta_2} \overline{L_x^{b_0} K_{xx}^{b_1} K_{yx}^{b_2}} - g k_x) \\
& + \overline{p}_y (q_y L_y^{\gamma_0} K_{xy}^{\gamma_1} K_{yy}^{\gamma_2} \overline{L_y^{c_0} K_{xy}^{c_1} K_{yy}^{c_2}} - g k_y) \\
& + \overline{w}_0 (L - L_x - L_y - L_c) + \overline{w}_x (k_x - K_{xx} - K_{xy} - K_{xc}) \\
& + \overline{w}_y (k_y - K_{yx} - K_{yy} - K_{yc}).
\end{aligned}
$$

Here $\overline{p}_x, \overline{p}_y, \overline{w}_0, \overline{w}_x$ and $\overline{w}_y$ the Lagrange multipliers which will represent the utility prices of the capital goods x and y, the rental rates of capital goods, and the wage rate of labor, all in terms of the price of the consumption good c. The first-order conditions, with respect to the inputs, are

$$
\begin{aligned}
\overline{w}_0 &= U' \alpha_0 q_c L_c^{\alpha_0 + a_0 - 1} K_{xc}^{\alpha_1 + a_1} K_{yc}^{\alpha_2 + a_2} \\
&= \overline{p}_x \beta_0 q_x L_x^{\beta_0 + b_0 - 1} K_{xx}^{\beta_1 + b_1} K_{yx}^{\beta_2 + b_2} \\
&= \overline{p}_y \gamma_0 q_y L_y^{\gamma_0 + c_0 - 1} K_{xy}^{\gamma_1 + c_1} K_{yy}^{\gamma_2 + c_2}, \\
\overline{w}_x &= U' \alpha_1 q_c L_c^{\alpha_0 + a_0} K_{xc}^{\alpha_1 + a_1 - 1} K_{yc}^{\alpha_2 + a_2} \\
&= \overline{p}_x \beta_1 q_x L_x^{\beta_0 + b_0} K_{xx}^{\beta_1 + b_1 - 1} K_{yx}^{\beta_2 + b_2} \\
&= \overline{p}_y \gamma_1 q_y L_y^{\gamma_0 + c_0} K_{xy}^{\gamma_1 + c_1 - 1} K_{yy}^{\gamma_2 + c_2}, \\
\overline{w}_y &= U' \alpha_2 q_c L_c^{\alpha_0 + a_0} K_{xc}^{\alpha_1 + a_1} K_{yc}^{\alpha_2 + a_2 - 1} \\
&= \overline{p}_x \beta_2 q_x L_x^{\beta_0 + b_0} K_{xx}^{\beta_1 + b_1} K_{yx}^{\beta_2 + b_2 - 1} \\
&= \overline{p}_y \gamma_2 q_y L_y^{\gamma_0 + c_0} K_{xy}^{\gamma_1 + c_1} K_{yy}^{\gamma_2 + c_2 - 1}.
\end{aligned}
$$

If we define

$$
\overline{w}_0 = U' w_0, \quad \overline{w}_x = U' w_x, \quad \overline{w}_y = U' w_y, \quad \overline{p}_x = U' p_x, \quad \overline{p}_y = U' p_y,
$$

then the first-order conditions become:

$$
\begin{aligned}
w_0 &= \alpha_0 q_c L_c^{\alpha_0 + a_0 - 1} K_{xc}^{\alpha_1 + a_1} K_{yc}^{\alpha_2 + a_2} \\
&= p_x \beta_0 q_x L_x^{\beta_0 + b_0 - 1} K_{xx}^{\beta_1 + b_1} K_{yx}^{\beta_2 + b_2} \\
&= p_y \gamma_0 q_y L_y^{\gamma_0 + c_0 - 1} K_{xy}^{\gamma_1 + c_1} K_{yy}^{\gamma_2 + c_2},
\end{aligned}
\tag{13.18}
$$

$$\begin{aligned} w_x &= \alpha_1 q_c L_c^{\alpha_0+a_0} K_{xc}^{\alpha_1+a_1-1} K_{yc}^{\alpha_2+a_2} \\ &= p_x \beta_1 q_x L_x^{\beta_0+b_0} K_{xx}^{\beta_1+b_1-1} K_{yx}^{\beta_2+b_2} \\ &= p_y \gamma_1 q_y L_y^{\gamma_0+c_0} K_{xy}^{\gamma_1+c_1-1} K_{yy}^{\gamma_2+c_2}, \end{aligned} \tag{13.19}$$

$$\begin{aligned} w_y &= \alpha_2 q_c L_c^{\alpha_0+a_0} K_{xc}^{\alpha_1+a_1} K_{yc}^{\alpha_2+a_2-1} \\ &= p_x \beta_2 q_x L_x^{\beta_0+b_0} K_{xx}^{\beta_1+b_1} K_{yx}^{\beta_2+b_2-1} \\ &= p_y \gamma_2 q_y L_y^{\gamma_0+c_0} K_{xy}^{\gamma_1+c_1} K_{yy}^{\gamma_2+c_2-1}. \end{aligned} \tag{13.20}$$

The first-order conditions with respect to L, after combining with the others, give the labor market equilibrium condition:

$$c^{(1-\sigma)} \alpha_0 L_c^{-1} = L^{\nu}. \tag{13.21}$$

If we assume constant returns at the social level, we have

$$\begin{aligned} \alpha_0 + \alpha_1 + \alpha_2 + a_0 + a_1 + a_2 &= \beta_0 + \beta_1 + \beta_2 + b_0 + b_1 + b_2 \\ &= \gamma_0 + \gamma_1 + \gamma_2 + c_0 + c_1 + c_2 = 1. \end{aligned}$$

The four equations of motion for the system are given by

$$\begin{pmatrix} \frac{dk_x}{dt} \\ \frac{dk_y}{dt} \end{pmatrix} = \begin{pmatrix} x \\ y \end{pmatrix} - g \begin{pmatrix} k_x \\ k_y \end{pmatrix}, \tag{13.22}$$

$$\begin{pmatrix} \frac{d(U' p_x)}{dt} \\ \frac{d(U' p_y)}{dt} \end{pmatrix} = rU'(c) \begin{pmatrix} p_x \\ p_y \end{pmatrix} - U'(c) \begin{pmatrix} w_x \\ w_y \end{pmatrix}. \tag{13.23}$$

To simplify (13.23), we define the following two matrices:

$$\left[\frac{\partial c}{\partial p}\right] = \begin{bmatrix} \frac{\partial c}{\partial p_x} & \frac{\partial c}{\partial p_y} \\ \frac{\partial c}{\partial p_x} & \frac{\partial c}{\partial p_y} \end{bmatrix}, \quad \left[\frac{\partial c}{\partial k}\right] = \begin{bmatrix} \frac{\partial c}{\partial k_x} & \frac{\partial c}{\partial k_y} \\ \frac{\partial c}{\partial k_x} & \frac{\partial c}{\partial k_y} \end{bmatrix}.$$

Evaluated at the steady state, where quantities and prices are stationary, (13.23) can be written as

$$\begin{aligned} \begin{pmatrix} \frac{dp_x}{dt} \\ \frac{dp_y}{dt} \end{pmatrix} &= r \begin{pmatrix} p_x \\ p_y \end{pmatrix} - \begin{pmatrix} w_x \\ w_y \end{pmatrix} - \left(\frac{U''(c)c}{U'(c)} \right) \begin{bmatrix} \frac{p_x}{c} & 0 \\ 0 & \frac{p_y}{c} \end{bmatrix} \\ &\quad \times \left(\left[\frac{\partial c}{\partial p}\right] \begin{pmatrix} \frac{dp_x}{dt} \\ \frac{dp_y}{dt} \end{pmatrix} + \left[\frac{\partial c}{\partial k}\right] \begin{pmatrix} \frac{dk_x}{dt} \\ \frac{dk_y}{dt} \end{pmatrix} \right) \end{aligned}$$

$$= \left[I - \sigma \begin{bmatrix} \frac{p_x}{c} & 0 \\ 0 & \frac{p_y}{c} \end{bmatrix} \left[\frac{\partial c}{\partial p} \right] \right]^{-1} \tag{13.24}$$

$$\cdot \left(r \begin{pmatrix} p_x \\ p_y \end{pmatrix} - \begin{pmatrix} w_x \\ w_y \end{pmatrix} + \sigma \begin{bmatrix} \frac{p_x}{c} & 0 \\ 0 & \frac{p_y}{c} \end{bmatrix} \left[\frac{\partial c}{\partial k} \right] \left(\begin{pmatrix} (x - gk_x) \\ (y - gk_y) \end{pmatrix} \right) \right),$$

where $\sigma = (-U''(c)c/U'(c))$. With logarithmic utility of consumption, we have, of course, $\sigma = 1$. The first-order conditions given by (13.18)–(13.21), and the equations of motion given by (13.22) and (13.24), completely describe the system.

13.3.2 Three-Sector Dynamics

We now linearize the dynamical system given by (13.22) and (13.24), and we evaluate the associated Jacobian $[JN]$ at the steady state. Let

$$[JN] = \begin{bmatrix} \left[\frac{\partial \dot{k}}{\partial k} \right] & \left[\frac{\partial \dot{k}}{\partial p} \right] \\ \left[\frac{\partial \dot{p}}{\partial k} \right] & \left[\frac{\partial \dot{p}}{\partial p} \right] \end{bmatrix}$$

where

$$\left[\frac{\partial \dot{k}}{\partial k} \right] = [[YK] - gI],$$

$$\left[\frac{\partial \dot{k}}{\partial p} \right] = [YP],$$

$$\left[\frac{\partial \dot{p}}{\partial k} \right] = [D]^{-1} \sigma \begin{bmatrix} \frac{p_x}{c} & 0 \\ 0 & \frac{p_y}{c} \end{bmatrix} \left[\frac{\partial c}{\partial k} \right] [[YK] - gI],$$

$$\left[\frac{\partial \dot{p}}{\partial p} \right] = [D]^{-1} \left[\sigma \begin{bmatrix} \frac{p_x}{c} & 0 \\ 0 & \frac{p_y}{c} \end{bmatrix} \left[\frac{\partial c}{\partial k} \right] [YP] + [rI - [WP]] \right],$$

and where

$$[D] = \left[I - \sigma \begin{bmatrix} \frac{p_x}{c} \\ 0 \end{bmatrix} \left[\frac{\partial c}{\partial p} \right] \right], \tag{13.25}$$

$$[WP] = \begin{bmatrix} \left[\frac{\partial w_x}{\partial p_x} \right] & \left[\frac{\partial w_x}{\partial p_y} \right] \\ \left[\frac{\partial w_y}{\partial p_x} \right] & \left[\frac{\partial w_y}{\partial p_y} \right] \end{bmatrix}, \tag{13.26}$$

$$[YK] = \begin{bmatrix} \left[\frac{\partial x}{\partial k_x}\right] & \left[\frac{\partial x}{\partial k_y}\right] \\ \left[\frac{\partial y}{\partial k_x}\right] & \left[\frac{\partial y}{\partial k_y}\right] \end{bmatrix}, \tag{13.27}$$

$$[YP] = \begin{bmatrix} \left[\frac{\partial x}{\partial p_x}\right] & \left[\frac{\partial x}{\partial p_y}\right] \\ \left[\frac{\partial y}{\partial p_x}\right] & \left[\frac{\partial y}{\partial p_y}\right] \end{bmatrix}, \tag{13.28}$$

$$\left[\frac{\partial c}{\partial p}\right] = \begin{bmatrix} \left[\frac{\partial c}{\partial p_x}\right] & \left[\frac{\partial c}{\partial p_y}\right] \\ \left[\frac{\partial c}{\partial p_x}\right] & \left[\frac{\partial c}{\partial p_y}\right] \end{bmatrix}, \tag{13.29}$$

$$\left[\frac{\partial c}{\partial k}\right] = \begin{bmatrix} \left[\frac{\partial c}{\partial k_x}\right] & \left[\frac{\partial c}{\partial k_y}\right] \\ \left[\frac{\partial c}{\partial k_x}\right] & \left[\frac{\partial c}{\partial k_y}\right] \end{bmatrix}. \tag{13.30}$$

In the appendix we show how the elements of the matrices $[WP]$, $[YK]$, $[YP]$, $[\partial c/\partial p]$, and $[\partial c/\partial k]$ can be evaluated using the steady state output elasticities and the steady-state values of the prices and quantities, all of which can be expressed in terms of the parameters of the economy. It is therefore possible to evaluate the roots of the Jacobian $[JN]$ at the steady state and check for indeterminacy, that is, to check for parameter values that yield more than two roots of $[JN]$ with negative real parts. The production parameters given below easily generate indeterminacy for our three sector economy, where the discount rate is $r = 0.05$, the population growth rate is $g = 0.01$, the intertemporal elasticity of substitution in consumption is $\sigma = 1$ (which implies logarithmic utility in consumption), and the inverse labor supply elasticity is $v = 1$.

Parameters for the Consumption Good c and Investment Goods x and y.

$q_c = 1$, $\alpha_0 = 0.66$; $\alpha_1 = 0.24$, $\alpha_2 = 0.1$; $a_0 = 0.00$; $a_1 = 0.00$, $a_2 = 0.00$;
$q_x = 1$, $\beta_0 = 0.64$; $\beta_1 = 0.20$, $\beta_2 = 0.1$; $b_0 = 0.00$; $b_1 = 0.06$; $b_2 = 0.00$;
$q_y = 1$, $\gamma_0 = 0.61$; $\gamma_1 = 0.23$, $\gamma_2 = 0.1$; $c_0 = 0.00$; $c_1 = 0.00$, $c_2 = 0.06$.

Roots of the Jacobian $[JN]$. $-10.8767, -0.0599, -0.8579, 0.1138$

Clearly, the parameters above are entirely standard, and the external effects, which are present only for the capital input x in the production of x, and for the capital input y in the production of y, both of which are set to 0.06, are extremely small. The production functions of the three goods differ only slightly, with the labor share in each one of them given by α_0, β_0, and γ_0, all at roughly equal to $\frac{2}{3}$. There are constant returns to scale at the social level and at the private level the agents face very slight diminishing returns to scale in producing x and y. Let us emphasize that the degree of perceived private decreasing returns in x and y is indeed negligibly small, which from the social perspective the Cobb-Douglas exponents add up to 1 and from the private perspective they add up to 0.94 for both x and y. Furthermore, indeterminacy seems very robust: Small variations in σ, v, r, g, or in the production parameters do not change the values of the roots by much, nor do they change their sign pattern. Eliminating the external effects completely, however, does eliminate indeterminacy, as expected, because the private and social optimum coincide in that case.[9]

13.4 The Stochastic Discrete Time Model and Calibration

The discrete time problem can be defined as

$$V(k_x, k_y, z) = Max\left(\frac{1}{1-\sigma}\right)(z_c q_c L_c^{\alpha_0} K_{xc}^{\alpha 1} K_{yc}^{\alpha_2} \overline{L_c^{a_0} K_{xc}^{a_1} K_{yc}^{a_2}})^{(1-\sigma)}$$

$$- (1+v)^{-1} L^{(1+v)} + pEV((1-g_x)k_x + x, (1-g_y)k_y + y, z'),$$

$$x = z_x q_x L_x^{\beta_0} K_{xx}^{\beta_1} K_{yx}^{\beta_2} \overline{L_x^{b_0} K_{xx}^{b_2} K_{yx}^{b_2}}, \tag{13.31}$$

$$y = z_y q_y L_y^{\gamma_0} K_{xy}^{\gamma_1} K_{yy}^{\gamma_2} \overline{L_y^{c_0} K_{xy}^{c_1} K_{yy}^{c_2}}, \tag{13.32}$$

where $\rho = (1 + (r - g))^{-1}$ is the discount factor, g_x and g_y are depreciation rates, $z = (z_c, z_x, z_y)$, z_i is a technology shock for $i = c, x, y$, where $\ln z_i = \zeta_i$, and

$$\zeta_{i,t+1} = \lambda_i \zeta_{i,t} + \widehat{\varepsilon}_{i,t+1}, \quad 0 \leqslant \lambda_i \leqslant 1, \tag{13.33}$$

$i = c, x, y$, and $\widehat{\varepsilon}_{i,t+1}$ is iid, normally distributed, and has means zero. z' is the value attained by z in the subsequent period. Note that we can write the consumption output as

$$c = z_c q_c \cdot (L - L_x - L_y)^{\alpha_0 + a_0}(k_x - K_{xx} - K_{xy})^{\alpha_1 + a_1}(k_y - K_{yx} - K_{yy})^{\alpha_2 + a_2}.$$

The first-order conditions, after simple substitutions, are

[9] It is also easy to construct examples of systems without externalities generating closed orbits or cycles as optimal paths, as in Benhabib and Nishimura (1979). For example if we set $r = 0.05$, $g = 0.01$, $v = 1$, $\sigma = 0.001$, $\alpha_1 = 0.0017$, $\alpha_2 = 0.459$, $\alpha_0 = 1 - \alpha_1 - \alpha_2$, $\beta_1 = 0.0265$, $\beta_2 = 0.0012$, $\beta_0 = 1 - \beta_1 - \beta_2$, $\gamma_1 = 0.5635$, $\gamma_2 = 0.423$, $\gamma_0 = 1 - \gamma_1 - \gamma_2$, then at $\sigma \approx 0.020386$, the Jacobian $[JN]$ has two complex roots with zero real parts which become negative for higher σ, and positive for lower σ, satisfying the conditions of the Hopf Bifurcation Theorem for existence of closed orbits. We note that the family of cycles as a function of σ in the example occur for low discount rates, but for a utility function of consumption that is close to linear.

$$c_t^{-\sigma} p_{x,t} = \rho E \left(c_{t+1}^{-\sigma} p_{x,t+1} \left(\frac{w_{x,t+1}}{p_{x,t+1}} + (1 - g_x) \right) \right), \tag{13.34}$$

$$c_t^{-\sigma} p_{y,t} = \rho E \left(c_{t+1}^{-\sigma} p_{y,t+1} \left(\frac{w_{y,t+1}}{p_{y,t+1}} + (1 - g_y) \right) \right), \tag{13.35}$$

where

$$\frac{w_x}{p_x} = \beta_1 q_x L_x^{\beta_0 + b_0} K_{xx}^{\beta_1 + b_1 - 1} K_{yx}^{\beta_2 + b_2}, \tag{13.36}$$

$$\frac{w_y}{p_y} = \gamma_2 q_y L_y^{\gamma_0 + c_0} K_{xy}^{\gamma_1 + c_1} K_{yy}^{\gamma_2 + c_2 - 1}, \tag{13.37}$$

$$p_x = \frac{\alpha_1 q_c L_c^{\alpha_0 + a_0} K_{xc}^{\alpha_1 + a_1 - 1} K_{yc}^{\alpha_2 + a_2}}{\beta_1 q_x L_x^{\beta_0 + b_0} K_{xx}^{\beta_1 + b_1 - 1} K_{yx}^{\beta_2 + b_2}}, \tag{13.38}$$

$$p_y = \frac{\alpha_1 q_c L_c^{\alpha_0 + a_0} K_{xc}^{\alpha_1 + a_1 - 1} K_{yc}^{\alpha_2 + a_2}}{\gamma_1 q_y L_y^{\gamma_0 + c_0} K_{xy}^{\gamma_1 + c_1 - 1} K_{yy}^{\gamma_2 + c_2}}. \tag{13.39}$$

The equations for accumulation are given by

$$k_{x,t+1} = (1 - g_x) k_{x,t} + x_t, \tag{13.40}$$

$$k_{y,t+1} = (1 - g_y) k_{y,t} + y_t. \tag{13.41}$$

The computations for the analysis and calibration of this model are presented in Appendix II. Here we proceed directly to study the local dynamics. The linearized dynamics of the model are

$$\begin{pmatrix} \widehat{k}_{x,t+1} \\ \widehat{k}_{y,t+1} \\ \widehat{p}_{x,t+1} \\ \widehat{p}_{y,t+1} \\ \widehat{z}_{c,t+1} \\ \widehat{z}_{x,t+1} \\ \widehat{z}_{y,t+1} \end{pmatrix} = [Q]^{-1} [R] \begin{pmatrix} \widehat{k}_{x,t} \\ \widehat{k}_{y,t} \\ \widehat{p}_{x,t} \\ \widehat{p}_{y,t} \\ \widehat{z}_{c,t} \\ \widehat{z}_{x,t} \\ \widehat{z}_{y,t} \end{pmatrix} + [Q]^{-1} \begin{pmatrix} 0 \\ 0 \\ -\widehat{s}_{x,t+1} \\ -\widehat{s}_{y,t+1} \\ \widehat{\varepsilon}_{c,t+1} \\ \widehat{\varepsilon}_{x,t+1} \\ \widehat{\varepsilon}_{y,t+1} \end{pmatrix}, \tag{13.42}$$

where $\widehat{s}_{i,t}$, $i = x, y$, are iid sunspot shocks with zero mean, acting on the "Euler" equations for the two capital stocks. The matrices $[R]$ and $[Q]$ are defined in Appendix II. Their elements are functions of parameters of the system, and of steady state quantities which are also functions of the parameters. We can therefore evaluate the roots of $[Q]^{-1}[R]$ to check for the possibility of indeterminacy. When externality parameters are set to zero, four of the roots of the Jacobian matrix come in pairs of $(\mu, 1/p\mu)$, and the other three roots are the autoregressive coefficients of the technology shocks.[10] For very modest externalities however indeterminacy arises, as it does in the continuous time case. The four roots no longer split with half inside and half outside the unit circle. We find that indeterminacy can easily occur for a large set of parameter values. The example below illustrates this point.

We calibrate the model along the lines of a standard RBC model. We set the quarterly discount factor to $r = 0.036$ and the deprecation rate to $g = 0.025$, so that quarterly net discount is $(r - g) = 0.011$. The instantaneous utility of consumption is logarithmic, so that $\sigma = 1$. Labor supply is taken to be quite elastic, although not infinitely elastic as is often the case in the real business cycle literature: We set $v = 0.2$, implying a labor supply elasticity of 5. The persistence parameters for the technology shocks, λ_c, λ_x, and λ_y are each set to 0.95. The production parameters and the resulting roots of the Jacobian $\left[[Q]^{-1}[R]\right]$ are as follows:

Parameters for the Consumption Good c and Investment Goods x and y.

$$q_c = 1,\ \alpha_0 = 0.58;\ \alpha_1 = 0.15,\ \alpha_2 = 0.20;\ a_0 = 0.00;\ a_1 = 0.07,\ a_2 = 0.00;$$
$$q_x = 1,\ \beta_0 = 0.50;\ \beta_1 = 0.22,\ \beta_2 = 0.21;\ b_0 = 0.00;\ b_1 = 0.07;\ b_2 = 0.00;$$
$$q_y = 1,\ \gamma_0 = 0.51;\ \gamma_1 = 0.26,\ \gamma_2 = 0.15;\ c_0 = 0.00;\ c_1 = 0.00,\ c_2 = 0.08.$$

Roots of Jacobian $[[Q]^{-1}[R]]$. (0.251, 1.057, 0.967, 0.425, 0.950, 0.950, 0.950).

The last three roots are simply the persistence parameters λ_c, λ_x, and λ_y. Of the remaining four roots, three are within the unit circle, which implies indeterminacy since there are two capital stocks and two prices.[11] Many other parametrization giving indeterminacy are also possible, but the one above is the parametrization that we use in the calibrations below.

To calibrate the model we set the standard deviations of sunspot shock $\widehat{s}_{x,t}$, and the innovations to technology shocks $\widehat{\varepsilon}_{i,t+1}$, $i = c, x, y$, all of which we take to be normally distributed, to 0.0039.[12] They are also set to imply a standard deviation for GNP of 1.76 to match the US data. In the simulations we take the technology shocks to be perfectly correlated, and the sunspot shock to be independent.

---

[10]Here again, since $\rho < 1$, there is the possibility of too many roots crossing and falling outside the unit circle even without external effects, making the steady state unstable and creating cycles on invariant circles via a Hopf bifurcation, or cycles via flip bifurcation. Our concern here however is with too many roots inside the unit circle, a situation that implies indeterminacy.

[11]The same parameters in the deterministic version of the model in continuous time would yield the four roots: $-0.72, 0.06, -0.03, -1.47$. The three negative roots imply indeterminacy.

[12]Of course both sunspot shocks cannot be independently chosen: There is a joint restriction on the properties of the sunspot shocks and the innovations to technology shocks that is needed to guarantee that the solution remains stationary, and that the effect of the root outside the unit modulus is nullified. We choose the sunspot shock $\widehat{s}_{y,t}$ as a linear combination of the innovations to technology shocks and the sunspot shock $\widehat{s}_{x,t}$ in order to satisfy this restriction. We note that, alternatively, we could have picked $\widehat{s}_{y,t}$ independently and $\widehat{s}_{x,t}$ to satisfy the restriction, or simply chosen them jointly.

Experimenting with independent technology shocks or with technology shocks correlated with the sunspot does not change the simulation results by much. The results of our calibration exercise are given in the table below.

	GNP	Consumption	Investment	Labor
St. Dev.	1.00	0.74	3.32	0.70
		(0.73)	(3.20)	(1.16)
Corr. with GNP	1.00	0.53	0.83	0.71
		(0.82)	(0.90)	(0.86)
AR1 Coeff.	0.93	0.97	0.92	0.80
	(0.90)	(0.84)	(0.76)	(0.90)

Standard deviations of the variables in the table are relative to those of GNP, and the numbers in parentheses are the same ratios for Hodrick-Prescott filtered US data. Investment corresponds to its aggregated value, evaluated at the current relative prices of x and y. GNP contains consumption, c, and investment, with the price of the consumption good normalized to unity each period. Individual components of GNP, or of investment, tend to be much more highly volatile than the aggregated series. We find however that this is the case for standard RBC calibrations, irrespective of whether the chosen parameter values generate determinate or indeterminate equilibria.[13]

The data generated by the model match US data reasonable well. Consumption is more weakly correlated with output for the data generated by the model than it is for actual US data: This in part may be because positive technology shocks initially lead to strong expansions in investment at the expense of consumption (see Fig. 13.2 and the preceding discussion). In addition, labor data from the model are less volatile and less correlated with output than they are for actual US data. One possible reason for this, as we pointed out earlier, is that we used a labor supply elasticity of 5, compared to the infinite labor supply elasticities used in much of the RBC literature.

Figure 13.1 gives a typical simulation with indeterminate equilibria and sunspots, calibrated to the parameters given above, for GNP, investment, and consumption. Clearly, investment displays oscillations of the largest amplitude while consumption is fairly smooth and GNP is in the intermediate range.

Figure 13.2 shows the impulse responses for consumption, investment, and GNP generated by an aggregate productivity shock impacting the three sectors simultaneously. The aggregative shock leads to a surge of investment, initially at the expense of consumption. Again we find that this feature, that is, the initial negative response of consumption to the aggregative technology shock, typically arises for standard RBC calibrations of multisector models that do not have any

[13] In multisector RBC models, in order to reduce volatility at the sectoral level and to insure that all sectors are procyclical, it may be necessary to introduce adjustment costs for the movement of factors across sectors. See Huffman and Wynne (1996).

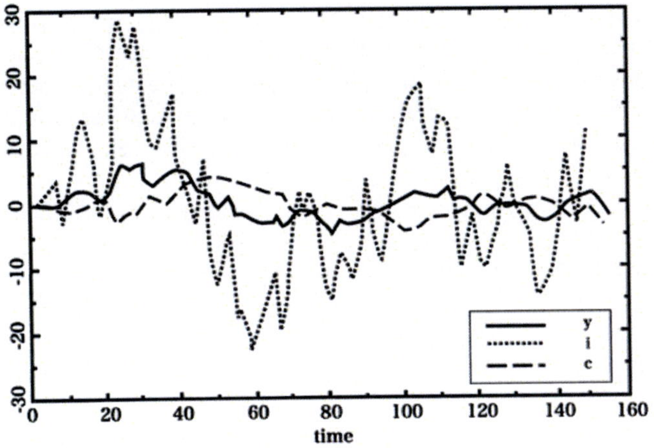

Fig. 13.1 Simulated data

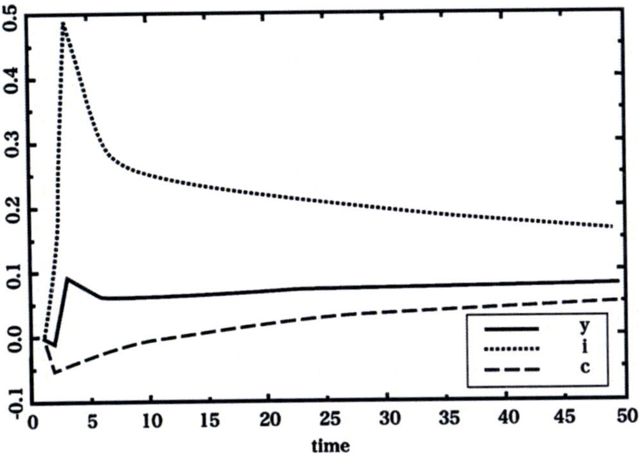

Fig. 13.2 Impulse response for GNP, C, and I

external effects and therefore exhibit determinate equilibria.[14] GNP also drops by a small amount when the shock hits, but rises immediately afterward as investment surges, and then subsides to generate the hump-shaped response found in the data.

While we have by no means performed an extensive search, the model with the above parameters that generates sunspot equilibria can provide a reasonable match to the various moments of actual data. There exist many other reasonable parameter combinations that give a good match, and yet there are still others that give a very poor match to the data. Furthermore, some of the moments generated by

[14]The impulse response function of consumption to a technology shock also exhibits the same behavior in the multisector model of Weder (1996) in this issue.

the model can be sensitive to parameter changes in certain regions of the parameter set. This is true whether we have external effects and indeterminacy, or whether we restrict ourselves to standard parametrizations of the model without externalities and indeterminacy. Another feature, shared with calibrated multisector models without external effects or market distortions that have determinate equilibria, is that prices and outputs of the individual investment goods tend to be more volatile than the aggregated value of investment, with some sectors even exhibiting counter-cyclical behavior (see for example Benhabib et al. 1997). This counter-factual observation about calibrated multisector models in the context of a determinate economy has led Huffman and Wynne (1996) to introduce adjustment costs for the sectoral reallocations of factors of production. It seems therefore that multisector RBC models, with or without indeterminacy and sunspots, raise some new issues for the RBC literature.[15] More information concerning the moments of individual output series must be considered to identify the best parametrization and to assess how good the match is between the data and the simulations. Further disaggregation may be necessary to identify the sectors of the model with the actual sectors of the economy for which data is available. On the other hand it also seems likely that increasing the number of sectors will expand the range of parameters yielding indeterminacy, much as going from one to two or three sectors does. We view the above calibration exercise only as suggestive of interesting possibilities that can expand the scope of the RBC literature.

13.5 Appendix I: The Continuous Time Case

In this appendix we will derive the expressions necessary to evaluate the steady-state Jacobian of the linearized dynamics of the three-sector model in continuous time.

13.5.1 The Static Structure

From (13.33)–(13.35) we can write:

$$\omega_{12} = \frac{w_x}{w_y} = \frac{\alpha_1 K_{yc}}{\alpha_2 K_{xc}} = \frac{\beta_1 K_{yx}}{\beta_2 K_{xx}} = \frac{\gamma_1 K_{yy}}{\gamma_2 K_{xy}}, \tag{13.43}$$

$$\omega_{10} = \frac{w_x}{w_0} = \frac{\alpha_1 L_c}{\alpha_0 K_{xc}} = \frac{\beta_1 L_x}{\beta_0 K_{xx}} = \frac{\gamma_1 L_y}{\gamma_0 K_{xy}}, \tag{13.44}$$

$$\omega_{20} = \frac{w_y}{w_0} = \frac{\alpha_2 L_c}{\alpha_0 K_{yc}} = \frac{\beta_2 L_x}{\beta_0 K_{yx}} = \frac{\gamma_2 L_y}{\gamma_0 K_{yy}}. \tag{13.45}$$

[15] For an exploration of these issues in a three-sector model without external effects, market distortions, or indeterminacies, see Benhabib et al. (1997).

If we denote $\widehat{x} = dx/x$, then logarithmic differentiation yields the following:

$$\widehat{\omega}_{12} = \widehat{K}_{yc} - \widehat{K}_{xc} = \widehat{K}_{yx} - \widehat{K}_{xx} = \widehat{K}_{yy} - \widehat{K}_{xy}, \tag{13.46}$$

$$\widehat{\omega}_{10} = \widehat{L}_c - \widehat{K}_{xc} = \widehat{L}_x - \widehat{K}_{xx} = \widehat{L}_y - \widehat{K}_{xy}, \tag{13.47}$$

$$\widehat{\omega}_{20} = \widehat{L}_c - \widehat{K}_{yc} = \widehat{L}_x - \widehat{K}_{yx} = \widehat{L}_y - \widehat{K}_{yy}. \tag{13.48}$$

Note that $(w_{ij}) = (w_{ji})^{-1}$ and that $(\omega_{ij})(\omega_{jh}) = (\omega_{ih})$. Now, combining equations (13.19), (13.20), and (13.43)–(13.45), and noting that at a steady state

$$w_x = rp_x, \qquad w_y = rp_y, \tag{13.49}$$

we have

$$\begin{aligned} w_x &= p_x \left(q_x \beta_1 \left(\frac{\beta_0}{\beta_1}\right)^{\beta_0+b_0} \left(\frac{\beta_2}{\beta_1}\right)^{\beta_2+b_2} \right) (\omega_{10})^{\beta_0+b_0} (\omega_{12})^{\beta_2+b_2} \\ &= rp_x, \end{aligned} \tag{13.50}$$

$$\begin{aligned} w_y &= p_y \left(q_y \gamma_2 \left(\frac{\gamma_0}{\gamma_2}\right)^{\gamma_0+c_0} \left(\frac{\gamma_1}{\gamma_2}\right)^{\gamma_2+c_2} \right) (\omega_{10})^{\gamma_0+c_0} (\omega_{12})^{-\gamma_1-c_1-\gamma_0-c_0} \\ &= rp_y. \end{aligned} \tag{13.51}$$

The exponential term $(-\gamma_1 - c_1 - \gamma_0 - c_0)$ appears in (13.51) because the factor price ratios ω_{21} and ω_{20} in the equation are replaced by $(\omega_{12})^{-1}$ and $(\omega_{10})(\omega_{12})^{-1}$. Taking logs in (13.50) and (13.51) we can write them as:

$$\left(\begin{pmatrix} \ln r - \ln \left(q_x \beta \left(\frac{\beta_0}{\beta_1}\right)^{\beta_0+b_0} \left(\frac{\beta_2}{\beta_1}\right)^{\beta_2+b_2} \right) \\ \ln r - \ln \left(q_y \gamma_2 \left(\frac{\gamma_0}{\gamma_2}\right)^{\gamma_0+c_0} \left(\frac{\gamma_1}{\gamma_2}\right)^{\gamma_2+c_2} \right) \end{pmatrix} \right) = M \begin{pmatrix} \ln \omega_{10} \\ \ln \omega_{12} \end{pmatrix}, \tag{13.52}$$

where

$$[M] = \begin{bmatrix} \beta_0 + b_0 & \beta_2 + b_2 \\ \gamma_0 + c_0 & -\gamma_1 - c_1 - \gamma_0 - c_0 \end{bmatrix}. \tag{13.53}$$

Equation 13.52 determines the steady-state values of ω_{10} and ω_{12}. We now solve, using (13.19), for the prices p_x and p_y. We have

$$\begin{aligned} \alpha_1 q_c \left(\frac{K_{yc}}{K_{xc}}\right)^{(\alpha_2+a_2)} \left(\frac{L_c}{K_{xc}}\right)^{(\alpha_0+a_0)} &= p_x \beta_1 q_x \left(\frac{K_{yx}}{K_{xx}}\right)^{(\beta_2+b_2)} \left(\frac{L_x}{K_{xx}}\right)^{(\beta_0+b_0)} \\ &= p_y \gamma_1 q_y \left(\frac{K_{yy}}{K_{xy}}\right)^{(\gamma_2+c_2)} \left(\frac{L_y}{K_{xy}}\right)^{(\gamma_0+c_0)}. \end{aligned}$$

Substituting from (13.43)–(13.45), we obtain

$$\alpha_1 q_c \left(\left(\frac{\alpha_2}{\alpha_1}\right)\omega_{12}\right)^{(\alpha_2+a_2)}\left(\left(\frac{\alpha_0}{\alpha_1}\right)\omega_{10}\right)^{(\alpha_0+a_0)}$$
$$= p_x\beta_1 q_x \left(\left(\frac{\beta_2}{\beta_1}\right)\omega_{12}\right)^{(\beta_2+b_2)}\left(\left(\frac{\beta_0}{\beta_1}\right)\omega_{10}\right)^{(\beta_0+b_0)}$$
$$= p_y\gamma_1 q_y \left(\left(\frac{\gamma_2}{\gamma_1}\right)\omega_{12}\right)^{(\gamma_2+c_2)}\left(\left(\frac{\gamma_0}{\gamma_1}\right)\omega_{10}\right)^{(\gamma_0+c_0)}. \tag{13.54}$$

Taking log we then have

$$\begin{pmatrix}\ln p_x\\ \ln p_y\end{pmatrix} = \begin{pmatrix}\ln\left(\left(\frac{q_c\alpha_1}{q_x\beta_1}\right)\left(\frac{\alpha_2}{\alpha_1}\right)^{(\alpha_2+a_2)}\left(\frac{\alpha_0}{\alpha_1}\right)^{(\alpha_0+a_0)}\left(\frac{\beta_2}{\beta_1}\right)^{-(\beta_2+b_2)}\left(\frac{\beta_0}{\beta_1}\right)^{-(\beta_0+b_0)}\right)\\ \ln\left(\left(\frac{q_c\alpha_1}{q_x\gamma_1}\right)\left(\frac{\alpha_2}{\alpha_1}\right)^{(\alpha_2+a_2)}\left(\frac{\alpha_0}{\alpha_1}\right)^{(\alpha_0+a_0)}\left(\frac{\gamma_2}{\gamma_1}\right)^{-(\gamma_2+c_2)}\left(\frac{\gamma_0}{\gamma_1}\right)^{-(\gamma_0+c_0)}\right)\end{pmatrix}$$
$$+[N]\begin{pmatrix}\ln\omega_{10}\\ \ln\omega_{12}\end{pmatrix} \tag{13.55}$$

where

$$[N] = \begin{bmatrix}\alpha_0 + a_0 - \beta_0 - b_0 & \alpha_2 + a_2 - \beta_2 - b_2\\ \alpha_0 + a_0 - \gamma_0 - c_0 & \alpha_2 + a_2 - \gamma_2 - c_2\end{bmatrix}. \tag{13.56}$$

Equation 13.55 allows us to solve for p_x and p_y, in terms of ω_{10} and ω_{12}, and then, using (13.52) and (13.49), for the steady-state values of the prices p_x and p_y and capital rentals w_x and w_y; we will need them to evaluate the Jacobian matrix describing the local dynamics around the steady state. Furthermore, using (13.55) we obtain

$$\begin{pmatrix}\widehat{\omega}_{10}\\ \widehat{\omega}_{12}\end{pmatrix} = [N]^{-1}\begin{pmatrix}\widehat{p}_x\\ \widehat{p}_y\end{pmatrix}. \tag{13.57}$$

Now, taking logarithmic derivatives of (13.50) and (13.51), we get

$$\begin{pmatrix}\widehat{w}_x\\ \widehat{w}_y\end{pmatrix} = \begin{bmatrix}\beta_0 + b_0 & \beta_2 + b_2\\ \gamma_0 + c_0 & \gamma_2 + c_2\end{bmatrix}\begin{pmatrix}\widehat{\omega}_{10}\\ \widehat{\omega}_{12}\end{pmatrix} + \begin{pmatrix}\widehat{p}_x\\ \widehat{p}_y\end{pmatrix}$$
$$= \left(\begin{bmatrix}\beta_0 + b_0 & \beta_2 + b_2\\ \gamma_0 + c_0 & \gamma_2 + c_2 - 1\end{bmatrix}[N]^{-1} + I\right)\begin{pmatrix}\widehat{p}_x\\ \widehat{p}_y\end{pmatrix}$$
$$= [G]\begin{pmatrix}\widehat{p}_x\\ \widehat{p}_y\end{pmatrix}. \tag{13.58}$$

Note that (13.58) defines the matrix $[G]$.

13.5.1.1 The Static Structure for the Two-Sector Case

In a two-sector economy without the capital good y, either as an input or an output, the above expression for the elasticity of $\widehat{w}_x$ with respect to $\widehat{p}_x$ can be simplified. The matrix $[N]$ becomes a scalar and, since at a steady state $w_x = rp_x$, we obtain

$$\frac{\partial w_x}{\partial p_x} = r\frac{\widehat{w}_x}{\widehat{p}_x} = r\frac{\alpha_0 + a_0}{\alpha_0 + a_0 - \beta_0 - b_0}. \tag{13.59}$$

We use this expression in Sect. 13.2 above.

13.5.2 *Unit Input Coefficients*

Computing unit input coefficients is straightforward. Taking logs of the production function for capital good x and using (13.43)–(13.45), we have

$$\begin{bmatrix} \beta_1 + b_1 & \beta_2 + b_2 & \beta_0 + b_0 \\ -1 & 1 & 0 \\ -1 & 0 & 1 \end{bmatrix} \begin{bmatrix} \ln K_{xx} \\ \ln K_{yx} \\ \ln L_x \end{bmatrix} = \begin{bmatrix} \ln x - \ln q_x \\ \ln w_x - \ln w_y + \ln \beta_2 - \ln \beta_1 \\ \ln w_x - \ln w_0 + \ln \beta_0 - \ln \beta_1 \end{bmatrix}.$$

Solving, we get

$$a_{01} = \frac{L_x}{x} = \left(\frac{1}{q_x}\right)\left(\frac{\beta_0 w_x}{\beta_1 w_0}\right)^{\beta_1+b_1}\left(\frac{\beta_0 w_y}{\beta_2 w_0}\right)^{\beta_2+b_2}, \tag{13.60}$$

$$a_{11} = \frac{K_{xx}}{x} = \left(\frac{1}{q_x}\right)\left(\frac{\beta_1 w_0}{\beta_0 w_x}\right)^{\beta_0+b_0}\left(\frac{\beta_1 w_y}{\beta_2 w_x}\right)^{\beta_2+b_2}, \tag{13.61}$$

$$a_{21} = \frac{K_{yx}}{x} = \left(\frac{1}{q_x}\right)\left(\frac{\beta_2 w_0}{\beta_0 w_y}\right)^{\beta_0+b_0}\left(\frac{\beta_2 w_x}{\beta_1 w_y}\right)^{\beta_1+b_1}. \tag{13.62}$$

Similarly, for the consumption and second capital good we obtain

$$a_{00} = \frac{L_c}{c} = \left(\frac{1}{q_c}\right)\left(\frac{\alpha_0 w_x}{\alpha_1 w_0}\right)^{\alpha_1+a_1}\left(\frac{\alpha_0 w_y}{\alpha_2 w_0}\right)^{\alpha_2+a_2}, \tag{13.63}$$

$$a_{10} = \frac{K_{xc}}{x} = \left(\frac{1}{q_c}\right)\left(\frac{\alpha_1 w_0}{\alpha_0 w_x}\right)^{\alpha_0+a_0}\left(\frac{\alpha_1 w_y}{\alpha_2 w_x}\right)^{\alpha_2+a_2}, \tag{13.64}$$

$$a_{20} = \frac{K_{yc}}{c} = \left(\frac{1}{q_c}\right)\left(\frac{\alpha_2 w_0}{\alpha_0 w_y}\right)^{\alpha_0+a_0}\left(\frac{\alpha_2 w_x}{\alpha_1 w_y}\right)^{\alpha_1+a_1}, \tag{13.65}$$

$$a_{02} = \frac{L_y}{y} = \left(\frac{1}{q_y}\right)\left(\frac{\gamma_0 w_x}{\gamma_1 w_0}\right)^{\gamma_1+c_1}\left(\frac{\gamma_0 w_y}{\gamma_2 w_0}\right)^{\gamma_2+c_2}, \tag{13.66}$$

$$a_{12} = \frac{K_{xy}}{y} = \left(\frac{1}{q_y}\right)\left(\frac{\gamma_1 w_0}{\gamma_0 w_x}\right)^{\gamma_0+c_0}\left(\frac{\gamma_1 w_y}{\gamma_2 w_x}\right)^{\gamma_2+c_2}, \tag{13.67}$$

$$a_{22} = \frac{K_{yy}}{y} = \left(\frac{1}{q_y}\right)\left(\frac{\gamma_0 w_x}{\gamma_1 w_0}\right)^{\gamma_1+c_1}\left(\frac{\gamma_0 w_y}{\gamma_2 w_0}\right)^{\gamma_2+c_2}. \tag{13.68}$$

Note that the input coefficients are functions of the ratios of factor rentals and can be written in terms of ω_{10} and ω_{12}, remembering that $(\omega_{ij}) = (\omega_{ji})^{-1}$ and that $(\omega_{ih}) = (\omega_{ij})(\omega_{jh})$.

13.5.3 Steady-State Quantities

At a steady state we have

$$x = gk_x; \quad y = gk_y. \tag{13.69}$$

Full employment then requires

$$\begin{bmatrix} a_{00}\ a_{01}\ a_{02} \\ a_{10}\ a_{11}\ a_{12} \\ a_{20}\ a_{21}\ a_{22} \end{bmatrix}\begin{pmatrix} c \\ gk_x \\ gk_y \end{pmatrix} = \begin{pmatrix} L \\ k_x \\ k_y \end{pmatrix}. \tag{13.70}$$

We can solve for k_x and k_y as

$$\begin{pmatrix} k_x \\ k_y \end{pmatrix} = \left[I - g\begin{bmatrix} a_{11}\ a_{12} \\ a_{21}\ a_{22} \end{bmatrix}\right]^{-1}\begin{pmatrix} a_{10} \\ a_{20} \end{pmatrix} c. \tag{13.71}$$

Let $\nu = (1/\upsilon)$. Then using (13.21), (13.69), and (13.70), we can solve, for steady-state c,

$$\left(a_{00} + g(a_{01}\ a_{02})\left[I - g\begin{bmatrix} a_{11}\ a_{12} \\ a_{21}\ a_{22} \end{bmatrix}\right]^{-1}\begin{pmatrix} a_{10} \\ a_{20} \end{pmatrix}\right) c = \alpha_0^{\nu} c^{(1-\sigma)\nu} L_c^{-\nu},$$

or, since $L_c = (a_{00})c$,

$$c = \left(\left(\frac{\alpha_0}{a_{00}}\right)^{(-\nu)}\left(a_{00} + g(a_{01}\ a_{02})\left[I - g\begin{bmatrix} a_{11}\ a_{12} \\ a_{21}\ a_{22} \end{bmatrix}\right]^{-1}\begin{pmatrix} a_{10} \\ a_{20} \end{pmatrix}\right)\right)^{\mu}, \tag{13.72}$$

where $\mu = (-1/(\sigma\nu + 1))$. Using (13.72) and (13.71) we can solve for steady-state k_x and k_y. Since unit input coefficients are functions of ω_{10} and ω_{12}, whose steady-state values are given by (13.52), the steady-state stocks k_x and k_y can be computed in terms of the parameters of the model. The steady-state outputs then are given by (13.69) and (13.72). We can express the steady-state factor inputs in the production function as

$$\begin{aligned} L_c &= a_{00}c;\ K_{xc} = a_{10}c;\ K_{yc} = a_{20}c; \\ L_x &= a_{01}x;\ K_{xx} = a_{11}x;\ K_{yx} = a_{21}x; \\ L_y &= a_{02}y;\ K_{xy} = a_{12}y;\ K_{yy} = a_{22}y; \end{aligned}$$

where, by construction

$$\begin{aligned} L &= L_c + L_x + L_y, \\ k_x &= K_{xc} + K_{xx} + K_{xy}, \\ k_y &= K_{yc} + K_{yx} + K_{yy}. \end{aligned}$$

13.5.4 Output Elasticities

First we compute the elasticities of inputs with respect to ω_{10} and ω_{12}. From the first-order condition for labor given by (13.21) we have

$$\widehat{L} = x_0\widehat{L}_c + x_1\widehat{K}_{xc} + x_2\widehat{K}_{yc}, \tag{13.73}$$

where

$$\begin{aligned} x_0 &= \frac{(\alpha_0 + a_0)(1-\sigma) - 1}{\nu}; \\ x_1 &= \frac{(\alpha_1 + a_1)(1-\sigma)}{\nu}; \\ x_2 &= \frac{(\alpha_2 + a_2)(1-\sigma)}{\nu}. \end{aligned}$$

Using (13.46)–(13.48), it follows that

$$\widehat{L} = (x_0 + x_1 + x_2)\widehat{K}_{yc} + x_0\widehat{\omega}_{20} - x_1\widehat{\omega}_{12}. \tag{13.74}$$

The following identity,

$$\begin{bmatrix} \left(\frac{L_c}{L}\right)\widehat{L}_c & \left(\frac{L_x}{L}\right)\widehat{L}_x & \left(\frac{L_y}{L}\right)\widehat{L}_y \\ \left(\frac{K_{xc}}{k_x}\right)\widehat{K}_{xc} & \left(\frac{K_{xx}}{k_x}\right)\widehat{K}_{xx} & \left(\frac{K_{xy}}{k_x}\right)\widehat{K}_{xy} \\ \left(\frac{K_{yc}}{k_y}\right)\widehat{K}_{yc} & \left(\frac{K_{yx}}{k_y}\right)\widehat{K}_{yx} & \left(\frac{K_{yy}}{k_y}\right)\widehat{K}_{yy} \end{bmatrix} \begin{pmatrix} 1 \\ 1 \\ 1 \end{pmatrix} = \begin{pmatrix} \widehat{L} \\ \widehat{k}_x \\ \widehat{k}_y \end{pmatrix} \tag{13.75}$$

may now be rewritten, after some substitutions using (13.74), (13.47), (13.46), and (13.48), as

$$\begin{bmatrix} \left(\frac{L_c}{L}+\frac{\sigma}{v}\right) & \left(\frac{L_x}{L}\right) & \left(\frac{L_y}{L}\right) \\ \left(\frac{K_{xc}}{k_x}\right) & \left(\frac{K_{xx}}{k_x}\right) & \left(\frac{K_{xy}}{k_x}\right) \\ \left(\frac{K_{yc}}{k_y}\right) & \left(\frac{K_{yx}}{k_y}\right) & \left(\frac{K_{yy}}{k_y}\right) \end{bmatrix} \begin{pmatrix} \widehat{K}_{yc} \\ \widehat{K}_{yx} \\ \widehat{K}_{yy} \end{pmatrix} = \begin{pmatrix} (\widehat{\omega}_{12}-\widehat{\omega}_{10})(1-x_0)-(\widehat{\omega}_{12})\,x_1 \\ \widehat{\omega}_{12}+\widehat{k}_x \\ \widehat{k}_y \end{pmatrix}. \tag{13.76}$$

In particular, to derive the first equation of (13.76) we use

$$(\widehat{\omega}_{20}) = (\widehat{\omega}_{10}) - (\widehat{\omega}_{12})$$

and

$$x_0 + x_1 + x_2 = -\frac{\sigma}{v}.$$

The equations given by (13.76) allow us to express the steady-state elasticities of K_{yc}, K_{yx}, and K_{yy} with respect to ω_{10}, ω_{12}, k_x, and k_y. From the production functions, on the other hand, we have

$$\begin{aligned} \widehat{c} &= (\alpha_0+a_0)(\widehat{K}_{yc}-\widehat{\omega}_{12}+\widehat{\omega}_{10}) + (\alpha_1+a_1)(\widehat{K}_{yc}-\widehat{\omega}_{12}) + (\alpha_2+a_2)(\widehat{K}_{yc}) \\ &= \widehat{K}_{yc} + (\alpha_0+a_0)(-\widehat{\omega}_{12}+\widehat{\omega}_{10}) - (\alpha_1+a_1)(\widehat{\omega}_{12}), \end{aligned} \tag{13.77}$$

$$\widehat{x} = \widehat{K}_{yx} + (\beta_0+b_0)(-\widehat{\omega}_{12}+\widehat{\omega}_{10}) - (\beta_1+b_1)(\widehat{\omega}_{12}), \tag{13.78}$$

$$\widehat{y} = \widehat{K}_{yy} + (\gamma_0+c_0)(-\widehat{\omega}_{12}+\widehat{\omega}_{10}) - (\gamma_1+c_1)(\widehat{\omega}_{12}). \tag{13.79}$$

Now notice that since $\widehat{K}_{yc}$, $\widehat{K}_{yx}$, and $\widehat{K}_{yy}$ depend on $\widehat{\omega}_{10}$, $\widehat{\omega}_{12}$, $\widehat{k}_x$, and $\widehat{k}_y$, and $\widehat{\omega}_{10}$, $\widehat{\omega}_{12}$ in turn depend on $\widehat{p}_x$ and $\widehat{p}_y$, and we can now express the output elasticities of c, x, and y with respect to p_x, p_y, k_x, and k_y. We define the elasticity function of an m-vector q with respect to an n-vector p as an $m \times n$ matrix as $E(q\colon p)$. Let the matrix in (13.76) be defined as

$$[F] = \begin{bmatrix} \left(\frac{L_c}{L}+\frac{\sigma}{v}\right) & \left(\frac{L_x}{L}\right) & \left(\frac{L_y}{L}\right) \\ \left(\frac{K_{xc}}{k_x}\right) & \left(\frac{K_{xx}}{k_x}\right) & \left(\frac{K_{xy}}{k_x}\right) \\ \left(\frac{K_{yc}}{k_y}\right) & \left(\frac{K_{yx}}{k_y}\right) & \left(\frac{K_{yy}}{k_y}\right) \end{bmatrix}.$$

Then we have

$$[E_{KK}] \equiv E((K_{yc}, K_{yx}, K_{yy})\colon (k_x, k_y)) = [F]^{-1} \begin{bmatrix} 0 & 0 \\ 1 & 0 \\ 0 & 1 \end{bmatrix}.$$

Now define $[S] = [N]^{-1}$. The price elasticities are

$$[E_{KP}] \equiv E((K_{yc}, K_{yx}, K_{yy})\colon (p_x, p_y))$$
$$= [F]^{-1}\begin{bmatrix} (-S_{11}+S_{21})(1-x_0)-x_1S_{21} & (-S_{12}+S_{22})(1-x_0)-x_1S_{22} \\ S_{21} & S_{22} \\ 0 & 0 \end{bmatrix}.$$

where the elements of the matrix $[S]$ are denoted by S_{ij}. We will denote the ijth element of $[E_{KK}]$ and $[E_{KP}]$ by $E_{KK}(i,j)$ and $E_{KP}(i,j)$. Substituting the elements of the matrices $[S]$, $[E_{KK}]$, and $[E_{KP}]$ into the production equations (13.77)–(13.79), we obtain the output elasticities:

$$\begin{aligned}
E(c\colon p_x) &= E_{KP}(1,1) + (\alpha_0 + a_0)((S_{11} - S_{21})) - (\alpha_1 + a_1)S_{21},\\
E(c\colon p_y) &= E_{KP}(2,1) + (\alpha_0 + a_0)((S_{12} - S_{22})) - (\alpha_1 + a_1)S_{22},\\
E(c\colon k_x) &= E_{KK}(1,1),\\
E(c\colon k_y) &= E_{KK}(2,1),\\
E(x\colon p_x) &= E_{KP}(2,1) + (\beta_0 + b_0)((S_{11} - S_{21})) - (\beta_1 + b_1)S_{21},\\
E(x\colon p_y) &= E_{KP}(2,2) + (\beta_0 + b_0)((S_{12} - S_{22})) - (\beta_1 + b_1)S_{22},\\
E(x\colon k_x) &= E_{KK}(2,1),\\
E(x\colon k_y) &= E_{KK}(2,2),\\
E(y\colon p_x) &= E_{KP}(3,1) + (\gamma_0 + c_0)((S_{11} - S_{21})) - (\gamma_1 + c_1)S_{21},\\
E(y\colon p_y) &= E_{KP}(3,2) + (\gamma_0 + c_0)((S_{12} - S_{22})) - (\gamma_1 + c_1)S_{22},\\
E(y\colon k_x) &= E_{KK}(3,1),\\
E(y\colon k_y) &= E_{KK}(3,2).
\end{aligned}$$

13.5.4.1 Output Elasticities for the Two-Sector Case

Simpler expressions can be obtained for the case of a two-sector model. We will derive an expression only for $(\partial x/\partial k_x)$ in the two-sector case, to be used in Sect. 13.2 above.[16] Setting $\widehat{\omega}_{10}$ and $\widehat{\omega}_{12}$ to zero and using (13.44) and (13.47), the matrix equation (13.76) can be modified and written as:

$$\begin{bmatrix} \left(\frac{L_c}{L} + \frac{\sigma}{v}\right) & \frac{L_x}{L} \\ \left(\frac{K_{xc}}{k_x}\right) & \left(\frac{K_{xx}}{k_x}\right) \end{bmatrix} \begin{pmatrix} \widehat{K}_{xc} \\ \widehat{K}_{xx} \end{pmatrix} = \begin{pmatrix} 0 \\ \widehat{k}_x \end{pmatrix}. \tag{13.80}$$

[16] A full characterization of the two-sector case is available from the authors.

Solving for $\widehat{K}_{xx}$, L_x we have,

$$\widehat{K}_{xx} = \left(\frac{kL}{Q}\right)\left(\frac{L_c}{L} - \left(\frac{\sigma}{\nu}\right)\right)\widehat{k}_x$$

where

$$Q = L_c K_x - L_x K_c + \left(\frac{\sigma}{\nu}\right) K_x L. \tag{13.81}$$

Since $\widehat{\omega}_{10}$ has been set to zero, from (13.47), we get

$$\widehat{L}_x = \widehat{K}_{xx} - \omega_{10} = \left(\frac{kL}{Q}\right)\left(\frac{L_c}{L} - \left(\frac{\sigma}{\nu}\right)\right)\widehat{k_x}.$$

Using the input coefficients given by (13.63), (13.64), (13.67), and (13.68), but modified for the two-sector model, we obtain

$$\begin{aligned}\left(\frac{Q}{L_c L_x}\right) &= \left(\frac{K_x}{L_x}\left(1 + \left(\frac{\sigma}{\nu}\right)\frac{L_c}{L}\right) - \frac{K_c}{L_c}\right)\\ &= \frac{\omega}{\alpha_0\beta_0}\left(\beta_1\left(1 - a_0 - a_1\right) - \alpha_1\left(1 - b_0 - b_1\right) + \beta_1\alpha_0\left(\frac{\sigma}{\nu}\right)\frac{L}{L_c}\right).\end{aligned}$$

Now the growth of the output x will be

$$\begin{aligned}\widehat{x} &= (\beta_1 + b_1)\widehat{K}_{xx} + (1 - \beta_1 - b_1)\widehat{L}_x\\ &= \left(\frac{k_x L}{Q}\right)\left(\frac{L_c}{L} + \left(\frac{\sigma}{\nu}\right)\right)\widehat{k}_x\\ &= \left(\frac{k_x L}{L_c L_x}\right)\left(\frac{\alpha_0\beta_0}{\omega}\right)\\ &\quad\times\left(\frac{\frac{L_c}{L} + \left(\frac{\sigma}{\nu}\right)}{\left(\beta_1\left(1 - a_0 - a_1\right) - \alpha_1\left(1 - b_0 - b_1\right) + \beta_1\alpha_0\left(\frac{\sigma}{\nu}\right)\frac{L}{L_c}\right)}\right)\widehat{k}_x\end{aligned}$$

and we have

$$\left(\frac{\partial x}{\partial k}\right) = \left(\frac{xL}{L_c L_x}\right)\left(\frac{\alpha_0\beta_0}{\omega}\right)\left(\frac{\left(\frac{L_c}{L} + \left(\frac{\sigma}{\nu}\right)\right)}{\left(\beta_1\left(1 - a_0 - a_1\right) - \alpha_1\left(1 - b_0 - b_1\right) + \beta_1\alpha_0\left(\frac{\sigma}{\nu}\right)\frac{L}{L_c}\right)}\right).$$

However, noting that at the steady state

$$w = rp = p\beta q_x\left(\frac{\beta_0}{\beta_1\omega}\right)^{(\beta_0+b_0)} = p\beta q_x\left(\frac{\beta_1\omega}{\beta_0}\right)^{(\beta_1+b_1)}\left(\frac{\beta_0}{\beta_1\omega}\right)$$

and

$$\left(\frac{x}{L_x}\right) = q_x \left(\frac{\beta_1 \omega}{\beta_0}\right)^{(\beta_1 + b_1)} = \left(\frac{r\omega}{\beta_0}\right),$$

we obtain

$$\left(\frac{\partial x}{\partial k}\right) = r\left(\frac{L}{L_c}\right)\left(\frac{\alpha_0\left(\frac{L}{L_c} + \left(\frac{\sigma}{v}\right)\right)}{\left(\beta_1\left(1 - a_0 - a_1\right) - \alpha_1\left(1 - b_0 - b_1\right) + \beta_1 \alpha_0 \left(\frac{\sigma}{v}\right)\frac{L}{L_c}\right)}\right)$$

$$= \left(\frac{r\alpha_0\left(1 + \left(\frac{L}{L_c}\right)\left(\frac{\sigma}{v}\right)\right)}{\left(\beta_1\left(1 - a_0 - a_1\right) - \alpha_1\left(1 - b_0 - b_1\right) + \beta_1 \alpha_0 \left(\frac{\sigma}{v}\right)\frac{L}{L_c}\right)}\right). \tag{13.82}$$

we use the expression $(\partial x / \partial k)$ in Sect. 13.2.

13.6 Appendix II: The Discrete Time Case

The discrete time problem can be defined as

$$V(k_x, k_y, z) = Max\left(\frac{1}{1-\sigma}\right)\left(z_c q_c L_c^{\alpha_0} K_{xc}^{\alpha_1} K_{yc}^{\alpha_2} \overline{L_c^{a_0} K_{xc}^{a_1} K_{yc}^{a_2}}\right)^{(1-\sigma)}$$

$$- (1+v)^{-1} L^{(1+v)} + \rho E V((1-g_x)k_x + x, (1-g_y)k_y + y, z'),$$

$$x = z_x q_x L_x^{\beta_0} K_{xx}^{\beta_1} K_{yx}^{\beta_2} \overline{L_x^{b_0} K_{xx}^{b_1} K_{yx}^{b_2}}, \tag{13.83}$$

$$y = z_y q_y L_y^{\gamma_0} K_{xy}^{\gamma_1} K_{yy}^{\gamma_2} \overline{L_y^{c_0} K_{xy}^{c_1} K_{yy}^{c_2}}, \tag{13.84}$$

where the discount factor is $\rho = (1 + (r - g))^{-1}$, g_x, g_y are depreciation rates, $z = (z_c, z_x, z_y)$, z_i is a technology shock where $\ln z_i = \zeta_i$,

$$\zeta_{i,t+1} = \lambda_i \zeta_{i,t} + \varepsilon_{i,t+1}; \quad 0 \leqslant \lambda_i \leqslant 1, \tag{13.85}$$

$i = c, x, y$, and $\varepsilon_{i,t+1}$ is iid, normally distributed, and has mean zero, z' is the value attained by z in the subsequent period. Note that we can write the consumption output as:

$$c = z_c q_c (L - L_x - L_y)^{\alpha_0 + a_0} (k_x - K_{xx} - K_{xy})^{\alpha_1 + a_1} (k_y - K_{yx} - K_{yy})^{\alpha_2 + a_2}.$$

The first-order conditions, after simple substitutions, are

$$c_t^{-\sigma} p_{x,t} = \rho E\left(c_{t+1}^{-\sigma} p_{x,t+1}\left(\frac{w_{x,t+1}}{p_{x,t+1}} + (1-g_x)\right)\right), \tag{13.86}$$

$$c_t^{-\sigma} p_{y,t} = \rho E\left(c_{t+1}^{-\sigma} p_{y,t+1}\left(\frac{w_{y,t+1}}{p_{y,t+1}} + (1-g_y)\right)\right), \tag{13.87}$$

and the equations for accumulation are given by

$$k_{x,t+1} = (1-g_x)k_{x,t} + x_t, \tag{13.88}$$

$$k_{y,t+1} = (1-g_y)k_{y,t} + y_t. \tag{13.89}$$

Before analyzing the dynamics it is easy to show that the steady state of the dynamic system, (13.86)–(13.89), and (13.85), with the random variables z_c, z_x, z_y, set to their long-run means, is identical to the steady state of the deterministic continuous time system if $g_x = g_y = g$. To see this set $\rho = (1+(r-g))^{-1}$ and note that at a steady state this implies

$$1 + r - g = \frac{w_x}{p_x} + (1-g) = \frac{w_y}{p_y} + (1-g).$$

We can define $q_i = \overline{q}_i \overline{z}_i$ where $\overline{z}_i$ is the long run mean of z_i for $i = c, x, y$. The steady-state values of variables computed in the previous sections therefore remain as before. Quantity and price variables will now depend on the realizations of shocks as well, so to study the dynamics around the steady state we now have to compute the elasticities of input coefficients and of outputs that will incorporate the effects of the stochastic shocks.

First we note that (13.43)–(13.45) as well as (13.76) remain unchanged. Equation 13.57 has to be slightly modified:

$$\begin{pmatrix} \widehat{\omega}_{10} \\ \widehat{\omega}_{12} \end{pmatrix} = [N]^{-1} \begin{pmatrix} \widehat{p}_x + \widehat{z}_x - \widehat{z}_c \\ \widehat{p}_y + \widehat{z}_y - \widehat{z}_c \end{pmatrix}. \tag{13.90}$$

In terms of elasticities, this implies that

$$E(\omega_{10}, \omega_{12}: z_x, z_y) = E(\omega_{10}, \omega_{12}: p_x, p_y),$$

$$E(\omega_{10}: \widehat{z}_c) = -E(\omega_{10}: p_x),$$

$$E(\omega_{12}: \widehat{z}_c) = -E(\omega_{12}: p_y),$$

If follows therefore that

$$E(c, x, y: z_x, z_y) = \begin{bmatrix} 0\,1\,0 \\ 0\,0\,1 \end{bmatrix} + E(c, x, y: p_x, p_y), \tag{13.91}$$

$$E(c, x, y: z_c) = (1\,0\,0) - E(c, x, y: p_x) - E(c, x, y: p_y). \tag{13.92}$$

Note that the elasticities of (w_x/p_x) and (w_y/p_y) can be obtained from (13.58) after a slight rearrangement to incorporate the technology shock. Let $r_x = (w_x/p_x)$ and $r_y = (w_y/p_y)$. Then

$$\begin{pmatrix} \widehat{r}_x \\ \widehat{r}_y \end{pmatrix} = \left(\begin{bmatrix} \beta_0 + b_0 & \beta_2 + b_2 \\ \gamma_0 + c_0 & \gamma_2 + c_2 - 1 \end{bmatrix} [N]^{-1} \right) \begin{pmatrix} \widehat{p}_x + \widehat{z}_x - \widehat{z}_c \\ \widehat{p}_y + \widehat{z}_y - \widehat{z}_c \end{pmatrix} + \begin{pmatrix} \widehat{z}_x \\ \widehat{z}_y \end{pmatrix}.$$

In terms of elasticities, this implies that

$$E(r_x, r_y: z_x, z_y) = I + E(r_x, r_y: p_x, p_y), \tag{13.93}$$

$$E(r_x, r_y: z_c) = -E(r_x, r_y: p_x) - E(r_x, r_y: p_y). \tag{13.94}$$

Note that $E(r_x, r_y: p_x, p_y)$ is simply given by the elements of the matrix

$$\left(\begin{bmatrix} \beta_0 + b_0 & \beta_2 + b_2 \\ \gamma_0 + c_0 & \gamma_2 + c_2 - 1 \end{bmatrix} [N]^{-1} \right).$$

Now we have all the elements to evaluate the Jacobian corresponding to the linear system evaluated at the steady-state. The relevant partial derivatives can be computed from the associated elasticities using steady-state values. To compute the appropriate Jacobian we note first that at the steady-state, $\rho((w_{y,t+1})/(p_{y,t+1}) + (1 - g_y)) = 1$. A similar equation holds for the linearization of (13.87). We can express the linearized dynamics as percentage deviations from the steady state with the help of the following matrices:

$$[R_{11}] = \begin{bmatrix} \left(\frac{k_x((\partial x/\partial k_x)+1-g)}{x+(1-g)k_x}\right) & \left(\frac{k_y(\partial x/\partial k_y)}{x+(1-g)k_x}\right) \\ \left(\frac{k_x(\partial y/\partial k_x)}{y+(1-g)k_y}\right) & \left(\frac{k_y((\partial y/\partial k_y)+1-g)}{y+(1-g)k_y}\right) \end{bmatrix},$$

$$[R_{12}] = \begin{bmatrix} \left(\frac{p_x(\partial x/\partial p_x)}{z_x+(1-g)k_x}\right) & \left(\frac{p_y((\partial x/\partial p_y)}{z_x+(1-g)k_x}\right) \\ \left(\frac{p_x(\partial y/\partial p_x)}{y+(1-g)k_y}\right) & \left(\frac{p_y(\partial y/\partial p_y)}{y+(1-g)k_y}\right) \end{bmatrix},$$

$$[R_{21}] = \begin{bmatrix} -\sigma\left(\frac{k_x}{c}\right)\left(\frac{\partial c}{\partial k_x}\right) & -\sigma\left(\frac{k_y}{c}\right)\left(\frac{\partial c}{\partial k_y}\right) \\ -\sigma\left(\frac{k_x}{c}\right)\left(\frac{\partial c}{\partial k_x}\right) & -\sigma\left(\frac{k_y}{c}\right)\left(\frac{\partial c}{\partial k_y}\right) \end{bmatrix},$$

$$[R_{22}] = \begin{bmatrix} -\sigma\left(\frac{p_x}{c}\right)\left(\frac{\partial c}{\partial p_x}\right)+1 & -\sigma\left(\frac{p_y}{c}\right)\left(\frac{\partial c}{\partial p_y}\right) \\ -\sigma\left(\frac{p_x}{c}\right)\left(\frac{\partial c}{\partial p_x}\right) & -\sigma\left(\frac{p_y}{c}\right)\left(\frac{\partial c}{\partial p_y}\right)+1 \end{bmatrix},$$

$$[R_{13}] = \begin{bmatrix} \left(\frac{z_c(\partial x/\partial z_c)}{x+(1-g)k_x}\right) & \left(\frac{z_x(\partial x/\partial z_x)}{x+(1-g)k_x}\right) & \left(\frac{z_y((\partial x/\partial z_y)}{x+(1-g)k_x}\right) \\ \left(\frac{z_c(\partial y/\partial z_c)}{y+(1-g)k_y}\right) & \left(\frac{z_x(\partial y/\partial z_x)}{y+(1-g)k_y}\right) & \left(\frac{z_y(\partial y/\partial z_y)}{y+(1-g)k_y}\right) \end{bmatrix},$$

$$[R_{23}] = \begin{bmatrix} -\sigma\left(\frac{z_c}{c}\right)\left(\frac{\partial_c}{\partial z_c}\right) & -\sigma\left(\frac{z_x}{c}\right)\left(\frac{\partial_c}{\partial z_x}\right) & -\sigma\left(\frac{z_y}{c}\right)\left(\frac{\partial_c}{\partial z_y}\right) \\ -\left(\frac{z_c}{c}\right)\left(\frac{\partial_c}{\partial z_c}\right) & -\sigma\left(\frac{z_x}{c}\right)\left(\frac{\partial_c}{\partial z_x}\right) & -\sigma\left(\frac{z_y}{c}\right)\left(\frac{\partial_c}{\partial z_y}\right) \end{bmatrix},$$

$$[R_{31}] = [R_{32}] = \begin{bmatrix} 0 & 0 \\ 0 & 0 \\ 0 & 0 \end{bmatrix},$$

$$[R_{33}] = \begin{bmatrix} \lambda_c & 0 & 0 \\ 0 & \lambda_x & 0 \\ 0 & 0 & \lambda_y \end{bmatrix},$$

$$[R] = \begin{bmatrix} [R_{11}] & [R_{12}] & [R_{13}] \\ [R_{21}] & [R_{22}] & [R_{23}] \\ [R_{31}] & [R_{32}] & [R_{33}] \end{bmatrix},$$

$$[Q_{11}] = \begin{bmatrix} 1 & 0 \\ 0 & 1 \end{bmatrix},$$

$$[Q_{12}] = \begin{bmatrix} 0 & 0 \\ 0 & 0 \end{bmatrix},$$

$$[Q_{13}] = \begin{bmatrix} 0 & 0 & 0 \\ 0 & 0 & 0 \end{bmatrix},$$

$$[Q_{21}] = \begin{bmatrix} \left(\frac{-\sigma k_x}{c}\right)\left(\frac{\partial c}{\partial k_x}\right) & \left(\frac{-\sigma k_y}{c}\right)\left(\frac{\partial c}{\partial k_y}\right) \\ \left(\frac{-\sigma k_x}{c}\right)\left(\frac{\partial c}{\partial k_x}\right) & \left(\frac{-\sigma k_y}{c}\right)\left(\frac{\partial c}{\partial k_y}\right) \end{bmatrix},$$

$$[Q_{22}] = \begin{bmatrix} \rho p_x\left(\frac{\partial r_x}{\partial p_x}\right)-\left(\frac{\sigma p_x}{c}\right)\left(\frac{\partial c}{\partial p_x}\right)+1 & \rho p_y\left(\frac{\partial r_x}{\partial p_y}\right)-\left(\frac{\sigma p_y}{c}\right)\left(\frac{\partial c}{\partial p_y}\right) \\ \rho p_x\left(\frac{\partial r_x}{\partial p_x}\right)-\left(\frac{\sigma p_x}{c}\right)\left(\frac{\partial c}{\partial p_x}\right) & \rho p_y\left(\frac{\partial r_x}{\partial p_y}\right)-\left(\frac{\sigma p_y}{c}\right)\left(\frac{\partial c}{\partial p_y}\right)+1 \end{bmatrix},$$

$$[Q_{23}] = \begin{bmatrix} \rho z_c \left(\frac{\partial r_x}{\partial z_c}\right) - \left(\frac{\sigma z_c}{c}\right)\left(\frac{\partial c}{\partial z_c}\right) & \rho z_x \left(\frac{\partial r_x}{\partial z_z}\right) - \left(\frac{\sigma z_x}{c}\right)\left(\frac{\partial c}{\partial z_x}\right) & \rho z_y \left(\frac{\partial r_x}{\partial z_y}\right) - \left(\frac{\sigma z_y}{c}\right)\left(\frac{\partial c}{\partial z_y}\right) \\ \rho z_c \left(\frac{\partial r_y}{\partial z_c}\right) - \left(\frac{\sigma z_c}{c}\right)\left(\frac{\partial c}{\partial z_c}\right) & \rho z_x \left(\frac{\partial r_y}{\partial z_x}\right) - \left(\frac{\sigma z_x}{c}\right)\left(\frac{\partial c}{\partial z_x}\right) & \rho z_y \left(\frac{\partial r_y}{\partial z_y}\right) - \left(\frac{\sigma z_y}{c}\right)\left(\frac{\partial c}{\partial z_y}\right) \end{bmatrix},$$

$$[Q_{31}] = [Q_{32}] = [R_{31}],$$

$$[Q_{33}] = \begin{bmatrix} 1 & 0 & 0 \\ 0 & 1 & 0 \\ 0 & 0 & 1 \end{bmatrix},$$

$$[Q] = \begin{bmatrix} [Q_{11}] & [Q_{12}] & [Q_{13}] \\ [Q_{21}] & [Q_{22}] & [Q_{23}] \\ [Q_{31}] & [Q_{32}] & [Q_{33}] \end{bmatrix}.$$

The linearized dynamics are then:

$$\begin{pmatrix} \widehat{k}_{x,t+1} \\ \widehat{k}_{y,t+1} \\ \widehat{p}_{x,t+1} \\ \widehat{p}_{y,t+1} \\ \widehat{z}_{c,t+1} \\ \widehat{z}_{x,t+1} \\ \widehat{z}_{y,t+1} \end{pmatrix} = [Q]^{-1}[R] \begin{pmatrix} \widehat{k}_{x,t} \\ \widehat{k}_{y,t} \\ \widehat{p}_{x,t} \\ \widehat{p}_{y,t} \\ \widehat{z}_{c,t} \\ \widehat{z}_{x,t} \\ \widehat{z}_{y,t} \end{pmatrix} + [Q]^{-1} \begin{pmatrix} 0 \\ 0 \\ -\widehat{s}_{x,t+1} \\ -\widehat{s}_{y,t+1} \\ \widehat{\varepsilon}_{c,t+1} \\ \widehat{\varepsilon}_{x,t+1} \\ \widehat{\varepsilon}_{y,t+1} \end{pmatrix} \tag{13.95}$$

where $\widehat{s}_{i,t}$, $i = x, y$ is an iid sunspot shock with zero mean, acting on the "Euler" equations for the two capital stocks. Note that the elements of the matrices $[R]$ and $[Q]$ are functions of the parameters of the system, and also of the steady-state quantities which are functions of the parameters as well. We can therefore evaluate the roots of $[Q]^{-1}[R]$ to check for the possibility of indeterminacy. When externality parameters are set to zero, as is well-known, the four of the roots of the Jacobian matrix come in pairs of $(\mu, 1/\rho\mu)$, and the other three are the autoregressive coefficients of the technology shocks. For modest externalities, however, it is easy to find large parameter regions for which there exist indeterminate equilibria, as the calibrated example in Sect. 13.4 illustrates.

13.6.1 The Calibration

Equation 13.95 can easily be used to simulate or assess the stochastic properties of our dynamic system. In order to then obtain series for outputs c, x, and y, as well as their inputs, we must first express them as functions of (p_x, p_y, k_x, k_y). Using (13.43)–(13.45), and (13.73), we can set up the matrix equation

$$\begin{pmatrix} \widehat{\omega}_{10} \\ \widehat{\omega}_{12} \\ \widehat{\omega}_{10} \\ \widehat{\omega}_{12} \\ \widehat{\omega}_{10} \\ \widehat{\omega}_{12} \\ \widehat{k}_x \\ \widehat{k}_y \\ 0 \end{pmatrix} = \begin{pmatrix} [N]^{-1} \begin{pmatrix} \widehat{p}_x \\ \widehat{p}_y \end{pmatrix} \\ [N]^{-1} \begin{pmatrix} \widehat{p}_x \\ \widehat{p}_y \end{pmatrix} \\ [N]^{-1} \begin{pmatrix} \widehat{p}_x \\ \widehat{p}_y \end{pmatrix} \\ \widehat{k}_x \\ \widehat{k}_y \\ 0 \end{pmatrix}$$

$$= \begin{pmatrix} 0 & 0 & 0 & -1 & 0 & 0 & 1 & 0 & 0 \\ 1 & 0 & 0 & -1 & 0 & 0 & 0 & 0 & 0 \\ 0 & 0 & 0 & 0 & -1 & 0 & 0 & 1 & 0 \\ 0 & 1 & 0 & 0 & -1 & 0 & 0 & 0 & 0 \\ 0 & 0 & 0 & 0 & 0 & -1 & 0 & 0 & 1 \\ 0 & 0 & 1 & 0 & 0 & -1 & 0 & 0 & 0 \\ 0 & 0 & 0 & \frac{K_{xc}}{k_x} & \frac{K_{xx}}{k_x} & \frac{K_{xy}}{k_x} & 0 & 0 & 0 \\ \frac{K_{yc}}{k_y} & \frac{K_{yx}}{k_y} & \frac{K_{yy}}{k_y} & 0 & 0 & 0 & 0 & 0 & 0 \\ -x_2 & 0 & 0 & -x_1 & 0 & 0 & \frac{L_c}{L} - x_0 & \frac{L_x}{L} & \frac{L_y}{L} \end{pmatrix} \begin{pmatrix} \widehat{K}_{yc} \\ \widehat{K}_{yx} \\ \widehat{K}_{yy} \\ \widehat{K}_{xc} \\ \widehat{K}_{xx} \\ \widehat{K}_{xy} \\ \widehat{L}_c \\ \widehat{L}_x \\ \widehat{L}_y \end{pmatrix}, \tag{13.96}$$

where x_0, x_1, x_2 are as in (13.73). Equation 13.96 , after inverting will solve for $\widehat{K}_{xc}$ and $\widehat{K}_{yc}$ in terms of $(\widehat{p}_x, \widehat{p}_y, \widehat{k}_x, \widehat{k}_y)$, so that we can obtain the associated elasticities. We now have all the elements of the Jacobian and we can analytically compute the variance-covariance matrix of the variables in (13.95), for contemporaneous as well as lagged values. Furthermore, using these we can easily compute a larger variance-covariance matrix that includes linear functions of the original variables, like the outputs of the three goods, and the aggregate value of investment or GNP.

References

Basu, S. and J. G. Fernald (1994a), Are Apparent Productive Spillovers a Figment of Specification Error?, International Finance Discussion Paper 463, Board of Governors of the Federal Reserve System.

Basu, S. and J. G. Fernald (1994b), Constant Returns and Small Markups in U.S. Manufacturing, International Finance Discussion Paper 483, Board of Governors of the Federal Reserve System.

Baxter, M. and R. King (1991), Productive Externalities and Business Cycles, Discussion Paper 53, Institute for Empirical Macroeconomics, Federal Reserve Bank of Minneapolis.

Beaudry, P. and M. Devereux (1993), Monopolistic Competition, Price Setting and the Effects of Real and Monetary Shocks, mineo.

Benhabib, J. and K. Nishimura (1979), "The Hopf Bifurcation and the Existence and Stability of Closed Orbits in Multisector Models of Optimal Economic Growth," *Journal of Economic Theory*, 21, 421-444.

Benhabib, J. and R. Farmer (1994), "Indeterminacy and Growth," *Journal of Economic Theory*, 63, 19-41.

Benhabib, J. and R. Perli (1994), "Uniqueness and Indeterminacy: Transitional Dynamics in a Model of Endogenous Growth," *Journal of Economic Theory*, 63, 113-142.

Benhabib, J., and A. Rustichini (1994), "Introduction to the Symposium on Growth, Fluctuations and Sunspots," *Journal of Economic Theory*, 63, 1-19.

Benhabib, J., Perli, R. and D. Xie (1994), "Monopolistic Competition, Indeterminacy and Growth," *Recherche Economique*, 48, 279-298.

Benhabib, J. and R. E. Farmer (1996a), "Indeterminacy and Sector Specific Externalities," *Journal of Monetary Economics*, 37, 397-419.

Benhabib, J. and R. E. Farmer (1996b), The Monetary Transmission Mechanism, C. V. Starr Center of Applied Economics, Working paper 96-13, New York University.

Benhabib, J., Perli, R. and P. Sakellaris (1997), Persistence of Business Cycles in Multisector RBC Models, C. V. Starr Working Paper 97-19, New York University.

Black, F. (1974), "Uniqueness of the Price Level in a Monetary Growth Model with Rational Expectations," *Journal of Economic Theory*, 7, 53-65.

Boldrin, M. and A. Rustichini (1994), "Indeterminacy of Equilibria in Models with Infinitely-Lived Agents and External Effects," *Econometrica*, 62, 323-342.

Burnside, C., Eichenbaum, M., and S. Rebelo (1995), "Capacity Utilization and Returns to Scale," *NBER Macroeconomics Annual*, 10, 67-110.

Burnside, C. (1996), "Production Function Regressions, Returns to Scale, and Externalities," *Journal of Monetary Economics*, 37, 177-201.

Chatterjee, S. and R. Cooper (1989), "Multiplicity of Equilibria and Fluctuations in Dynamic Imperfectly Competitive Economies," *American Economic Review*, 79, 353-357.

Christiano, L. J. and S. G. Harrison (1996), Chaos, Sunspots and Automatic Stabilizers, NBER Working paper 5703.

Farmer R. E. A. and J. T. Guo (1994), "Real Business Cycles and the Animal Spirits Hypothesis," *Journal of Economic Theory*, 63, 42-73.

Farmer R. E. A. and J. T. Guo (1995), "The Econometrics of Indeterminacy: An Applied Study," *Carnegie-Rochester Conference Series on Public Policy*, 43, 225-272.

Gali, J. (1994), "Monopolistic Competition, Business Cycles, and the Composition of Aggregate Demand," *Journal of Economic Theory*, 63, 73-96.

Gali, J. and F. Zilibotti (1995), "Endogenous Growth and Poverty Traps in a Cournotian Model," *Annals of Economics and Statistics*, 37/38, 197-213.

Hall, R. E. (1988), "The Relation between Price and Marginal Cost in U.S. Industry," *Journal of Political Economy*, 96, 921-948.

Hall, R. E. (1990), Invariance Properties of Solow's Productivity Residual, in *Growth, Productivity, Unemployment*, ed. by Diamond, P., MIT Press, Cambridge, MA., 71-112.

Huffman, G. W. and M. A. Wynne (1996), The Role of Intertemporal Adjustment Costs in a Multi-Sector Economy, working paper, Southern Methodist University.

Kydland F. E. and E. C. Prescott (1990), "Business Cycles: Real Facts and a Monetary Myth," *Federal Reserve Bank of Minneapolis Quarterly Review*, Spring, 3-18.

Perli, R. (1994), Indeterminacy, Home Production and the Business Cycle: A Calibration Analysis, New York University Working Paper.

Rotemberg, J. J. and M. Woodford (1992), "Oligopolistic Pricing and the Effects of Aggregate Demand on Economic Activity," *Journal of Political Economy*, 100, 1153-1207.

Schmitt-Grohé, S. (1997), "Comparing Four Models of Aggregate Fluctuations due to Self-Fulfilling Expectations," *Journal of Economic Theory*, 72, 96-146.

Schmitt-Grohé, S. and M. Uribe (1996), Balanced Budget Rules, Distortionary Taxes, and Aggregate Instability, working paper, Board of Governors of the Federal Reserve System, Washington.

Weder, M. (1996), Indeterminacy, Business Cycles and Modest Increasing Returns, Humbolt University Working Paper.

Xie, Dangyang (1994), "Divergence in Economic Performance: Transitional Dynamics with Multiple Equilibria," *Journal of Economic Theory*, 63, 97-112.

Chapter 14
Trade and Indeterminacy in a Dynamic General Equilibrium Model*

Kazuo Nishimura and Koji Shimomura**

14.1 Introduction

This study investigates the dynamic behavior of multiple countries' economic activities in a two-good, two-factor model in which factors are internationally immobile and countries' technologies are subject to sector-specific external effects.

It has been demonstrated that in a perfect foresight model with many consumers, a competitive equilibrium path behaves like an optimal growth path; see Becker (1980), Bewley (1982), Yano (1984), and Epstein (1987). As these results suggest, a perfect foresight equilibrium path may exhibit the same behavior as in a single consumer model even in a many consumer model such as a large-country trade model. Nishimura and Yano (1993a,b) studied the interlinkage of business cycles between large countries in the discrete time model. Atkeson and Kehoe (2000) applied a large-country Heckscher-Ohlin (H-O) model to the analysis of development patterns. Bond et al. (2000) characterized an integrated world equilibrium path in a dynamic H-O model. Ghiglino and Olszak-Duquenne (2001) investigated economic fluctuations in a two-country dynamic general equilibrium model. However, indeterminacy in a disaggregated model of world economy has not been characterized in the existing literature. Indeterminacy means that there

*Journal of Economic Theory 105, 244–260, 2002.

**We thank Christian Ghiglino, Murray Kemp, and Takumi Naito for their useful comments on the earlier version of the paper.

K. Nishimura (✉)
Institute of Economic Research, Kyoto University, Japan
e-mail: nishimura@kier.kyoto-u.ac.jp

K. Shimomura
Research Institute for Economics & Business Administration, Kobe University, Japan

J. Stachurski et al. (eds.), *Nonlinear Dynamics in Equilibrium Models*,
DOI 10.1007/978-3-642-22397-6_14,

exists a continuum of equilibria starting from the same initial condition, all of which converge to a steady state.

Recently there has been a growing literature on the existence of indeterminate equilibria in dynamic general equilibrium economies. While the earlier results on indeterminacy relied on relatively large increasing returns, Benhabib and Nishimura (1998) and Benhabib et al. (2000) showed that in multisector models indeterminacy can arise with constant social returns to scale if there is a small wedge between private and social returns.

The present paper extends the H-O model by introducing sector-specific externalities. Given a two (-country) by two (-good) by two (-factor) model of international trade in which production technologies are subject to constant returns to scale and preferences are homothetic, and in which the difference between the two countries is only in the factor endowment ratio, the H-O Theorem tells us that each country exports such a good that the abundantly endowed factor of production is intensively used for producing it.

The H-O Theorem is a result in a static framework. Formulating a two-sector dynamic general equilibrium model in which capital accumulation is taken into account, Chen (1992) studied a dynamic version of the H-O model.

In the present paper we introduce sector-specific externalities in the dynamic general equilibrium model and show that indeterminacy of the equilibrium path in the world market can occur. It follows that there are multiple equilibrium paths from the same initial distribution of capital in the world market, and the distribution of capital in the limit differs among equilibrium paths, and one equilibrium path converges to a long-run equilibrium in which the international ranking of factor endowment ratios differs from the initial ranking, whereas another equilibrium path maintains the initial ranking and converges to another long-run equilibrium. Since the path realized is indeterminate, so is the long-run trade pattern. Therefore the Long-Run H-O prediction is vulnerable to the introduction of externality.

In Sect. 14.2 we present the dynamic H-O model in a continuous time model with factor-generated externalities. In Sect. 14.3 we show the existence of a continuum of long-run equilibria. In Sect. 14.4 we derive indeterminacy results and discuss implications for the distribution of capital in the long-run equilibrium. Section 14.5 concludes.

14.2 The Dynamic Two-Country Model

We shall formulate the continuous-time version of the dynamic two-country model.

14.2.1 The Production Side

Two goods, a consumption good and an investment good, are produced using two factors of production, capital, and labor. The home and the foreign countries are endowed with the same fixed amount of labor l and capital stocks k and k^*, respectively. The first good is the investment good and the second good is the consumption good.

Following Benhabib et al. (2000), we assume Cobb-Douglas technologies which are constant returns to scale from the social perspective but decreasing returns to scale from the private perspective due to factor-generated externalities. That is, the industry production function of good $i, i = 1, 2$, is

$$y_i = l_i^{a_i} k_i^{b_i} \overline{l_i}^{\alpha_i} \overline{k_i}^{\beta_i}, \tag{14.1}$$

where $a_i + \alpha_i + b_i + \beta_i = 1$ and all the parameters are positive. $\overline{l_i}^{\alpha_i} \overline{k_i}^{\beta_i}$, $i = 1, 2$, are externality terms. Defining $\theta_i \equiv a_i + \alpha_i$ and under $l_i = \overline{l}_i$ and $k_i = \overline{k}_i$, profit maximization of each firm implies

$$w = a_1 l_1^{\theta_1 - 1} k_1^{1-\theta_1}, \quad r = b_1 l_1^{\theta_1} k_1^{-\theta_1} \tag{14.2a}$$

$$w = pa_2 l_1^{\theta_2 - 1} k_1^{1-\theta_2}, \quad r = pb_2 l_2^{\theta_2} k_2^{-\theta_2}. \tag{14.2b}$$

Let

$$W_i \equiv \frac{\theta_i w}{a_i} \text{ and } R_i \equiv \frac{(1-\theta_i) r}{b_i} \quad i = 1, 2. \tag{14.3}$$

From (14.2) and (14.3) we can derive the "virtual" average cost = price condition for each good.

$$1 = \overline{c}^1(W_1, R_1) \tag{14.4a}$$

$$p = \overline{c}^2(W_2, R_2) \tag{14.4b}$$

$$\overline{c}_W^1(W_1, R_1) y_1 + \overline{c}_W^2(W_2, R_2) y_2 = l \tag{14.4c}$$

$$\overline{c}_R^1(W_1, R_1) y_1 + \overline{c}_R^2(W_2, R_2) y_2 = k, \tag{14.4d}$$

where $\overline{c}_W^i(W_i, R_i) \equiv \frac{\partial}{\partial W_i} \overline{c}^i(W_i, R_i)$ and $\overline{c}_R^i(W_i, R_i) \equiv \frac{\partial}{\partial R_i} \overline{c}^i(W_i, R_i)$, $i = 1, 2$. Note that

$$\frac{W_i \overline{c}_W^i(W_i, R_i)}{\overline{c}^i(W_i, R_i)} = \theta_i$$

and

$$\frac{R_i \overline{c}_R^i(W_i, R_i)}{\overline{c}^i(W_i, R_i)} = 1 - \theta_i.$$

Equations 14.3 and 14.4 determine w, r, y_1, and y_2 for given p and k. Under Cobb-Douglas technologies and substituting (14.3) into (14.4), we see that, as far as $\theta_1 \neq \theta_2$, (14.4a) and (14.4b) have a unique pair $(w(p), r(p))$ for any given p. Logarithmically differentiating the pair with respect to p, we obtain the Stolper-Samuelson properties

$$\frac{pw'(p)}{w(p)} = -\frac{1-\theta_1}{\theta_1-\theta_2} \quad \text{and} \quad \frac{pr'(p)}{r(p)} = \frac{\theta_1}{\theta_1-\theta_2}, \tag{14.5}$$

where $w'(p) \equiv \frac{d}{dp}w(p)$ and $r'(p) \equiv \frac{d}{dp}r(p)$. If $\theta_1 > (<)\theta_2$, we say that the consumption good is capital (labor) intensive from the social perspective. On the other hand, if $\Delta \equiv a_1b_2 - a_2b_1 > (<)0$, then we say that the consumption good is capital (labor) intensive from the private perspective. The model becomes the standard dynamic Heckscher-Ohlin model if $\alpha_i = \beta_i = 0, i = 1, 2$.

Remark 1. As technologies are internationally identical, the same domestic price $(w(p), r(p))$ prevails in the foreign country as well. Thus, factor price equalization concerning still holds in a model with externalities as long as the social returns are constant. This is an extension of factor price equalization in the literature.[1]

14.2.2 The Consumption Side

The demand for the consumption good, z, is based on the dynamic optimization problem the representative household in each country faces. The home household is assumed to maximize the discounted sum of its utilities

$$\int_0^\infty \frac{z^{1-\eta}}{1-\eta}e^{-\rho t}dt, \quad 0 < \eta < 1, \rho > 0. \tag{14.6}$$

Note that in the present case the total factor income $wl + rk$ is not necessarily equal to $y_1 + py_2$, because of the presence of externalities. The gap $\Pi \equiv (y_1 + py_2) - (wl + rk)$ can be interpreted as profits or the remuneration for sector-specific factor of production.[2] Thus, the home household maximizes the discounted sum of its utility (14.6) subject to the flow budget constraint

$$\begin{aligned}\dot{k} &= y_1 + py_2 - pz - \delta k \\ &= (wl + rk) + \Pi - pz - \delta k,\end{aligned} \tag{14.7}$$

where $\delta(> 0)$ is the rate of capital depreciation. The Hamiltonian associated with the problem is

---

[1]We thank Murray Kemp for pointing this out to us. See Kemp and Okawa (1998) concerning recent issues on factor price equalization. Note that factor price equalization holds only for interindustry-mobile factors of production.

[2]Benhabib and Nishimura (1998) assume a fixed cost of entry which makes profits possible. Alternatively, Nishimura and Shimomura (2001) assume that there exist sector-specific factors of production in both sectors and externalities may be negative and study the case that production technology in each industry is subject to constant returns to scale from both the private and social perspectives. Under the latter assumption, Π is interpreted as the remuneration of sector-specific factors of production, compatible with free entry and exit.

$$H = \frac{z^{1-\eta}}{1-\eta} + \lambda[wl + rk + \Pi - pz - \delta k]. \tag{14.8}$$

The necessary conditions for optimality are the first-order condition

$$z^{-\eta} = \lambda p, \tag{14.9}$$

the differential equation of the co-state variable

$$\dot{\lambda} = \lambda[\rho + \delta - r(p)], \tag{14.10}$$

and the transversality condition

$$\lim_{t \to \infty} k(t)\lambda(t)e^{-\rho t} = 0. \tag{14.11}$$

Note that the transversality condition is a necessary condition in the present model.[3] We assume that both countries have the same technologies and preferences. Thus, the foreign household also faces the same dynamic optimization problem, and we obtain

$$\begin{aligned} \dot{k}^* &= y_1^* + py_2^* - pz^* - \delta k^* \\ &= w(p)l^* + r(p)k^* + \Pi^* - pz^* - \delta k^* && (14.12a) \\ [z^*]^{-b} &= \lambda^* p && (14.12b) \\ \dot{\lambda}^* &= \lambda^*[\rho + \delta - r(p)] && (14.12c) \\ \lim_{t \to \infty} k^*(t)\lambda^*(t)e^{-\rho t} &= 0. && (14.12d) \end{aligned}$$

The asterisk (*) indicates the variables belonging to the foreign country.

Finally, the world market-clearing condition for the consumption good is

$$z + z^* = y_2 + y_2^*. \tag{14.13}$$

The market-clearing condition for the investment good is obtained from (14.7), (14.12a) and (14.13).

14.3 A Long-Run Equilibrium Under Incomplete Specialization

We shall prove that the stationary states of the system, (14.7), and (14.9)–(14.13), exist under incomplete specialization. For this purpose we will assume the following.

[3] See Kamihigashi (2001).

Assumption 5 *The consumption good is labor intensive from the social perspective but capital intensive from the private perspective, i.e.,* $\theta_1 < \theta_2$ *and* $\Delta > 0$.

Assumption 6 $1/\eta > \max[1, 1/\widetilde{\eta}]$, *where*

$$\widetilde{\eta}^{-1} \equiv \frac{\theta_1 a_1 b_2(\rho+\delta) + (1-\theta_1)b_1\{\rho a_2 + \delta a_1 b_2 + (1-b_1)a_2\delta\}}{\Delta(\theta_2-\theta_1)\{\rho+\delta(1-b_1)\}}.$$

Benhabib and Nishimura (1998) derived indeterminacy in their closed economy model under Assumption 5. Note that Assumption 6 is satisfied for a sufficiently small η, which means that the felicity function is close to be linear in consumption.[4]

Since the pair of factor prices $(w(p), r(p))$ is established in both countries, we also see from (14.4c) and (14.4d) that as long as production is incompletely specialized,[5]

$$y_1 + py_2 = \frac{r(p)k(a_1-a_2) - w(p)l(b_1-b_2)}{\Delta}, \tag{14.14a}$$

$$y_1^* + py_2^* = \frac{r(p)k^*(a_1-a_2) - w(p)l(b_1-b_2)}{\Delta}, \tag{14.14b}$$

$$y_2 = \frac{r(p)a_1 k - w(p)b_1 l}{p\Delta}, \tag{14.14c}$$

$$y_2^* = \frac{r(p)a_1 k^* - w(p)b_1 l}{p\Delta}. \tag{14.14d}$$

Making use of (14.14), the dynamic general equilibrium model can be written as

$$\dot{k} = \frac{r(p)k(a_1-a_2) - w(p)l(b_1-b_2)}{\Delta} - p[\lambda p]^{-\frac{1}{\eta}} - \delta k \tag{14.15a}$$

$$\dot{k}^* = \frac{r(p)k^*(a_1-a_2) - w(p)l(b_1-b_2)}{\Delta} - p[\lambda^* p]^{-\frac{1}{\eta}} - \delta k^* \tag{14.15b}$$

$$\dot{\lambda} = \lambda[\rho + \delta - r(p)] \tag{14.15c}$$

$$\dot{\lambda}^* = \lambda^*[\rho + \delta - r(p)] \tag{14.15d}$$

$$[\lambda p]^{-\frac{1}{\eta}} + [\lambda^* p]^{-\frac{1}{\eta}} = \frac{r(p)a_1(k+k^*) - 2lw(p)b_1}{p\Delta}. \tag{14.15e}$$

If a solution path $(k(t), k^*(t), \lambda(t), \lambda^*(t), p(t))$ of (14.15) satisfies the transversality conditions (14.11) and (14.12d), we say that $(k(t), k^*(t))$ is called an

[4] Benhabib and Nishimura (1998) assumed a linear felicity function.

[5] See Appendix 14.6 for the derivation of (14.14).

equilibrium from $(k(0), k^*(0))$. We shall analyze the system in the following way. First, let $K \equiv k + k^*$ and $L \equiv 2l$. Summing (14.15a) and (14.15b), we have

$$\dot{K} = \frac{b_2 w(p)L - a_2 r(p)K}{\Delta} - \delta K. \tag{14.16}$$

(14.15c) and (14.15d) mean that $\frac{\lambda^*}{\lambda} \equiv m$, the ratio of the home and foreign marginal utilities of wealth, is constant over time. Therefore, for any given positive m, the solution to the system that is composed of (14.15a), (14.15c), (14.15e), $K = k + k^*$, (14.16) and $\lambda^* = m\lambda$ is equivalent to the system (14.15).

Theorem 1. *(i) In the long-run equilibrium the total capital and the price* (K^e, p^e) *are uniquely determined. (ii) In the long-run equilibrium there exists a continuum of countries' capital* (k^e, k^{e*}) *at which both economies are incompletely specialized.*

Proof. (i) Consider the system of equations

$$\frac{b_2 w(p)L - a_2 r(p)K}{\Delta} - \delta K = 0 \tag{14.17a}$$

$$\rho + \delta = r(p). \tag{14.17b}$$

First, as far as $\theta_1 \neq \theta_2$, for any positive ρ and δ, (14.17b) uniquely determines the stationary-state p, say p^e. Second, substituting p^e into (14.17a) and solving for K, we obtain the unique stationary-state K, K^e. That is,

$$p^e = r^{-1}(\rho + \delta), \tag{14.18a}$$

where $r^{-1}(.)$ is the inverse function of $r(.)$, and

$$K^e = \frac{b_2 w(p^e)L}{\rho a_2 + \delta a_1 b_2 + (1 - b_1)a_2\delta} > 0. \tag{14.18b}$$

The long-run equilibrium in a world trade market is independent of m.

(ii) Given $m > 0$, (14.15a) and (14.15e) yield the following

$$\{(\rho + \delta)(a_1 - a_2) - \delta\Delta\}k = w(p)l(b_1 - b_2) + p[\lambda p]^{-\frac{1}{\eta}}\Delta \tag{14.19a}$$

$$[\lambda p]^{-\frac{1}{\eta}} + [m\lambda p]^{-\frac{1}{\eta}} = \frac{r(p)a_1 K - w(p)b_1 L}{p\Delta}. \tag{14.19b}$$

By substituting (14.18) into (14.19), we have

$$k^e = \frac{w(p^e)l(b_1 - b_2) + p^e[\lambda^e p^e]^{-\frac{1}{\eta}}\Delta}{(\rho + \delta)(a_1 - a_2) - \delta\Delta} \tag{14.20}$$

$$[\lambda^e p^e]^{-\frac{1}{\eta}}(1+m^{-\frac{1}{\eta}}) = \frac{r(p^e)a_1K^e - w(p^e)b_1L}{p^e\Delta}$$
$$= \frac{w(p^e)L\{\rho+\delta(1-b_1)\}}{p^e\{\rho a_2+\delta a_1 b_2+(1-b_1)a_2\delta\}}. \tag{14.21}$$

Also, from $\lambda^* = m\lambda$ and $k^{e*} = K^e - k^e$

$$k^{e*} = \frac{w(p^e)l(b_1-b_2) + p^e[m\lambda^e p^e]^{-\frac{1}{\eta}}\Delta}{(\rho+\delta)(a_1-a_2)-\delta\Delta}. \tag{14.22}$$

Fact 1. If $m = 1$,

$$k^e = k^{e*} = \frac{b_2 w(p^e)l}{\rho a_2+\delta a_1 b_2+(1-b_1)a_2\delta}. \tag{14.23}$$

(See Appendix 14.6 for the proof.)

From Fact 1, (14.20) and (14.22) the following Fact holds, and it completes the proof.

Fact 2. There exists $\overline{m} > 1$. Such that for $\overline{m}^{-1} < m < \overline{m}$, the stationary state under incomplete specialization $(K^e, k^e, \lambda^e, p^e)$ uniquely exists.

Note that λ^e and k^e do depend on the value of m. Subtracting (14.22) from (14.20), we have

$$k^e - k^{e*} = \frac{(\lambda^e)^{-\frac{1}{\eta}}(p^e)^{1-\frac{1}{\eta}}[1-m^{-\frac{1}{\eta}}]\Delta}{(\rho+\delta)(a_1-a_2)-\delta\Delta}.$$

It follows from Assumption 5 that if $(\rho+\delta)(a_1-a_2)-\delta\Delta < 0$,

$$k^e \begin{Bmatrix} < \\ = \\ > \end{Bmatrix} k^{e*} \text{ if and only if } m \begin{Bmatrix} > \\ = \\ < \end{Bmatrix} 1. \tag{14.24}$$

See Fig. 14.1. Equation 14.24 means that under $(\rho+\delta)(a_1-a_2)-\delta\Delta<0$, if $m =$ (respectively $>$ and $<$)1, then the stationary state is on (respectively, below and above) the 45°-line. In the rest of this paper, we make the assumption.

Assumption 7 $(\rho+\delta)(a_1-a_2)-\delta\Delta < 0$.

14.4 Indeterminacy

Under Assumptions 5–7, we can obtain the local indeterminacy result.

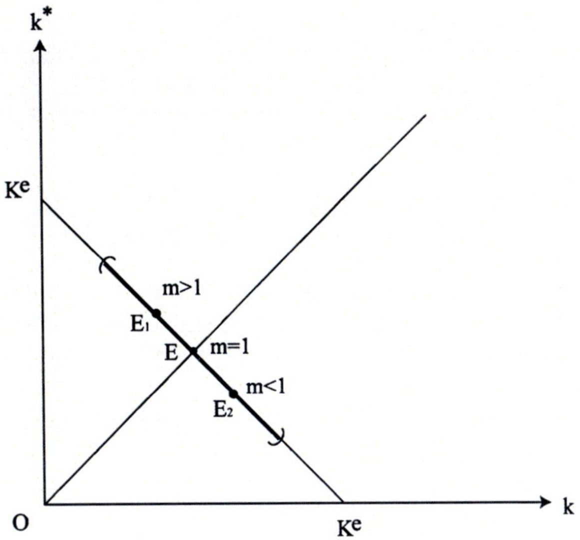

Fig. 14.1
$(\rho + \delta)(a_1 - a_2) - \delta\Delta < 0$

Theorem 2. *There exists a neighborhood of a long-run equilibrium such that from any initial distribution of capital $(k(0), k^*(0))$ in that neighborhood there exists a continuum of equilibrium paths. Moreover, different equilibrium paths converge to different long-run equilibria.*

We consider the dynamic general equilibrium model.

$$\dot{K} = \frac{b_2 w(p)L - a_2 r(p)K}{\Delta} - \delta K \tag{14.25a}$$

$$\dot{\lambda} = \lambda[\rho + \delta - r(p)] \tag{14.25b}$$

$$\dot{k} = \frac{r(p)k(a_1 - a_2) - w(p)l(b_1 - b_2)}{\Delta} - p[\lambda p]^{-\frac{1}{\eta}} - \delta k \tag{14.25c}$$

$$1 + m^{-\frac{1}{\eta}} = \frac{(r(p)a_1 K - w(p)b_1 L)p^{\frac{1-\eta}{\eta}}\lambda^{\frac{1}{\eta}}}{\Delta}. \tag{14.25d}$$

We first prove that for any m in $\left(\overline{m}^{-1}, \overline{m}\right)$ the long-run equilibrium of the system (14.25) is locally stable.

Lemma 1. *Given $m \in \left(\overline{m}^{-1}, \overline{m}\right)$ there is a three-dimensional stable manifold of the system (14.25) in a neighborhood of the stationary state (K^e, k^e, λ^e) such that the solution to the above system converges to the stationary state. $\lambda(0)$ can be arbitrarily chosen.*

Proof. First, totally differentiating (14.15e) with respect to p, K and λ, we have

$$\frac{dp}{\Gamma} = -\frac{r(p^e)a_1(p^e)^{\frac{1}{\eta}}(\lambda^e)^{\frac{1}{\eta}}}{(1+m^{-\frac{1}{\eta}})\Delta}dK - \frac{p^e}{\eta\lambda^e}d\lambda, \tag{14.26}$$

where

$$\Gamma \equiv \left[\frac{1-\eta}{\eta} + \frac{p^e\{r'a_1K^e - w'b_1L\}}{\{ra_1K^e - wb_1L\}}\right]^{-1}.$$

By substituting (14.5), (14.17b), and (14.18a), we get $\Gamma = [\eta^{-1} - \widetilde{\eta}^{-1}]^{-1} > 0$ from Assumption 6. Linearizing the above system around the stationary state and making use of (14.26), we can obtain the characteristic equation

$$F(x) \equiv \begin{vmatrix} x + \left[\frac{ra_2}{\Delta} + \delta + \frac{\Gamma\Omega ra_1(p^e)^{\frac{1}{\eta}}(\lambda^e)^{\frac{1}{\eta}}}{(1+m^{-\frac{1}{\eta}})\Delta^2}\right] & \frac{\Gamma\Omega p^e}{\eta\lambda^e\Delta} & 0 \\ -\frac{\Gamma\lambda^e r' ra_1(p^e)^{\frac{1}{\eta}}(\lambda^e)^{\frac{1}{\eta}}}{(1+m^{-\frac{1}{\eta}})\Delta} & x - \frac{\Gamma r' p^e}{\eta} & 0 \\ * & * & x + \left\{\delta - \frac{r(a_1-a_2)}{\Delta}\right\} \end{vmatrix}$$

$$= \left[x + \frac{\delta\Delta - (\rho+\delta)(a_1-a_2)}{\Delta}\right]$$

$$\times \left[x^2 + \left\{\frac{(\rho+\delta)a_2}{\Delta} + \delta + \frac{\Gamma\Omega ra_1(p^e)^{\frac{1}{\eta}}(\lambda^e)^{\frac{1}{\eta}}}{(1+m^{-\frac{1}{\eta}})\Delta^2} - \frac{\Gamma r' p^e}{\eta}\right\}x\right.$$

$$\left. - \frac{\Gamma r' p^e}{\eta}\left\{\frac{(\rho+\delta)a_2}{\Delta} + \delta\right\}\right]$$

$$= 0, \tag{14.27}$$

where $\Omega \equiv b_2Lw' - a_2K^er'$. In the light of Assumptions 5 and 7, the characteristic equation $F(x) = 0$ has one negative real root

$$-\frac{\delta\Delta - (\rho+\delta)(a_1-a_2)}{\Delta}.$$

Since (14.5) and Assumption 5 imply $r' < 0$ and $\Omega > 0$, we see that

$$\frac{(\rho+\delta)a_2}{\Delta} + \delta + \frac{\Gamma\Omega ra_1(p^e)^{\frac{1}{\eta}}(\lambda^e)^{\frac{1}{\eta}}}{(1+m^{-\frac{1}{\eta}})\Delta^2} - \frac{\Gamma r' p^e}{\eta} > 0 \tag{14.28}$$

and

$$-\frac{\Gamma r' p^e}{\eta}\left\{\frac{(\rho+\delta)a_2}{\Delta} + \delta\right\} = -\frac{\Gamma r' p^e}{\eta\Delta}[\rho a_2 + \delta a_1 b_2 + \delta a_2(1-b_1)] > 0. \tag{14.29}$$

Therefore, the other two roots also have negative real parts. It follows that the characteristic equation $F(x) = 0$ has three roots with negative real parts. This fact implies the Lemma. □

Proof of Theorem 2. From Lemma 1, even if m is given, for any choice of λ an equilibrium path converges to a long-run equilibrium. Moreover, if the value of m varies, an equilibrium path from given initial stock $(k\,(0)\,, k^*\,(0))$ and, from (14.20) and (14.22), the long-run equilibrium (k^e, k^{e*}) it converges to also varies. Since m is a ratio of the two jump variables λ and λ^*, there is no economic mechanism in the decentralized world economy which chooses a particular value of m. This completes the proof of Theorem 2. □

14.5 Dynamic Heckscher-Ohlin Theorem

Let us examine the long-run trade pattern in the long-run equilibrium. First, (14.21) ensures us that if m is larger, λ^e is smaller. Second, differentiating the long-run home demand for the consumption good,

$$D_2 = [\lambda^e p^e]^{-\frac{1}{\eta}}$$

and the stationary-state supply of it

$$S_2 = \frac{r(p^e)a_1k^e - w(p^e)b_1l}{p^e\Delta}$$

with respect to λ^e, and making use of (14.20), we have

$$\begin{aligned}
&\frac{d}{d\lambda^e}\left[[\lambda^e p^e]^{-\frac{1}{\eta}} - \frac{(\rho+\delta)a_1k^e - w(p^e)b_1l}{p^e\Delta}\right] \\
&= \left(-\frac{1}{\eta}\right)(p^e)^{-\frac{1}{\eta}}(\lambda^e)^{-\frac{1}{\eta}-1} - \frac{(\rho+\delta)a_1}{p^e\Delta}\frac{dk^e}{d\lambda^e} \\
&= \frac{(p^e)^{-\frac{1}{\eta}}(\lambda^e)^{-\frac{1}{\eta}-1}\{\rho a_2 + \delta a_1b_2 + \delta a_2(1-b_1)\}}{\eta\{(\rho+\delta)(a_1-a_2) - \delta\Delta\}},
\end{aligned} \tag{14.30}$$

which is negative from Assumption 7.

Lemma 2. *In the long-run equilibrium the capital-abundant country exports (respectively, imports) the consumption good.*

Proof. First, notice that if $m = 1$, the long-run home excess demand for the consumption good, $D_2 - S_2$, is equal to zero. Second, Assumption 7 ensures us that (14.30) is negative, which means that $D_2 - S_2 <$ (respectively $>$)0 if $m <$ (respectively $>$)1. For, (14.21) tells us that m is negatively related to λ^e. Finally,

Fig. 14.2 $EK^e/OK^e = 1/2$ and $(\rho + \delta)(a_1 - a_2) - \delta\Delta < 0$

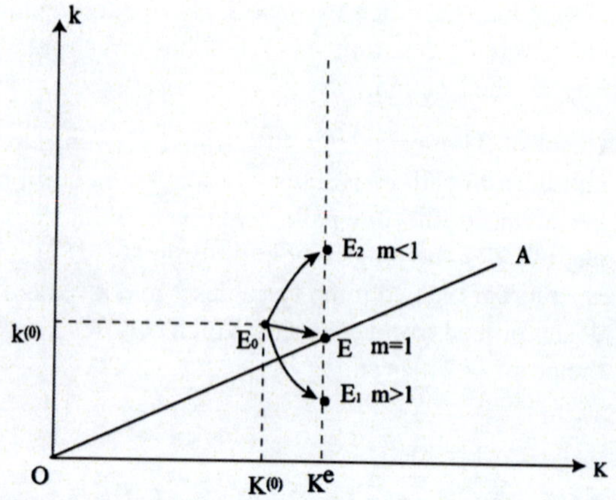

note that (14.24) implies that $m <$ (respectively $>$)1 means that $k^e >$ (respectively $<$)k^{e*}, i.e., the home (respectively, foreign) country is more capital abundant than the foreign (respectively, home) country and exports the consumption good.

See Fig. 14.2. If $m = 1$, the stationary state is point E on the line OA with slope $1/2$. Theorem 2 asserts that, starting from any initial point E_0 in a neighborhood of the long-run equilibrium E, the world economy converges to E, in which the two countries have identical factor endowments and no country has comparative advantage to any good. If $m <$ (respectively $>$)1, the stationary state is above (respectively, below) the line OA like E_1 (respectively E_2). The world economy starting from any initial point E_0 in a neighborhood of the stationary state, converges to E_1 (respectively E_2), in which the home country is more (respectively, less) capital abundant, and, following Lemma 2, it exports (respectively, imports) the consumption good. □

Thus, the long-run trade pattern crucially depends on the sign of $(\rho + \delta)(a_1 - a_2) - \delta\Delta$ and whether m is larger or smaller than 1. However, as we have already mentioned, there is no economic mechanism in the decentralized world economy which chooses a particular value of m. Thus, we arrive at the third theorem.

Theorem 3. *There exists a neighborhood of the long-run equilibrium* (k^e, k^{*e}) *with* $m = 1$ *such that from any initial distribution of capital* $(k(0), k^*(0))$ *in that neighborhood there exists an equilibrium path converging to a long-run equilibrium with* $m < 1$ *and another equilibrium converging to a long-run equilibrium with* $m > 1$. *It follows that the initial world distribution of capital does not determine the long-run trade pattern.*

Remark 2. Using a two-country dynamic Heckscher-Ohlin model without externalities, Chen (1992) showed that the long-run pattern of trade is determined by the initial world distribution of capital. Our results mean that his long-run Heckscher-Ohlin Theorem does not hold if externalities are introduced.

14.6 Concluding Remarks

Let us add a couple of concluding remarks.

First, the indeterminacy results obtained in this paper crucially depend on the presence of externalities. Suppose that there is no externality. Then, $\Delta = a_1 - a_2 = \theta_1 - \theta_2$, which implies that one root of the characteristic equation, $F(x) = 0$, is

$$-\frac{\delta\Delta - (\rho + \delta)(a_1 - a_2)}{\Delta} = \rho > 0,$$

and that $F(0)(=(14.29))$ is negative. It follows that the characteristic equation (14.27) has at most one negative root, which means that for any given positive m the stable manifold is at most one-dimensional. It follows that in order for the solution to the aforementioned dynamic general equilibrium model to converge to a stationary state, we have to choose appropriate $\lambda(0)$ and m. Thus we have a continuum of dynamic general equilibrium paths which converge to different stationary states, where there is no indeterminacy as far as the world initial distribution of capital is given. That is substantially what Chen (1992) and Shimomura (1992) discussed.

Second, a large extent of externalities is not necessary for indeterminacy. Let us consider the following numerical example:

$$a_1 = 0.3, \quad \alpha_1 = 0, \quad b_1 = 0.69, \quad \beta_1 = 0.01$$
$$a_2 = 0.301, \quad \alpha_2 = 0, \quad b_2 = 0.699, \quad \beta_2 = 0.$$

Since $\theta_1 - \theta_2 = -0.001 < 0$, $\Delta = 0.0201 > 0$ and $a_1 - a_2 = -0.001$, this example satisfies Assumptions 5 and 7. It follows that only a small externality of capital in the production of the consumption good is sufficient for indeterminacy in this paper.

Appendix 1: The Derivation of (14.14)

Solving (14.4) for y_1 and y_2, we have

$$y_1 = \frac{l\bar{c}_R^2 - k\bar{c}_W^2}{\bar{c}_W^1\bar{c}_R^2 - \bar{c}_W^2\bar{c}_R^1} \quad \text{and} \quad y_2 = \frac{kc_W^1 - l\bar{c}_R^1}{\bar{c}_W^1\bar{c}_R^2 - \bar{c}_W^2\bar{c}_R^1}. \tag{A.1}$$

Using (14.3), y_1 can be written as

$$y_1 = \frac{l\left(\frac{R_2\bar{c}_R^2}{\bar{c}^2}\right)\left(\frac{1}{R_2}\right) - k\left(\frac{W_2\bar{c}_W^2}{\bar{c}^2}\right)\left(\frac{1}{W_2}\right)}{\left(\frac{W_1\bar{c}_W^1}{\bar{c}^1}\right)\left(\frac{R_2\bar{c}_R^2}{\bar{c}^2}\right)\left(\frac{1}{W_1R_2}\right) - \left(\frac{W_2\bar{c}_W^2}{\bar{c}^2}\right)\left(\frac{R_1\bar{c}_R^1}{\bar{c}^1}\right)\left(\frac{1}{W_2R_1}\right)}$$

$$= \frac{\frac{l(1-\theta_2)b_2}{r(1-\theta_2)} - \frac{k\theta_2 a_2}{w\theta_2}}{\frac{\theta_1(1-\theta_2)a_1 b_2}{\theta_1(1-\theta_2)wr} - \frac{\theta_2(1-\theta_1)a_1 b_2}{\theta_2(1-\theta_1)wr}} = \frac{b_2 lw - a_2 kr}{\Delta}. \tag{A.2}$$

Making similar calculations, we obtain

$$y_2 = \frac{a_1 kr - b_1 lw}{p\Delta}. \tag{A.3}$$

(14.14) directly follows from (A.2) and (A.3).

Appendix 2: The Proof of Fact 1

If $m = 1$, (14.21) becomes

$$[\lambda^e p^e]^{-\frac{1}{\eta}} = \frac{\delta w(p^e) l \{\rho + \delta(1 - b_1)\}}{p^e \{\rho a_2 + \delta a_1 b_2 + (1 - b_1) a_2 \delta\}}. \tag{A.4}$$

The substitution of (A.4) into (14.20) and (14.22) yields

$$\begin{aligned} k^e = k^{*e} &= \frac{w(p^e) l (b_1 - b_2) + \frac{w(p^e) l \{\rho + \delta(1-b_1)\}\Delta}{\{\rho a_2 + \delta a_1 b_2 + (1-b_1) a_2 \delta\}}}{\{(\rho + \delta)(a_1 - a_2) - \delta\Delta\}} \\ &= \frac{w(p^e) l [(b_1 - b_2)\{\rho a_2 + \delta a_1 b_2 + (1 - b_1) a_2 \delta\} + \{\rho + \delta(1 - b_1)\}\Delta]}{\{(\rho + \delta)(a_1 - a_2) - \delta\Delta\}\{\rho a_2 + \delta a_1 b_2 + (1 - b_1) a_2 \delta\}} \\ &= \frac{w(p^e) l \Theta}{\{(\rho + \delta)(a_1 - a_2) - \delta\Delta\}\{\rho a_2 + \delta a_1 b_2 + (1 - b_1) a_2 \delta\}}, \end{aligned}$$

where

$$\begin{aligned} \Theta &\equiv (b_1 - b_2)(\rho + \delta) a_2 + (b_1 - b_2)\delta\Delta + \{\rho + \delta(1 - b_1)\}\Delta \\ &= (b_1 - b_2)(\rho + \delta) a_2 - b_2 \delta\Delta + (\rho + \delta)\Delta \\ &= (\rho + \delta)\{(b_1 - b_2) a_2 + (a_1 b_2 - a_2 b_1)\} - b_2 \delta\Delta \\ &= (\rho + \delta)(a_1 - a_2) b_2 - b_2 \delta\Delta \\ &= b_2\{(\rho + \delta)(a_1 - a_2) - \delta\Delta\}. \end{aligned}$$

Therefore, we have

$$k^e = k^{*e} = \frac{b_2 w(p^e) l}{\rho a_2 + \delta a_1 b_2 + (1 - b_1) a_2 \delta},$$

as was to be proved.

References

Atkeson, A. and P. Kehoe (2000), "Paths of Development for Early- and Late-boomers in a Dynamic Heckscher-Ohlin Model," *Research Staff Report* 256, Federal Reserve Bank of Minneapolis.

Becker, R. (1980), "On the Long-run Steady State in a Simple Dynamic Model of Equilibrium with Heterogeneous Households," *Quarterly Journal of Economics*, 95, 375–382 .

Benhabib, J., Meng, Q. and K. Nishimura (2000), "Indeterminacy under Constant Returns to Scale in Multisector Economies," *Econometrica*, 68, 1541–1548.

Benhabib, J. and K. Nishimura (1998), "Indeterminacy and Sunspots with Constant Returns," *Journal of Economic Theory*, 81, 58–96.

Bewley, T. (1982), "An Integration of Equilibrium Theory and Turnpike Theory," *Journal of Mathematical Economics*, 10, 233–267.

Bond, E., Trask, K. and P. Wang (2000), Factor Accumulation and Trade; Dynamic Comparative Advantage with Endogenous Physical and Human Capital, WP 00-W31 , Dept. of Economics, Vanderbilt University, Nashville.

Chen, Z. (1992), "Long-Run Equilibria in a Dynamic Heckscher-Ohlin Model," *Canadian Journal of Economics*, 25, 923–943.

Epstein, L. (1987), "A Simple Dynamic General Equilibrium Model," *Journal of Economic Theory*, 41, 68–95.

Ghiglino, C. and M. Olszak-Duquenne (2001), "Inequalities and Fluctuations in a Dynamic General Equilibrium Model," *Economic Theory*, 17, 1–24.

Kamihigashi, T. (2001) "Necessity of Transversality Conditions for Infinite Horizon Problems," *Econometrica*, 69, 995–1012.

Kemp, M. and M. Okawa (1998), "Market Structure and Factor Price Equalization," *Japanese Economic Review*, 49, 335–339.

Nishimura, K. and K. Shimomura (2001), "Indeterminacy in a Dynamic Small Open Economy," *Journal of Economic Dynamics and Control*, 27, 271–281

Nishimura, K. and M. Yano (1993a), "Interlinkage in the Endogenous Real Business Cycles of International Economies," *Economic Theory*, 3, 151–168.

Nishimura, K. and M. Yano (1993b), Endogenous Real Business Cycles and International Specialization, in *Theory, Policy and Dynamics in International Trade*, ed. by Ethier, W., Helpman, E. and P. Neary, Cambridge University Press, Cambridge, UK.

Shimomura, K. (1992), A Two-Sector Dynamic General Equilibrium Model of Distribution, in *Dynamic Economic Models and Optimal Control*, ed. by Feichtinger, G., North-Holland, Amsterdam.

Yano, M. (1984) "The Turnpike of Dynamic General Equilibrium Paths and its Insensitivity to Initial Conditions," *Journal of Mathematical Economics*, 13, 235–254.

Chapter 15
Indeterminacy in Discrete-Time Infinite-Horizon Models with Non-linear Utility and Endogenous Labor*

Kazuo Nishimura and Alain Venditti**

15.1 Introduction

Over the last decade, a large number of papers have established the fact that locally indeterminate equilibria and sunspots fluctuations may arise in infinite-horizon growth models with external effects in production.[1] These contributions also show that there exist significant differences between one-sector and two-sector models. As initially established by Benhabib and Farmer (1994), in one-sector models local indeterminacy requires some increasing returns based on externalities coming from capital and more importantly labor, a strongly elastic labor supply, and a large enough elasticity of intertemporal substitution in consumption which may however remain within plausible intervals.[2] On the contrary, as proved in Benhabib and Nishimura (1998), in two-sector models with sector specific externalities, local indeterminacy is compatible with constant returns at the social level, does not require some elastic labor supply and is mainly based on a technological mechanism coming from the broken duality between the Rybczynski and Stolper-Samuelson

***Journal of Mathematical Economics* 43**, 446–476, 2007.

**This paper benefited from comments received during a presentation at the International Conference on "Instability and Fluctuations in Intertemporal Equilibrium Models", Marseille, June 2005. We thank all the participants and in particular J. Gali and K. Shimomura. We are also grateful to an anonymous referee for useful comments and suggestions.

[1] See Benhabib and Farmer (1999) for a survey.

[2] See also Lloyd-Braga et al. (2006) and Pintus (2006).

K. Nishimura (✉)
Institute of Economic Research, Kyoto University, Japan
e-mail: nishimura@kier.kyoto-u.ac.jp

A. Venditti
National Center for Scientific Research, CNRS-GREQAM and EDHEC, Marseille, France
e-mail: alain.venditti@univmed.fr

J. Stachurski et al. (eds.), *Nonlinear Dynamics in Equilibrium Models*,
DOI 10.1007/978-3-642-22397-6_15, © Elsevier B.V. 2007. Published by
Springer-Verlag Berlin Heidelberg 2012. All Rights Reserved

effects.[3] However, most of the contributions on two-sector models are based on the assumption of a linear utility function, i.e. an infinite elasticity of intertemporal substitution in consumption. Of course, indeterminacy remains possible by continuity when the utility function is sufficiently close to linear.

When we compare with the results characterizing one-sector models mentioned above, at least two open questions on two-sector models remain:

1. Finding a lower bound on the elasticity of intertemporal substitution in consumption above which local indeterminacy occurs;
2. Characterizing the role of an elastic labor supply in the existence of local indeterminacy.

In order to provide answers to these questions, we study in this paper the interactions between the technological mechanism from which local indeterminacy originates, and the preference mechanisms based on the consumption-leisure trade-off and the intertemporal substitution of consumption. We shall then assume that the two sectors are characterized by CES technologies with asymmetric elasticities of capital-labor substitution and that the preferences of the representative agent are given by a CES additively separable utility function defined over consumption and leisure. The cases with Cobb-Douglas technologies, a linear utility function with respect to consumption, or inelastic labor will be considered as particular specifications.

Considering in a first step the limit case of an infinitely elastic labor supply, we show that the steady state is always a saddle-point for any value of the elasticity of intertemporal substitution in consumption and any amount of sector-specific external effects. This result is drastically different from the one obtained in one-sector models in which local indeterminacy is mainly obtained under the assumption of an infinitely elastic labor supply.[4] While this configuration makes the occurrence of local indeterminacy easier in one-sector models, it necessarily implies saddle-point stability in two-sector models. This conclusion also echoes one of the results provided by Bosi et al. (2005). They show that in a two-sector optimal growth model with leisure and additive separable preferences, the steady-state is saddle-point stable when the elasticity of labor supply is infinite.

In a second step, we consider an infinite elasticity of intertemporal substitution in consumption. We then show that the local stability properties of the steady state do not depend on the elasticity of the labor supply. Under the assumption of a capital intensive consumption good at the private level, we also give precise conditions on the elasticities of capital-labor substitution for the existence of local indeterminacy which are compatible with Cobb-Douglas technologies in both sectors.

[3] Benhabib and Nishimura (1998) consider a continuous-time model with Cobb-Douglas technologies. The extension to the discrete-time case is given by Benhabib et al. (2002). Two-sector models with CES technologies and non-unitary elasticities of capital-labor substitution are considered in Nishimura and Venditti (2004a,b).

[4] See, for instance, Benhabib and Farmer (1994), Lloyd-Braga et al. (2006) or Pintus (2006).

Building on these results, we introduce in a third step a non-linear utility in consumption but under the assumption of an inelastic labor supply. This restriction allows to precisely isolate the effect of the intertemporal substitutability. We use a geometrical method provided by Grandmont et al. (1998). It is based on a particular property characterizing the product and sum of characteristic roots, which allows to provide a complete characterization of the local stability properties of the steady state and of the occurrence of bifurcations. We then show under some simple conditions on the CES coefficients of technologies that the steady state is locally indeterminate if and only if the elasticity of intertemporal substitution in consumption is large enough. Moreover, we prove that period-two cycles always occur as the elasticity of intertemporal substitution in consumption is decreased and the steady state becomes saddle-point stable.

Finally, we consider the general model with a finite elasticity of intertemporal substitution in consumption and a finite elasticity of the labor supply. Using again the geometrical method of Grandmont et al. (1998), we show that contrary to one-sector models, when the elasticity of intertemporal substitution in consumption is sufficiently large, the steady state is locally indeterminate if and only if the elasticity of the labor supply is low enough. Moreover, period-two cycles always occur as the elasticity of the labor supply is increased and the steady state becomes saddle-point stable.

The paper is organized as follows: Sect. 15.2 presents the basic model, the intertemporal equilibrium and the steady state. Preliminary results are provided in Sect. 15.3 while Sect. 15.4 contains the main contributions of the paper. Some concluding comments are provided in Sect. 15.5 and all the proofs are gathered in a final Appendix.

15.2 Model

15.2.1 The Basic Structure

We consider a discrete-time two-sector economy having an infinitely lived representative agent with single period CES utility function defined over consumption c and leisure $\mathscr{L} = 1 - \ell$:

$$u(c) = \frac{c^{1-\sigma}}{1-\sigma} - \frac{\ell^{1+\gamma}}{1+\gamma}$$

with $\sigma \geq 0$ and $\gamma \geq 0$. The elasticity of intertemporal substitution in consumption is thus given by $\epsilon_c = 1/\sigma$ while the elasticity of the labor supply is given by $\epsilon_\ell = 1/\gamma$.

There are two goods: the pure consumption good, c, and the pure capital good, k. Each good is assumed to be produced with a CES technology which contains some sector specific externalities. We denote by c and y the outputs of sectors c and k, and by e_c and e_y the corresponding external effects. The *private production functions* are thus defined as:

$$c = \left(\alpha_1 K_c^{-\rho_c} + \alpha_2 L_c^{-\rho_c} + e_c\right)^{-\frac{1}{\rho_c}}, \quad y = \left(\beta_1 K_y^{-\rho_y} + \beta_2 L_y^{-\rho_y} + e_y\right)^{-\frac{1}{\rho_y}}$$

with $\rho_c, \rho_y > -1$ and $\gamma_c = 1/(1+\rho_c) \geq 0$, $\gamma_y = 1/(1+\rho_y) \geq 0$ the elasticities of capital-labor substitution in each sector. The externalities e_c and e_y depend on $\bar{K}_i$ and $\bar{L}_i$, which denote the average use of capital and labor in sector $i = c, y$, and will be equal to

$$e_c = a_1 \bar{K}_c^{-\rho_c} + a_2 \bar{L}_c^{-\rho_c}, \quad e_y = b_1 \bar{K}_y^{-\rho_y} + b_2 \bar{L}_y^{-\rho_y} \tag{15.1}$$

with $a_i, b_i \geq 0$, $i = 1, 2$. We assume that these economy-wide averages are taken as given by individual firms. At the equilibrium, all firms of sector $i = c, y$ being identical, we have $\bar{K}_i = K_i$ and $\bar{L}_i = L_i$. Denoting $\hat{\alpha}_i = \alpha_i + a_i$, $\hat{\beta}_i = \beta_i + b_i$, the *social production functions* are defined as

$$c = \left(\hat{\alpha}_1 K_c^{-\rho_c} + \hat{\alpha}_2 L_c^{-\rho_c}\right)^{-1/\rho_c}, \quad y = \left(\hat{\beta}_1 K_y^{-\rho_y} + \hat{\beta}_2 L_y^{-\rho_y}\right)^{-1/\rho_y}. \tag{15.2}$$

The returns to scale are therefore constant at the social level, and decreasing at the private level. We will assume in the following that $\hat{\alpha}_1 + \hat{\alpha}_2 = 1$ and $\hat{\beta}_1 + \hat{\beta}_2 = 1$ so that the production functions collapse to Cobb-Douglas in the particular case $\rho_c = \rho_y = 0$.

Total labor is given by $L_c + L_y = \ell$, and the total stock of capital is given by $K_c + K_y = k$. We assume complete depreciation of capital in one period so that the capital accumulation equation is $y_t = k_{t+1}$.[5] Optimal factor allocations across sectors are obtained by solving the following program:

$$\begin{aligned} \max_{K_{ct}, L_{ct}, K_{yt}, L_{yt}} \quad & \left(\alpha_1 K_{ct}^{-\rho_c} + \alpha_2 L_{ct}^{-\rho_c} + e_{ct}\right)^{-1/\rho_c} \\ \text{s.t.} \quad & k_{t+1} = \left(\beta_1 K_{yt}^{-\rho_y} + \beta_2 L_{yt}^{-\rho_y} + e_{yt}\right)^{-1/\rho_y} \\ & \ell_t = L_{ct} + L_{yt} \\ & k_t = K_{ct} + K_{yt} \\ & e_{ct}, e_{yt} \quad \text{given.} \end{aligned}$$

Denote by p_t, w_t and r_t, respectively, the price of the capital good, the wage rate of labor and the rental rate of the capital good at time $t \geq 0$, all in terms of the price of the consumption good. The Lagrangian is:

[5] Full depreciation is introduced in order to simplify the analysis and to focus on the role of preferences. Of course, local indeterminacy may also occur under partial depreciation (see Nishimura and Venditti (2004c) for some results with a linear utility function).

$$\mathcal{L}_t = \frac{\left(\alpha_1 K_{ct}^{-\rho_c} + \alpha_2 L_{ct}^{-\rho_c} + e_{ct}\right)^{-\frac{1-\sigma}{\rho_c}}}{1-\sigma} - \frac{\ell_t^{1+\gamma}}{1+\gamma} + w_t\left(\ell_t - L_{ct} - L_{yt}\right) \quad (15.3)$$

$$+ r_t\left(k_t - K_{ct} - K_{yt}\right) + p_t\left[\left(\beta_1 K_{yt}^{-\rho_y} + \beta_2 L_{yt}^{-\rho_y} + e_{yt}\right)^{-\frac{1}{\rho_y}} - k_{t+1}\right].$$

For any given (k_t, k_{t+1}, ℓ_t), solving the first order conditions gives input demand functions $\tilde{K}_c = K_c(k_t, k_{t+1}, \ell_t, e_{ct}, e_{yt})$, $\tilde{L}_c = L_c(k_t, k_{t+1}, \ell_t, e_{ct}, e_{yt})$, $\tilde{K}_y = K_y(k_t, k_{t+1}, \ell_t, e_{ct}, e_{yt})$, $\tilde{L}_y = L_y(k_t, k_{t+1}, \ell_t, e_{ct}, e_{yt})$. We then define the production frontier as

$$c_t = T(k_t, k_{t+1}, \ell_t, e_{ct}, e_{yt}) = \left(\alpha_1 \tilde{K}_c^{-\rho_c} + \alpha_2 \tilde{L}_c^{-\rho_c} + e_{ct}\right)^{-1/\rho_c}.$$

Using the envelope theorem we derive:

$$\begin{aligned} r_t &= T_1(k_t, k_{t+1}, \ell_t, e_{ct}, e_{yt}) \\ p_t &= -T_2(k_t, k_{t+1}, \ell_t, e_{ct}, e_{yt}) \\ w_t &= T_3(k_t, k_{t+1}, \ell_t, e_{ct}, e_{yt}) \end{aligned} \quad (15.4)$$

where $T_1 = \frac{\partial T}{\partial k_t}$, $T_2 = \frac{\partial T}{\partial k_{t+1}}$ and $T_3 = \frac{\partial T}{\partial \ell_t}$.

The representative consumer's optimization program is then given by

$$\begin{aligned} \max_{\{k_t, \ell_t\}_{t=0}^{\infty}} \quad & \sum_{t=0}^{\infty} \delta^t \left[\frac{T(k_t, k_{t+1}, \ell_t, e_{ct}, e_{yt})^{1-\sigma}}{1-\sigma} - \frac{\ell_t^{1+\gamma}}{1+\gamma}\right] \\ \text{s.t.} \quad & k_0, (e_{ct})_{t=0}^{+\infty}, (e_{yt})_{t=0}^{+\infty} \quad \text{given} \end{aligned}$$

with $\delta \in (0,1)$ the discount factor. The first order conditions give the Euler equation and the equation describing the consumption/leisure trade-off

$$\begin{aligned} p_t c_t^{-\sigma} &= \delta r_{t+1} c_{t+1}^{-\sigma} \\ w_t c_t^{-\sigma} &= \ell_t^{\gamma}. \end{aligned} \quad (15.5)$$

From the input demand functions together with the external effects (15.1) considered at the equilibrium we may define the equilibrium factors demand functions $\hat{K}_i = \hat{K}_i(k_t, k_{t+1}, \ell_t)$, $\hat{L}_i = \hat{L}_i(k_t, k_{t+1}, \ell_t)$ so that $\hat{e}_c = \hat{e}_c(k_t, k_{t+1}, \ell_t) = a_1 \hat{K}_c^{-\rho_c} + a_2 \hat{L}_c^{-\rho_c}$ and $\hat{e}_y = \hat{e}_y(k_t, k_{t+1}, \ell_t) = b_1 \hat{K}_y^{-\rho_y} + b_2 \hat{L}_c^{-\rho_y}$. From (15.4) prices now satisfy

$$\begin{aligned} r(k_t, k_{t+1}, \ell_t) &= T_1(k_t, k_{t+1}, \ell_t, \hat{e}_c(k_t, k_{t+1}, \ell_t), \hat{e}_y(k_t, k_{t+1}, \ell_t)) \\ p(k_t, k_{t+1}, \ell_t) &= -T_2(k_t, k_{t+1}, \ell_t, \hat{e}_c(k_t, k_{t+1}, \ell_t), \hat{e}_y(k_t, k_{t+1}, \ell_t)) \\ w(k_t, k_{t+1}, \ell_t) &= T_3(k_t, k_{t+1}, \ell_t, \hat{e}_c(k_t, k_{t+1}, \ell_t), \hat{e}_y(k_t, k_{t+1}, \ell_t)) \end{aligned} \quad (15.6)$$

and the consumption level at time t is given by

$$c(k_t, k_{t+1}, \ell_t) = T(k_t, k_{t+1}, \ell_t, \hat{e}_c(k_t, k_{t+1}, \ell_t), \hat{e}_y(k_t, k_{t+1}, \ell_t)) \tag{15.7}$$

We then get (15.5) evaluated at $\hat{e}_c$ and $\hat{e}_y$:

$$\begin{aligned} p(k_t, k_{t+1}, \ell_t)c(k_t, k_{t+1}, \ell_t)^{-\sigma} &= \delta r(k_{t+1}, k_{t+2}, \ell_{t+1})c(k_{t+1}, k_{t+2}, \ell_{t+1})^{-\sigma} \\ w(k_t, k_{t+1}, \ell_t)c(k_t, k_{t+1}, \ell_t)^{-\sigma} &= \ell_t^{\gamma}. \end{aligned} \tag{15.8}$$

Any solution $\{(k_t, \ell_t)\}_{t=0}^{+\infty}$ which also satisfies the transversality condition

$$\lim_{t \to +\infty} \delta^t c(k_t, k_{t+1}, \ell_t)^{-\sigma} p(k_t, k_{t+1}, \ell_t) k_{t+1} = 0$$

is called an equilibrium path.

15.2.2 Steady State and Characteristic Polynomial

A steady state is defined by $k_t = k^*$, $y_t = y^* = k^*$, $\ell_t = \ell^*$ and is determined as a solution of

$$\begin{aligned} \delta r(k, k, \ell) - p(k, k, \ell) &= 0 \\ w(k, k, \ell)c(k, k, \ell)^{-\sigma} - \ell^{\gamma} &= 0. \end{aligned} \tag{15.9}$$

The methodology used in this paper consists in approximating (15.8) in order to compute the steady state and the characteristic polynomial.

We will assume the following restriction on parameters' values:

Assumption 1 $\rho_y \in (-1,\ \hat{\rho}_y)$ *with* $\hat{\rho}_y \equiv \ln \hat{\beta}_1 / \left[\ln(\delta\beta_1) - \ln \hat{\beta}_1\right] > 0$.

Such a restriction is quite standard with CES technologies. Indeed when the elasticity of capital-labor substitution is less than 1, the Inada conditions are not satisfied and corner solutions cannot be a priori ruled out. Assumption 1 precisely guarantees positiveness and interiority of all the steady state values for the input demand functions K_c, K_y, L_c and L_y.

We start by proving existence and uniqueness of the steady state (k^*, ℓ^*). As mentioned in Remark 2 in Appendix 15.6.2, the production frontier, i.e. (15.7), when evaluated along a stationary solution, is homogeneous of degree 1. It follows that the price functions (15.6) are homogeneous of degree 0. Denoting $\kappa^* = k^*/\ell^*$, a steady state may thus be characterized as a pair (κ^*, ℓ^*) solution of:

$$\begin{aligned} \delta r(\kappa, \kappa, 1) - p(\kappa, \kappa, 1) &= 0 \\ w(\kappa, \kappa, 1)c(\kappa, \kappa, 1)^{-\sigma} - \ell^{\gamma+\sigma} &= 0. \end{aligned} \tag{15.10}$$

Notice from the first equation that κ^* is only determined by technological characteristics and does not depend on preferences. Once κ^* is obtained, the second equation gives the steady state value of the labor supply ℓ^*.

Proposition 1. *Under Assumption 1, there exists a unique steady state* $(\kappa^*, \ell^*) > 0$*, such that:*

$$\kappa^* = \frac{\left(\frac{\alpha_1\beta_2}{\alpha_2\beta_1}\right)^{\frac{1}{1+\rho_c}} \left(\frac{(\delta\beta_1)^{\frac{\rho_y}{1+\rho_y}} - \hat{\beta}_1}{\hat{\beta}_2}\right)^{\frac{1+\rho_y}{\rho_y(1+\rho_c)}}}{1-(\delta\beta_1)^{\frac{1}{1+\rho_y}}\left[1-\left(\frac{\alpha_1\beta_2}{\alpha_2\beta_1}\right)^{\frac{1}{1+\rho_c}}\left(\frac{(\delta\beta_1)^{\frac{\rho_y}{1+\rho_y}} - \hat{\beta}_1}{\hat{\beta}_2}\right)^{\frac{\rho_y-\rho_c}{\rho_y(1+\rho_c)}}\right]}$$

$$\ell^* = \left\{\frac{\frac{\alpha_1\beta_2}{\beta_1}\left(\frac{(\delta\beta_1)^{\frac{\rho_y}{1+\rho_y}} - \hat{\beta}_1}{\hat{\beta}_2}\right)^{\frac{1+\rho_y}{\rho_y}}}{\left[1-(\delta\beta_1)^{\frac{1}{1+\rho_y}}\right]^{\sigma}\left[\hat{\alpha}_1 + \hat{\alpha}_2\left(\frac{\alpha_1\beta_2}{\alpha_2\beta_1}\right)^{\frac{\rho_c}{1+\rho_c}}\left(\frac{(\delta\beta_1)^{\frac{\rho_y}{1+\rho_y}} - \hat{\beta}_1}{\hat{\beta}_2}\right)^{\frac{\rho_c(1+\rho_y)}{\rho_y(1+\rho_c)}}\right]^{\frac{1+\rho_c-\sigma}{\rho_c}} \kappa^{*\sigma}}\right\}^{\frac{1}{\gamma+\sigma}}.$$

Proof. See Appendix 15.6.1.

Consider the following notations:

$$\mathcal{T}_{i1}(k_t, k_{t+1}, \ell_t) = \partial T_i(k_t, k_{t+1}, \ell_t, \hat{e}_c(k_t, k_{t+1}, \ell_t), \hat{e}_y(k_t, k_{t+1}, \ell_t))/\partial k_t$$

$$\mathcal{T}_{i2}(k_t, k_{t+1}, \ell_t) = \partial T_i(k_t, k_{t+1}, \ell_t), \hat{e}_c(k_t, k_{t+1}, \ell_t), \hat{e}_y(k_t, k_{t+1}, \ell_t))/\partial k_{t+1}$$

$$\mathcal{T}_{i3}(k_t, k_{t+1}, \ell_t) = \partial T_i(k_t, k_{t+1}, \ell_t, \hat{e}_c(k_t, k_{t+1}, \ell_t), \hat{e}_y(k_t, k_{t+1}, \ell_t))/\partial \ell_t$$

for $i = 1, 2, 3$. The linearization of the Euler equation around (κ^*, ℓ^*) gives the following characteristic polynomial:

Theorem 1. *Under Assumption 1, the characteristic polynomial is*

$$\mathcal{P}(x) = \gamma\frac{T_3^*}{\mathcal{T}_{32}^*\ell^*}\left\{\left(x+\frac{\mathcal{T}_{11}^*}{\mathcal{T}_{12}^*}\right)\left(x+\frac{\mathcal{T}_{22}^*}{\delta\mathcal{T}_{12}^*}\right)+\sigma\frac{T_1^{*2}}{c^*\mathcal{T}_{12}^*}(\delta\mathcal{B}x-\mathcal{A})(x-1)\right\}$$
$$-\sigma\frac{T_1^{*2}}{c^*\mathcal{T}_{12}^*}(1+\mathcal{C})^2\mathcal{D}\left[x-\hat{\beta}_1(\delta\beta_1)^{\frac{-\rho_y}{1+\rho_y}}\right]\left[x(\delta\beta_1)^{\frac{1}{1+\rho_y}}-1\right]$$

with

$$\frac{\mathcal{T}_{11}^*}{\mathcal{T}_{12}^*} = -\left\{(\delta\beta_1)^{\frac{1}{1+\rho_y}}\left[1-\left(\frac{\alpha_1\beta_2}{\alpha_2\beta_1}\right)^{\frac{1}{1+\rho_c}}\left(\frac{(\delta\beta_1)^{\frac{\rho_y}{1+\rho_y}}-\hat{\beta}_1}{\hat{\beta}_2}\right)^{\frac{\rho_y-\rho_c}{\rho_y(1+\rho_c)}}\right]\right\}^{-1}$$

$$\frac{\mathscr{T}_{22}^*}{\delta \mathscr{T}_{12}^*} = -\hat{\beta}_1 (\delta\beta_1)^{\frac{-\rho_y}{1+\rho_y}} \left[1 - \frac{\hat{\alpha}_1 \hat{\beta}_2}{\hat{\alpha}_2 \hat{\beta}_1} \left(\frac{\alpha_2 \beta_1}{\alpha_1 \beta_2} \right)^{\frac{\rho_c}{1+\rho_c}} \left(\frac{(\delta\beta_1)^{\frac{\rho_y}{1+\rho_y}} - \hat{\beta}_1}{\hat{\beta}_2} \right)^{\frac{\rho_y - \rho_c}{\rho_y(1+\rho_c)}} \right]$$

$$\mathscr{A} = \frac{\hat{\alpha}_1}{\alpha_1} \frac{(1+\rho_y)(\delta\beta_1)^{\frac{\rho_y}{1+\rho_y}} \left(1 - (\delta\beta_1)^{\frac{1}{1+\rho_y}} \right) + \delta\beta_1 (1+\rho_c) \frac{\hat{\alpha}_2 \hat{\beta}_1}{\hat{\alpha}_1 \hat{\beta}_2} \frac{\alpha_1 \beta_2}{\alpha_2 \beta_1} \left(1 + \frac{\mathscr{T}_{22}^*}{\delta \mathscr{T}_{12}^*} \right)}{(1+\rho_y)(\delta\beta_1)^{\frac{\rho_y}{1+\rho_y}} \left[1 - (\delta\beta_1)^{\frac{1}{1+\rho_y}} \right] + (1+\rho_c) \left[\delta\beta_1 + \hat{\beta}_1 \frac{\mathscr{T}_{12}^*}{\mathscr{T}_{11}^*} \right]}$$

$$\mathscr{B} = \frac{\hat{\alpha}_2}{\alpha_2} \frac{\beta_2}{\hat{\beta}_2} \frac{\left[\begin{array}{l} (1+\rho_y) \left[1 - (\delta\beta_1)^{\frac{1}{1+\rho_y}} \right] \left((\delta\beta_1)^{\frac{\rho_y}{1+\rho_y}} - \hat{\beta}_1 \left(1 - \frac{\hat{\alpha}_1 \hat{\beta}_2}{\hat{\alpha}_2 \hat{\beta}_1} \frac{\alpha_2 \beta_1}{\alpha_1 \beta_2} \right) \right) \\ +(1+\rho_c)\delta\beta_1 \left(1 + \frac{\mathscr{T}_{22}^*}{\delta \mathscr{T}_{12}^*} \right) \end{array} \right]}{(1+\rho_y)(\delta\beta_1)^{\frac{\rho_y}{1+\rho_y}} \left[1 - (\delta\beta_1)^{\frac{1}{1+\rho_y}} \right] + (1+\rho_c) \left[\delta\beta_1 + \hat{\beta}_1 \frac{\mathscr{T}_{12}^*}{\mathscr{T}_{11}^*} \right]}$$

$$\mathscr{C} = \frac{\hat{\alpha}_2}{\hat{\alpha}_1} \left(\frac{\alpha_1 \beta_2}{\alpha_2 \beta_1} \right)^{\frac{\rho_c}{1+\rho_c}} \left(\frac{(\delta\beta_1)^{\frac{\rho_y}{1+\rho_y}} - \hat{\beta}_1}{\hat{\beta}_2} \right)^{\frac{\rho_c(1+\rho_y)}{\rho_y(1+\rho_c)}} > 0$$

$$\mathscr{D} = -\frac{\mathscr{T}_{11}^*}{\mathscr{T}_{12}^*} \frac{\hat{\alpha}_1}{\alpha_1} \left(\frac{\alpha_1 \beta_2}{\alpha_2 \beta_1} \right)^{\frac{1}{1+\rho_c}} \left(\frac{(\delta\beta_1)^{\frac{\rho_y}{1+\rho_y}} - \hat{\beta}_1}{\hat{\beta}_2} \right)^{\frac{1+\rho_y}{\rho_y(1+\rho_c)}}$$

$$\frac{T_1^{*2}}{c^* \mathscr{T}_{12}^*} = -\frac{\alpha_1}{\hat{\alpha}_1} \frac{\mathscr{T}_{11}^*}{\mathscr{T}_{12}^*} \frac{(1+\rho_y)(\delta\beta_1)^{\frac{\rho_y}{1+\rho_y}} \left[1 - (\delta\beta_1)^{\frac{1}{1+\rho_y}} \right] + (1+\rho_c) \left[\delta\beta_1 + \hat{\beta}_1 \frac{\mathscr{T}_{12}^*}{\mathscr{T}_{11}^*} \right]}{(1+\rho_c)(1+\rho_y)(\delta\beta_1)^{\frac{\rho_y}{1+\rho_y}} \left[1 - (\delta\beta_1)^{\frac{1}{1+\rho_y}} \right] \mathscr{C}}$$

$$\frac{T_3^*}{\mathscr{T}_{32}^* \ell^*} = -\frac{T_1^{*2}}{c^* \mathscr{T}_{12}^*} \left[1 - (\delta\beta_1)^{\frac{1}{1+\rho_y}} \right] \kappa^* \frac{\hat{\alpha}_1}{\alpha_1} \mathscr{C}(1+\mathscr{C}).$$

Proof. See Appendix 15.6.2.

We have now to study the stability properties of the steady state depending on the value of preference and technological parameters. As in Nishimura and Venditti (2004a), if the elasticities of capital-labor substitution are identical across sectors, the consumption good is capital intensive at the private level if and only if $\alpha_1\beta_2 - \alpha_2\beta_1 > 0$ while it is capital intensive at the social level if and only if $\hat{\alpha}_1\hat{\beta}_2 - \hat{\alpha}_2\hat{\beta}_1 > 0$. However, with asymmetric elasticities of capital-labor substitution, the capital intensity differences between sectors also depend on the prices and the parameters ρ_c and ρ_y. As shown in Nishimura and Venditti (2004b), we have the following characterization at the steady state:

Proposition 2. *Under Assumption 1, at the steady state:*

(i) The consumption (investment) good sector is capital intensive from the private perspective if and only if

$$\left(\frac{(\delta\beta_1)^{\frac{\rho_y}{1+\rho_y}} - \hat{\beta}_1}{\hat{\beta}_2}\right)^{\frac{\rho_c-\rho_y}{\rho_y(1+\rho_c)}} < (>) \left(\frac{\alpha_1\beta_2}{\alpha_2\beta_1}\right)^{\frac{1}{1+\rho_c}} \quad (15.11)$$

(ii) The consumption (investment) good sector is capital intensive from the social perspective if and only if

$$\left(\frac{(\delta\beta_1)^{\frac{\rho_y}{1+\rho_y}} - \hat{\beta}_1}{\hat{\beta}_2}\right)^{\frac{\rho_c-\rho_y}{\rho_y(1+\rho_c)}} < (>) \left(\frac{\hat{\alpha}_1\hat{\beta}_2}{\hat{\alpha}_2\hat{\beta}_1}\right)^{\frac{1}{1+\rho_c}} \left(\frac{\hat{\beta}_2\beta_1}{\hat{\beta}_1\beta_2}\right)^{\frac{\rho_c-\rho_y}{(1+\rho_y)(1+\rho_c)}}. \quad (15.12)$$

Remark 1. If both technologies are Cobb-Douglas with $\rho_c = \rho_y = 0$, we get the following expressions:

$$\frac{\mathcal{T}_{11}^*}{\mathcal{T}_{12}^*} = -\left\{\delta\beta_1\left(1 - \frac{\alpha_1\beta_2}{\alpha_2\beta_1}\right)\right\}^{-1}, \quad \frac{\mathcal{T}_{22}^*}{\delta\mathcal{T}_{12}^*} = -\hat{\beta}_1\left(1 - \frac{\hat{\alpha}_1\hat{\beta}_2}{\hat{\alpha}_2\hat{\beta}_1}\right).$$

Condition (15.11) becomes $1 < (>)\alpha_1\beta_2/\alpha_2\beta_1$ and condition (15.12) becomes $1 < (>)\hat{\alpha}_1\hat{\beta}_2/\hat{\alpha}_2\hat{\beta}_1$. It follows that $\mathcal{T}_{11}^*/\mathcal{T}_{12}^*$ is positive if and only if the consumption good is capital intensive at the private level while $\mathcal{T}_{22}^*/\mathcal{T}_{12}^*$ is positive if and only if the consumption good is capital intensive at the social level. Moreover, notice from (15.11) and the expression $\mathcal{T}_{11}^*/\mathcal{T}_{12}^*$ in Theorem 1 that as in the Cobb-Douglas formulation, $\mathcal{T}_{11}^*/\mathcal{T}_{12}^*$ is positive if and only if the consumption good is capital intensive at the private level. On the contrary, when $\rho_c \neq \rho_y \neq 0$, the sign of $\mathcal{T}_{22}^*/\mathcal{T}_{12}^*$ does not directly depend on the sign of the capital intensity difference across sectors at the social level.

In order to simplify the exposition in the rest of the paper, we will discuss the local stability properties of the steady state depending on the sign of the differences $\alpha_1\beta_2 - \alpha_2\beta_1$ and $\hat{\alpha}_1 - \hat{\beta}_1$,[6] and the values of the elasticities of substitution in both sectors. We will only refer to capital intensities when the results are economically interpreted. We now introduce the following standard definition:

Definition 1. A steady state k^* is called locally indeterminate if there exists $\epsilon > 0$ such that from any k_0 belonging to $(k^* - \epsilon, k^* + \epsilon)$ there are infinitely many equilibrium paths converging to the steady state.

[6] We have indeed $\hat{\alpha}_1\hat{\beta}_2 - \hat{\alpha}_2\hat{\beta}_1 = \hat{\alpha}_1 - \hat{\beta}_1$.

If both roots of the characteristic equation have modulus less than 1 then the steady state is locally indeterminate. If a steady state is not locally indeterminate, then we call it locally determinate.

15.3 Preliminary Results

Before analyzing the general configuration with $\sigma, \gamma > 0$, we start by considering three polar cases: in the first one we assume a linear utility function with respect to labor, i.e. an infinite elasticity of the labor supply. The second extreme configuration will be based on the assumption of linear utility with respect to consumption, i.e. an infinite elasticity of intertemporal substitution in consumption. A third particular case concerns the model with inelastic labor and non-linear utility function with respect to consumption. From these polar cases, we will be able to study the general formulation.

15.3.1 *Infinite Elasticity of the Labor Supply: $\sigma > 0$, $\gamma = 0$*

Consider the case of a linear utility function with respect to labor, i.e. $\gamma = 0$. The characteristic polynomial then reduces to

$$\mathscr{P}(x) = \left(x - \hat{\beta}_1 (\delta\beta_1)^{\frac{-\rho y}{1+\rho y}}\right)\left[x (\delta\beta_1)^{\frac{1}{1+\rho y}} - 1\right] = 0$$

and the characteristic roots can be explicitly computed, namely

$$x_1 = \hat{\beta}_1 (\delta\beta_1)^{\frac{-\rho y}{1+\rho y}}, \quad x_2 = (\delta\beta_1)^{\frac{-1}{1+\rho y}} > 1.$$

Assumption 1 also implies $x_1 \in (0, 1)$ so that we get:

Theorem 2. *Under Assumption 1, if $\gamma = 0$ the steady state is saddle-point stable.*

This result implies that for any intertemporal elasticity of substitution in consumption and any amount of external effects, the steady state is locally determinate. This conclusion may be compared to one of the main results provided by Bosi et al. (2005). They show that in a two-sector optimal growth model with leisure and additive separable preferences, the steady-state is saddle-point stable when the elasticity of the labor supply is infinite (utility is linear in labor).

This conclusion points out a role of the labor supply which is the complete opposite of the one obtained in one-sector models. Indeed, as shown in Lloyd-Braga et al. (2006) and Pintus (2006), the occurrence of local indeterminacy in aggregate models requires the consideration of large elasticities of labor supply. Such a drastic difference is explained by the fact that there exists a discontinuity between one-sector and two-sector models. Indeed, if the capital intensity difference at the private level is equal to zero, then $\mathcal{T}_{11}^*/\mathcal{T}_{12}^* = \infty$ and the characteristic polynomial is no longer well-defined.

15.3.2 Infinite Elasticity of Intertemporal Substitution in Consumption: $\sigma = 0$, $\gamma > 0$

Consider the case of a linear utility function with respect to consumption, i.e. $\sigma = 0$, in which the elasticity of intertemporal substitution in consumption is infinite. The characteristic polynomial then reduces to

$$\mathcal{P}(x) = \left(x + \frac{\mathcal{T}_{11}^*}{\mathcal{T}_{12}^*}\right)\left(\delta x + \frac{\mathcal{T}_{22}^*}{\mathcal{T}_{12}^*}\right).$$

Again the characteristic roots can be explicitly computed, namely

$$x_1 = -\frac{\mathcal{T}_{11}^*}{\mathcal{T}_{12}^*}, \quad x_2 = -\frac{\mathcal{T}_{22}^*}{\delta \mathcal{T}_{12}^*}.$$

This case has been analyzed by Nishimura and Venditti (2004b). However, a new conclusion is exhibited here:

Proposition 3. *Under Assumption 1, let $\sigma = 0$. Then the local stability properties of the steady state only depend on the CES coefficients $(\alpha_i, \beta_i, a_i, b_i, \rho_i)$, $i = c, y$, and do not depend on the elasticity of the labor supply $\epsilon_\ell = 1/\gamma$.*

When the utility function is additively separable and the elasticity of intertemporal substitution in consumption is infinite, the local stability properties are completely characterized by the production side of the model. A similar conclusion is obtained by Bosi et al. (2005) in a discrete-time two-sector optimal growth model with leisure and additively separable preferences.

As shown in Nishimura and Venditti (2004b), local indeterminacy cannot hold when the investment good is capital intensive at the private level. We then need to introduce the following assumption:

Assumption 2 *The consumption good is capital intensive at the private level.*

We will focus in the rest of the paper on configurations with $\alpha_1\beta_2 > \alpha_2\beta_1$. With Cobb-Douglas technologies this condition is equivalent to Assumption 2. Indeed, as shown in Benhabib et al. (2002), local indeterminacy in a Cobb-Douglas economy with linear preferences is obtained if and only if $\alpha_1\beta_2 - \alpha_2\beta_1 > \alpha_2/\delta$ and $\hat{\alpha}_2 > \hat{\alpha}_1 - \hat{\beta}_1$. The first inequality is then considered as a new Assumption:

Assumption 3 $\alpha_1\beta_2 - \alpha_2\beta_1 > \alpha_2/\delta$.

A slightly more restrictive condition than the second inequality will be introduced to get clear-cut conclusions. We will however exclusively focus on parameter configurations of CES technologies compatible with a Cobb-Douglas production function in both sectors.[7]

Theorem 3. *Under Assumptions 1–3, let $\hat{\alpha}_2 > \hat{\alpha}_1$. Then there exist $\underline{\rho}_c \in (-1, 0)$ and $\bar{\rho}_c \in (0, +\infty)$ with the following property: for any given $\rho_c \in (\underline{\rho}_c, \bar{\rho}_c)$, there exists $\bar{\rho}_y \in (0, \hat{\rho}_y)$ such that the steady state is locally indeterminate for all $\rho_y \in (-1, \bar{\rho}_y)$.*

Proof. See Appendix 15.6.3.

In order to give more precise economic interpretations, assume first that the technologies in both sectors are Cobb-Douglas, i.e. $\rho_c = \rho_y = 0$. Condition $\hat{\alpha}_2 > \hat{\alpha}_1$ implies that the aggregate share of labor in the consumption good sector is higher than the aggregate share of capital. This inequality, which also implies $\hat{\alpha}_2 > \hat{\alpha}_1 - \hat{\beta}_1$, may be satisfied when, at the social level, the investment good is capital intensive, i.e. $\hat{\alpha}_1 - \hat{\beta}_1 < 0$, and is also compatible with a consumption good which is capital intensive at the social level, i.e. $\hat{\alpha}_1 - \hat{\beta}_1 > 0$.

Consider now general asymmetric elasticities of capital-labor substitution. Theorem 3 shows that local indeterminacy still occurs with elasticities of capital-labor substitution which are arbitrarily large in the investment good sector. This cannot be the case with symmetric elasticities.

15.3.3 *Inelastic Labor Supply: $\sigma > 0$, $\gamma = +\infty$*

Assuming inelastic labor is equivalent to considering a zero elasticity of the labor supply, i.e. $\gamma = +\infty$. In such a case the utility function becomes

$$u(c) = \frac{c^{1-\sigma}}{1-\sigma}$$

and the roots of the characteristic polynomial may be approximated by the roots of the following equation:

$$\left(x + \frac{\mathcal{T}_{11}^*}{\mathcal{T}_{12}^*}\right)\left(x + \frac{\mathcal{T}_{22}^*}{\delta \mathcal{T}_{12}^*}\right) + \sigma \frac{T_1^{*2}}{c^* \mathcal{T}_{12}^*} \left(\delta \mathcal{B} x - \mathcal{A}\right)(x-1) = 0. \tag{15.13}$$

[7] We provide here a more precise formulation for some results exhibited in Nishimura and Venditti (2004b). We give indeed conditions for local indeterminacy which are valid for some intervals of values for both ρ_c and ρ_y. In Nishimura and Venditti (2004b) on the contrary, for a fixed value of ρ_c, we provide some interval for ρ_y in which local indeterminacy holds.

A similar formulation has been studied by Nishimura et al. (2006) but under the assumption that both sectors have the same elasticity of capital-labor substitution, i.e. $\rho_c = \rho_y = \rho$. We provide here a much more detailed analysis using the geometrical method provided by Grandmont et al. (1998). It is based on a particular property characterizing the product ($\mathscr{D}et$) and sum ($\mathscr{T}r$) of characteristic roots, which allow to provide a complete characterization of the local stability properties of the steady state and of the occurrence of bifurcations. The characteristic equation (15.13) is written as

$$x^2 - \mathscr{T}r(\sigma)x + \mathscr{D}et(\sigma) = 0 \tag{15.14}$$

with

$$\mathscr{D}et(\sigma) = \frac{\frac{\mathscr{T}_{11}^*}{\mathscr{T}_{12}^*}\frac{\mathscr{T}_{22}^*}{\delta\mathscr{T}_{12}^*} + \sigma\frac{T_1^{*2}}{c^*\mathscr{T}_{12}^*}\mathscr{A}}{1 + \sigma\delta\frac{T_1^{*2}}{c^*\mathscr{T}_{12}^*}\mathscr{B}}, \tag{15.15}$$

$$\mathscr{T}r(\sigma) = \frac{-\frac{\mathscr{T}_{11}^*}{\mathscr{T}_{12}^*} - \frac{\mathscr{T}_{22}^*}{\delta\mathscr{T}_{12}^*} + \sigma(\mathscr{A} + \delta\mathscr{B})\frac{T_1^{*2}}{c^*\mathscr{T}_{12}^*}}{1 + \sigma\delta\frac{T_1^{*2}}{c^*\mathscr{T}_{12}^*}\mathscr{B}}.$$

We analyze the local stability of the steady state by studying the variations of $\mathscr{T}r(\sigma)$ and $\mathscr{D}et(\sigma)$ in the $(\mathscr{T}r, \mathscr{D}et)$ plane when the inverse of the elasticity of intertemporal substitution in consumption σ varies continuously. Solving the two equations in (15.15) with respect to σ shows that when σ covers the interval $[0, +\infty)$, $\mathscr{D}et(\sigma)$ and $\mathscr{T}r(\sigma)$ vary along a line, called in what follows Δ_∞, which is defined by the following equation:

$$\mathscr{D}et = \mathscr{S}_\infty \mathscr{T}r + \frac{\mathscr{A}\left(\frac{\mathscr{T}_{11}^*}{\mathscr{T}_{12}^*} + \frac{\mathscr{T}_{22}^*}{\delta\mathscr{T}_{12}^*}\right) + (\mathscr{A} + \delta\mathscr{B})\frac{\mathscr{T}_{11}^*}{\mathscr{T}_{12}^*}\frac{\mathscr{T}_{22}^*}{\delta\mathscr{T}_{12}^*}}{\delta\mathscr{B}\left(1 + \frac{\mathscr{T}_{11}^*}{\mathscr{T}_{12}^*} + \frac{\mathscr{T}_{22}^*}{\delta\mathscr{T}_{12}^*} + \frac{\mathscr{A}}{\delta\mathscr{B}}\right)} \tag{15.16}$$

with

$$\mathscr{S}_\infty = \frac{\frac{\mathscr{A}}{\delta\mathscr{B}} - \frac{\mathscr{T}_{11}^*}{\mathscr{T}_{12}^*}\frac{\mathscr{T}_{22}^*}{\delta\mathscr{T}_{12}^*}}{1 + \frac{\mathscr{T}_{11}^*}{\mathscr{T}_{12}^*} + \frac{\mathscr{T}_{22}^*}{\delta\mathscr{T}_{12}^*} + \frac{\mathscr{A}}{\delta\mathscr{B}}} \tag{15.17}$$

the slope of Δ_∞.

Since $\gamma = +\infty$, labor is inelastic and $\ell^* = 1$. Moreover, κ^* does not depend on σ or γ, as shown in Proposition 1. The steady state then remains the same along the line Δ_∞. Figure 15.1 provides a possible illustration of Δ_∞.

We also introduce three other relevant lines: line AC ($\mathscr{D}et = \mathscr{T}r - 1$) along which one characteristic root is equal to 1 in (15.14), line AB ($\mathscr{D}et = -\mathscr{T}r - 1$) along which one characteristic root is equal to -1 in (15.14) and segment BC

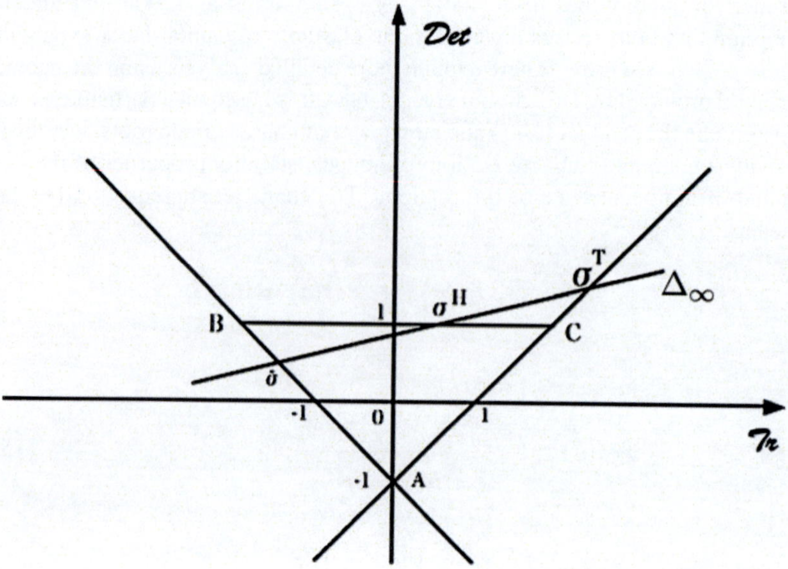

Fig. 15.1 Stability triangle and Δ_∞ line

$(\mathscr{D}et = 1, |\mathscr{T}r| < 2)$ along which the characteristic roots in (15.14) are complex conjugate with modulus equal to 1. These lines divide the space $(\mathscr{T}r, \mathscr{D}et)$ into three different types of regions according to the number of characteristic roots with modulus less than 1.

When $(\mathscr{T}r, \mathscr{D}et)$ belongs to the interior of triangle ABC, the steady state is locally indeterminate. Let $\bar{\sigma}$, σ^H and σ^T in $(0, +\infty)$ be the values of σ at which the Δ_∞ line, respectively, crosses segments AB, BC and AC. Then as σ, respectively, goes through $\bar{\sigma}$, σ^H or σ^T, a flip, Hopf or transcritical bifurcation is generically expected to occur. We have shown that a unique steady state always exists so that if the critical value σ^T exists, it will be only associated with a loss of stability of the steady state.

We compute the starting and end points of the pair $(\mathscr{T}r(\sigma), \mathscr{D}et(\sigma))$ on the line Δ_∞ as σ moves from 0 to $+\infty$. Consider first the starting point when $\sigma = 0$. Straightforward computations from (15.15) give

$$\mathscr{D}et(0) = \frac{\mathscr{T}_{11}^*}{\mathscr{T}_{12}^*}\frac{\mathscr{T}_{22}^*}{\delta \mathscr{T}_{12}^*}, \quad \mathscr{T}r(0) = -\frac{\mathscr{T}_{11}^*}{\mathscr{T}_{12}^*} - \frac{\mathscr{T}_{22}^*}{\delta \mathscr{T}_{12}^*}.$$

$\mathscr{D}et(0)$ and $\mathscr{T}r(0)$ correspond to the values of the characteristic roots obtained in Sect. 15.3.2 under an infinite elasticity of intertemporal substitution in consumption. Theorem 3 gives sufficient conditions for $(\mathscr{T}r(0), \mathscr{D}et(0))$ to belong to the triangle

ABC. If $\hat{\alpha}_2 > \hat{\alpha}_1$, there exist $\underline{\rho}_c \in (-1, 0)$ and $\bar{\rho}_c \in (0, +\infty)$ with the following property: for any given $\rho_c \in (\underline{\rho}_c, \bar{\rho}_c)$, we can define a function $\bar{\rho}_y(\rho_c) \in (0, \hat{\rho}_y)$ such that local indeterminacy occurs for all $\rho_y \in (-1, \bar{\rho}_y(\rho_c))$. We state these conditions as an assumption in order to focus on the case with local indeterminacy when utility is linear in consumption with such values $\underline{\rho}_c$, $\bar{\rho}_c$ and a function $\bar{\rho}_y(\rho_c)$.

Assumption 4 *$\hat{\alpha}_2 > \hat{\alpha}_1$, $\rho_c \in \left(\underline{\rho}_c, \bar{\rho}_c\right)$ and $\rho_y \in (-1, \bar{\rho}_y(\rho_c))$ with $\underline{\rho}_c \in (-1, 0)$, $\bar{\rho}_c \in (0, +\infty]$ and $\bar{\rho}_y(\rho_c) \in (0, \hat{\rho}_y)$.*

Under this assumption, the starting point of Δ_∞ with $\sigma = 0$ is within the triangle *ABC* as indicated in Fig. 15.2.

Consider now the end point of the locus $(\mathscr{T}r(\sigma), \mathscr{D}et(\sigma))$ when $\sigma = +\infty$. From (15.15) we get

$$\mathscr{D}et(+\infty) = \frac{\mathscr{A}}{\delta\mathscr{B}}, \quad \mathscr{T}r(+\infty) = 1 + \frac{\mathscr{A}}{\delta\mathscr{B}}. \tag{15.18}$$

Notice that

$$1 - \mathscr{T}r(+\infty) + \mathscr{D}et(+\infty) = 0 \tag{15.19}$$

so that the point $(\mathscr{T}r(+\infty), \mathscr{D}et(+\infty))$ lies on the line AC. It follows that the critical value σ^T mentioned on Fig. 15.1 is equal to $+\infty$.

Based on these results, in order to locate the line Δ_∞ we finally need to study the slope $\mathscr{S}_\infty$ and how $\mathscr{D}et(\sigma)$ and $\mathscr{T}r(\sigma)$ vary with σ.

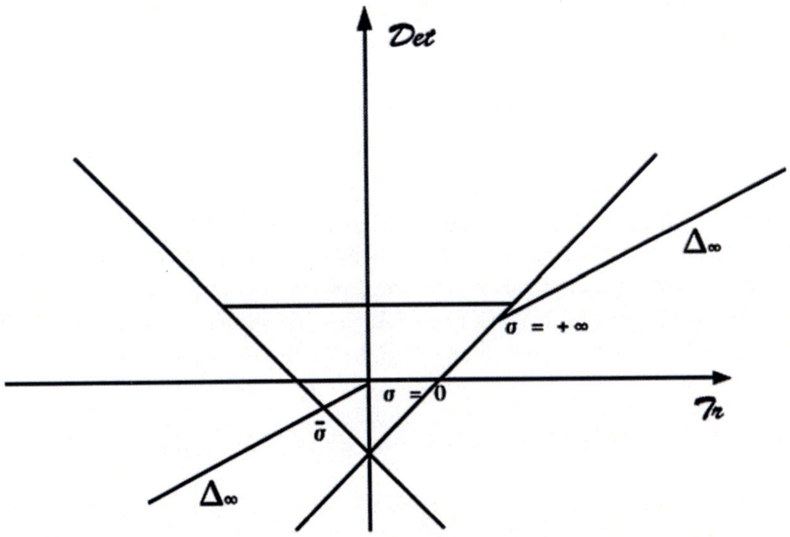

Fig. 15.2 Local indeterminacy with low σ and $\gamma = +\infty$

Lemma 1. *Under Assumptions 1–4, there exists* $\sigma^* \in (0, +\infty)$ *such that* $\lim_{\sigma \to \sigma^*_-} \mathscr{D}et(\sigma) = -\infty$, $\lim_{\sigma \to \sigma^*_+} \mathscr{D}et(\sigma) = +\infty$, *and* $\mathscr{D}et(\sigma)$ *and* $\mathscr{T}r(\sigma)$ *are monotonically decreasing in* σ *over each of the intervals* $[0, \sigma^*)$ *and* $(\sigma^*, +\infty)$. *Moreover,* $\mathscr{S}_\infty \in (0, 1)$.

Proof. See Appendix 15.6.4.

The critical value σ^* in Lemma 1 is defined from (15.15) and is equal to

$$\sigma^* = -\frac{c^* \mathscr{T}_{12}^*}{\delta \mathscr{B} T_1^{*2}}. \tag{15.20}$$

Notice also that, as shown in Sect. 15.3.2, when $\sigma = 0$ the characteristic roots are real. Under Assumption 4, since $\mathscr{S}_\infty \in (0, 1)$, we conclude that complex characteristic roots cannot occur and any real root cannot be equal to 1 for finite values of σ. Therefore, starting from $(\mathscr{T}r(0), \mathscr{D}et(0))$ into the triangle ABC, when σ increases, the point $(\mathscr{T}(\sigma), \mathscr{D}(\sigma))$ decreases along a Δ_∞ line as $\sigma \in (0, \sigma^*)$, goes through $-\infty$ when $\sigma = \sigma^*$ and finally decreases from $+\infty$ as $\sigma > \sigma^*$ until it reaches the end point $(\mathscr{T}r(+\infty), \mathscr{D}et(+\infty))$ which is located on the line AC, as shown in the following Fig. 15.2.

We derive from Theorem 3 and Fig. 15.2.

Theorem 4. *Let Assumptions 1–4 hold. Then there exists* $\bar{\sigma} \in (0, +\infty)$ *such that the steady state is locally indeterminate when* $\sigma \in [0, \bar{\sigma})$ *and saddle-point stable when* $\sigma > \bar{\sigma}$. *Moreover, the steady state undergoes a flip bifurcation when* $\sigma = \bar{\sigma}$ *so that locally indeterminate (respectively, saddle-point stable) period-two cycles generically exist in left (respectively, right) neighbourhood of* $\bar{\sigma}$.

The bifurcation value $\bar{\sigma}$ in Lemma 1 is defined as the solution of $1 + \mathscr{T}r + \mathscr{D}et = 0$ and is equal to

$$\bar{\sigma} = -\left(1 - \frac{\mathscr{T}_{11}^*}{\mathscr{T}_{12}^*}\right)\left(1 - \frac{\mathscr{T}_{22}^*}{\delta \mathscr{T}_{12}^*}\right) \frac{c^* \mathscr{T}_{12}^*}{2(\mathscr{A} + \delta \mathscr{B}) T_1^{*2}}. \tag{15.21}$$

Moreover, as shown in Fig. 15.2, we necessarily have $\bar{\sigma} < \sigma^*$.

Theorem 4 provides conditions for the occurrence of local indeterminacy which are compatible with the standard following intuition formulated, for instance, in Benhabib and Nishimura (1998): expectations-driven fluctuations are more likely under a high elasticity of intertemporal substitution in consumption $\epsilon_c = 1/\sigma$. When the elasticity of intertemporal substitution is decreased and crosses $\bar{\epsilon}_c = 1/\bar{\sigma}$, the steady state becomes saddle-point stable through a flip bifurcation and endogenous equilibrium cycles are generated in a neighbourhood of $\bar{\epsilon}_c$.

15.4 Main Results: $\sigma > 0$ and $\gamma > 0$

We finally consider the general configuration with $\sigma, \gamma > 0$. From the three polar cases previously analyzed, we will be able to provide a clear picture of the local stability properties of equilibria. Indeed, as in Sect. 15.3.3, a simple geometrical method based on the variations of $\mathcal{T}r$ and $\mathcal{D}et$ in the $(\mathcal{T}r, \mathcal{D}et)$ plane can be applied. We easily derive from the characteristic polynomial in Theorem 1 the expressions of $\mathcal{D}et$ and $\mathcal{T}r$:

$$\mathcal{D}et(\sigma,\gamma) = \frac{\gamma \frac{T_3^*}{\mathcal{T}_{32}^* \ell^*}\left(\frac{\mathcal{T}_{11}^*}{\mathcal{T}_{12}^*}\frac{\mathcal{T}_{22}^*}{\delta \mathcal{T}_{12}^*} + \sigma \frac{T_1^{*2}}{c^* \mathcal{T}_{12}^*}\mathcal{A}\right) - \sigma \frac{T_1^{*2}}{c^* \mathcal{T}_{12}^*}(1+\mathcal{C})^2 \mathcal{D}\hat{\beta}_1 (\delta\beta_1)^{\frac{-\rho y}{1+\rho y}}}{\gamma \frac{T_3^*}{\mathcal{T}_{32}^* \ell^*}\left(1 + \sigma \frac{T_1^{*2}}{c^* \mathcal{T}_{12}^*}\delta\mathcal{B}\right) - \sigma \frac{T_1^{*2}}{c^* \mathcal{T}_{12}^*}(1+\mathcal{C})^2 \mathcal{D}(\delta\beta_1)^{\frac{1}{1+\rho y}}}$$

$$\mathcal{T}r(\sigma,\gamma) = -\frac{\gamma \frac{T_3^*}{\mathcal{T}_{32}^* \ell^*}\left(\frac{\mathcal{T}_{11}^*}{\mathcal{T}_{12}^*} + \frac{\mathcal{T}_{22}^*}{\delta \mathcal{T}_{12}^*} - \sigma(\mathcal{A}+\delta\mathcal{B})\frac{T_1^{*2}}{c^* \mathcal{T}_{12}^*}\right) + \sigma \frac{T_1^{*2}}{c^* \mathcal{T}_{12}^*}(1+\mathcal{C})^2 \mathcal{D}\left(1 + \hat{\beta}_1 (\delta\beta_1)^{\frac{1-\rho y}{1+\rho y}}\right)}{\gamma \frac{T_3^*}{\mathcal{T}_{32}^* \ell^*}\left(1 + \sigma \frac{T_1^{*2}}{c^* \mathcal{T}_{12}^*}\delta\mathcal{B}\right) - \sigma \frac{T_1^{*2}}{c^* \mathcal{T}_{12}^*}(1+\mathcal{C})^2 \mathcal{D}(\delta\beta_1)^{\frac{1}{1+\rho y}}}. \tag{15.22}$$

Solving the two equations in (15.22) with respect to γ shows that when the inverse of the elasticity of labor supply γ covers the interval $[0, +\infty)$, $\mathcal{D}et(\sigma,\gamma)$ and $\mathcal{T}r(\sigma,\gamma)$ also vary along a line, denoted Δ_γ, which is defined as

$$\begin{aligned}\mathcal{D}et &= \mathcal{S}_\gamma(\sigma)\mathcal{T}r \\ &+ \frac{\left(1 + \hat{\beta}_1 (\delta\beta_1)^{\frac{1-\rho y}{1+\rho y}}\right)\left(\frac{\mathcal{T}_{11}^*}{\mathcal{T}_{12}^*}\frac{\mathcal{T}_{22}^*}{\delta \mathcal{T}_{12}^*} + \sigma \frac{T_1^{*2}}{c^* \mathcal{T}_{12}^*}\mathcal{A}\right) + \hat{\beta}_1(\delta\beta_1)^{\frac{-\rho y}{1+\rho y}}\left(\frac{\mathcal{T}_{11}^*}{\mathcal{T}_{12}^*} + \frac{\mathcal{T}_{22}^*}{\delta \mathcal{T}_{12}^*} - \sigma(\mathcal{A}+\delta\mathcal{B})\frac{T_1^{*2}}{c^* \mathcal{T}_{12}^*}\right)}{(\delta\beta_1)^{\frac{1}{1+\rho y}}\left(\frac{\mathcal{T}_{11}^*}{\mathcal{T}_{12}^*} + \frac{\mathcal{T}_{22}^*}{\delta \mathcal{T}_{12}^*} - \sigma(\mathcal{A}+\delta\mathcal{B})\frac{T_1^{*2}}{c^* \mathcal{T}_{12}^*}\right) + \left(1 + \hat{\beta}_1 (\delta\beta_1)^{\frac{1-\rho y}{1+\rho y}}\right)\left(1 + \sigma \frac{T_1^{*2}}{c^* \mathcal{T}_{12}^*}\delta\mathcal{B}\right)}\end{aligned} \tag{15.23}$$

with

$$\mathcal{S}_\gamma(\sigma) = \frac{\hat{\beta}_1 (\delta\beta_1)^{\frac{-\rho y}{1+\rho y}}\left(1 + \sigma \frac{T_1^{*2}}{c^* \mathcal{T}_{12}^*}\delta\mathcal{B}\right) - (\delta\beta_1)^{\frac{1}{1+\rho y}}\left(\frac{\mathcal{T}_{11}^*}{\mathcal{T}_{12}^*}\frac{\mathcal{T}_{22}^*}{\delta \mathcal{T}_{12}^*} + \sigma \frac{T_1^{*2}}{c^* \mathcal{T}_{12}^*}\mathcal{A}\right)}{(\delta\beta_1)^{\frac{1}{1+\rho y}}\left(\frac{\mathcal{T}_{11}^*}{\mathcal{T}_{12}^*} + \frac{\mathcal{T}_{22}^*}{\delta \mathcal{T}_{12}^*} - \sigma(\mathcal{A}+\delta\mathcal{B})\frac{T_1^{*2}}{c^* \mathcal{T}_{12}^*}\right) + \left(1 + \hat{\beta}_1 (\delta\beta_1)^{\frac{1-\rho y}{1+\rho y}}\right)\left(1 + \sigma \frac{T_1^{*2}}{c^* \mathcal{T}_{12}^*}\delta\mathcal{B}\right)} \tag{15.24}$$

the slope of Δ_γ. We compute the starting and end points of the pair $(\mathcal{T}r(\sigma,\gamma), \mathcal{D}et(\sigma,\gamma))$ on Δ_γ as γ moves from 0 to $+\infty$. From (15.22) we get

$$\begin{aligned}\mathcal{D}et(\sigma,0) &= \frac{\hat{\beta}_1}{\delta\beta_1}, \\ \mathcal{T}r(\sigma,0) &= (\delta\beta_1)^{\frac{-1}{1+\rho y}} + \hat{\beta}_1 (\delta\beta_1)^{\frac{-\rho y}{1+\rho y}}, \\ \mathcal{D}et(\sigma,+\infty) &= \frac{\frac{\mathcal{T}_{11}^*}{\mathcal{T}_{12}^*}\frac{\mathcal{T}_{22}^*}{\delta \mathcal{T}_{12}^*} + \sigma \frac{T_1^{*2}}{c^* \mathcal{T}_{12}^*}\mathcal{A}}{1 + \sigma\delta \frac{T_1^{*2}}{c^* \mathcal{T}_{12}^*}\mathcal{B}},\end{aligned} \tag{15.25}$$

$$\mathcal{T}r(\sigma, +\infty) = \frac{\sigma(\mathcal{A} + \delta\mathcal{B})\frac{T_1^{*2}}{c^* \mathcal{T}_{12}^*} - \frac{\mathcal{T}_{11}^*}{\mathcal{T}_{12}^*} - \frac{\mathcal{T}_{22}^*}{\delta \mathcal{T}_{12}^*}}{1 + \sigma\delta\frac{T_1^{*2}}{c^* \mathcal{T}_{12}^*}\mathcal{B}}.$$

Of course, $\mathcal{D}et(\sigma, 0)$ and $\mathcal{T}r(\sigma, 0)$ correspond to the values of the characteristic roots obtained in Sect. 15.3.1 under an infinite elasticity of the labor supply while $\mathcal{D}et(\sigma, +\infty)$ and $\mathcal{T}r(\sigma, +\infty)$ correspond to the values of the characteristic roots obtained in Sect. 15.3.3 under inelastic labor. Notice that under Assumption 1 the end point $(\mathcal{T}r(\sigma, 0), \mathcal{D}et(\sigma, 0))$ satisfies

$$1 - \mathcal{T}r(\sigma, 0) + \mathcal{D}et(\sigma, 0) = -\left(1 - (\delta\beta_1)^{\frac{1}{1+\rho_y}}\right)\left((\delta\beta_1)^{\frac{\rho_y}{1+\rho_y}} - \hat{\beta}_1\right)(\delta\beta_1)^{-1} < 0$$

so that it lies below the line AC with $\mathcal{D}et(\sigma, 0) > 1$ and $\mathcal{T}r(\sigma, 0) > 2$.

Geometrically, for some given value of σ, the end point $(\mathcal{T}r(\sigma, +\infty), \mathcal{D}et(\sigma, +\infty))$ on the line Δ_γ belongs to the line Δ_∞ which is obtained by setting γ to be $+\infty$ and which is analyzed in Sect. 15.3.3. Based on these results, in order to locate the line Δ_γ we finally need to study the slope $\mathcal{S}_\gamma(\sigma)$ and how $\mathcal{D}et(\sigma, \gamma)$ and $\mathcal{T}r(\sigma, \gamma)$ vary with γ. We introduce an additional assumption to get clear-cut results:

Assumption 5 $\alpha_1\beta_2/\alpha_2\beta_1 > \hat{\alpha}_1\hat{\beta}_2/\hat{\alpha}_2\hat{\beta}_1$.

Assumption 3 implies $\alpha_1\beta_2/\alpha_2\beta_1 > 1$. Assumption 5 is then compatible with $\hat{\alpha}_1\hat{\beta}_2/\hat{\alpha}_2\hat{\beta}_1 > 1$ and $\hat{\alpha}_1\hat{\beta}_2/\hat{\alpha}_2\hat{\beta}_1 < 1$, i.e. with both capital intensity differences at the social level when the technologies are Cobb-Douglas.

Lemma 2. *Under Assumptions 1–5, consider the critical value σ^* as defined by (15.20). Then there exist $\sigma_1 \in (\sigma^*, +\infty)$ and $\sigma_2 \in (\sigma^*, +\infty)$ such that the following results hold*[8]*:*

(i) $\mathcal{D}et_2(\sigma, \gamma) > 0$ *when* $\sigma \in (0, \sigma_1)$, $\mathcal{D}et_2(\sigma, \gamma) = 0$ *when* $\sigma = \sigma_1$ *and* $\mathcal{D}et_2(\sigma, \gamma) < 0$ *when* $\sigma > \sigma_1$*;*
(ii) $\mathcal{T}r_2(\sigma, \gamma) > 0$ *when* $\sigma \in (0, \sigma_2)$, $\mathcal{T}r_2(\sigma, \gamma) = 0$ *when* $\sigma = \sigma_2$ *and* $\mathcal{T}r_2(\sigma, \gamma) < 0$ *when* $\sigma > \sigma_2$*;*
(iii) $\mathcal{S}_\gamma(\sigma) > 0$ *when* $\sigma \in (0, \min\{\sigma_1, \sigma_2\}) \cup (\max\{\sigma_1, \sigma_2\}, +\infty)$ *and* $\mathcal{S}_\gamma(\sigma) < 0$ *when* $\sigma \in (\min\{\sigma_1, \sigma_2\}, \max\{\sigma_1, \sigma_2\})$.

Proof. See Appendix 15.6.5.

Building on our previous results, we are looking for conditions for the occurrence of local indeterminacy which are compatible with the case of inelastic labor.

[8]The partial derivatives of $\mathcal{D}et(\sigma, \gamma)$ and $\mathcal{T}r(\sigma, \gamma)$ with respect to γ are, respectively, denoted $\mathcal{D}et_2(\sigma, \gamma)$ and $\mathcal{T}r_2(\sigma, \gamma)$.

According to Theorem 4, we thus introduce a restriction on the elasticity of intertemporal substitution in consumption which ensures that $(\mathcal{T}r(\sigma, +\infty), \mathcal{D}et(\sigma, +\infty))$ belongs to the triangle *ABC*.

Assumption 6 $\sigma \in [0, \bar{\sigma})$ *with* $\bar{\sigma} > 0$ *as defined by (15.21).*

Since $\bar{\sigma} < \sigma^*$, Assumption 6 implies $\sigma < \min\{\sigma_1, \sigma_2\}$ so that $\mathcal{D}et(\sigma, \gamma)$ and $\mathcal{T}r(\sigma, \gamma)$ are increasing functions of γ with $\mathcal{S}_\gamma(\sigma) > 0$. Moreover, we get the following result.

Lemma 3. *Under Assumptions 1–6, for any given* $\sigma \in [0, \bar{\sigma})$, *there exists* $\gamma^* \in (0, +\infty)$ *such that* $\lim_{\gamma \to \gamma^*_+} \mathcal{D}et(\sigma, \gamma) = -\infty$, $\lim_{\gamma \to \gamma^*_-} \mathcal{D}et(\sigma, \gamma) = +\infty$, *and* $\mathcal{S}_\gamma(\sigma) \in (0, 1)$ *for all* $\gamma > 0$.

Proof. See Appendix 15.6.6.

All these results may be summarized with the following geometrical representation:

When γ is large enough, i.e. when the elasticity of labor supply is low enough, the steady state is locally indeterminate. We finally derive from Theorem 4, Lemmas 2–3 and Fig. 15.3:

Theorem 5. *Let Assumptions 1–5 hold. Then, for any given* σ *satisfying Assumption 6, there exists* $\underline{\gamma} \in (0, +\infty)$ *such that the steady state is locally indeterminate when* $\gamma > \underline{\gamma}$ *and becomes saddle-point stable when* $\gamma < \underline{\gamma}$. *Moreover, the steady state undergoes a flip bifurcation when* $\gamma = \underline{\gamma}$ *so that locally indeterminate*

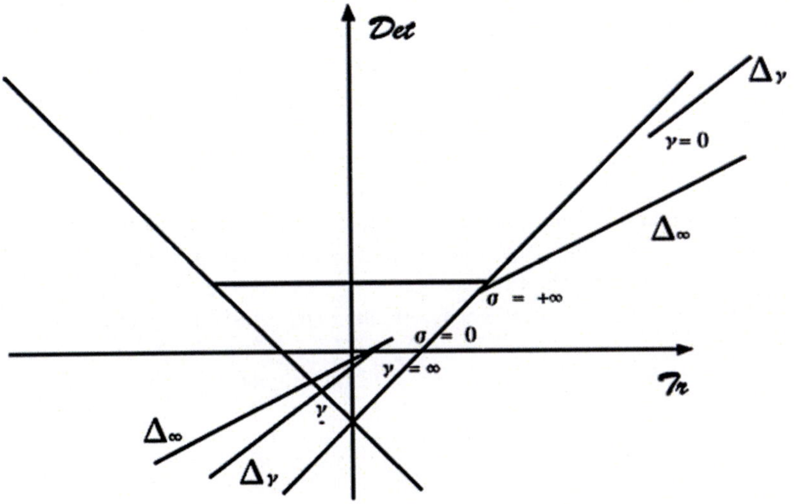

Fig. 15.3 Local indeterminacy with low σ and large γ

(respectively, saddle-point stable) period-two cycles generically exist in a right (respectively, left) neighbourhood of $\underline{\gamma}$.

Theorem 5 shows that the steady state is locally indeterminate provided that the elasticity of intertemporal substitution in consumption is large enough while the elasticity of labor supply is low. We also immediately derive from Lemma 2 that the consideration of endogenous labor does not provide any additional room for the occurrence of local indeterminacy. Indeed, the existence of multiple equilibria again requires σ to be lower than the bound $\bar{\sigma}$ exhibited in the case of inelastic labor.[9]

15.5 Concluding Comments

The main objective of this paper has been to discuss jointly the roles of the elasticity of intertemporal substitution in consumption and the elasticity of the labor supply on the local indeterminacy properties of the long-run equilibrium. We have considered a discrete-time two-sector model with sector specific externalities in which the technologies are given by CES functions with asymmetric elasticities of capital-labor substitution, and the preferences of the representative agent are given by a CES additively separable utility function defined over consumption and leisure.

Different specific configurations have been studied in order to identify the influence of each precise parameter. In a first step we have clearly showed that when inelastic labor is considered, the existence of local indeterminacy is obtained if and only if the elasticity of intertemporal substitution in consumption is large enough. This result confirms basic intuitions obtained within one-sector models. In a second step, we have proved that the consideration of endogenous labor within a two-sector model does not introduce any additional dimensions for the occurrence of local indeterminacy.

Indeed we have shown on one hand that when the labor supply is infinitely elastic, local indeterminacy is ruled out and the steady state is always saddle-point stable, no matter what the elasticity of intertemporal substitution in consumption and the size of externalities are. On the other hand, we have proved that when the elasticity of intertemporal substitution in consumption is large, local indeterminacy requires a low enough elasticity of the labor supply. These results show that the effects of the elasticity of the labor supply appear to be in complete opposition with the main conclusions obtained within one-sector models.

---

[9]Indeed, when $\sigma \in (\bar{\sigma}, \sigma_1)$ and $\mathscr{D}et_2(\sigma, \gamma) > 0$, or when $\sigma > \sigma_1$ and $\mathscr{D}et_2(\sigma, \gamma) < 0$, the pair $(\mathscr{T}r, \mathscr{D}et)$ is necessarily outside of the triangle ABC for any $\gamma > 0$.

15.6 Appendix

15.6.1 Proof of Proposition 1

From the Lagrangian (15.3) we derive the first order conditions:

$$\alpha_1 \left(c/K_c\right)^{1+\rho_c} - r = 0 \tag{15.26}$$

$$\alpha_2 \left(c/L_c\right)^{1+\rho_c} - w = 0 \tag{15.27}$$

$$p\beta_1 \left(y/K_y\right)^{1+\rho_y} - r = 0 \tag{15.28}$$

$$p\beta_2 \left(y/L_y\right)^{1+\rho_y} - w = 0. \tag{15.29}$$

Using $K_c = k_0 - K_y$, $L_y = 1 - L_c$, and manipulating (15.26)–(15.29) give

$$L_y = \left((y^{-\rho_y} - \hat{\beta}_1 K_y^{-\rho_y})/\hat{\beta}_2\right)^{-1/\rho_y} \tag{15.30}$$

$$\left(\frac{\alpha_1\beta_2}{\alpha_2\beta_1}\right) = \left(\frac{k_0 - K_y}{1 - L_y}\right)^{1+\rho_c} \left(\frac{L_y}{K_y}\right)^{1+\rho_y}. \tag{15.31}$$

By solving (15.30)–(15.31) with respect to K_y and substituting $y = k_1$, we get $K_y = g(k_0, k_1, \ell)$. From (15.26) and (15.30) we get

$$r = \alpha_1 \left\{ \hat{\alpha}_1 + \hat{\alpha}_2 \left(\frac{\alpha_1\beta_2}{\alpha_2\beta_1}\right)^{\frac{\rho_c}{1+\rho_c}} \left(\frac{(g/y)^{\rho_y} - \hat{\beta}_1}{\hat{\beta}_2}\right)^{\frac{\rho_c(1+\rho_y)}{\rho_y(1+\rho_c)}} \right\}^{-\frac{1+\rho_c}{\rho_c}}. \tag{15.32}$$

Moreover, we have from (15.28) and (15.27)

$$p = \frac{w}{\beta_1}\left(\frac{g}{y}\right)^{1+\rho_y}, \quad w = r\frac{\beta_2}{\beta_1}\left(\frac{(g/y)^{\rho_y} - \hat{\beta}_1}{\hat{\beta}_2}\right)^{\frac{1+\rho_y}{\rho_y}}.$$

From the envelope theorem we finally conclude that $T_1 = r$, $T_2 = -p$ and $T_3 = w$. A steady state (k^*, ℓ^*) is a solution of (15.9) with $y^* = k^*$. Using the derivatives of T in the definition of k^* gives

$$g(k^*, k^*, \ell^*) = (\delta\beta_1)^{\frac{1}{1+\rho_y}} k^*. \tag{15.33}$$

Constant returns to scale at the social level imply that $g(k, k, \ell)$ is homogeneous of degree 1 (see Remark 2 in Appendix 15.6.2). Denoting $\kappa = k/\ell$ and $\bar{g} = g(\kappa, \kappa, 1)$, we may then derive from (15.33)

$$\bar{g}^* = (\delta\beta_1)^{\frac{1}{1+\rho_y}} \kappa^*. \tag{15.34}$$

When $k_0 = k^*$, (15.30)–(15.31) with the fact that $g/y = \bar{g}/\kappa$ give

$$K_c^* = \ell\kappa^* \left(1 - (\delta\beta_1)^{\frac{1}{1+\rho_y}}\right), \quad L_y^* = \ell\kappa^* (\delta\beta_1)^{\frac{1}{1+\rho_y}} \left(\frac{(\delta\beta_1)^{\frac{\rho_y}{1+\rho_y}} - \hat{\beta}_1}{\hat{\beta}_2}\right)^{-\frac{1}{\rho_y}} \tag{15.35}$$

and $L_c^* = \ell - L_y^*$. Under Assumption 1, $k^* > K_c^* > 0$ and $\ell > L_y^* > 0$. Substituting these input demand functions into (15.31) we get the expression of κ^*. From (15.2), (15.34) and (15.35) we compute

$$c^* = \frac{\kappa^* \ell^* \left(1 - (\delta\beta_1)^{\frac{1}{1+\rho_y}}\right)}{\left[\hat{\alpha}_1 + \hat{\alpha}_2 \left(\frac{\alpha_1\beta_2}{\alpha_2\beta_1}\right)^{\frac{\rho_c}{1+\rho_c}} \left(\frac{(\delta\beta_1)^{\frac{\rho_y}{1+\rho_y}} - \hat{\beta}_1}{\hat{\beta}_2}\right)^{\frac{\rho_c(1+\rho_y)}{\rho_y(1+\rho_c)}}\right]^{\frac{1}{\rho_c}}}. \tag{15.36}$$

Substituting this expression into $wc^{-\sigma} - \ell^{\gamma} = 0$ finally gives ℓ^*. ■

15.6.2 Proof of Theorem 1

Various preliminary lemmas are necessary to prove Theorem 1. Denote in what follows g_1, g_2 and g_3 the partial derivatives of g with respect to k_0, k_1 and ℓ.

Lemma A.1. *Under Assumption 1, at the steady state the following hold:*

$$g_1 = \frac{1+\rho_c}{K_c \Delta}, \quad g_2 = \frac{\left[(1+\rho_y) + (1+\rho_c)\left(L_y/L_c\right)\right]\left(L_y/y\right)^{\rho_y}}{\hat{\beta}_2 y \Delta}, \quad g_3 = -\frac{1+\rho_c}{L_c \Delta}$$

with g, K_c, L_y and L_c, respectively, given by (15.33)–(15.34) and

$$\Delta = \frac{1+\rho_y}{g} + \frac{1+\rho_c}{K_c} + \left((1+\rho_y) + (1+\rho_c)\frac{L_y}{L_c}\right)\frac{\hat{\beta}_1}{\hat{\beta}_2 g}\left(\frac{L_y}{g}\right)^{\rho_y}.$$

Proof. From (15.31) we get

$$\frac{\alpha_1\beta_2}{\alpha_2\beta_1} = g^{-1-\rho_y}(k_0 - g)^{1+\rho_c}\left(\frac{y^{-\rho_y} - \hat{\beta}_1 g^{-\rho_y}}{\hat{\beta}_2}\right)^{-\frac{1+\rho_y}{\rho_y}} \times \left\{\ell - \left(\frac{y^{-\rho_y} - \hat{\beta}_1 g^{-\rho_y}}{\hat{\beta}_2}\right)^{-\frac{1}{\rho_y}}\right\}^{-1-\rho_c}.$$

Totally differentiating this expression gives after simplification

$$\begin{bmatrix} (1+\rho_y)g^{-1} + (1+\rho_c)(k_0-g)^{-1} + (1+\rho_y)\left(\frac{y^{-\rho_y}-\hat{\beta}_1 g^{-\rho_y}}{\hat{\beta}_2}\right)^{-1} \frac{\hat{\beta}_1}{\hat{\beta}_2} g^{-1-\rho_y} \\ +(1+\rho_c)\left\{1-\left(\frac{y^{-\rho_y}-\hat{\beta}_1 g^{-\rho_y}}{\hat{\beta}_2}\right)^{-\frac{1}{\rho_y}}\right\}^{-1} \left(\frac{y^{-\rho_y}-\hat{\beta}_1 g^{-\rho_y}}{\hat{\beta}_2}\right)^{-\frac{1+\rho_y}{\rho_y}} \frac{\hat{\beta}_1}{\hat{\beta}_2} g^{-1-\rho_y} \end{bmatrix} dg$$

$$= (1+\rho_c)(k_0-g)^{-1} dk_0 + (1+\rho_y)\left(\frac{y^{-\rho_y}-\hat{\beta}_1 g^{-\rho_y}}{\hat{\beta}_2}\right)^{-1} \frac{y^{-1-\rho_y}}{\hat{\beta}_2} dy$$

$$+(1+\rho_c)\left\{1-\left(\frac{y^{-\rho_y}-\hat{\beta}_1 g^{-\rho_y}}{\hat{\beta}_2}\right)^{-\frac{1}{\rho_y}}\right\}^{-1} \left(\frac{y^{-\rho_y}-\hat{\beta}_1 g^{-\rho_y}}{\hat{\beta}_2}\right)^{-\frac{1+\rho_y}{\rho_y}} \frac{y^{-1-\rho_y}}{\hat{\beta}_2} dy.$$

Notice from (15.30) and (15.31) that

$$\frac{(g/y)^{\rho_y}}{\hat{\beta}_2} - \frac{\hat{\beta}_1}{\hat{\beta}_2} = \left(\frac{g}{L_y}\right)^{\rho_y} \quad \text{and} \quad \left(\frac{\alpha_1\beta_2}{\alpha_2\beta_1}\right)^{\frac{1}{1+\rho_c}} \left(\frac{K_y}{L_y}\right)^{\frac{1+\rho_y}{1+\rho_c}} = \frac{K_c}{L_c}. \tag{15.37}$$

Substituting (15.37) into the previous total differentiation and considering $dy = dk_1$ together with

$$\Delta = \frac{1+\rho_y}{g} + \frac{1+\rho_c}{K_c} + \left((1+\rho_y) + (1+\rho_c)\frac{L_y}{L_c}\right) \frac{\hat{\beta}_1}{\hat{\beta}_2 g} \left(\frac{L_y}{g}\right)^{\rho_y}$$

we derive

$$\Delta dg = dk_0 \frac{1+\rho_c}{K_c} + dk_1 \left[(1+\rho_y) + (1+\rho_c)\frac{L_y}{L_c}\right] \frac{1}{\hat{\beta}_2 y} \left(\frac{L_y}{y}\right)^{\rho_y} - d\ell \frac{1+\rho_c}{L_c}.$$

This completes the proof.

Lemma A.2. *Under Assumption 1, at the steady state the following holds:*

$$g_1 = \frac{g - g_2 y}{g}\left(1 - \frac{K_c}{L_c}\frac{L_y}{g}\right)^{-1}, \quad g_3 = -g_1 \frac{K_c}{L_c}$$

with g, K_c, L_y and L_c, respectively, given by (15.33)–(15.35).

Proof. *From Lemma A.1 and (15.30) we have*

$$g - g_2 y = \frac{(1+\rho_c)}{K_c \Delta} g \left(1 - \frac{K_c}{L_c}\frac{L_y}{g}\right).$$

The result follows.

Remark 2. From Lemma A.2 we derive that at the steady state the function $g(k, y, \ell)$ is homogeneous of degree 1 since $g(k, k, \ell) = g_1 k + g_2 k + g_3 \ell$. It follows from (15.36) that the production frontier, i.e. (15.7), when evaluated along a stationary solution, is also homogeneous of degree 1.

Consider the following notations:

$$\mathcal{T}_{i1}(k_0, k_1, \ell) = \partial T_i(k_0, k_1, \ell, \hat{e}_c(k_0, k_1, \ell), \hat{e}_y(k_0, k_1, \ell))/\partial k_0$$

$$\mathcal{T}_{i2}(k_0, k_1, \ell) = \partial T_i(k_0, k_1, \ell, \hat{e}_c(k_0, k_1, \ell), \hat{e}_y(k_0, k_1, \ell))/\partial k_1$$

$$\mathcal{T}_{i3}(k_0, k_1, \ell) = \partial T_i(k_0, k_1, \ell, \hat{e}_c(k_0, k_1, \ell), \hat{e}_y(k_0, k_1, \ell))/\partial \ell$$

and $\mathcal{T}_{ij}(k^*, k^*, \ell^*) \equiv \mathcal{T}_{ij}^*$ for $i, j = 1, 2, 3$.

Lemma A.3. *Under Assumption 1, at the steady state, $k_0 = k_1 = y = k^*$ and the following hold:*

$$\frac{\mathcal{T}_{11}(k^*, k^*, l)}{\mathcal{T}_{12}(k^*, k^*, l)} = -\frac{y}{g}\left(1 - \frac{K_c}{L_c}\frac{L_y}{g}\right)^{-1} = \frac{\mathcal{T}_{31}(k^*, k^*, l)}{\mathcal{T}_{32}(k^*, k^*, l)}$$

$$\frac{\mathcal{T}_{22}(k^*, k^*, l)}{\mathcal{T}_{12}(k^*, k^*, l)} = \frac{g}{\beta_1 y}\left[\frac{\alpha_2\beta_1}{\alpha_1\beta_2}\frac{\hat{\alpha}_1\hat{\beta}_2}{\hat{\alpha}_2}\frac{K_c}{L_c}\frac{L_y}{g} + \hat{\beta}_2\left(\frac{g}{L_y}\right)^{\rho_y} - \left(\frac{g}{y}\right)^{\rho_y}\right]$$

$$\frac{\mathcal{T}_{21}(k^*, k^*, l)}{\mathcal{T}_{12}(k^*, k^*, l)} = \frac{\mathcal{T}_{22}(k^*, k^*, l)}{\mathcal{T}_{12}(k^*, k^*, l)}\frac{\mathcal{T}_{11}(k^*, k^*, l)}{\mathcal{T}_{12}(k^*, k^*, l)}$$

$$\frac{\mathcal{T}_{13}(k^*, k^*, l)}{\mathcal{T}_{12}(k^*, k^*, l)} = \frac{K_c}{L_c}\left(1 - \frac{K_c}{L_c}\frac{L_y}{g}\right)^{-1} = \frac{\mathcal{T}_{33}(k^*, k^*, l)}{\mathcal{T}_{32}(k^*, k^*, l)}$$

$$\frac{\mathcal{T}_{23}(k^*, k^*, l)}{\mathcal{T}_{12}(k^*, k^*, l)} = \frac{\mathcal{T}_{22}(k^*, k^*, l)}{\mathcal{T}_{12}(k^*, k^*, l)}\frac{\mathcal{T}_{13}(k^*, k^*, l)}{\mathcal{T}_{12}(k^*, k^*, l)}$$

with g, K_c, L_y and L_c, respectively, given by (15.33)–(15.35).

Proof. By definition we have

$$\mathcal{T}_{11}^* = \frac{\partial r}{\partial k_0}, \quad \mathcal{T}_{21}^* = -\frac{\partial p}{\partial k_0}, \quad \mathcal{T}_{31}^* = \frac{\partial w}{\partial k_0}$$

$$\mathcal{T}_{12}^* = \frac{\partial r}{\partial k_1}, \quad \mathcal{T}_{22}^* = -\frac{\partial p}{\partial k_1}, \quad \mathcal{T}_{32}^* = \frac{\partial w}{\partial k_1}$$

$$\mathcal{T}_{13}^* = \frac{\partial r}{\partial \ell}, \quad \mathcal{T}_{23}^* = -\frac{\partial p}{\partial \ell}, \quad \mathcal{T}_{33}^* = \frac{\partial w}{\partial \ell}.$$

Simple computations give

$$\frac{\partial r}{\partial k_0} = -(1+\rho_y)\left(\frac{r}{\alpha_1}\right)^{\frac{\rho_c}{1+\rho_c}} r\frac{\hat{\alpha}_2}{\hat{\beta}_2}\left(\frac{\alpha_1\beta_2}{\alpha_2\beta_1}\right)^{\frac{\rho_c}{1+\rho_c}}\left(\frac{(\delta\beta_1)^{\frac{\rho_y}{1+\rho_y}}-\hat{\beta}_1}{\hat{\beta}_2}\right)^{\frac{\rho_c-\rho_y}{\rho_y(1+\rho_c)}} \times\left(\frac{g}{y}\right)^{\rho_y}\frac{g_1}{g}$$

$$\frac{\partial p}{\partial k_0} = \frac{1}{\beta_1}\left(\frac{g}{y}\right)^{1+\rho_y}\left[\frac{\partial r}{\partial k_0}+(1+\rho_y)r\frac{g_1}{g}\right]$$

$$\frac{\partial w}{\partial k_0} = \frac{\beta_2}{\beta_1}\left(\frac{(\delta\beta_1)^{\frac{\rho_y}{1+\rho_y}}-\hat{\beta}_1}{\hat{\beta}_2}\right)^{\frac{1}{\rho_y}} \times\left[\frac{\partial r}{\partial k_0}\left(\frac{(\delta\beta_1)^{\frac{\rho_y}{1+\rho_y}}-\hat{\beta}_1}{\hat{\beta}_2}\right)+(1+\rho_y)r\left(\frac{g}{y}\right)^{\rho_y}\frac{1}{\hat{\beta}_2}\frac{g_1}{g}\right]$$

$$\frac{\partial r}{\partial k_1} = -(1+\rho_y)\left(\frac{r}{\alpha_1}\right)^{\frac{\rho_c}{1+\rho_c}} r\frac{\hat{\alpha}_2}{\hat{\beta}_2}\left(\frac{\alpha_1\beta_2}{\alpha_2\beta_1}\right)^{\frac{\rho_c}{1+\rho_c}}\left(\frac{(\delta\beta_1)^{\frac{\rho_y}{1+\rho_y}}-\hat{\beta}_1}{\hat{\beta}_2}\right)^{\frac{\rho_c-\rho_y}{\rho_y(1+\rho_c)}} \times\left(\frac{g}{y}\right)^{\rho_y}\frac{g_2y-g}{gy}$$

$$\frac{\partial p}{\partial k_1} = \frac{1}{\beta_1}\left(\frac{g}{y}\right)^{1+\rho_y}\left[\frac{\partial r}{\partial k_1}+(1+\rho_y)r\frac{g_2y-g}{yg}\right]$$

$$\frac{\partial w}{\partial k_1} = \frac{\beta_2}{\beta_1}\left(\frac{(\delta\beta_1)^{\frac{\rho_y}{1+\rho_y}}-\hat{\beta}_1}{\hat{\beta}_2}\right)^{\frac{1}{\rho_y}} \times\left[\frac{\partial r}{\partial k_1}\left(\frac{(\delta\beta_1)^{\frac{\rho_y}{1+\rho_y}}-\hat{\beta}_1}{\hat{\beta}_2}\right)+(1+\rho_y)r\left(\frac{g}{y}\right)^{\rho_y}\frac{1}{\hat{\beta}_2}\frac{g_2y-g}{gy}\right]$$

$$\frac{\partial r}{\partial \ell} = -(1+\rho_y)\left(\frac{r}{\alpha_1}\right)^{\frac{\rho_c}{1+\rho_c}} r\frac{\hat{\alpha}_2}{\hat{\beta}_2}\left(\frac{\alpha_1\beta_2}{\alpha_2\beta_1}\right)^{\frac{\rho_c}{1+\rho_c}}\left(\frac{(\delta\beta_1)^{\frac{\rho_y}{1+\rho_y}}-\hat{\beta}_1}{\hat{\beta}_2}\right)^{\frac{\rho_c-\rho_y}{\rho_y(1+\rho_c)}} \times\left(\frac{g}{y}\right)^{\rho_y}\frac{g_3}{g}$$

$$\frac{\partial p}{\partial \ell} = \frac{1}{\beta_1}\left(\frac{g}{y}\right)^{1+\rho_y}\left[\frac{\partial r}{\partial \ell}+(1+\rho_y)r\frac{g_3}{g}\right]$$

$$\frac{\partial w}{\partial \ell} = \frac{\beta_2}{\beta_1}\left(\frac{(\delta\beta_1)^{\frac{\rho_y}{1+\rho_y}} - \hat{\beta}_1}{\hat{\beta}_2}\right)^{\frac{1}{\rho_y}}$$

$$\times \left[\frac{\partial r}{\partial \ell}\left(\frac{(\delta\beta_1)^{\frac{\rho_y}{1+\rho_y}} - \hat{\beta}_1}{\hat{\beta}_2}\right) + (1+\rho_y) r \left(\frac{g}{y}\right)^{\rho_y} \frac{1}{\hat{\beta}_2}\frac{g_3}{g}\right].$$

Substituting (r/α_1) from (15.32), and using (15.37) we get

$$\mathscr{T}_{11}^* = -(1+\rho_y) r \left(\frac{g}{y}\right)^{\rho_y} \frac{g_1}{g}\left[\frac{\hat{\alpha}_1\hat{\beta}_2}{\hat{\alpha}_2}\left(\frac{\alpha_2\beta_1}{\alpha_1\beta_2}\right)\frac{K_c}{L_c}\frac{L_y}{g} + \hat{\beta}_2\left(\frac{g}{L_y}\right)^{\rho_y}\right]^{-1}. \tag{15.38}$$

Let us denote

$$\mathscr{E} = \left(\frac{g}{y}\right)^{\rho_y}\left[\frac{\hat{\alpha}_1\hat{\beta}_2}{\hat{\alpha}_2}\left(\frac{\alpha_2\beta_1}{\alpha_1\beta_2}\right)\frac{K_c}{L_c}\frac{L_y}{g} + \hat{\beta}_2\left(\frac{g}{L_y}\right)^{\rho_y}\right]^{-1}.$$

Following the same procedure we obtain

$$\mathscr{T}_{21}^* = -(1+\rho_y)\frac{r}{\beta_1}\left(\frac{g}{y}\right)^{1+\rho_y}\frac{g_1}{g}(1-\mathscr{E}),$$

$$\mathscr{T}_{12}^* = -(1+\rho_y) r \frac{g_2 y - g}{yg}\mathscr{E}$$

$$\mathscr{T}_{22}^* = -(1+\rho_y)\frac{r}{\beta_1}\left(\frac{g}{y}\right)^{1+\rho_y}\frac{g_2 y - g}{yg}(1-\mathscr{E}),$$

$$\mathscr{T}_{13}^* = -(1+\rho_y) r \frac{g_3}{g}\mathscr{E}$$

$$\mathscr{T}_{23}^* = -(1+\rho_y)\frac{r}{\beta_1}\left(\frac{g}{y}\right)^{1+\rho_y}\frac{g_3}{g}(1-\mathscr{E}).$$

Let us now denote

$$\mathscr{F} = \left(\frac{g}{L_y}\right)^{1+\rho_y}\left[\frac{1}{\hat{\beta}_2}\left(\frac{L_y}{y}\right)^{\rho_y} - \mathscr{E}\right].$$

We also get

$$\mathscr{T}_{31}^* = (1+\rho_y) r \frac{\beta_2}{\beta_1}\frac{g_1}{g}\mathscr{F},\ \mathscr{T}_{32}^* = (1+\rho_y) r \frac{\beta_2}{\beta_1}\frac{g_2 y - g}{yg}\mathscr{F},\ \mathscr{T}_{33}^* = (1+\rho_y) r \frac{\beta_2}{\beta_1}\frac{g_3}{g}\mathscr{F}$$

Then we conclude

$$\frac{\mathcal{T}_{11}^*}{\mathcal{T}_{12}^*} = \frac{g_1 y}{g_2 y - g} = \frac{\mathcal{T}_{31}^*}{\mathcal{T}_{32}^*}, \quad \frac{\mathcal{T}_{22}^*}{\mathcal{T}_{12}^*}$$

$$= \frac{g}{\beta_1 y}\left[\frac{\hat{\alpha}_1\hat{\beta}_2}{\hat{\alpha}_2}\left(\frac{\alpha_2\beta_1}{\alpha_1\beta_2}\right)\frac{K_c}{L_c}\frac{L_y}{g} + \hat{\beta}_2\left(\frac{g}{L_y}\right)^{\rho_y} - \left(\frac{g}{y}\right)^{\rho_y}\right]$$

$$\frac{\mathcal{T}_{21}^*}{\mathcal{T}_{12}^*} = \frac{\mathcal{T}_{22}^*}{\mathcal{T}_{12}^*}\frac{\mathcal{T}_{11}^*}{\mathcal{T}_{12}^*}, \quad \frac{\mathcal{T}_{13}^*}{\mathcal{T}_{12}^*} = \frac{g_3 y}{g_2 y - g} = \frac{\mathcal{T}_{33}^*}{\mathcal{T}_{32}^*}, \quad \frac{\mathcal{T}_{23}^*}{\mathcal{T}_{12}^*} = \frac{\mathcal{T}_{13}^*}{\mathcal{T}_{12}^*}\frac{\mathcal{T}_{22}^*}{\mathcal{T}_{12}^*}.$$

Considering Lemmas A.1 and A.2 completes the proof.

We may finally prove Theorem 1. We first evaluate $\mathcal{T}_{11}^*/\mathcal{T}_{12}^*$ and $\mathcal{T}_{22}^*/\mathcal{T}_{12}^*$. Substituting (15.33) and (15.37) into the formulas given in Lemma A.3 gives the expressions provided in Theorem 1. We need now to compute the characteristic polynomial. From (15.7), consider the following notations:

$$c_1^* = \partial c(k^*, k^*, \ell^*)/\partial k_t, \quad c_2^* = \partial c(k^*, k^*, \ell^*)/\partial k_{t+1}, \quad c_3^* = \partial c(k^*, k^*, \ell^*)/\partial \ell_t.$$

From Lemma A.1, we get

$$c = \left[\hat{\alpha}_1(k-g)^{-\rho_c} + \hat{\alpha}_2\left(1 - \left(\frac{y^{-\rho_y} - \hat{\beta}_1 g^{-\rho_y}}{\hat{\beta}_2}\right)^{-1/\rho_y}\right)^{-\rho_c}\right]^{-1/\rho_c}. \tag{15.39}$$

It follows that

$$c_1 = \hat{\alpha}_1\left(\frac{c}{K_c}\right)^{1+\rho_c}(1-g_1) + \frac{\hat{\alpha}_2\hat{\beta}_1}{\hat{\beta}_2}\left(\frac{c}{L_c}\right)^{1+\rho_c}\left(\frac{L_y}{g}\right)^{1+\rho_y} g_1$$

$$c_2 = -\hat{\alpha}_1\left(\frac{c}{K_c}\right)^{1+\rho_c} g_2 - \frac{\hat{\alpha}_2}{\hat{\beta}_2}\left(\frac{c}{L_c}\right)^{1+\rho_c}\left[\left(\frac{L_y}{y}\right)^{1+\rho_y} - \hat{\beta}_1\left(\frac{L_y}{g}\right)^{1+\rho_y} g_2\right]$$

$$c_3 = -\hat{\alpha}_1\left(\frac{c}{K_c}\right)^{1+\rho_c} g_3 + \hat{\alpha}_2\left(\frac{c}{L_c}\right)^{1+\rho_c}\left[1 + \frac{\hat{\beta}_1}{\hat{\beta}_2}\left(\frac{L_y}{g}\right)^{1+\rho_y} g_3\right].$$

Notice now that the first order conditions (15.26)–(15.29) imply

$$\left(\frac{c}{K_c}\right)^{1+\rho} = \frac{T_1}{\alpha_1}, \quad \left(\frac{c}{L_c}\right)^{1+\rho} = \frac{w}{\alpha_2} = \frac{T_1}{\alpha_1}\left(\frac{K_c}{L_c}\right)^{1+\rho}.$$

Finally, using (15.31) we get after simplifications

$$c_1^* = T_1^*\frac{\hat{\alpha}_1}{\alpha_1}\left[1 - g_1\left(1 - \frac{\hat{\alpha}_2\hat{\beta}_1}{\hat{\alpha}_1\hat{\beta}_2}\frac{\alpha_1\beta_2}{\alpha_2\beta_1}\right)\right] \equiv T_1^*\mathcal{A}$$

$$c_2^* = -\delta T_1^* \frac{\hat{\alpha}_2}{\alpha_2} \frac{\beta_2}{\hat{\beta}_2} \left[1 - \frac{g_2}{\delta} \frac{\hat{\beta}_1}{\beta_1} \left(1 - \frac{\hat{\alpha}_1 \hat{\beta}_2}{\hat{\alpha}_2 \hat{\beta}_1} \frac{\alpha_2 \beta_1}{\alpha_1 \beta_2} \right) \right] \equiv -\delta T_1^* \mathscr{B}$$

$$c_3^* = T_1^* \frac{\hat{\alpha}_1}{\alpha_1} \left[\frac{\hat{\alpha}_2}{\hat{\alpha}_1} \left(\frac{K_c}{L_c} \right)^{1+\rho_c} - g_3 \left(1 - \frac{\hat{\alpha}_2 \hat{\beta}_1}{\hat{\alpha}_1 \hat{\beta}_2} \frac{\alpha_1 \beta_2}{\alpha_2 \beta_1} \right) \right] \equiv T_1^* \mathscr{G}.$$

Considering Lemma A.3, total differentiation of (15.8) gives

$$\begin{aligned} dk_{t+2}\delta \left[1 + \sigma\delta \frac{T_1^{*2}}{c^* \mathscr{T}_{12}^*} \mathscr{B} \right] + dk_{t+1} \left[\frac{\mathscr{T}_{22}^*}{\mathscr{T}_{12}^*} + \delta \frac{\mathscr{T}_{11}^*}{\mathscr{T}_{12}^*} - \sigma\delta \frac{T_1^{*2}}{c^* \mathscr{T}_{12}^*} (\delta\mathscr{B} + \mathscr{A}) \right] \\ + dk_t \left[\frac{\mathscr{T}_{22}^*}{\mathscr{T}_{12}^*} \frac{\mathscr{T}_{11}^*}{\mathscr{T}_{12}^*} + \sigma\delta \frac{T_1^{*2}}{c^* \mathscr{T}_{12}^*} \mathscr{A} \right] \\ + d\ell_{t+1}\delta \left[\frac{\mathscr{T}_{13}^*}{\mathscr{T}_{12}^*} - \sigma \frac{T_1^{*2}}{c^* \mathscr{T}_{12}^*} \mathscr{G} \right] + d\ell_t \left[\frac{\mathscr{T}_{22}^*}{\mathscr{T}_{12}^*} \frac{\mathscr{T}_{13}^*}{\mathscr{T}_{12}^*} + \sigma\delta \frac{T_1^{*2}}{c^* \mathscr{T}_{12}^*} \mathscr{G} \right] = 0 \end{aligned}$$

$$\begin{aligned} dk_{t+1} \left[1 + \sigma\delta \frac{T_1^* T_3^*}{c^* \mathscr{T}_{32}^*} \mathscr{B} \right] + dk_t \left[\frac{\mathscr{T}_{11}^*}{\mathscr{T}_{12}^*} - \sigma \frac{T_1^* T_3^*}{c^* \mathscr{T}_{32}^*} \mathscr{A} \right] \\ + d\ell_t \left[\frac{\mathscr{T}_{13}^*}{\mathscr{T}_{12}^*} - \sigma \frac{T_1^* T_3^*}{c^* \mathscr{T}_{32}^*} \mathscr{G} - \gamma \frac{T_3^*}{\mathscr{T}_{32}^* \ell^*} \right] = 0. \end{aligned}$$

Solving the first equation with respect to dk_{t+2} and substituting the result into the second equation considered one period forward gives

$$\begin{aligned} dk_{t+1} \Bigg\{ \left[1 + \sigma\delta \frac{T_1^* T_3^*}{c^* \mathscr{T}_{32}^*} \mathscr{B} \right] \left[\frac{\mathscr{T}_{22}^*}{\mathscr{T}_{12}^*} - \sigma\delta^2 \frac{T_1^{*2}}{c^* \mathscr{T}_{12}^*} \mathscr{B} \right] \\ + \sigma\delta \frac{T_1^{*2}}{c^* \mathscr{T}_{12}^*} \left[\frac{T_3^*}{T_1^*} \frac{\mathscr{T}_{12}^*}{\mathscr{T}_{32}^*} - 1 \right] \left[\mathscr{A} + \delta \frac{\mathscr{T}_{11}^*}{\mathscr{T}_{12}^*} \mathscr{B} \right] \Bigg\} \\ + dk_t \left[1 + \sigma\delta \frac{T_1^* T_3^*}{c^* \mathscr{T}_{32}^*} \mathscr{B} \right] \left[\frac{\mathscr{T}_{22}^*}{\mathscr{T}_{12}^*} \frac{\mathscr{T}_{11}^*}{\mathscr{T}_{12}^*} + \sigma\delta \frac{T_1^{*2}}{c^* \mathscr{T}_{12}^*} \mathscr{A} \right] \\ + d\ell_{t+1} \Bigg\{ \left[1 + \sigma\delta \frac{T_1^{*2}}{c^* \mathscr{T}_{12}^*} \mathscr{B} \right] \delta\gamma \frac{T_3^*}{\mathscr{T}_{32}^* \ell^*} \\ + \sigma\delta \frac{T_1^{*2}}{c^* \mathscr{T}_{12}^*} \left[\frac{T_3^*}{T_1^*} \frac{\mathscr{T}_{12}^*}{\mathscr{T}_{32}^*} - 1 \right] \left[\mathscr{G} + \delta \frac{\mathscr{T}_{13}^*}{\mathscr{T}_{12}^*} \mathscr{B} \right] \Bigg\} \\ + d\ell_t \left[1 + \sigma\delta \frac{T_1^* T_3^*}{c^* \mathscr{T}_{32}^*} \mathscr{B} \right] \left[\frac{\mathscr{T}_{22}^*}{\mathscr{T}_{12}^*} \frac{\mathscr{T}_{13}^*}{\mathscr{T}_{12}^*} + \sigma\delta \frac{T_1^{*2}}{c^* \mathscr{T}_{12}^*} \mathscr{G} \right] = 0. \end{aligned}$$

Tedious computations finally give the characteristic polynomial

$$\begin{aligned}
\mathscr{P}(x) &= \gamma \frac{T_3^*}{\mathscr{T}_{32}^* \ell^*} \left\{ \left(x + \frac{\mathscr{T}_{11}^*}{\mathscr{T}_{12}^*} \right) \left(\delta x + \frac{\mathscr{T}_{22}^*}{\mathscr{T}_{12}^*} \right) + \sigma \delta \frac{T_1^{*2}}{c^* \mathscr{T}_{12}^*} (\delta \mathscr{B} x - \mathscr{A})(x-1) \right\} \\
&\quad + \sigma \frac{T_1^{*2}}{c^* \mathscr{T}_{12}^*} \left[\delta x \left(\frac{T_3^*}{T_1^*} \frac{\mathscr{T}_{12}^*}{\mathscr{T}_{32}^*} - 1 \right) + \frac{T_3^*}{T_1^*} \frac{\mathscr{T}_{22}^*}{\mathscr{T}_{32}^*} + \delta \right] \\
&\quad \times \left[x \left(\mathscr{G} + \delta \frac{\mathscr{T}_{13}^*}{\mathscr{T}_{12}^*} \mathscr{B} \right) + \mathscr{G} \frac{\mathscr{T}_{11}^*}{\mathscr{T}_{12}^*} - \mathscr{A} \frac{\mathscr{T}_{13}^*}{\mathscr{T}_{12}^*} \right].
\end{aligned}$$

Considering Lemmas A.1–A.3 and the definitions of $\mathscr{A}$, $\mathscr{B}$ and $\mathscr{G}$, we can show that

$$\frac{T_3^*}{T_1^*} \frac{\mathscr{T}_{12}^*}{\mathscr{T}_{32}^*} = -\frac{\hat{\alpha}_2}{\hat{\alpha}_1} \left(\frac{\alpha_1 \beta_2}{\alpha_2 \beta_1} \right)^{\frac{\rho_c}{1+\rho_c}} \left(\frac{(\delta \beta_1)^{\frac{\rho_y}{1+\rho_y}} - \hat{\beta}_1}{\hat{\beta}_2} \right)^{\frac{\rho_c(1+\rho_y)}{\rho_y(1+\rho_c)}} \equiv -\mathscr{C} \tag{15.40}$$

$$\frac{T_3^*}{T_1^*} \frac{\mathscr{T}_{22}^*}{\mathscr{T}_{32}^*} + \delta = -\mathscr{C} \frac{\mathscr{T}_{22}^*}{\mathscr{T}_{12}^*} + \delta = \delta \hat{\beta}_1 (\delta \beta_1)^{\frac{-\rho_y}{1+\rho_y}} (1 + \mathscr{C})$$

$$\begin{aligned}
\mathscr{G} + \delta \frac{\mathscr{T}_{13}^*}{\mathscr{T}_{12}^*} \mathscr{B} &= \left[\mathscr{A} \frac{\mathscr{T}_{13}^*}{\mathscr{T}_{12}^*} - \mathscr{G} \frac{\mathscr{T}_{11}^*}{\mathscr{T}_{12}^*} \right] (\delta \beta_1)^{\frac{1}{1+\rho_y}} \\
&= -\frac{\mathscr{T}_{11}^*}{\mathscr{T}_{12}^*} \frac{\hat{\alpha}_1}{\alpha_1} \left(\frac{\alpha_1 \beta_2}{\alpha_2 \beta_1} \right)^{\frac{1}{1+\rho_c}} \left(\frac{(\delta \beta_1)^{\frac{\rho_y}{1+\rho_y}} - \hat{\beta}_1}{\hat{\beta}_2} \right)^{\frac{1+\rho_y}{\rho_y(1+\rho_c)}} (1 + \mathscr{C})(\delta \beta_1)^{\frac{1}{1+\rho_y}} \\
&\equiv \mathscr{D}(1 + \mathscr{C}) (\delta \beta_1)^{\frac{1}{1+\rho_y}}.
\end{aligned}$$

A final substitution of all these expressions into $\mathscr{P}(x)$ gives the formulation of the characteristic polynomial provided in Theorem 1.

Consider now Lemma A.1. Using (15.33)–(15.35) gives the following expressions:

$$g_1 = \frac{(1+\rho_c)(\delta \beta_1)^{\frac{1}{1+\rho_y}} \left[(\delta \beta_1)^{\frac{\rho_y}{1+\rho_y}} - \hat{\beta}_1 \right]}{(1+\rho_y)(\delta \beta_1)^{\frac{\rho_y}{1+\rho_y}} \left[1 - (\delta \beta_1)^{\frac{1}{1+\rho_y}} \right] + (1+\rho_c) \left[\delta \beta_1 + \hat{\beta}_1 \frac{\mathscr{T}_{12}^*}{\mathscr{T}_{11}^*} \right]}$$

$$g_2 = \frac{\delta \beta_1 \left((1+\rho_y) \left[1 - (\delta \beta_1)^{\frac{1}{1+\rho_y}} \right] + (1+\rho_c) \left[\delta \beta_1 + \hat{\beta}_1 \frac{\mathscr{T}_{12}^*}{\mathscr{T}_{11}^*} \right] \right)}{(1+\rho_y)(\delta \beta_1)^{\frac{\rho_y}{1+\rho_y}} \left[1 - (\delta \beta_1)^{\frac{1}{1+\rho_y}} \right] + (1+\rho_c) \left[\delta \beta_1 + \hat{\beta}_1 \frac{\mathscr{T}_{12}^*}{\mathscr{T}_{11}^*} \right]}.$$

Substituting g_1 and g_2 into $\mathscr{A}$ and $\mathscr{B}$, and considering the values of $\mathscr{T}_{11}^*/\mathscr{T}_{12}^*$ and $\mathscr{T}_{22}^*/\mathscr{T}_{12}^*$ then gives the expressions provided in Theorem 1. We finally have to compute $T_1^{*2}/c^* \mathscr{T}_{12}^*$ and $T_3^*/\mathscr{T}_{32}^* \ell^*$. From (15.39) and (15.35) we get after simplifications

$$c^* = \left[1 - (\delta\beta_1)^{\frac{1}{1+\rho_y}}\right] \ell\kappa^* \left(\frac{T_1^*}{\alpha_1}\right)^{\frac{1}{1+\rho_c}}. \tag{15.41}$$

Using Lemma A.1 it follows therefore that

$$\frac{T_1^* k^*}{c^*} = \left[1 - (\delta\beta_1)^{\frac{1}{1+\rho_y}}\right]^{-1} (1 + \mathscr{C})^{-1}.$$

Moreover, from (15.38) we obtain after simplifications

$$\frac{\mathscr{T}_{11}^* k^*}{T_1^*} = -\frac{(1+\rho_c)(1+\rho_y)(\delta\beta_1)^{\frac{\rho_y}{1+\rho_y}}\mathscr{C}}{(1+\mathscr{C})\left[(1+\rho_y)(\delta\beta_1)^{\frac{\rho_y}{1+\rho_y}}\left[1-(\delta\beta_1)^{\frac{1}{1+\rho_y}}\right]+(1+\rho_c)\left[\delta\beta_1+\hat{\beta}_1\frac{\mathscr{T}_{12}^*}{\mathscr{T}_{11}^*}\right]\right]}.$$

Then we get from the two previous expressions

$$\frac{T_1^{*2}}{c^* \mathscr{T}_{11}^*} = -\frac{\alpha_1}{\hat{\alpha}_1} \frac{(1+\rho_y)(\delta\beta_1)^{\frac{\rho_y}{1+\rho_y}}\left[1-(\delta\beta_1)^{\frac{1}{1+\rho_y}}\right] + (1+\rho_c)\left[\delta\beta_1 + \hat{\beta}_1 \frac{\mathscr{T}_{12}^*}{\mathscr{T}_{11}^*}\right]}{(1+\rho_c)(1+\rho_y)(\delta\beta_1)^{\frac{\rho_y}{1+\rho_y}}\left[1-(\delta\beta_1)^{\frac{1}{1+\rho_y}}\right]\mathscr{C}}.$$

The result is derived from the fact that

$$\frac{T_1^{*2}}{c^* \mathscr{T}_{12}^*} = \frac{T_1^{*2}}{c^* \mathscr{T}_{11}^*} \frac{\mathscr{T}_{11}^*}{\mathscr{T}_{12}^*}. \tag{15.42}$$

Finally, to compute $T_3^*/\mathscr{T}_{32}^* \ell^*$, consider (15.32) and (15.36) with (15.40). We get

$$\frac{T_3^*}{\mathscr{T}_{32}^* \ell^*} = -\frac{T_1^*}{\mathscr{T}_{12}^* \ell^*}\mathscr{C} = -\frac{T_1^{*2}}{c^* \mathscr{T}_{12}^*} \frac{c^*}{T_1^* \ell^*}\mathscr{C} = -\frac{T_1^{*2}}{c^* \mathscr{T}_{12}^*}\left[1-(\delta\beta_1)^{\frac{1}{1+\rho_y}}\right]\frac{\kappa^* \hat{\alpha}_1 \mathscr{C}(1+\mathscr{C})}{\alpha_1}. \tag{15.43}$$

Notice that (15.42) and (15.43) do not depend on the preference parameters σ and γ. ■

15.6.3 Proof of Theorem 3

Our strategy consists in considering a fixed value for ρ_c and varying ρ_y in order to find intervals of values in which local indeterminacy occurs. To simplify the analysis, we have to impose some restrictions on the parameters ρ_c and ρ_y such that the following function:

$$g(\rho_y) = \left(\frac{(\delta\beta_1)^{\frac{\rho_y}{1+\rho_y}} - \hat{\beta}_1}{\hat{\beta}_2} \right)^{\frac{\rho_y - \rho_c}{\rho_y(1+\rho_c)}} \tag{15.44}$$

is monotone decreasing.

Lemma A.4. *Under Assumption 1, there is a* $\tilde{\rho}_c \in (\hat{\rho}_y, +\infty)$ *such that for any given* $\rho_c \in (-1, \tilde{\rho}_c)$ *there exists* $\tilde{\rho}_y \in (0, \hat{\rho}_y]$ *such that* $g(\rho_y)$ *is a monotone decreasing function for all* $\rho_y \in (-1, \tilde{\rho}_y)$.

Proof. Available upon request.

We may now start the proof of Theorem 3. Consider first the root x_1. To simplify the exposition we will study its inverse:

$$x_1^{-1} = (\delta\beta_1)^{\frac{1}{1+\rho_y}} \left[1 - \left(\frac{\alpha_1\beta_2}{\alpha_2\beta_1} \right)^{\frac{1}{1+\rho_c}} \left(\frac{(\delta\beta_1)^{\frac{\rho_y}{1+\rho_y}} - \hat{\beta}_1}{\hat{\beta}_2} \right)^{\frac{\rho_y - \rho_c}{\rho_y(1+\rho_c)}} \right].$$

We have shown in Nishimura and Venditti (2004b) that if $x_1^{-1} > 0$ then necessarily $x_1^{-1} \in (0, 1)$ so that local indeterminacy cannot occur. We have therefore to find conditions to get $x_1^{-1} < -1$. We know that $\rho_y \in (-1, \hat{\rho}_y)$ with $\hat{\rho}_y > 0$. Notice that L'Hôpital rule gives:

$$\lim_{\rho_y \to 0} g(\rho_y) = (\delta\beta_1)^{\frac{-\rho_c}{(1+\rho_c)\hat{\beta}_2}}. \tag{15.45}$$

It follows that

$$\lim_{\rho_y \to 0} x_1^{-1} = \delta\beta_1 \left[1 - \left(\frac{\alpha_1\beta_2}{\alpha_2\beta_1} \right)^{\frac{1}{1+\rho_c}} (\delta\beta_1)^{\frac{-\rho_c}{(1+\rho_c)\hat{\beta}_2}} \right].$$

Under Assumption 3, we then get

$$\lim_{\rho_y \to 0} x_1^{-1}|_{\rho_c \to -1} = \delta\beta_1$$

$$\lim_{\rho_y \to 0} x_1^{-1}|_{\rho_c = 0} < -1.$$

Therefore, we derive from Lemma A.4 that there exist $\rho_c^1 \in (-1, 0)$ and $\rho_c^2 \in (0, \tilde{\rho}_c]$ such that $\lim_{\rho_y \to 0} x_1^{-1} < -1$ for any $\rho_c \in (\rho_c^1, \rho_c^2)$. Consider now

$$\lim_{\rho_y \to -1} x_1^{-1} = -(\delta\beta_1)^{\frac{1}{1+\rho_c}} \left(\frac{\alpha_1\beta_2}{\alpha_2\beta_1} \right)^{\frac{1}{1+\rho_c}} \frac{1}{\hat{\beta}_2}$$

It follows that $\lim_{\rho_y \to -1} x_1^{-1} < -1$ if and only if $\hat{\beta}_2^{1+\rho_c} < (\delta\alpha_1\beta_2/\alpha_2)$. Notice then that Assumption 3 can be equivalently written as

$$\frac{\alpha_1\beta_2}{\alpha_2\beta_1} > 1 + \frac{1}{\delta\beta_1} > \frac{1}{\delta\beta_1} \text{ and thus } \delta\frac{\alpha_1\beta_2}{\alpha_2} > 1.$$

It follows that under Assumption 3 the inequality $\hat{\beta}_2^{1+\rho_c} < (\delta\alpha_1\beta_2/\alpha_2)$ holds for any $\rho_c > -1$. Finally we have

$$\lim_{\rho_y \to \hat{\rho}_y} x_1^{-1} = \begin{cases} (\delta\beta_1)^{\frac{1}{1+\hat{\rho}_y}}, & \text{if } \rho_c < \hat{\rho}_y \\ (\delta\beta_1)^{\frac{1}{1+\hat{\rho}_y}} \left[1 - \left(\frac{\alpha_1\beta_2}{\alpha_2\beta_1}\right)^{\frac{1}{1+\hat{\rho}_y}}\right], & \text{if } \rho_c = \hat{\rho}_y \\ -\infty, & \text{if } \rho_c > \hat{\rho}_y. \end{cases}$$

We derive from all this and Lemma A.4 the following results: under Assumption 3,

(a) For a given $\rho_c \in (-1, \rho_c^1)$, there exists $\rho_y^1 \in (-1, 0)$ such that $x_1^{-1} < -1$ for any $\rho_y \in (-1, \rho_y^1)$.
(b) For a given $\rho_c \in (\rho_c^1, \hat{\rho}_y)$, there exists $\rho_y^2 \in (0, \hat{\rho}_y)$ such that $x_1^{-1} < -1$ for any $\rho_y \in (-1, \rho_y^2)$.

Consider now the root x_2:

$$x_2 = \hat{\beta}_1 (\delta\beta_1)^{\frac{-\rho_y}{1+\rho_y}} \left[1 - \frac{\hat{\alpha}_1\hat{\beta}_2}{\hat{\alpha}_2\hat{\beta}_1}\left(\frac{\alpha_2\beta_1}{\alpha_1\beta_2}\right)^{\frac{\rho_c}{1+\rho_c}} \left(\frac{(\delta\beta_1)^{\frac{\rho_y}{1+\rho_y}} - \hat{\beta}_1}{\hat{\beta}_2}\right)^{\frac{\rho_y-\rho_c}{\rho_y(1+\rho_c)}}\right].$$

We have shown in Nishimura and Venditti (2004b) (see p. 146) that if $x_2 > 0$ then necessarily $x_2 \in (0, 1)$. Since we are looking for the occurrence of local indeterminacy, we have therefore to find conditions to get $x_2 > -1$. Using again (15.45), we get

$$\lim_{\rho_y \to 0} x_2 = \hat{\beta}_1 \left[1 - \frac{\hat{\alpha}_1\hat{\beta}_2}{\hat{\alpha}_2\hat{\beta}_1}\left(\frac{\alpha_2\beta_1}{\alpha_1\beta_2}\right)^{\frac{\rho_c}{1+\rho_c}} (\delta\beta_1)^{\frac{-\rho_y}{(1+\rho_y)\hat{\beta}_2}}\right].$$

Under Assumption 3, we derive

$$\lim_{\rho_y \to 0} x_2|_{\rho_c \to -1} = \begin{cases} -\infty, \text{ if and only if } \frac{\alpha_1\beta_2}{\alpha_2\beta_1}(\delta\beta_1)^{1/\hat{\beta}_2} > 1 \\ \hat{\beta}_1, \text{ if and only if } \frac{\alpha_1\beta_2}{\alpha_2\beta_1}(\delta\beta_1)^{1/\hat{\beta}_2} < 1 \end{cases} \tag{15.46}$$

$$\lim_{\rho_y \to 0} x_2|_{\rho_c = 0} = \hat{\beta}_1 \left[1 - \frac{\hat{\alpha}_1\hat{\beta}_2}{\hat{\alpha}_2\hat{\beta}_1}\right] > -1, \text{ if and only if } \hat{\alpha}_2 > \hat{\alpha}_1 - \hat{\beta}_1.$$

Therefore, assuming that $\hat{\alpha}_2 > \hat{\alpha}_1 - \hat{\beta}_1$, there exist $\rho_c^3 \in [-1, 0)$ and $\rho_c^4 \in (0, \tilde{\rho}_c]$ such that for any $\rho_c \in (\rho_c^3, \rho_c^4)$, $\lim_{\rho_y \to 0} x_2 > -1$. Consider now

$$\lim_{\rho_y \to -1} x_2 = -(\delta\beta_1)^{\frac{-\rho_c}{1+\rho_c}} \frac{\hat{\alpha}_1}{\hat{\alpha}_2} \left(\frac{\alpha_1\beta_2}{\alpha_2\beta_1}\right)^{\frac{\rho_c}{1+\rho_c}} > -1 \Leftrightarrow \left(\frac{\alpha_1\beta_2}{\alpha_2}\delta\right)^{\frac{\rho_c}{1+\rho_c}} > \frac{\hat{\alpha}_1}{\hat{\alpha}_2}.$$

If we assume that $\hat{\alpha}_2 > \hat{\alpha}_1$, Assumption 3 implies that there exists $\rho_c^5 \in (-1, 0)$ such that the above inequality will be satisfied for any $\rho_c \in (\rho_c^5, \tilde{\rho}_c)$. Finally under Assumption 1 we have

$$\begin{aligned} \lim_{\rho_y \to \hat{\rho}_y} x_2 &\in (0, 1), && \text{if } \rho_c < \hat{\rho}_y \\ &= \left[1 - \frac{\hat{\alpha}_1\hat{\beta}_2}{\hat{\alpha}_2\hat{\beta}_1}\left(\frac{\alpha_2\beta_1}{\alpha_1\beta_2}\right)^{\frac{\hat{\rho}_y}{1+\hat{\rho}_y}}\right], && \text{if } \rho_c = \hat{\rho}_y \\ &= -\infty, && \text{if } \rho_c \in \left(\hat{\rho}_y, \tilde{\rho}_c\right). \end{aligned}$$

Therefore there exists $\rho_c^6 \in (0, \tilde{\rho}_c)$ such that $\lim_{\rho_y \to \hat{\rho}_y} x_2 > -1$ for any $\rho_c < \rho_c^6$.

We derive from all this and Lemma A.4 the following results: under Assumption 3 and $\hat{\alpha}_2 > \hat{\alpha}_1$,

(c) For any given $\rho_c \in (\max\{\rho_c^3, \rho_c^5\}, \rho_c^4)$, there exists $\rho_y^4 \in (0, \hat{\rho}_y)$ such that $x_2 > -1$ for all $\rho_y \in (-1, \rho_y^4)$.

(d) For any given $\rho_c \in (\max\{\rho_c^3, \rho_c^5\}, \min\{\rho_c^4, \rho_c^6\})$, $x_2 > -1$ for all $\rho_y \in (-1, \hat{\rho}_y)$.

The final result is derived from (a)-(c) considering $\underline{\rho}_c = \max\{\rho_c^1, \rho_c^3, \rho_c^5\}$, $\bar{\rho}_c = \min\{\rho_c^2, \rho_c^4\}$ and $\bar{\rho}_y = \min\{\rho_y^2, \rho_y^3, \rho_y^4\}$. ■

15.6.4 Proof of Lemma 1

It is easy to derive from Theorem 1 that

$$1 + \frac{\mathcal{T}_{22}^*}{\delta\mathcal{T}_{12}^*} = (\delta\beta_1)^{\frac{-\rho_y}{1+\rho_y}} \left[(\delta\beta_1)^{\frac{\rho_y}{1+\rho_y}} - \hat{\beta}_1\right] \frac{1+\mathcal{C}}{\mathcal{C}}$$

$$\delta\beta_1 + \hat{\beta}_1 \frac{\mathcal{T}_{12}^*}{\mathcal{T}_{11}^*} = (\delta\beta_1)^{\frac{1}{1+\rho_y}} \left(\frac{\alpha_1\beta_2}{\alpha_2\beta_1}\right)^{\frac{1}{1+\rho_c}} \left(\frac{(\delta\beta_1)^{\frac{\rho_y}{1+\rho_y}} - \hat{\beta}_1}{\hat{\beta}_2}\right)^{\frac{\rho_y - \rho_c}{\rho_y(1+\rho_c)}}$$

$$\times \left[\hat{\beta}_1 + \frac{\hat{\alpha}_1\hat{\beta}_2}{\hat{\alpha}_2} \frac{\alpha_2\beta_1}{\alpha_1\beta_2}\mathcal{C}\right].$$

It follows that $\mathscr{A} > 0$ and $\mathscr{B} > 0$. Moreover, direct computations give

$$\frac{\mathscr{A}}{\delta\mathscr{B}} = \frac{\hat{\beta}_1}{\delta\beta_1}\frac{(1+\rho_y)(\delta\beta_1)^{\frac{\rho_y}{1+\rho_y}}\left[1-(\delta\beta_1)^{\frac{1}{1+\rho_y}}\right]+\delta\beta_1(1+\rho_c)\frac{\hat{\alpha}_2\hat{\beta}_1}{\hat{\alpha}_1\hat{\beta}_2}\frac{\alpha_1\beta_2}{\alpha_2\beta_1}\left(1+\frac{\mathscr{T}_{22}^*}{\delta\mathscr{T}_{12}^*}\right)}{(1+\rho_y)(\delta\beta_1)^{\frac{\rho_y}{1+\rho_y}}\left[1-(\delta\beta_1)^{\frac{1}{1+\rho_y}}\right]\mathscr{X}+\delta\beta_1(1+\rho_c)\frac{\hat{\alpha}_2\hat{\beta}_1}{\hat{\alpha}_1\hat{\beta}_2}\frac{\alpha_1\beta_2}{\alpha_2\beta_1}\left(1+\frac{\mathscr{T}_{22}^*}{\delta\mathscr{T}_{12}^*}\right)} \tag{15.47}$$

with

$$\mathscr{X} = \frac{\hat{\alpha}_2\hat{\beta}_1}{\hat{\alpha}_1\hat{\beta}_2}\frac{\alpha_1\beta_2}{\alpha_2\beta_1}\left[1-\hat{\beta}_1(\delta\beta_1)^{\frac{-\rho_y}{1+\rho_y}}\right]+\hat{\beta}_1(\delta\beta_1)^{\frac{-\rho_y}{1+\rho_y}}.$$

Since $\hat{\beta}_1/\delta\beta_1 > 1$, a sufficient condition for $\mathscr{A}/\delta\mathscr{B}$ to be greater than 1 is $\mathscr{X} < 1$. On the contrary, $\mathscr{X} > 1$ is a necessary condition for $\mathscr{A}/\delta\mathscr{B}$ to be less than 1. Notice that $\mathscr{X} > 1$ if and only if

$$\frac{\hat{\alpha}_1\hat{\beta}_2}{\hat{\alpha}_2\hat{\beta}_1} < \frac{\alpha_1\beta_2}{\alpha_2\beta_1}. \tag{15.48}$$

Considering again Theorem 1, let us now denote

$$\frac{T_1^{*2}}{c^*\mathscr{T}_{12}^*} \equiv -\frac{\mathscr{T}_{11}^*}{\mathscr{T}_{12}^*}\mathscr{H}.$$

We derive from the above results that $\mathscr{H} > 0$. Straightforward computations from (15.15) also give

$$\mathscr{D}et'(\sigma) = -\frac{\mathscr{T}_{11}^*}{\mathscr{T}_{12}^*}\frac{\mathscr{H}\delta\mathscr{B}\left(\frac{\mathscr{A}}{\delta\mathscr{B}}-\frac{\mathscr{T}_{11}^*}{\mathscr{T}_{12}^*}\frac{\mathscr{T}_{22}^*}{\delta\mathscr{T}_{12}^*}\right)}{\left[1+\sigma\delta\frac{T_1^{*2}}{c^*\mathscr{T}_{12}^*}\mathscr{B}\right]^2}, \tag{15.49}$$

$$\mathscr{T}r'(\sigma) = -\frac{\mathscr{T}_{11}^*}{\mathscr{T}_{12}^*}\frac{\mathscr{H}(\mathscr{A}+\delta\mathscr{B})\left(1+\frac{\mathscr{T}_{11}^*}{\mathscr{T}_{12}^*}+\frac{\mathscr{T}_{22}^*}{\delta\mathscr{T}_{12}^*}\right)}{\left[1+\sigma\delta\frac{T_1^{*2}}{c^*\mathscr{T}_{12}^*}\mathscr{B}\right]^2}.$$

Recall that when $\sigma = 0$, the characteristic roots are

$$x_1 = -\frac{\mathscr{T}_{11}^*}{\mathscr{T}_{12}^*}, \quad x_2 = -\frac{\mathscr{T}_{22}^*}{\delta\mathscr{T}_{12}^*}.$$

Under Assumption 4, we derive from the proof of Theorem 3 that $x_1 \in (-1, 0)$ and $x_2 \in (-1, 1)$. It follows that $\mathscr{T}_{11}^*/\mathscr{T}_{12}^* > 0$ while $\mathscr{T}_{22}^*/\delta\mathscr{T}_{12}^*$ may be positive or negative. If $\mathscr{T}_{22}^*/\delta\mathscr{T}_{12}^* > 0$ then we get $1 + \mathscr{T}_{11}^*/\mathscr{T}_{12}^* + \mathscr{T}_{22}^*/\delta\mathscr{T}_{12}^* > 0$. If

$\mathscr{T}_{22}^*/\delta\mathscr{T}_{12}^* < 0$, then we necessarily have $\mathscr{T}_{22}^*/\delta\mathscr{T}_{12}^* > -1$ and we find again $1+\mathscr{T}_{11}^*/\mathscr{T}_{12}^* + \mathscr{T}_{22}^*/\delta\mathscr{T}_{12}^* > 0$. Therefore, we get $\mathscr{T}r'(\sigma) < 0$.

The sign of $\mathscr{D}et'(\sigma)$ is given by the sign of the following expression:

$$-\left(\frac{\mathscr{A}}{\delta\mathscr{B}} - \frac{\mathscr{T}_{11}^*}{\mathscr{T}_{12}^*}\frac{\mathscr{T}_{22}^*}{\delta\mathscr{T}_{12}^*}\right). \tag{15.50}$$

Under Assumption 4, if $x_2 \in (0,1)$, then $\mathscr{T}_{22}^*/\delta\mathscr{T}_{12}^* < 0$ and (15.50) is negative. If on the contrary $x_2 \in (-1,0)$, i.e. $\mathscr{T}_{22}^*/\delta\mathscr{T}_{12}^* > 0$, assume first that $\mathscr{A}/\delta\mathscr{B} > 1$. Since $\mathscr{T}_{11}^*\mathscr{T}_{22}^*/\delta\mathscr{T}_{12}^{*2} \in (0,1)$, we find again that (15.50) is negative. Assume finally that $\mathscr{A}/\delta\mathscr{B} < 1$, i.e. (15.48) holds and $\mathscr{X} > 1$. Consider (15.47) and the following notations:

$$\frac{\mathscr{A}}{\delta\mathscr{B}} \equiv \frac{\hat{\beta}_1}{\delta\beta_1}\frac{\mathscr{M}+\mathscr{N}}{\mathscr{M}\mathscr{X}+\mathscr{N}}$$

with $\mathscr{M} > 0$ and $\mathscr{N} > 0$. It follows that

$$\frac{\mathscr{A}}{\delta\mathscr{B}} - \frac{\mathscr{T}_{11}^*}{\mathscr{T}_{12}^*}\frac{\mathscr{T}_{22}^*}{\delta\mathscr{T}_{12}^*} = \frac{\mathscr{M}\left[\hat{\beta}_1 - \delta\beta_1\mathscr{X}\frac{\mathscr{T}_{11}^*}{\mathscr{T}_{12}^*}\frac{\mathscr{T}_{22}^*}{\delta\mathscr{T}_{12}^*}\right] + \mathscr{N}\left[\hat{\beta}_1 - \delta\beta_1\frac{\mathscr{T}_{11}^*}{\mathscr{T}_{12}^*}\frac{\mathscr{T}_{22}^*}{\delta\mathscr{T}_{12}^*}\right]}{\delta\beta_1(\mathscr{M}\mathscr{X}+\mathscr{N})}.$$

Straightforward computations then give

$$\hat{\beta}_1 - \delta\beta_1\mathscr{X}\frac{\mathscr{T}_{11}^*}{\mathscr{T}_{12}^*}\frac{\mathscr{T}_{22}^*}{\delta\mathscr{T}_{12}^*} = \hat{\beta}_1\frac{(1-\mathscr{X}) - \left(\frac{\alpha_1\beta_2}{\alpha_2\beta_1}\right)^{\frac{1}{1+\rho_c}}\left(\frac{(\delta\beta_1)^{\frac{\rho_y}{1+\rho_y}} - \hat{\beta}_1}{\hat{\beta}_2}\right)^{\frac{\rho_y-\rho_c}{\rho_y(1+\rho_c)}}\left[1-\mathscr{X}\frac{\frac{\hat{\alpha}_1\hat{\beta}_2}{\hat{\alpha}_2\hat{\beta}_1}}{\frac{\alpha_1\beta_2}{\alpha_2\beta_1}}\right]}{1-\left(\frac{\alpha_1\beta_2}{\alpha_2\beta_1}\right)^{\frac{1}{1+\rho_c}}\left(\frac{(\delta\beta_1)^{\frac{\rho_y}{1+\rho_y}} - \hat{\beta}_1}{\hat{\beta}_2}\right)^{\frac{\rho_y-\rho_c}{\rho_y(1+\rho_c)}}}$$

with

$$1-\mathscr{X}\frac{\frac{\hat{\alpha}_1\hat{\beta}_2}{\hat{\alpha}_2\hat{\beta}_1}}{\frac{\alpha_1\beta_2}{\alpha_2\beta_1}} = \hat{\beta}_1(\delta\beta_1)^{\frac{-\rho_y}{1+\rho_y}}\left[1-\frac{\frac{\hat{\alpha}_1\hat{\beta}_2}{\hat{\alpha}_2\hat{\beta}_1}}{\frac{\alpha_1\beta_2}{\alpha_2\beta_1}}\right] < 0$$

and $1-\mathscr{X} < 0$ while Assumption 4 implies that the denominator of the above expression is negative. It follows that

$$\hat{\beta}_1 - \delta\beta_1\mathscr{X}\frac{\mathscr{T}_{11}^*}{\mathscr{T}_{12}^*}\frac{\mathscr{T}_{22}^*}{\delta\mathscr{T}_{12}^*} > 0 \text{ and } \hat{\beta}_1 - \delta\beta_1\frac{\mathscr{T}_{11}^*}{\mathscr{T}_{12}^*}\frac{\mathscr{T}_{22}^*}{\delta\mathscr{T}_{12}^*} > 0.$$

Therefore, (15.50) is again negative and we conclude that $\mathscr{D}et'(\sigma) < 0$.

Consider now the expression of $\mathscr{D}et(\sigma)$ as given by (15.22). Under Assumption 4, $T_1^{*2}/c^*\mathscr{T}_{12}^* < 0$ and there exists a critical value σ^* given by (15.20) such that the denominator of $\mathscr{D}et(\sigma)$ is equal to zero when $\sigma = \sigma^*$. Moreover, when $\sigma = \sigma^*$, the numerator of $\mathscr{D}et(\sigma)$ is equal to expression (15.50) which is negative. It follows that $\lim_{\sigma \to \sigma^*_-} \mathscr{D}et(\sigma) = -\infty$ and $\lim_{\sigma \to \sigma^*_+} \mathscr{D}et(\sigma) = +\infty$.

Consider finally the slope $\mathscr{S}_\infty$ given by (15.17). All the previous arguments show that under Assumption 4, $\mathscr{S}_\infty > 0$. Then $\mathscr{S}_\infty < 1$ is equivalent to

$$\left(1 + \frac{\mathscr{T}_{11}^*}{\mathscr{T}_{12}^*}\right)\left(1 + \frac{\mathscr{T}_{22}^*}{\delta \mathscr{T}_{12}^*}\right) > 0.$$

This inequality is then derived from the same argument that has been used to show $\mathscr{T}r'(\sigma) < 0$. ■

15.6.5 Proof of Lemma 2

From (15.22) we derive:

$$\mathscr{D}et_2(\sigma, \gamma) = -\frac{\frac{T_3^*}{\mathscr{T}_{32}^* \ell^*} \sigma \frac{T_1^{*2}}{c^* \mathscr{T}_{12}^*} (1 + \mathscr{C})^2 \mathscr{D} (\delta\beta_1)^{\frac{1}{1+\rho_Y}} \left[\frac{\mathscr{T}_{11}^*}{\mathscr{T}_{12}^*} \frac{\mathscr{T}_{22}^*}{\delta \mathscr{T}_{12}^*} - \frac{\hat{\beta}_1}{\delta\beta_1} + \sigma \frac{T_1^{*2}}{c^* \mathscr{T}_{12}^*} \left(\mathscr{A} - \frac{\hat{\beta}_1}{\beta_1} \mathscr{B}\right)\right]}{\left[\gamma \frac{T_3^*}{\mathscr{T}_{32}^* \ell^*} \left(1 + \sigma \frac{T_1^{*2}}{c^* \mathscr{T}_{12}^*} \delta \mathscr{B}\right) - \sigma \frac{T_1^{*2}}{c^* \mathscr{T}_{12}^*} (1 + \mathscr{C})^2 \mathscr{D} (\delta\beta_1)^{\frac{1}{1+\rho_Y}}\right]^2}$$

$$\begin{aligned}
\mathscr{T}r_2(\sigma, \gamma) = {} & \frac{T_3^*}{\mathscr{T}_{32}^* \ell^*} \sigma \frac{T_1^{*2}}{c^* \mathscr{T}_{12}^*} (1 + \mathscr{C})^2 \mathscr{D} \\
& \times \left[\gamma \frac{T_3^*}{\mathscr{T}_{32}^* \ell^*} \left(1 + \sigma \frac{T_1^{*2}}{c^* \mathscr{T}_{12}^*} \delta \mathscr{B}\right) - \sigma \frac{T_1^{*2}}{c^* \mathscr{T}_{12}^*} (1 + \mathscr{C})^2 \mathscr{D} (\delta\beta_1)^{\frac{1}{1+\rho_Y}}\right]^{-2} \\
& \times \left[1 + (\delta\beta_1)^{\frac{1}{1+\rho_Y}} \frac{\mathscr{T}_{11}^*}{\mathscr{T}_{12}^*} + \hat{\beta}_1 (\delta\beta_1)^{\frac{1-\rho_Y}{1+\rho_Y}} + (\delta\beta_1)^{\frac{1}{1+\rho_Y}} \frac{\mathscr{T}_{22}^*}{\delta \mathscr{T}_{12}^*}\right. \\
& \left. -\sigma \frac{T_1^{*2}}{c^* \mathscr{T}_{12}^*} (\delta\beta_1)^{\frac{1}{1+\rho_Y}} \left(\mathscr{A} - \delta \mathscr{B} (\delta\beta_1)^{\frac{-1}{1+\rho_Y}} + \delta \mathscr{B} (\delta\beta_1)^{\frac{-\rho_Y}{1+\rho_Y}} \left[(\delta\beta_1)^{\frac{\rho_Y}{1+\rho_Y}} - \hat{\beta}_1\right]\right)\right].
\end{aligned}$$

Under Assumptions 4 and 5, tedious but straightforward computations give

$$\frac{\mathscr{T}_{11}^*}{\mathscr{T}_{12}^*} \frac{\mathscr{T}_{22}^*}{\delta \mathscr{T}_{12}^*} - \frac{\hat{\beta}_1}{\delta\beta_1} < 0, \quad \frac{T_1^{*2}}{c^* \mathscr{T}_{12}^*} \mathscr{D} > 0$$

$$1 + (\delta\beta_1)^{\frac{1}{1+\rho_Y}} \frac{\mathscr{T}_{11}^*}{\mathscr{T}_{12}^*} + \hat{\beta}_1 (\delta\beta_1)^{\frac{1-\rho_Y}{1+\rho_Y}} + (\delta\beta_1)^{\frac{1}{1+\rho_Y}} \frac{\mathscr{T}_{22}^*}{\delta \mathscr{T}_{12}^*} > 0, \quad \frac{T_3^*}{\mathscr{T}_{32}^* \ell^*} > 0$$

$$\mathscr{A} - \delta \mathscr{B} (\delta\beta_1)^{\frac{-1}{1+\rho_Y}} + \delta \mathscr{B} (\delta\beta_1)^{\frac{-\rho_Y}{1+\rho_Y}} \left[(\delta\beta_1)^{\frac{\rho_Y}{1+\rho_Y}} - \hat{\beta}_1\right] < 0, \quad \mathscr{A} - \frac{\hat{\beta}_1}{\beta_1} \mathscr{B} < 0.$$

It follows that there exist $\sigma_1 > 0$ and $\sigma_2 > 0$, as defined by

$$\sigma_1 = -\frac{\frac{\mathscr{T}_{11}^*}{\mathscr{T}_{12}^*}\frac{\mathscr{T}_{22}^*}{\delta\mathscr{T}_{12}^*} - \frac{\hat{\beta}_1}{\delta\beta_1}}{\frac{T_1^{*2}}{c^*\mathscr{T}_{12}^*}\left(\mathscr{A} - \frac{\hat{\beta}_1}{\beta_1}\mathscr{B}\right)},$$

$$\sigma_2 = \frac{1 + (\delta\beta_1)^{\frac{1}{1+\rho_y}}\frac{\mathscr{T}_{11}^*}{\mathscr{T}_{12}^*} + \hat{\beta}_1(\delta\beta_1)^{\frac{1-\rho_y}{1+\rho_y}} + (\delta\beta_1)^{\frac{1}{1+\rho_y}}\frac{\mathscr{T}_{22}^*}{\delta\mathscr{T}_{12}^*}}{\frac{T_1^{*2}}{c^*\mathscr{T}_{12}^*}(\delta\beta_1)^{\frac{1}{1+\rho_y}}\left(\mathscr{A} - \delta\mathscr{B}(\delta\beta_1)^{\frac{-1}{1+\rho_y}} + \delta\mathscr{B}(\delta\beta_1)^{\frac{-\rho_y}{1+\rho_y}}\left[(\delta\beta_1)^{\frac{\rho_y}{1+\rho_y}} - \hat{\beta}_1\right]\right)}$$

such that $\mathscr{D}et_2(\sigma,\gamma) > 0$ if and only if $\sigma \in [0,\sigma_1)$ and $\mathscr{T}r_2(\sigma,\gamma) > 0$ if and only if $\sigma \in [0,\sigma_2)$. Consider the critical value σ^* as defined by (15.20). It is then easy to show that under Assumptions 4 and 5, $\sigma^* < \sigma_1$ and $\sigma^* < \sigma_2$. However, there is no clear condition to compare σ_1 and σ_2.

Consider finally the slope $\mathscr{S}_\gamma(\sigma)$ as defined by (15.24). We can show that

$$\mathscr{S}_\gamma(\sigma) = \frac{\mathscr{A} - \frac{\hat{\beta}_1}{\beta_1}\mathscr{B}}{\mathscr{A} - \delta\mathscr{B}(\delta\beta_1)^{\frac{-1}{1+\rho_y}} + \delta\mathscr{B}(\delta\beta_1)^{\frac{-\rho_y}{1+\rho_y}}\left[(\delta\beta_1)^{\frac{\rho_y}{1+\rho_y}} - \hat{\beta}_1\right]}\frac{\sigma - \sigma_1}{\sigma - \sigma_2}.$$

Then $\mathscr{S}_\gamma(\sigma) > 0$ when $\sigma \in (0, \min\{\sigma_1,\sigma_2\}) \cup (\max\{\sigma_1,\sigma_2\}, +\infty)$ and $\mathscr{S}_\gamma(\sigma) < 0$ when $\sigma \in (\min\{\sigma_1,\sigma_2\}, \max\{\sigma_1,\sigma_2\})$. ■

15.6.6 Proof of Lemma 3

Consider the expression of $\mathscr{D}et(\sigma,\gamma)$ as defined by (15.22). Assumption 4 implies that the denominator of $\mathscr{D}et(\sigma,\gamma)$ is equal to zero when γ is equal to the following value:

$$\gamma^* = \frac{\sigma\frac{T_1^{*2}}{c^*\mathscr{T}_{12}^*}(1+\mathscr{C})^2\mathscr{D}(\delta\beta_1)^{\frac{1}{1+\rho_y}}}{\frac{T_3^*}{\ell^*\mathscr{T}_{32}^*}\left(1 + \sigma\delta\frac{T_1^{*2}}{c^*\mathscr{T}_{12}^*}\mathscr{B}\right)}.$$

Moreover, under Assumption 6, when $\gamma = \gamma^*$, the numerator of $\mathscr{D}et(\sigma,\gamma)$ is equal to

$$\frac{\sigma^2\left(\frac{T_1^{*2}}{c^*\mathscr{T}_{12}^*}\right)^2(1+\mathscr{C})^2\mathscr{D}(\delta\beta_1)^{\frac{1}{1+\rho_y}}\left(\mathscr{A} - \frac{\hat{\beta}_1}{\beta_1}\mathscr{B}\right)}{1 + \sigma\delta\frac{T_1^{*2}}{c^*\mathscr{T}_{12}^*}\mathscr{B}}(\sigma - \sigma_1) < 0.$$

Therefore, $\lim_{\gamma\to\gamma_+^*}\mathscr{D}et(\sigma,\gamma) = -\infty$ and $\lim_{\gamma\to\gamma_-^*}\mathscr{D}et(\sigma,\gamma) = +\infty$.

Under Assumption 6, we know from Lemma 2 that $\mathscr{S}_\gamma(\sigma) > 0$. We have also proved that when γ goes to 0, the end point $(\mathscr{T}r(\sigma, 0), \mathscr{D}et(\sigma, 0))$ lies below the line AC and thus satisfies $\mathscr{D}et(\sigma, 0) > 1$ and $\mathscr{T}r(\sigma, 0) > 2$. As a result the line $\Delta_\gamma(\sigma)$ cannot cross the line AC, or equivalently $\mathscr{S}_\gamma(\sigma) \in (0, 1)$. ■

References

Benhabib, J. and R. Farmer (1994), "Indeterminacy and Increasing Returns," *Journal of Economic Theory*, 63, 19–41.

Benhabib, J. and R. Farmer (1999), Indeterminacy and Sunspots in Macroeconomics, in *Handbook of Macroeconomics*, ed. by Taylor, J.B. and M. Woodford, North-Holland, Amsterdam, 387–448.

Benhabib, J. and K. Nishimura (1998), "Indeterminacy and Sunspots with Constant Returns," *Journal of Economic Theory*, 81, 58–96.

Benhabib, J., Nishimura, K. and A. Venditti (2002), "Indeterminacy and Cycles in Two-Sector Discrete-Time Model," *Economic Theory*, 20, 217–235.

Bosi, S., Magris, F. and A. Venditti (2005), "Competitive Equilibrium Cycles with Endogenous Labor," *Journal of Mathematical Economics*, 41, 325–349.

Grandmont, J.-M., Pintus, P. and R. De Vilder (1998), "Capital-Labor Substitution and Competitive Nonlinear Endogenous Business Cycles," *Journal of Economic Theory*, 80, 14–59.

Lloyd-Braga, T., Nourry, C. and A. Venditti (2006), "Indeterminacy with Small Externalities: the Role of Non-Separable Preferences," *International Journal of Economic Theory*, 2, 217–239.

Nishimura, K. and A. Venditti (2004a), "Indeterminacy and the Role of Factor Substitutability", *Macroeconomic Dynamics*, 8, 436–465.

Nishimura, K. and A. Venditti (2004b), "Asymmetric Factor Substitutability and Indeterminacy", *Journal of Economics*, 83, 125–150.

Nishimura, K. and A. Venditti (2004c), "Capital Depreciation, Factors Substitutability and Indeterminacy," *Journal of Difference Equations and Applications*, 10, 1153–1169.

Nishimura, K., Takahashi, H. and A. Venditti (2006), "Endogenous Fluctuations in Two-Sector Models: the Role of Preferences," *Journal of Optimization Theory and Applications*, 128 (2).

Pintus, P (2006), "Indeterminacy with Almost Constant Returns to Scale: Capital-Labor Substitution Matters", *Economic Theory*, 28, 633–649.

Part VII
Indeterminacy in Endogenous Growth Models

Chapter 16
Indeterminacy Under Constant Returns to Scale in Multisector Economies*

Jess Benhabib, Qinglai Meng, and Kazuo Nishimura**

16.1 Introduction

Recently there has been a renewed interest in indeterminacy, or alternatively put, in the existence of a continuum of equilibria in dynamic economies that exhibit some market imperfections.[1] One of the primary concerns of this literature has been the empirical plausibility of indeterminacy, which arises in markets with external effects or with monopolistic competition, often coupled with some degree of increasing returns. While the early results on indeterminacy relied on relatively large increasing returns and high markups, more recently Benhabib and Farmer (1996) showed that indeterminacy can also occur in two-sector models with small sector-specific external effects and very mild increasing returns. Nevertheless, a number of empirical researchers, refining the earlier findings of Hall (1990) on

*Econometrica 68, 1541–1548, 2000

**We wish to thank a co-editor, Danyang Xie, and anonymous referees for very useful comments. Support from the C. V. Starr Center for Applied Economics at New York University is gratefully acknowledged.

[1] See, for example, Benhabib and Farmer (1994), Benhabib and Perli (1994), Benhabib et al. (1994), Boldrin and Rustichini (1994), Bond et al. (1996), Schmitt-Grohé (1997), or Xie (1994).

J. Benhabib (✉)
Department of Economics, New York University, USA
e-mail: jess.benhabib@nyu.edu

Q. Meng
Department of Economics, Oregon State University, USA
e-mail: meng2008@oregonstate.edu

K. Nishimura
Institute of Economic Research, Kyoto University, Japan
e-mail: nishimura@kier.kyoto-u.ac.jp

J. Stachurski et al. (eds.), *Nonlinear Dynamics in Equilibrium Models*,
DOI 10.1007/978-3-642-22397-6_16,

disaggregated US data, have concluded that returns to scale seem to be roughly constant, if not decreasing.[2] While one can argue whether the degree of increasing returns required for indeterminacy in Benhabib and Farmer (1996) falls within the standard errors of the recent empirical estimates, one may also inquire whether increasing returns are at all needed for indeterminacy to arise in a plausible manner.

In this paper we argue that in multisector models indeterminacy can arise as a type of coordination problem, even without increasing returns, if there is a small wedge between private and social returns. In simple one-sector models increasing returns, sustained in a market context by external effects or monopolistic competition, can create a coordination problem. In such a setting, if all agents were to simultaneously increase their investment in an asset above the level associated with the initial equilibrium, the rate of return on that asset would tend to increase, justifying the higher level of investment. In a multisector model however, the rates of return and marginal products depend not only on the stocks of assets, but also on the composition of output across sectors. The rate of return on an asset can increase with the stock of the asset even in the absence of increasing returns. For example, consider a two-sector model with a pure consumption and a pure capital good. Increasing the relative price and hence the output of the capital good by moving along the production possibility frontier will increase the marginal product of the capital good if it is relatively more capital intensive. When combined with market distortions and external effects, the consequent rise in the stock of capital may not be enough to offset the initial increase of its marginal product. Both the stock and the marginal product of the capital good would rise simultaneously, mimicking the effect of increasing returns in the one-sector model. It is therefore possible to have constant aggregate returns in all sectors at the social level, and still to obtain indeterminacy with minor or even negligible external effects in some of the sectors. We illustrate this in the next section in the context of a two-sector endogenous growth model and discuss some extensions.

16.2 An Endogenous Growth Model with Non-linear Utility

We consider an economy without fixed factors that exhibits unbounded growth. A representative agent optimizes an additively separable utility function with discount rate $(r - g) > 0$ and g is the depreciation rate. We have

$$\max \int_0^\infty U(c)\, e^{-(r-g)t} dt \qquad (16.1)$$

subject to

$$y_j = e_j \prod_{i=1}^{2} (x_{ij})^{\beta_{ij}} \qquad (j = 1, 2), \qquad (16.2)$$

[2] See Basu and Fernald (1997).

$$\frac{dx_1}{dt} = y_1 - gx_1 - c, \tag{16.3}$$

$$\frac{dx_2}{dt} = y_2 - gx_2, \tag{16.4}$$

$$\sum_{j=1}^{2} x_{ij} = x_i \qquad (i = 1, 2), \tag{16.5}$$

where we assume that $\beta_{ij} > 0$.[3] We specify the utility function as $U(c) = (1-\sigma)^{-1} c^{1-\sigma}$, $\sigma \geq 0$. Note that as in one sector growth models, we do not have a pure consumption good. The first good is both a factor of production and a consumption good. Production is subject to an external effect e_j, treated as a constant by the agent,

$$e_j = \prod_{i=1}^{2} x_{ij}{}^{b_{ij}} \qquad (j = 1, 2). \tag{16.6}$$

Therefore the true production functions are

$$y_j = \prod_{i=1}^{2} (x_{ij})^{\beta_{ij} + b_{ij}} \qquad (j = 1, 2), \tag{16.7}$$

where under social constant returns $\Sigma_{s=1}^{2} (\beta_{sj} + b_{sj}) = 1$.

The Hamiltonian associated with the problem given by (16.1)–(16.5) is

$$\begin{aligned} H = U(c) &+ p_1 \left(e_1 \prod_{i=1}^{2} (x_{i1})^{\beta_{i1}} - gx_1 - c \right) \\ &+ p_2 \left(e_2 \prod_{i=1}^{2} (x_{i2})^{\beta_{i2}} - gx_2 \right) \\ &+ \sum_{i=1}^{2} w_i \left(x_i - \sum_{j=1}^{2} x_{ij} \right). \end{aligned}$$

Here p_j and w_i are Lagrange multipliers, representing utility prices of the capital goods and their rentals, respectively. The static first order conditions for this problem are given by

$$U'(c) = c^{-\sigma} = p_1, \tag{16.8}$$

$$w_s = p_j \left(\beta_{sj} \prod_{i=1}^{2} (x_{ij})^{\beta_{ij} + b_{ij}} \right) (x_{sj})^{-1} \qquad (s, j = 1, 2). \tag{16.9}$$

[3]We assume $\beta_{ij} > 0$, which assures that all inputs are used in the production of all goods, for computational and analytical simplicity. It is not difficult to relax this assumption but the notation becomes cumbersome.

Note that to derive (16.9), we use the fact that in equilibrium the inputs x_{ij} generating external effects e_j are identical to the inputs chosen by the firm. Under constant returns the unit cost functions are independent of output levels and are invertible. Therefore factor rentals $w = (w_1, w_2)$ are uniquely determined by output prices $p = (p_1, p_2)$.

Let $x = (x_1, x_2)$. The laws of motion for problem (16.1) are given by (16.3), (16.4), where $y_i = y_i(x, p)$, and by

$$\frac{dp_i}{dt} = rp_i - w_i(p) \quad (i = 1, 2) \,. \tag{16.10}$$

Constant social returns coupled with small external effects imply that some sectors must have a small degree of decreasing returns at the private level. This is in contrast to models of indeterminacy with social increasing, but private constant returns to scale. An implication of decreasing private returns is positive profits. Unless the number of firms is fixed, we must assume that there is a fixed entry cost to determine the number of firms along the equilibrium trajectories. As is clear from Proposition 1 below, the external effects and the degree of decreasing returns required for indeterminacy may be arbitrarily small, and generate only a small amount of profits. If the discounted value of profits along equilibrium trajectories that converge to the balanced growth path is small, a small fixed cost of entry will be sufficient to deter new entrants.[4]

Let the growth rate of c and x_i along the balanced growth path be μ. It follows from (16.8) that prices must then decline at the rate $\sigma\mu$. We define discounted variables as

$$\chi_i = e^{-\mu t} x_i, \quad \pi_i = e^{\sigma\mu t} p_i, \quad \psi_i = e^{-\mu t} y_i, \quad \omega_i = e^{\sigma\mu t} w_i \tag{16.11}$$

for $i = 1, 2$. Note that $e^{-\mu t} c = e^{-\mu t} (p_1)^{-\frac{1}{\sigma}} = (\pi_1)^{-\frac{1}{\sigma}}$. Since there are no fixed factors, outputs y are homogenous of degree one in the stocks x, and homogenous of degree zero in prices p, and the factor prices w are homogenous of degree one in prices. Then the (16.3), (16.4) and (16.10) can be written as

$$\frac{d\chi_1}{dt} = \psi_1(\chi, \pi) - (g + \mu)\chi_1 - (\pi_1)^{-\frac{1}{\sigma}}, \tag{16.12}$$

$$\frac{d\chi_2}{dt} = \psi_2(\chi, \pi) - (g + \mu)\chi_2, \tag{16.13}$$

$$\frac{d\pi_i}{dt} = (r + \sigma\mu)\pi_i - \omega_i(\pi) \quad (i = 1, 2) \,. \tag{16.14}$$

[4]The existence of a balanced growth path is easily proved, with small modifications to allow for external effects, along the lines of the proof in Bond et al. (1996). To assure positive prices and quantities, a lower bound to the discount rate is required, as shown in footnote 5 below.

The balanced growth path corresponds to the stationary point $(\chi_1^*, \chi_2^*, \pi_1^*, \pi_2^*)$ of the above system. The Jacobian of this system is given by

$$J = \begin{bmatrix} \left[\frac{\partial \psi}{\partial \chi}\right] - (g+\mu) I & \left[\frac{\partial \psi}{\partial \pi}\right] - Z \\ \mathbf{0} & (r+\sigma\mu) I - \left[\frac{\partial \omega}{\partial \pi}\right] \end{bmatrix}$$

where Z is a matrix with zeros except for the element of the first row and the first column, which is $(1/\sigma)\,\pi_1^{-\frac{1}{\sigma}-1}$.

Using (16.9) and the social constant returns restriction $\Sigma_{s=1}^2(\beta_{sj} + b_{sj}) = 1$, we find that output prices satisfy

$$p_j = \prod_{s=1}^{2} \left(\frac{w_s}{\beta_{sj}}\right)^{\beta_{sj}+b_{sj}} \quad (j = 1, 2), \tag{16.15}$$

or

$$\pi_j = \prod_{s=1}^{2} \left(\frac{\omega_s}{\beta_{sj}}\right)^{\beta_{sj}+b_{sj}} \quad (j = 1, 2) \tag{16.16}$$

and unit input coefficients are

$$a_{ij} = \frac{\beta_{ij} p_j}{w_i} \tag{16.17}$$

$$= \frac{\beta_{ij}}{w_i} \prod_{s=1}^{2} \left(\frac{w_s}{\beta_{sj}}\right)^{\beta_{sj}+b_{sj}} \tag{16.18}$$

$$= \frac{\beta_{ij}}{\omega_i} \prod_{s=1}^{2} \left(\frac{\omega_s}{\beta_{sj}}\right)^{\beta_{sj}+b_{sj}} \quad (i, j = 1, 2)\,. \tag{16.19}$$

Let A be the input coefficient matrix with elements a_{ij}. Full employment of factors implies $A\psi = \chi$. Differentiating, we have

$$A d\psi + S\psi = d\chi \tag{16.20}$$

where the elements of the matrix S are $\left[\sum_{s=1}^{2}(\partial a_{ij}/\partial \omega_s) d\omega_s\right]$. In order to obtain the dual relationship of price and output in the context of externalities, we express the price function in terms of input coefficients and Cobb-Douglas exponents. Let

$$\hat{a}_{ij} = a_{ij}(\beta_{ij} + b_{ij})/\beta_{ij} \quad (i, j = 1, 2) \tag{16.21}$$

and define $\hat{A} = \left[\hat{a}_{ij}\right]$. Prices satisfy (16.16). Differentiating (16.16) and using (16.19) we obtain

$$d\pi = \left[\hat{A}\right]' d\omega. \tag{16.22}$$

Using (16.20) and (16.22) the Jacobian matrix J becomes

$$J = \begin{bmatrix} [A]^{-1} - (g+\mu)\, I & \left[\frac{\partial \psi}{\partial \omega}\right] - Z \\ \mathbf{0} & (r + \sigma\mu)\, I - \left[\hat{A}'\right]^{-1} \end{bmatrix} \tag{16.23}$$

provided A and $\hat{A}$ are nonsingular (see Assumption 1 below). On a balanced growth path the full employment of factors, $A\psi = \chi$, implies

$$[I - A\,(g+\mu)]\, \chi^* = z \tag{16.24}$$

where $z = \left(a_{11}\pi_1^*,\ a_{21}\pi_1^*\right)$. The price equations, $\pi = \left[\hat{A}\right]' \omega$, imply that on a balanced growth path,

$$\left[\left[\hat{A}\right]' (r + \sigma\mu) - I\right] \pi^* = 0. \tag{16.25}$$

Note that the above relation implies that the matrix $\left[\left[\hat{A}\right]' (r + \sigma\mu) - I\right]$ must be singular and it corresponds to the lower right submatrix of the Jacobian J. As expected this always yields a root for J that is identically zero. The vector π will be determined up to a multiplicative constant, while (16.24) will determine χ. The vector χ pins down not the level of stocks x, but their discounted values, as is clear from (16.11). The same is true for π which does not pin down the prices p, but their upcounted values. Thus on the balanced growth path quantities x, y, c grow at the rate μ, while prices p decline at the rate $\sigma\mu$.

To determine the growth rate μ, we note from (16.25) that the quantity $(r + \sigma\mu)$ corresponds to the inverse of the Frobenius root of the nonnegative matrix $\hat{A}'$. This is the only root that is associated with the positive eigenvector π^*.[5]

The signs of the roots of J are the same as those of the roots of $\left[[A]^{-1} - (g+\mu)\, I\right]$ and $\left[(r + \sigma\mu)\, I - \left[\hat{A}'\right]^{-1}\right]$. The system will be locally determinate if J has one negative root and two roots with positive real parts. Then we

[5]Since the matrix $\hat{A}'$ is indecomposable, π^* is the unique nonnegative eigenvector of $\hat{A}'$ and the associated positive Frobenius root $\hat{\lambda} = (r + \sigma\mu)^{-1}$ is the largest root in absolute value. Since with positive externalities the elements of $\hat{A}$ are at least as large as those of A, the Frobenius root of $\hat{A}$ will be at least as large as that of A. From these observations it also follows that if $\hat{\lambda}^{-1} > g + \mu$, the inverse of $[I - A\,(g+\mu)]$ will be a positive matrix and assure, from (16.24) that $\chi > 0$. The restriction on the discount rate is then $(1-\sigma)\left(\hat{\lambda}^{-1} - g\right) < r - g$. If some externalities are negative, the elements of $\hat{A}$ may be smaller than those of A. Then, if the Frobenius root of A is less than μ^{-1}, we can choose g less than r so that $[I - (g+\mu)\, A]^{-1}$ is a positive matrix.

can choose the initial prices so that initial conditions lie on the two dimensional manifold spanned by the one dimensional center manifold corresponding to the balanced growth path, and the one dimensional stable manifold corresponding to the negative root.[6] If J has two roots with negative real parts however, the system will be indeterminate. In this case, depending on initial stocks, the initial prices can be chosen on the three dimensional space spanned by the one dimensional center manifold corresponding to the balanced growth path, and the two dimensional stable manifold corresponding to the two negative roots. For example, if $[A]^{-1}$ has one root with negative real part and $\hat{A}'$ has at least two real positive roots, the system will be indeterminate. In the multisector version of the model, the system will be locally indeterminate if $[A]^{-1}$ has $(n-1)$ roots with negative real parts and $\hat{A}'$ has at least two real positive roots (see Proposition 2 below).

Let $B = [\beta_{si}]$ and $\hat{B} = [\beta_{si} + b_{si}]$. We make the following assumption:

Assumption 1 *The matrices B and $\hat{B}$ are strictly positive and nonsingular.*

Let Ω denote the 2×2 diagonal matrix with diagonal elements ω_i, $i = 1, 2$ and zero off-diagonal elements. Similarly let Π denote the 2×2 diagonal matrix with diagonal elements π_i, $i = 1, 2$ and zero off-diagonal elements. Note from (16.9) that $a_{ij} = p_j \beta_{ij} / w_i$, and therefore we have $\hat{a}_{ij} = p_j \left(\beta_{ij} + b_{ij}\right) / w_i$. It follows that $A = \Omega^{-1} B \Pi$ and $\hat{A} = \Omega^{-1} \hat{B} \Pi$, and from Assumption 1 that A and $\hat{A}$ will be nonsingular.

Lemma 1. *Along the balanced growth path the sign pattern of roots of B is the same as that of $A^{-1} = \Pi^{-1} B^{-1} \Omega$, and the sign pattern of roots of $\widehat{B}$ is the same as that of $A^{-1} = \Omega \left[\widehat{B}'\right]^{-1} \Pi^{-1}$.*

Proof. Along the balanced growth path $\omega_i^* = (r + \sigma\mu)\, \pi_i^*$. The lemma follows from noting that $|A^{-1}| = |\Pi^{-1} B^{-1} \Omega| = (r + \sigma\mu)^2 \, |B^{-1}|$ and that every principal minor of $\left[\Pi^{-1} B^{-1} \Omega\right]$ of order i will be given by the corresponding principal minor of B^{-1} multiplied by $(r + \sigma\mu)^i$. If the characteristic equation of B^{-1} is $f(\lambda) = (-\lambda)^2 + b_1(-\lambda) + b_0 = 0$, the coefficients b_{n-i} will be the sum of principal minors of order i. Therefore, the characteristic polynomial of $\left[\Pi^{-1} B^{-1} \Omega\right]$ will have coefficients $(r + \sigma\mu)^i \, b_{2-i}$. If the characteristic equation of $\left[\Pi^{-1} B^{-1} \Omega\right]$ is given by $g(\nu) = 0$, then

$$
\begin{aligned}
(r + \sigma\mu)^{-2} g(\nu) &= (r + \sigma\mu)^{-2} (-\nu)^2 + (r + \sigma\mu)^{-1} b_1 (-\nu) + b_0 \\
&= f\left(\nu / (r + \sigma\mu)\right).
\end{aligned}
$$

[6] A standard alternative method to working with a system that has an identically zero root is to reduce the dimension of the system using ratios of stocks, x_i / x_1, as in Mulligan and Sala-i-Martin (1993) or Benhabib and Farmer (1994).

Therefore if λ is a root of B^{-1}, then $\lambda/(r+\sigma\mu)$ is a root of $\left[\Pi^{-1}B^{-1}\Omega\right]$ and the sign pattern of the roots of B and B^{-1} is the same as that of $\left[\Pi^{-1}B^{-1}\Omega\right]$. The proof that the inertia of $\hat{B}'$ is the same as that of $\left[\Omega\left[\widehat{B}'\right]^{-1}\Pi^{-1}\right]$ is identical. □

Proposition 1 below gives conditions for indeterminacy that are independent of the utility function. From Lemma 1 the factor intensity difference $a_{22}/a_{12}-a_{21}/a_{11}$ is directly related to $\beta_{22}/\beta_{12}-\beta_{21}/\beta_{11}$. Therefore we may say that the consumable capital good (first good) is intensive in the pure capital good (second good) *from the private perspective* if $\beta_{22}\beta_{11}-\beta_{21}\beta_{12}<0$, but that it is intensive in itself *from the social perspective* if $(\beta_{22}+b_{22})(\beta_{11}+b_{11})-(\beta_{21}+b_{21})(\beta_{12}+b_{12})>0$. The proposition follows from noting the signs of the determinant and trace of the matrices B and $\hat{B}$.

Proposition 1. *In the two-sector endogenous growth model, if the consumable capital good is intensive in the pure capital good from the private perspective, but it is intensive in itself from the social perspective, then the balanced growth path is indeterminate.*

Proof. If the consumable capital good is intensive in the pure capital good from the private perspective, B has negative determinant. This implies that B^{-1} has negative determinant and one negative root. In this case $[A]^{-1}-(g+\mu)I$ has at least one negative root.

If the consumable capital good is intensive in itself from the social perspective, $\widehat{B}$ has positive trace and positive determinant. In this case $\widehat{B}$, and hence $\hat{A}'$ have two positive roots. One of the positive roots of $\hat{A}'$ is the Frobenius root $(r+\sigma\mu)^{-1}$, which has to be real, and the other one is smaller in modulus. Therefore the positive real root of $\left[\hat{A}'\right]^{-1}$ other than the inverse of the Frobenius root of $\hat{A}'$ will dominate $(r+\sigma\mu)$. On the other hand as the inverse of the Frobenius root, $(r+\sigma\mu)$, is the root of $\left[\hat{A}'\right]^{-1}$, $\left[\hat{A}'\right]^{-1}-(r+\sigma\mu)I$ has one zero root and one negative root. Therefore J has one zero root and at least two negative roots. □

The analysis of the model above can easily be recast in an n-sector framework. In a multisector model, the matrices B and $\hat{B}$, composed of the Cobb-Douglas exponents, will be of dimension higher than two, with $i,j=1,\ldots,n$. Indeterminacy will now follow if 2n-dimensional matrix J has less than n roots with positive real parts. Then the proposition above generalizes as follows:

Proposition 2. *In the multisector endogenous growth model, if the matrix B has $(n-1)$ roots with negative real parts and the matrix $\hat{B}$ has at least two roots with positive real parts, the system will be indeterminate.*

Proof. From Lemma 1, at the steady state the root structure of the n-dimensional input matrices A and $\hat{A}$ corresponds to the root structure of B and $\hat{B}$. The system will be indeterminate if the now 2n-dimensional matrix J has less than n positive roots. This will happen if $[A]^{-1}$ has $(n-1)$ roots with negative real parts and $\left[\hat{A}'\right]$ has at least two real positive roots. Since one of the positive real roots of $\hat{A}'$ is the Frobenius root $(r+\sigma\mu)^{-1}$, all the other roots are smaller in modulus. The real positive roots of $\left[\hat{A}'\right]^{-1}$ other than the Frobenius root will dominate $(r+\sigma\mu)$. Therefore J will have at least $n+1$ roots with negative real parts. □

The result of Proposition 1 holds exactly in a two-sector nonendogenous growth model with fixed labor supply if the first good is a pure consumption good, the second good is a pure capital good, and if utility is linear (see Benhabib and Nishimura 1998). The reason that we need to resort to linear utility in a two-sector model with fixed labor and a pure consumption good is simple. The existence of multiple equilibrium paths implies that for a given level of the capital stock, there is a continuum of ratios of initial investment to consumption that are consistent with equilibrium. However some curvature in the utility function may destroy the possibility of multiple equilibria if the cost of foregoing current consumption is large relative to future benefits that come from higher initial investment. We can regain some flexibility if the first good is both a consumption good and a capital good, as in the one sector model. In that case increasing the investment level in the second (pure capital) good does not solely come at the expense of consumption. However if we stick with such a setup, we would have two goods and three factors, one of which is labor. This would significantly complicate the analysis. Switching to an endogenous growth model without a fixed factor avoids this difficulty by making the number of goods and factors equal.

References

Basu, S. and J. G. Fernald (1997), "Returns to Scale in US Production: Estimates and Implications," *Journal of Political Economy*, 105, 249–283.

Benhabib, J. and R. Farmer (1994), "Indeterminacy and Increasing Returns," *Journal of Economic Theory*, 63, 19–41.

Benhabib, J. and R. Farmer (1996), "Indeterminacy and Sector Specific Externalities," *Journal of Monetary Economics*, 37, 397–419.

Benhabib, J. and K. Nishimura (1998), "Indeterminacy and Sunspots with Constant Returns," *Journal of Economic Theory*, 81, 58–96.

Benhabib, J. and R. Perli (1994), "Uniqueness and Indeterminacy: Transitional Dynamics in a Model of Endogenous Growth," *Journal of Economic Theory*, 63, 113–142.

Benhabib, J., Perli, R. and D. Xie (1994), "Monopolistic Competition, Indeterminacy and Growth," *Ricerche Economiche*, 48, 279–298.

Boldrin, M. and A. Rustichini (1994), "Indeterminacy of Equilibria in Models with Infinitely-lived Agents and External Effects," *Econometrica*, 62, 323–342.

Bond, E. W., Wang, P. and C. K. Yip (1996), "A General Two-Sector Model of Endogenous Growth with Human and Physical Capital: Balanced Growth and Transitional Dynamics," *Journal of Economic Theory*, 68, 149–173.

Hall, R. E. (1990), Invariance Properties of Solow's Productivity Residual, in *Growth, Productivity, Unemployment*, ed. by Diamond, P., Cambridge: MIT Press, 71–112.

Mulligan, C. B. and X. Sala-i-Martin (1993), "Transitional Dynamics in Two–Sector Models of Endogenous Growth," *Quarterly Journal of Economics*, 108, 739–773.

Schmitt-Grohé, S. (1997), "Comparing Four Models of Aggregate Fluctuations Due to Self-Fulfilling Expectations," *Journal of Economic Theory*, 72, 96–147.

Xie, D. (1994), "Divergence in Economic Performance: Transitional Dynamics with Multiple Equilibria," *Journal of Economic Theory*, 63, 97–11.

Chapter 17
Global Externalities, Endogenous Growth and Sunspot Fluctuations*

Kazuo Nishimura, Harutaka Takahashi, and Alain Venditti

17.1 Introduction

Over the last decade, a large literature has focussed on the existence of local indeterminacy, i.e. a continuum of equilibrium paths converging toward a steady state, and sunspot fluctuations in endogenous growth models. These studies provide a possible explanation of the facts that economic growth rates are volatile over time and dispersed across countries. Within infinite-horizon two-sector models, the occurrence of multiple equilibrium paths is related to the existence of market imperfections such that productive externalities. Two different strategies have been followed to modelize these learning-by-doing effects. A first set of contributions builds upon Benhabib and Farmer (1996) and consider sector-specific externalities with constant returns at the social level. Most of the papers are based on continuous-time models and local indeterminacy is shown to arise if the final good sector is human capital (labor) intensive at the private level but physical capital intensive at

* **Advances in Discrete Dynamical Systems**, ed. by Elaydi, S., Nishimura, K., Shishikura, M. and N. Tose, Mathematical Society Japan, Tokyo, 203–214, 2009

K. Nishimura (✉)
Institute of Economic Research, Kyoto University, Japan
e-mail: nishimura@kier.kyoto-u.ac.jp

H. Takahashi
Faculty of Economics, Meiji Gakuin University, Japan
e-mail: haru@eco.meijigakuin.ac.jp

A. Venditti
National Center for Scientific Research, CNRS-GREQAM and EDHEC, Marseille, France
e-mail: alain.venditti@univmed.fr

J. Stachurski et al. (eds.), *Nonlinear Dynamics in Equilibrium Models*,
DOI 10.1007/978-3-642-22397-6_17, © Mathematical Society Japan 2009.
Published by Springer-Verlag Berlin Heidelberg 2012. All Rights Reserved

the social level.[1] It is worth noticing however that Mino et al. (2008) have recently considered a discrete-time two-sector analogue model and show that the existence of local indeterminacy follows from a more complex set of conditions based on capital intensities differences at the private and social level, the rate of depreciation of capital and the discount factor.

A second set of contributions build upon the initial endogenous growth framework developed by Lucas (1988) and Romer (1986). Local indeterminacy is derived from the consideration of global externalities and increasing returns at the social level. Most of the papers are however based on specific functional forms with a Leontief technology in the investment good sector, and/or assume a degenerate allocation of human capital (labor) across sectors.[2] In such a framework, there is no clear condition in terms of capital intensities differences to ensure the occurrence of multiple equilibria. Considering a non-trivial allocation mechanism of labor between the two sectors, Drugeon et al. (2003) have shown within a general continuous-time model with labor-augmenting global external effects borrowed from Boldrin and Rustichini (1994) that the occurrence of local indeterminacy necessarily requires a physical capital intensive investment good sector at the private level. This condition appears to be the complete opposite to the one derived with sector-specific externalities. The aim of this paper is then to explore the robustness of this result by considering a discrete-time formulation.

Considering a discrete-time analogue of the Drugeon et al. (2003) model, Goenka and Poulsen (2004) suggest that local indeterminacy may arise under both configurations for the capital intensity difference at the private level. However, their conditions are quite complex as they involve the endogenous growth rate, and they do not show that a non-empty set of economies may satisfy these conditions. We then consider a discrete-time model with Cobb-Douglas technologies and labor-augmenting global externalities. We prove the existence of a balanced growth path and we give conditions on the Cobb-Douglas coefficients for the occurrence of sunspot fluctuations that are compatible with both types of capital intensity configuration at the private level provided the elasticity of intertemporal substitution in consumption admits intermediary values. However, the occurrence of period-2 cycles requires the consumption good to be physical capital intensive at the private level. All these results are finally illustrated through numerical examples.

The rest of the paper is organized as follows: Sect. 17.2 sets up the basic model. In Sect. 17.3 we study the existence of a balanced growth path. Section 17.4 provides the main results on local indeterminacy. Section 17.5 contains concluding comments. The proofs are gathered in a final Appendix.

[1] Benhabib et al. (2000), Mino (2001), and Nishimura and Venditti (2004). See also Bond et al. (1996) in which similar results are obtained from the consideration of distortionary factor taxes.

[2] See Benhabib and Perli (1994), Boldrin and Rustichini (1994), Boldrin et al. (2001), and Xie (1994).

17.2 Model

We consider a discrete-time two-sector economy having an infinitely-lived representative agent with a single period utility function given by

$$u(c) = \frac{c^{1-\sigma}}{1-\sigma}$$

where c is consumption and $\sigma > 0$ is the inverse of the elasticity of intertemporal substitution in consumption. The labor supply is inelastic. There are two goods: the consumption good, c, and the capital good, k. Each good is produced with a Cobb-Douglas technology. We assume that the production functions contain positive global externalities given by the average capital stock in the economy and which can be interpreted as a labor augmenting technical progress. We denote by y and c the outputs of sectors k and c:

$$c = K_c^{\alpha_1}(eL_c)^{\alpha_2}, \qquad y = AK_y^{\beta_1}(eL_y)^{\beta_2}$$

with $e = \bar{k}$ and $A > 0$ a normalization constant. Labor is normalized to one, $L_c + L_y = 1$ and the total stock of capital is $K_c + K_y = k$. We assume that the economy-wide average $\bar{k}$ is taken as given by individual firms. At the equilibrium, all firms of sector $i = c, y$ being identical, we have $\bar{k} = k$.

We assume constant returns to scale at the private level, i.e. $\alpha_1 + \alpha_2 = \beta_1 + \beta_2 = 1$. At the social level the returns to scale are therefore increasing. It can be easily shown that if $\beta_1/\beta_2 > (<)\alpha_1/\alpha_2$ the investment (consumption) good sector is capital intensive from the private perspective.

We assume complete depreciation of capital in one period so that the capital accumulation equation is $y_t = k_{t+1}$. Optimal factor allocations across sectors are obtained by solving the following program:

$$\begin{array}{ll} \max_{K_{ct},\, L_{ct},\, K_{yt},\, L_{yt}} & K_{ct}^{\alpha_1}(e_t L_{ct})^{\alpha_2} \\ \text{s.t.} & k_{t+1} = AK_{yt}^{\beta_1}(e_t L_{yt})^{\beta_2} \\ & 1 = L_{ct} + L_{yt} \\ & k_t = K_{ct} + K_{yt} \\ & e_t \ \text{ given.} \end{array} \tag{17.1}$$

Denote by p_t, w_t and r_t respectively the price of the capital good, the wage rate of labor and the rental rate of capital, all in terms of the price of the consumption good. For any (k_t, k_{t+1}, e_t), solving the associated first order conditions gives inputs as $\tilde{K}_c(k_t, k_{t+1}, e_t)$, $\tilde{L}_c(k_t, k_{t+1}, e_t)$, $\tilde{K}_y(k_t, k_{t+1}, e_t)$ and $\tilde{L}_y(k_t, k_{t+1}, e_t)$. We then define the efficient production frontier as

$$T(k_t, k_{t+1}, e_t) = \tilde{K}_c(k_t, k_{t+1}, e_t)^{\alpha_1}[e_t \tilde{L}_c(k_t, k_{t+1}, e_t)]^{\alpha_2}$$

which describes the standard trade-off between consumption and investment: for a given e_t, $T(k_t, k_{t+1}, e_t)$ is increasing with respect to k_t and decreasing with respect to k_{t+1}. Using the envelope theorem we then derive the equilibrium prices

$$p_t = -T_2\left(k_t, k_{t+1}, e_t\right), r_t = T_1\left(k_t, k_{t+1}, e_t\right) \tag{17.2}$$

where $T_1 = \frac{\partial T}{\partial k}$ and $T_2 = \frac{\partial T}{\partial y}$. The representative consumer's optimization program is finally given by

$$\begin{array}{ll} \max_{\{k_t\}_{t=0}^{+\infty}} & \sum_{t=0}^{\infty} \delta^t \frac{[T(k_t,\, k_{t+1},\, e_t)]^{1-\sigma}}{1-\sigma} \\ \text{s.t.} & k_0 = \hat{k}_0, \{e_t\}_{t=0}^{+\infty} \quad \text{given} \end{array} \tag{17.3}$$

with $\delta \in (0, 1)$ the discount factor. The corresponding Euler equation is

$$-c_t^{-\sigma} p_t + \delta c_{t+1}^{-\sigma} r_{t+1} = 0. \tag{17.4}$$

Let $\{k_t\}_{t=0}^{+\infty}$ denote a solution of (17.4) which obviously depends on $\{e_t\}_{t=0}^{\infty}$, i.e. $k_t = k(t, \{e_t\}_{t=0}^{\infty})$ for all $t \geq 0$. As we have assumed that $e_t = \bar{k}_t$ with $\bar{k}_t$ the economy-wide average capital stock, expectations are realized if there exists a solution of an infinite-dimensional fixed-point problem such that $e_t = k(t, \{e_t\}_{t=0}^{\infty})$ for any $t = 0, 1, 2, \ldots$. Assuming that such a solution exists,[3] prices may now be written as

$$r(k_t, k_{t+1}) = T_1(k_t, k_{t+1}, k_t), \;\; p(k_t, k_{t+1}) = -T_2(k_t, k_{t+1}, k_t) \tag{17.5}$$

and consumption at time t is given by a linear homogeneous function.[4]

$$c(k_t, k_{t+1}) = T(k_t, k_{t+1}, k_t) \tag{17.6}$$

Equation 17.4 finally becomes:

$$p(k_t, k_{t+1}) c(k_t, k_{t+1})^{-\sigma} = \delta r(k_{t+1}, k_{t+2}) c(k_{t+1}, k_{t+2})^{-\sigma} \tag{17.7}$$

Any solution $\{k_t\}_{t=0}^{+\infty}$ of (17.7) which also satisfies the transversality condition

$$\lim_{t \to +\infty} \delta^t c(k_t, k_{t+1})^{-\sigma} p(k_t, k_{t+1}) k_{t+1} = 0 \tag{17.8}$$

and the summability condition

[3] A detailed treatment of the existence of such a solution within a discrete-time version of the Lucas (1988) model is provided in Mitra (1998).

[4] See Proposition 1 in Drugeon and Venditti (1998) or Lemma 1 in Drugeon et al. (2003).

$$\sum_{t=0}^{\infty} \delta^t \frac{[c(k_t, k_{t+1})]^{1-\sigma}}{1-\sigma} < +\infty \tag{17.9}$$

is called an equilibrium path.[5]

17.3 Balanced Growth Path

We call a path $\{k_t\}_{t=0}^{+\infty}$ satisfying $k_0 = \hat{k}_0$ and $k_t \leq k_{t+1} \leq Ak_t$, a feasible path (from $k_0 = \hat{k}_0$). We call $\{k_t\}_{t=0}^{+\infty}$ a balanced growth path if it is in equilibrium and $k_{t+1}/k_t = \theta$ for $t = 0, 1, \ldots$.

Lemma 1. *Along an equilibrium path, prices satisfy*

$$r(k_t, k_{t+1}) = A\alpha_1 \left[\frac{\alpha_2\beta_1}{\alpha_1\beta_2}\left(\frac{k_{t+1}}{A\bar{g}(k_t, k_{t+1})}\right)^{1/\beta_2}\right]^{\alpha_2}, \quad p(k_t, k_{t+1}) = \frac{r_t \bar{g}(k_t, k_{t+1})}{\beta_1 k_{t+1}} \tag{17.10}$$

with

$$\bar{g}(k_t, k_{t+1}) = \left\{ K_y \in (0, Ak) / \frac{\alpha_1\beta_2}{\alpha_2\beta_1} = \frac{[k_t - K_y][(k_{t+1}/A)^{1/\beta_2} k_t^{-1} K_y^{-\beta_1/\beta_2}]}{[1 - (k_{t+1}/A)^{1/\beta_2} k_t^{-1} K_y^{-\beta_1/\beta_2}] K_y} \right\}$$

a linear homogeneous function.

Define $\theta_t = k_{t+1}/k_t (= c_{t+1}/c_t)$ the growth factor of capital (and thus consumption) at time t. Notice that $ln\theta_t$ is the growth rate of capital. By the feasibility condition $k_t \leq k_{t+1} \leq Ak_t$, it holds that $\theta_t \in (0, A)$. Denoting $g(\theta_t^{-1}) = \bar{g}(\theta_t^{-1}, 1)$ and using (17.10), the Euler equation (17.7) can be transformed into an implicit recursive equation as follows

$$\delta\beta_1 g(\theta_{t+1}^{-1})^{-\alpha_2/\beta_2} = g(\theta_t^{-1})^{1-(\alpha_2/\beta_2)} \theta_t^{\sigma} \tag{17.11}$$

A balanced growth factor is defined by $\theta_t = \theta_{t+1} = \theta$ and satisfies (17.11) together with conditions (17.8) and (17.9). Using the normalization constant A we will show that there exists such a normalized balanced growth factor.

Proposition 1. *Let* $\tilde{\sigma} = ln\beta_1/ln(\delta\beta_1) > 0$. *There exists a normalized balanced growth factor (NBGF)* $\theta^* = (\delta\beta_1)^{-1}$ *solution of the Euler equation (17.11) if* $\sigma > \tilde{\sigma}$ *and the normalization constant* A *is set at the following value*

[5] See Lemma 1 and Corollary 1 in Boldrin et al. (2001).

$$A^* = \left[\frac{\alpha_2\beta_1}{\alpha_1\beta_2} \frac{1-(\delta\beta_1)^{\sigma}\left(1-\frac{\alpha_1\beta_2}{\alpha_2\beta_1}\right)}{(\delta\beta_1)^{(1+\sigma)/\beta_2}} \right]^{\beta_2}. \quad (17.12)$$

Proposition 1 establishes the existence of a normalized balanced growth rate (NBGR) $ln\theta^*$ from which we define a normalized balanced growth path (NBGP), namely $k_t = \hat{k}_0\theta^{*t} = \hat{k}_0(\delta\beta_1)^{-t}$.

17.4 Local Indeterminacy

The local stability properties of the NBGP are obtained from the linearization of the Euler equation (17.11) around θ^*.

Assumption 1 $\alpha_1\beta_2 \neq \alpha_2\beta_1$.

Assumption 1, which is equivalent to a non-zero capital intensity difference at the private level, implies that the technologies are not identical.

Lemma 2. *Under Assumption 1, if $\sigma > \tilde{\sigma}$ and $A = A^*$ as defined in Proposition 1, the linearization of the Euler equation (17.11) around θ^* gives*

$$\frac{d\theta_{t+1}}{d\theta_t} = \sigma\beta_2 \frac{1-\frac{\alpha_1\beta_2}{\alpha_2\beta_1} + \frac{1-(\delta\beta_1)^{\sigma}\left(1-\frac{\alpha_1\beta_2}{\alpha_2\beta_1}\right)}{\beta_2(\delta\beta_1)^{\sigma}}}{\alpha_2\left(1-\frac{\alpha_1\beta_2}{\alpha_2\beta_1}\right)} - \beta_1\left(1-\frac{\alpha_1\beta_2}{\alpha_2\beta_1}\right). \quad (17.13)$$

Notice that the Euler equation (17.11) is forward looking since the initial growth rate $\theta_0 = k_1/\hat{k}_0$ is not pre-determined. As a result, if $d\theta_{t+1}/d\theta_t \in (-1, 1)$, then every initial point in the neighborhood of the NBGR has indeterminate equilibrium paths satisfying (17.8) and (17.9), and we say that the NBGP is locally indeterminate. If on the contrary $|d\theta_{t+1}/d\theta_t| > 1$, the NBGR is locally unstable and starting from $\hat{k}_0$ one possible equilibrium consists in jumping at $t = 1$ on the NBGP. In such a case the NBGP is locally determinate.

Proposition 2. *Let $\tilde{\sigma} = ln\beta_1/ln(\delta\beta_1)$. Under Assumption 1, if one of the following set of conditions is satisfied:*

(i) The investment good is capital intensive at the private level with

$$\frac{\tilde{\sigma}}{\beta_1}\left[1-\beta_1^2\left(1-\frac{\alpha_1\beta_2}{\alpha_2\beta_1}\right)\right] - \alpha_2\left(1-\frac{\alpha_1\beta_2}{\alpha_2\beta_1}\right)\left[1+\beta_1\left(1-\frac{\alpha_1\beta_2}{\alpha_2\beta_1}\right)\right] < 0 \quad (17.14)$$

(ii) The consumption good is capital intensive at the private level with

$$\frac{\tilde{\sigma}}{\beta_1}\left[1-\beta_1^2\left(1-\frac{\alpha_1\beta_2}{\alpha_2\beta_1}\right)\right]+\alpha_2\left(1-\frac{\alpha_1\beta_2}{\alpha_2\beta_1}\right)\left[1-\beta_1\left(1-\frac{\alpha_1\beta_2}{\alpha_2\beta_1}\right)\right]<0 \tag{17.15}$$

then there exist $\underline{\sigma} \geq \tilde{\sigma}$ and $\bar{\sigma} > \underline{\sigma}$ such that the NBGP $k_t = \hat{k}_0(\delta\beta_1)^{-t}$ is locally indeterminate when $\sigma \in (\underline{\sigma}, \bar{\sigma})$.

We easily derive from Proposition 2 that there exists some case in which the growth rate of capital exhibits period-2 cycles.

Corollary 1. *Let $\tilde{\sigma} = ln\beta_1/ln(\delta\beta_1)$. Under Assumption 1, if the consumption good is capital intensive at the private level and condition (17.15) holds, then there exist $\underline{\sigma} \geq \tilde{\sigma}$ and $\bar{\sigma} > \underline{\sigma}$ such that the NBGP is locally indeterminate when $\sigma \in (\underline{\sigma}, \bar{\sigma})$. Moreover, when σ crosses $\bar{\sigma}$ from below the NBGP becomes locally determinate and there exist locally indeterminate (determinate) period-2 growth cycles in a right (left) neighborhood of $\bar{\sigma}$.*

Corollary 1 shows that the existence of endogenous fluctuations requires a capital intensive consumption good. The intuition for this result, initially provided by Benhabib and Nishimura (1985), may be summarized as follows. Consider an instantaneous increase in the capital stock k_t. This results in two opposing forces:

- Since the consumption good is more capital intensive than the investment good, the trade-off in production becomes more favorable to the consumption good. The Rybczinsky theorem thus implies a decrease of the output of the capital good y_t. This tends to lower the investment and the capital stock in the next period k_{t+1}, and thus implies a decrease of the balanced growth factor $\theta_t = k_{t+1}/k_t$.
- In the next period the decrease of k_{t+1} implies again through the Rybczinsky effect an increase of the output of the capital good y_{t+1}. Indeed, the decrease of k_{t+1} improves the trade-off in production in favor of the investment good which is relatively less intensive in capital. Therefore this tends to lower the investment and the capital stock in period $t+2$, k_{t+2}, and thus implies an increase of the balanced growth factor $\theta_t = k_{t+2}/k_{t+1}$.

Boldrin et al. (2001) consider a similar two-sector model but assume a Leontief technology in the investment good sector with a degenerate allocation of labor across sectors as the investment good is only produced from physical capital. They prove the existence of global indeterminacy of equilibria and construct robust examples of both topological and ergodic chaos for the dynamics of balanced growth paths. We may expect similar results from Corollary 1. However, showing the existence of chaotic dynamics with a Cobb-Douglas technology in both sector is beyond the goal of the current paper and is left for future research.

Remark 1. Considering a discrete-time two-sector optimal endogenous growth model with two capital goods, one being consumable while the other is not, Drugeon

(2004) shows that the occurrence of period-2 cycles requires a non-unitary rate of capital depreciation in one sector at least. In a similar formulation but extended to include sector-specific externalities, Mino et al. (2008) show that the existence of local indeterminacy is based on the same necessary condition. We prove that the consideration of global externalities allows to get sunspot fluctuations even under full depreciation of capital.

Numerical illustration: In order to check whether all the conditions of Proposition 2 can be satisfied simultaneously, we perform some numerical simulations.

(i) Let $\delta = 0.98$, $\beta_1 = 0.85$ and $\alpha_1 = 0.55$. The investment good is thus capital intensive at the private level and the NBGF is equal to $\theta^* \approx 1.2$. The NBGP is locally indeterminate for any $\sigma \in (\tilde{\sigma}, \bar{\sigma})$ with $\tilde{\sigma} \approx 0.8894345$ and $\bar{\sigma} \approx 1.07066$.
(ii) Let $\delta = 0.95$, $\beta_1 = 0.35$ and $\alpha_1 = 0.88$. The consumption good is thus capital intensive at the private level and $\theta^* \approx 3.0075$. The NBGP is locally indeterminate for any $\sigma \in (\tilde{\sigma}, \bar{\sigma})$ with $\tilde{\sigma} \approx 0.953417$ and $\bar{\sigma} \approx 1.07$. Moreover, $\bar{\sigma}$ is a flip bifurcation value so that there exist locally indeterminate (resp. determinate) period-2 cycles in a right (resp. left) neighborhood of $\bar{\sigma}$.

17.5 Concluding Comments

We have considered a discrete-time two-sector endogenous growth model with Cobb-Douglas technologies augmented to include labor-augmenting global externalities. We have proved the existence of a normalized balanced growth path and we have shown that sunspot fluctuations arise under both types of capital intensity configuration at the private level provided the elasticity of intertemporal substitution in consumption admits intermediary values. Moreover, the dynamics of growth rates exhibits period-2 cycles if the consumption good is capital intensive at the private level.

17.6 Appendix

17.6.1 Proof of Lemma 1

The Lagrangian associated with program (17.1) is:

$$\mathcal{L} = K_{ct}^{\alpha_1}(e_t L_{ct})^{\alpha_2} + p_t[AK_{yt}^{\beta_1}(e_t L_{yt})^{\beta_2} - k_{t+1}] + \omega_t[1 - L_{ct} - L_{yt}]$$
$$+ r_t[k_t - K_{ct} - K_{yt}].$$

The first order conditions are:

$$\alpha_1 c_t / K_{cc} = p_t \beta_1 y_t / K_{yt} = r_t \tag{17.16}$$

$$\alpha_2 c_t / L_{ct} = p_t \beta_2 y_t / L_{yt} = w_t. \tag{17.17}$$

Solving $y_t = AK_{yt}^{\beta_1}(k_t L_{yt})^{\beta_2}$ with respect to L_{yt} gives

$$L_{yt} = (y_t/A)^{1/\beta_2} k_t^{-1} K_{yt}^{-\beta_1/\beta_2}. \tag{17.18}$$

Using $K_{ct} = k_0 - K_{yt}$, $L_{yt} = 1 - L_{ct}$, and merging (17.16)–(17.18) we get:

$$\frac{\alpha_1\beta_2}{\alpha_2\beta_1} = \frac{K_{ct}}{L_{ct}}\frac{L_{yt}}{K_{yt}} = \frac{k_t - K_{yt}}{1-(y_t/A)^{1/\beta_2}k_t^{-1}K_{yt}^{-\beta_1/\beta_2}}\frac{(y_t/A)^{1/\beta_2}k_t^{-1}K_{yt}^{-\beta_1/\beta_2}}{K_{yt}}. \tag{17.19}$$

The solution K_{yt}^* is obtained as an implicit linear homogeneous function $\bar{g}(k_t, y_t)$ as shown in Drugeon and Venditti (1998).[6] We then derive from (17.16)

$$r_t = A\alpha_1 \left(\frac{k_t L_{ct}}{K_{ct}}\right)^{\alpha_2}, \quad p_t = \frac{r_t K_{yt}}{\beta_1 y_t}.$$

Using (17.18) and (17.19) with $y_t = k_{t+1}$ gives the final results. □

17.6.2 *Proof of Proposition 1*

Consider the Euler equation (17.11):

$$\delta\beta_1 g(\theta_{t+1}^{-1})^{-\alpha_2/\beta_2} = g(\theta_t^{-1})^{1-(\alpha_2/\beta_2)}\theta_t^{\sigma}$$

Along a balanced growth path with $\theta_t = \theta_{t+1} = \theta$, we get $\delta\beta_1 = g(\theta^{-1})\theta^{\sigma}$. Now consider (17.19) with $K_y = \bar{g}(k_t, k_{t+1}) = k_{t+1}\bar{g}(k_t/k_{t+1}, 1) \equiv k_{t+1}g(\theta_t^{-1})$. We derive

$$\frac{\alpha_1\beta_2}{\alpha_2\beta_1} = \frac{\theta_t^{-1} - g(\theta_t^{-1})}{A^{1/\beta_2}\theta_t^{-1}g(\theta_t^{-1})^{1/\beta_2} - g(\theta_t^{-1})}. \tag{17.20}$$

If $\theta_t = \theta$ and thus $g(\theta^{-1}) = \delta\beta_1\theta^{-\sigma}$ we get after simplifications

$$\frac{\alpha_1\beta_2}{\alpha_2\beta_1} = \frac{1-\delta\beta_1\theta^{1-\sigma}}{A^{1/\beta_2}(\delta\beta_1\theta_t^{-\sigma})^{1/\beta_2} - \delta\beta_1\theta^{1-\sigma}}.$$

[6] See also Drugeon et al. (2003).

It follows that $\theta = (\delta\beta_1)^{-1} \equiv \theta^*$ is a solution of this equation if and only if the normalization constant A satisfies $A = A^*$ as defined by (17.12). Along the stationary balanced growth path $k_t = (\delta\beta_1)^{-t}\hat{k}_0$, using the fact that $c(k, y)$ is homogeneous of degree one and $p(k, y)$ is homogeneous of degree zero, the transversality condition (17.8) becomes

$$\delta^{-1}\hat{k}_0^{1-\sigma} c(\delta\beta_1, 1)^{-\sigma} p(\delta\beta_1, 1) \lim_{t\to+\infty} (\delta^{\sigma}\beta_1^{\sigma-1})^{t+1} = 0.$$

It will be satisfied if $\delta^{\sigma} < \beta_1^{1-\sigma}$ or equivalently $\sigma > \tilde{\sigma} = ln\beta_1/ln(\delta\beta_1) > 0$. Similarly the summability condition (17.9) becomes

$$\delta^{-1}\hat{k}_0^{1-\sigma} \frac{[c(\delta\beta_1, 1)]^{1-\sigma}}{1-\sigma} \sum_{t=0}^{\infty} (\delta^{\sigma}\beta_1^{\sigma-1})^{t+1} < +\infty$$

and is satisfied if $\delta^{\sigma} < \beta_1^{1-\sigma}$ or equivalently $\sigma > \tilde{\sigma} = ln\beta_1/ln(\delta\beta_1) > 0$. Finally, the feasibility condition requires $(\delta\beta_1)^{-1} \leq A^*$ or equivalently

$$\frac{\alpha_2\beta_1}{\alpha_1\beta_2} \frac{1-(\delta\beta_1)^{\sigma}\left(1-\frac{\alpha_1\beta_2}{\alpha_2\beta_1}\right)}{(\delta\beta_1)^{\sigma/\beta_2}} \geq 1$$

$$\Leftrightarrow 1-(\delta\beta_1)^{\sigma} + (\delta\beta_1)^{\sigma}\frac{\alpha_1\beta_2}{\alpha_2\beta_1}\left[1-(\delta\beta_1)^{(\sigma\beta_1)/\beta_2}\right] \geq 0.$$

This inequality is always satisfied. □

17.6.3 Proof of Lemma 2

Notice first that $g(\theta^{*-1}) = g(\delta\beta_1) = (\delta\beta_1)^{1+\sigma}$. Total differentiation of the Euler equation (17.11) around $\theta = \theta^*$ with $A = A^*$ gives

$$\frac{d\theta_{t+1}}{d\theta_t} = \frac{\sigma\beta_2(\delta\beta_1)^{\sigma}}{\alpha_2 g'(\theta^{*-1})} + \frac{\beta_2-\alpha_2}{\alpha_2}. \tag{17.21}$$

Consider now (17.20) and let us denote $X = \theta^{-1}$. We get:

$$\frac{\alpha_1\beta_2}{\alpha_2\beta_1} = \frac{X-g(X)}{A^{1/\beta_2}Xg(X)^{1/\beta_2}-g(X)}.$$

Total differentiation with respect to X gives

$$\frac{dg}{dX} = g'(X) = \frac{(\delta\beta_1)^{\sigma}\left(1-\frac{\alpha_1\beta_2}{\alpha_2\beta_1}\right)}{1-\frac{\alpha_1\beta_2}{\alpha_2\beta_1}+\frac{1-(\delta\beta_1)^{\sigma}\left(1-\frac{\alpha_1\beta_2}{\alpha_2\beta_1}\right)}{\beta_2(\delta\beta_1)^{\sigma}}}.$$

Substituting this expression in (17.21) and using the fact that $\beta_2 - \alpha_2 = -\alpha_2\beta_1(1 - \alpha_1\beta_2/\alpha_2\beta_1)$ give the final result. □

17.6.4 Proof of Proposition 2

Consider the expression (17.13) which may be expressed as follows

$$\frac{d\theta_{t+1}}{d\theta_t} = \frac{\sigma}{\alpha_2}\left[\frac{\left(1-\frac{\alpha_1\beta_2}{\alpha_2\beta_1}\right)^{-1}}{(\delta\beta_1)^{\sigma}} - \beta_1\right] - \beta_1\left(1-\frac{\alpha_1\beta_2}{\alpha_2\beta_1}\right). \tag{17.22}$$

Local indeterminacy of the NBGP will be obtained if and only if $d\theta_{t+1}/d\theta_t \in (-1, 1)$. We first obtain that

$$\lim_{\sigma\to+\infty}\frac{d\theta_{t+1}}{d\theta_t} = +\infty \Leftrightarrow 1-\frac{\alpha_1\beta_2}{\alpha_2\beta_1} > 0 \tag{17.23}$$
$$= -\infty \Leftrightarrow 1-\frac{\alpha_1\beta_2}{\alpha_2\beta_1} < 0.$$

Let $\sigma = \tilde{\sigma}$, or equivalently $(\delta\beta_1)^{\tilde{\sigma}} = \beta_1$, so that (17.13) becomes

$$\left.\frac{d\theta_{t+1}}{d\theta_t}\right|_{\sigma=\tilde{\sigma}} = \frac{1}{\alpha_2\left(1-\frac{\alpha_1\beta_2}{\alpha_2\beta_1}\right)}\left\{\frac{\tilde{\sigma}}{\beta_1}\left[1-\beta_1^2\left(1-\frac{\alpha_1\beta_2}{\alpha_2\beta_1}\right)\right] - \alpha_2\beta_1\left(1-\frac{\alpha_1\beta_2}{\alpha_2\beta_1}\right)^2\right\}.$$

(i) Consider first the case in which the investment good is capital intensive at the private level, i.e. $1 - \alpha_1\beta_2/\alpha_2\beta_1 > 0$. We get

$$\left.\frac{d\theta_{t+1}}{d\theta_t}\right|_{\sigma=\tilde{\sigma}} > -1$$
$$\Leftrightarrow \frac{\tilde{\sigma}}{\beta_1}\left[1-\beta_1^2\left(1-\frac{\alpha_1\beta_2}{\alpha_2\beta_1}\right)\right] + \alpha_2\left(1-\frac{\alpha_1\beta_2}{\alpha_2\beta_1}\right)\left[1-\beta_1\left(1-\frac{\alpha_1\beta_2}{\alpha_2\beta_1}\right)\right] > 0.$$

This inequality is always satisfied since $1 - \alpha_1\beta_2/\alpha_2\beta_1 < 1$. We also have

$$\left.\frac{d\theta_{t+1}}{d\theta_t}\right|_{\sigma=\tilde{\sigma}} < 1$$
$$\Leftrightarrow \frac{\tilde{\sigma}}{\beta_1}\left[1-\beta_1^2\left(1-\frac{\alpha_1\beta_2}{\alpha_2\beta_1}\right)\right] - \alpha_2\left(1+\frac{\alpha_1\beta_2}{\alpha_2\beta_1}\right)\left[1-\beta_1\left(1-\frac{\alpha_1\beta_2}{\alpha_2\beta_1}\right)\right] < 0.$$

This last inequality corresponds to condition (17.14). If it holds we conclude from (17.23) that there exists $\bar{\sigma} > \tilde{\sigma}$ such that the NBGP is locally indeterminate when $\sigma \in (\tilde{\sigma}, \bar{\sigma})$.

(ii) Consider now the case in which the consumption good is capital intensive at the private level, i.e. $1 - \alpha_1\beta_2/\alpha_2\beta_1 < 0$. We get

$$\left.\frac{d\theta_{t+1}}{d\theta_t}\right|_{\sigma=\tilde{\sigma}} > -1$$
$$\Leftrightarrow \frac{\tilde{\sigma}}{\beta_1}\left[1 - \beta_1^2\left(1 - \frac{\alpha_1\beta_2}{\alpha_2\beta_1}\right)\right] + \alpha_2\left(1 - \frac{\alpha_1\beta_2}{\alpha_2\beta_1}\right)\left[1 - \beta_1\left(1 - \frac{\alpha_1\beta_2}{\alpha_2\beta_1}\right)\right] < 0.$$

This last inequality corresponds to condition (17.15). If it holds we conclude from (17.23) that there exist $\underline{\sigma} \geq \tilde{\sigma}$ and $\bar{\sigma} > \underline{\sigma}$ such that the NBGP is locally indeterminate when $\sigma \in (\underline{\sigma}, \bar{\sigma})$. We may indeed have $\underline{\sigma} > \tilde{\sigma}$ since it may be the case that $d\theta_{t+1}/d\theta_t|_{\sigma=\tilde{\sigma}} > 1$. □

17.6.5 *Proof of Corollary 1*

Let the consumption good be capital intensive at the private level and condition (17.15) holds. Then $d\theta_{t+1}/d\theta_t|_{\sigma=\tilde{\sigma}} > -1$. Considering (17.23) we derive from Proposition 2 that $d\theta_{t+1}/d\theta_t|_{\sigma=\bar{\sigma}} = -1$ and $d\theta_{t+1}/d\theta_t < -1$ when $\sigma > \bar{\sigma}$. Therefore $\bar{\sigma}$ is a flip bifurcation value (see Ruelle 1989) and the result follows. □

References

Benhabib, J. and R. Farmer (1996), "Indeterminacy and Sector-specific Externalities," *Journal of Monetary Economics*, 37, 397–419.

Benhabib, J., Meng, Q. and K. Nishimura (2000), "Indeterminacy under Constant Returns to Scale in Multisector Economies," *Econometrica*, 68, 1541–1548.

Benhabib, J. and K. Nishimura (1985), "Competitive Equilibrium Cycles", *Journal of Economic Theory*, 35, 284–306.

Benhabib, J. and R. Perli (1994), "Uniqueness and Indeterminacy: Transitional Dynamics in a Model of Endogenous Growth," *Journal of Economic Theory*, 63, 113–142.

Boldrin, M., Nishimura, K., Shigoka, T. and M. Yano (2001), "Chaotic Equilibrium Dynamics in Endogenous Growth Models," *Journal of Economic Theory*, 96, 97–132.

Boldrin, M. and A. Rustichini (1994), "Growth and Indeterminacy in Dynamic Models with Externalities," *Econometrica*, 62, 323–342.

Bond, E., Wang, P. and C. Yip (1996), "A General Two-sector Model of Endogenous Growth with Human and Physical Capital: Balanced Growth and Transitional Dynamics," *Journal of Economic Theory*, 68, 149–173.

Drugeon, J.P. (2004), Optimal Cycles of Unbounded Growth, mimeo, EUREQua.

Drugeon, J.P. and A. Venditti (1998), Intersectoral External Effects, Multiplicities and Indeterminacies II. The Long-run Growth Case, GREQAM Working paper 98A20.

Drugeon, J.P., Poulsen, O. and A. Venditti (2003), "On Intersectoral Allocations, Factors Substitutability and Multiple Long-run Growth Paths," *Economic Theory*, 21, 175–183.
Goenka, A. and O. Poulsen (2004), "Indeterminacy and Labor Augmenting Externalities," *Journal of Economics*, 10, 61–82.
Lucas, R. (1988), "On the Mechanics of Economic Development," *Journal of Monetary Economics*, 22, 3–42.
Mino, K. (2001), "Indeterminacy and Endogenous Growth with Social Constant Returns," *Journal of Economic Theory*, 97, 203–222.
Mino, K., Nishimura, K., Shimomura, K. and P. Wang (2008), "Equilibrium Dynamics in Discrete-time Endogenous Growth Models with Social Constant Returns," *Economic Theory*, 34, 1–23.
Mitra, T. (1998), "On Equilibrium Dynamics under Externalities in a Model of Economic Development," *Japanese Economic Review*, 49, 85–107.
K., Nishimura and A. Venditti (2004), "Indeterminacy and the Role of Factor Substitutability," *Macroeconomic Dynamics*, 8, 436–465.
Romer, P. (1986), "Increasing Returns and Long-run Growth," *Journal of Political Economy*, 94, 1002–1037.
Ruelle, D. (1989), Elements of Differentiable Dynamics and Bifurcation Theory, Academic Press, San Diego.
Xie, D. (1994), "Divergence in Economic Performance: Transitional Dynamics with Multiple Equilibria," *Journal of Economic Theory*, 63, 97–112.

Chapter 18
Homoclinic Bifurcation and Global Indeterminacy of Equilibrium in a Two-Sector Endogenous Growth Model*

Paolo Mattana, Kazuo Nishimura, and Tadashi Shigoka

18.1 Introduction

If for a given endogenous growth model, there exists a continuum of equilibria in a small neighborhood of a balanced growth path (BGP) that continues to stay in this neighborhood, it is said that equilibrium is locally indeterminate. Here a small neighborhood of a BGP means the region in which results based on linear approximations of equilibrium dynamics on the BGP are qualitatively valid. If for a given endogenous growth model, there exists a continuum of equilibria outside a small neighborhood of a BGP, it is said that equilibrium is globally indeterminate. Local indeterminacy of equilibrium in the Lucas (1988) model and related systems is now well understood owing to Chamley (1993) and Benhabib and Perli (1994). In contrast, we know only a little about global indeterminacy of equilibrium in the Lucas model and related systems. This is especially the case, when there exist multiple balanced growth paths. The purpose of the present paper is to analyze global indeterminacy of equilibrium in an endogenous growth model that belongs to the same family as the Lucas model. In the Lucas model the external effect appears in the physical-goods sector, whereas in our model, it appears in the educational sector. In our model, this external effect yields multiple balanced growth paths. We show that our model undergoes a homoclinic bifurcation and that it exhibits global indeterminacy of equilibrium. In doing so, we shall use the result of Kopell and Howard (1975) that was introduced into the economics literature by Benhabib

*International Journal of Economic Theory 5, 25–47, 2009

P. Mattana (✉)
Department of Economics, University of Cagliari, Italy
e-mail: mattana@unica.it

K. Nishimura · T. Shigoka
Institute of Economic Research, Kyoto University, Japan
e-mail: nishimura@kier.kyoto-u.ac.jp

J. Stachurski et al. (eds.), *Nonlinear Dynamics in Equilibrium Models*, 427
DOI 10.1007/978-3-642-22397-6_18,

et al. (2001) and that has been utilized in analyses of a bounded growth model by Benhabib et al. (2008). Benhabib et al. (2001) treat a macroeconomic model with money, without capital. However, their analytical framework can also be used to analyze global indeterminacy of equilibrium in models with capital, when there exist either multiple steady states such as in Benhabib et al. (2008) or multiple balanced growth paths as in the present paper.

Consider the following endogenous growth model that generates the Uzawa model, the Lucas model and our model according to the choice of parameters.

$$\underset{C(t),u(t)}{\text{Max}} \int_0^\infty \frac{C^{1-\sigma}-1}{1-\sigma} e^{-\rho t} dt, \tag{P}$$

subject to

$$\begin{aligned}
\dot{K} &= AK^\beta (uH)^{1-\beta} H_a^\gamma - C \\
\dot{H} &= \delta H (1-u)^\alpha (1-u)_a^\eta \\
K(0) &= K_0 > 0 \\
H(0) &= H_0 > 0,
\end{aligned}$$

where C is consumption, K is physical capital, H is human capital, and u is the fraction of labor allocated to the production of physical capital with $0 \leq u \leq 1$. H_a is the average level of human capital, $(1-u)_a$ is the average level of investment in human capital, the last two lines are initial conditions for K and H, and $(A, \beta, \delta, \alpha, \rho, \sigma, \gamma, \eta)$ is a set of control parameters. ρ is a time preference rate, and σ is the inverse of the intertemporal elasticity of substitution. We assume that $(A, \delta, \rho) \in R^3_{++}$, $0 < \beta < 1$, $0 < \alpha \leq 1$, and $(\sigma, \gamma, \eta) \in R^3_+$, where R_{++} is a set of strictly positive real numbers, and R_+ is a set of non-negative real numbers. If $\gamma = \eta = 0$, the model reduces to an optimal endogenous growth model. If $\gamma > 0$, externalities arise between the *stocks* of human capital of different individuals, and when the representative agent solves problem (P), the agent treats H_a^γ as parametrically given, and $H_a = H$ holds in equilibrium. If $\eta > 0$, externalities arise between the flows of human capital accumulation by different individuals, and when the representative agent solves problem (P), the agent treats $(1-u)_a^\eta$ as parametrically given, and $(1-u)_a = (1-u)$ holds in equilibrium. We have the following results on global dynamics in this model.

1. Suppose $\gamma = \eta = \sigma = 0$ and $0 < \alpha < 1$. Then, equilibrium is determinate. Uzawa (1965) shows that if an optimal BGP exists, it is unique and globally stable, not asymptotically, but in the sense that the economy reaches the BGP during a finite length of time. He treats more general production functions without external effects.
2. Suppose $\gamma = \eta = 0$, $\sigma > 0$ and $0 < \alpha \leq 1$. Then, equilibrium is determinate. Caballe and Santos (1993) show that if an optimal BGP exists, it is unique and globally asymptotically stable. They treat more general production functions without external effects.

3. Suppose $\alpha = 1$, $\eta = 0$, $\sigma > 0$ and $\gamma > 0$. This is the Lucas (1988) model. If a BGP exists, it is unique. Benhabib and Perli (1994, p. 124) point out the possibility of a Hopf bifurcation in this model that implies global indeterminacy of equilibrium.[1] Mattana and Venturi (1999) and Mattana (2004) show that the model undergoes a supercritical Hopf bifurcation for some parameter values, and that it undergoes a subcritical Hopf bifurcation for other parameter values.
4. Suppose $\alpha = 1$, $\eta = 0$, $\sigma > 0$, and $\gamma > 0$. Xie (1994) shows that if $\gamma > \beta$, and $\delta < \rho < \delta(1 + \gamma - \beta)$, and if $\sigma = \beta$, a unique BGP exists, equilibrium is globally indeterminate and the economy converges to this BGP asymptotically.

Lucas (1988) assumes that $\alpha = 1, \sigma > 0, \gamma > 0$ and $\eta = 0$. In contrast, the present paper assumes that $\alpha = 1$, $\sigma > 0$, $\gamma = 0$ and $\eta > 0$. In the Lucas model, the external effect appears in the physical-goods sector, whereas in our model, it appears in the educational sector. In the Lucas model, if a BGP exists, it is unique, whereas in our model, the external effect on human capital accumulation yields multiple balanced growth paths. Chamley (1993) has already pointed out that this type of external effect can yield multiple balanced growth paths,[2] but he does not analyze global equilibrium dynamics. Since our specification of the external effect is simpler and more tractable than his specification, we are able to analyze the global equilibrium dynamics.

The remainder of the paper is organized as follows. Section 18.2 illustrates a model. Section 18.3 characterizes multiple balanced growth paths. Section 18.4 shows that our model undergoes a homoclinic bifurcation. Section 18.5 shows that our model exhibits global indeterminacy of equilibrium. Section 18.6 briefly concludes.

18.2 Model

Our model belongs to the same family as the Lucas model. In the Lucas model, externalities arise between the *stocks* of human capital of different individuals, whereas in our model, externalities arise between the *flows* of human capital accumulation by different individuals. Formally, the model is given as

$$\underset{C(t),u(t)}{\text{Max}} \int_0^\infty \frac{C^{1-\sigma} - 1}{1 - \sigma} e^{-\rho t} dt, \tag{P1}$$

[1] Kopell and Howard (1975, p. 307) suggest that the hypotheses of Hopf bifurcation are local and easily verifiable assumptions about a one-parameter family of equations but that the resulting conclusions are global: the existence of a periodic orbit outside a small neighborhood of a steady state.

[2] See also Sect. 3 of Benhabib and Perli (1994).

subject to

$$\dot{K} = AK^{\beta}(uH)^{1-\beta} - C$$
$$\dot{H} = \delta H(1-u)(1-u)_a^{\eta}$$
$$K(0) = K_0 > 0$$
$$H(0) = H_0 > 0,$$

where $0 \leq u \leq 1$, and $(1-u)_a^{\eta} > 0$. The meanings of the symbols are the same as in the previous section. $(A, \beta, \delta, \eta, \rho, \sigma)$ is a set of control parameters. $K(t) = K_t$ and $H(t) = H_t$ are predetermined state variables, and $C(t) = C_t$ and $u(t) = u_t$ are non-predetermined control variables. When the representative agent solves problem (P1), the agent treats $(1-u(t))_a^{\eta} = (1-u_t)_a^{\eta}$ as a parametrically given function of time, and $(1-u_t)_a = (1-u_t)$ holds in equilibrium.

Problem (P1) is solved by defining the current value Hamiltonian

$$\mathscr{H} = \frac{C^{1-\sigma}-1}{1-\sigma} + \lambda_1(AK^{\beta}(uH)^{1-\beta} - C) + \lambda_2\delta H(1-u)(1-u)_a^{\eta},$$

where we assume that an optimal solution satisfies $C > 0$, $0 < u < 1$, and $\beta(1-u) - \eta u \neq 0$, respectively. From the first order conditions, $C^{-\sigma} = \lambda_1$ and $(1-\beta)\lambda_1 AK^{\beta}u^{-\beta}H^{1-\beta} = \lambda_2\delta H(1-u)_a^{\eta}$, we can easily check the maximized Hamiltonian is linear in K and H, and so it is weakly concave in K and H. Thus it satisfies Arrow's condition.[3] Therefore the following system of equations constitutes the optimal solution of problem (P1) for the representative agent:

$$C^{-\sigma} = \lambda_1$$
$$\lambda_2\delta(1-u)_a^{\eta} = (1-\beta)\lambda_1 AK^{\beta}u^{-\beta}H^{-\beta}$$
$$\dot{K} = AK^{\beta}(uH)^{1-\beta} - C$$
$$\dot{H} = \delta H(1-u)(1-u)_a^{\eta}$$
$$\dot{\lambda}_1 = \rho\lambda_1 - \lambda_1\beta AK^{\beta-1}(uH)^{1-\beta}$$
$$\dot{\lambda}_2 = \rho\lambda_2 - \lambda_1(1-\beta)AK^{\beta}u^{1-\beta}H^{-\beta} - \lambda_2\delta(1-u)(1-u)_a^{\eta}$$
$$K(0) = K_0 > 0,\ H(0) = H_0 > 0.$$
$$\lim_{t\to\infty}\lambda_1(t)K(t)e^{-\rho t} = 0, \quad \lim_{t\to\infty}\lambda_2(t)H(t)e^{-\rho t} = 0.$$

K and H are predetermined variables whose boundary conditions are given by initial conditions. The last line represents the transversality conditions. λ_1 and λ_2 are non-predetermined variables whose boundary conditions are given by the transversality conditions.

[3] See Kamien and Schwartz (1991, pp. 221–222) for Arrow's condition. Under this condition, it is straightforward to prove the statement in the next sentence.

The market equilibrium condition is given by $(1-u)_a = (1-u)$. From the first six equations, eliminate λ_1 and λ_2 and substitute $(1-u)$ for $(1-u)_a$, to obtain the following four-dimensional autonomous ordinary differential equation[4]:

$$\begin{aligned}
\dot{K} &= AK^{\beta}(uH)^{1-\beta} - C \\
\dot{H} &= \delta H(1-u)^{1+\eta} \\
\dot{C} &= \frac{\beta}{\sigma}AK^{\beta-1}(uH)^{1-\beta}C - \frac{\rho}{\sigma}C \\
\dot{u} &= \frac{u(1-u)}{\beta(1-u)-\eta u}(\delta(1-u)^{\eta} - \beta\delta(1-u)^{1+\eta} - \beta\frac{C}{K}).
\end{aligned} \tag{18.1}$$

We also have the following relation that we shall use later when we treat the transversality condition:

$$\dot{\lambda}_2 = \rho\lambda_2 - \lambda_2\delta(1-u)^{\eta}. \tag{18.2}$$

In (18.1), K and H are predetermined variables, and C and u are non-predetermined variables. The equilibrium dynamics are described by the autonomous differential equation (18.1), the initial conditions and the transversality conditions. Suppose that a BGP exists, and that μ_K, μ_H, μ_C, and μ_u are balanced growth rates of K, H, C, and u, respectively. Then from (18.1), we have $\mu_K = \mu_H = \mu_C \geq 0$ and $\mu_u = 0$. Let u^* be the value of u corresponding to the BGP such that $0 < u^* < 1$ with $u^* \neq \frac{\beta}{\beta+\eta}$; then we have $\mu_K = \mu_H = \mu_C = \delta(1-u^*)^{1+\eta} > 0$.

To simplify the dynamics of (18.1), we introduce the following notation,

$$X = \frac{K}{H}, \quad Q = \frac{C}{K}, \quad \mathbf{B} = \begin{bmatrix} -1 & 1 & 0 \\ 0 & -1 & 1 \end{bmatrix},$$
$$k = \log K, \quad h = \log H, \quad c = \log C, \quad x = \log X, \quad q = \log Q, \tag{18.3}$$

where X is a predetermined variable and Q is a non-predetermined variable. Then, the four-dimensional autonomous ordinary differential equation (18.1) can be rewritten in the following way:

$$\begin{aligned}
\dot{X} &= AX^{\beta}u^{1-\beta} - QX - \delta X(1-u)^{1+\eta}, \\
\dot{u} &= \frac{u(1-u)}{\beta(1-u)-\eta u}(\delta(1-u)^{\eta} - \beta\delta(1-u)^{1+\eta} - \beta Q), \\
\dot{Q} &= \frac{\beta-\sigma}{\sigma}AX^{\beta-1}u^{1-\beta}Q - \frac{\rho}{\sigma}Q + Q^2,
\end{aligned} \tag{18.4}$$

---

[4]Note that we have assumed $\beta(1-u) - \eta u \neq 0$.

and

$$h(0) = h_0, \quad \dot{h} = \delta(1-u)^{1+\eta}, \quad \mathbf{B}\begin{bmatrix} h \\ k \\ c \end{bmatrix} = \begin{bmatrix} x \\ q \end{bmatrix}. \tag{18.5}$$

Therefore, the four-dimensional system (18.1) is decomposed into the three-dimensional stationary autonomous component (18.4) and a log-linear component with one equation of motion (18.5). By construction, if the stationary component (18.4) has an interior steady state, there exists a BGP in (18.1) corresponding to the steady state of (18.4). In the next section, we shall show that the transversality conditions are satisfied on a BGP, and we shall provide sufficient conditions to establish the existence of a BGP in System (18.1).

Suppose that $(X, u, Q) = (X^*, u^*, Q^*) \in R_{++} \times (0,1) \times R_{++}$ with $u^* \neq \frac{\beta}{\beta+\eta}$ is a steady state of System (18.4). Then the BGP in System (18.1) is given by the following one-dimensional manifold in R^3.

$$\left\{ \begin{bmatrix} h \\ k \\ c \end{bmatrix} \in R^3 : \mathbf{B}\begin{bmatrix} h \\ k \\ c \end{bmatrix} = \begin{bmatrix} x^* \\ q^* \end{bmatrix} \right\},$$

where $x^* = \log X^*$ and $q^* = \log Q^*$. Movements of System (18.1) on the BGP are completely characterized by the following equations.

$$h(0) = h_0, \quad \dot{h}_t = \delta(1-u^*)^{1+\eta}, \quad \mathbf{B}\begin{bmatrix} h_t \\ k_t \\ c_t \end{bmatrix} = \begin{bmatrix} x^* \\ q^* \end{bmatrix}. \tag{18.6}$$

Note that $\dot{h}_t = \delta(1-u^*)^{1+\eta}$ has a center eigenvalue, because the right-hand side of the equation is constant.

The transitional dynamics around the BGP are given by System (18.4). Note that this system includes one predetermined variable and two non-predetermined variables. We define local indeterminacy of equilibrium and global indeterminacy of equilibrium in the following way.

Definition 1. If System (18.1) equipped with the initial conditions and the transversality conditions has a continuum of solutions, it is said that this system exhibits indeterminacy of equilibrium. In this case, it is also said that the system composed of (18.4) and (18.5) exhibits indeterminacy of equilibrium.

(i) Suppose that $(X^*, u^*, Q^*) \in R_{++} \times (0,1) \times R_{++}$ with $u^* \neq \frac{\beta}{\beta+\eta}$ is a steady state of system (18.4), and consider the Jacobian of the right-hand side of (18.4) evaluated at (X^*, u^*, Q^*). If this Jacobian is hyperbolic, and if the number of stable roots is greater then one, it is said that equilibrium is locally indeterminate.

(ii) If equilibrium is indeterminate for a reason different from the case of local indeterminacy, it is said that equilibrium is globally indeterminate.

18.3 Multiple Balanced Growth Paths

The purpose of this section is to establish conditions under which there exist multiple balanced growth paths. Let (X^*, u^*, Q^*) be values of $(X, u, Q) \in R_{++} \times (0,1) \times R_{++}$ with $u \neq \frac{\beta}{\beta+\eta}$ such that $\dot{X} = \dot{u} = \dot{Q} = 0$. These values are found by setting the right-hand side of (18.4) to zero, where $(X^*, u^*, Q^*) \in R_{++} \times (0,1) \times R_{++}$ with $u^* \neq \frac{\beta}{\beta+\eta}$. Appendix I establishes that u^* satisfies the following equation:

$$\sigma(1-u^*)^{1+\eta} - (1-u^*)^{\eta} + \frac{\rho}{\delta} = 0. \tag{18.7}$$

For a balanced growth path to constitute a valid equilibrium, one must also establish that transversality conditions are satisfied; that is,

$$\lim_{t\to\infty} \lambda_1(t)K(t)e^{-\rho t} = 0, \tag{18.8}$$

$$\lim_{t\to\infty} \lambda_2(t)H(t)e^{-\rho t} = 0. \tag{18.9}$$

If we set

$$Z^* := AX^{*\beta-1}u^{*1-\beta} = \frac{\delta}{\beta}(1-u^*)^{\eta} \tag{18.10}$$

as in (18.27) of Appendix I, then it follows from (18.1), (18.2), (18.10) and the first equation of (18.26) in Appendix I that the following relations hold on the BGP.

$$\begin{aligned} -\rho + \frac{\dot{\lambda}_1}{\lambda_1} + \frac{\dot{K}}{K} &= -\rho - \sigma\frac{\dot{C}}{C} + \frac{\dot{K}}{K} = -\beta Z^* + \delta(1-u^*)^{1+\eta} \\ &= -\delta(1-u^*)^{\eta}u^* < 0. \\ -\rho + \frac{\dot{\lambda}_2}{\lambda_2} + \frac{\dot{H}}{H} &= -\rho + \rho - \delta(1-u^*)^{\eta} + \delta(1-u^*)^{1+\eta} \\ &= -\delta(1-u^*)^{\eta}u^* < 0. \end{aligned}$$

Since the left hand sides of these expressions represent the rates of change of (18.8) and (18.9), and since these rates of change are negative, we have established that the transversality conditions are satisfied on the BGP.

We now turn to the question: Do there exist multiple balanced growth paths? To answer this question, let $l = 1 - u$, and let $f : [0, 1] \to R$ be defined as

$$f(l) = \sigma l^{1+\eta} - l^{\eta} + \frac{\rho}{\delta}. \tag{18.11}$$

If $f(l^*) = 0$, $f(1) \neq 0$ and $l^* \neq \frac{\eta}{\beta+\eta}$, then $1 - l^*$ is a steady state value of u^* from (18.7) and (18.11). At a steady state, by construction, we have[5]

$$X^* = A^{\frac{1}{1-\beta}} u^* \left[\frac{\delta}{\beta}(1-u^*)^{\eta}\right]^{\frac{-1}{1-\beta}},$$

$$Q^* = \frac{\delta}{\beta}(1-u^*)^{\eta} - \delta(1-u^*)^{1+\eta}.$$

By differentiating the function $f(l)$ (defined in Relation (18.11)), under the assumption $0 < \eta \leq 1$, we can establish the following for each $l \in (0, 1]$:

$$f'(l) = l^{\eta-1}(\sigma(1+\eta)l - \eta)$$

$$f''(l) = \sigma\eta(1+\eta)l^{\eta-1} + \eta(1-\eta)l^{\eta-2}.$$

$f''(l)$ is positive for each $l \in (0, 1]$, and $f(l)$ is continuous at $l = 0$. Therefore, $f = f(l)$ is a convex function. If $0 < \frac{\eta}{\sigma(1+\eta)} < 1$, then $f(l)$ takes the global minimum value at $l = \frac{\eta}{\sigma(1+\eta)}$ given these facts. $f\left(\frac{\eta}{\beta+\eta}\right) = \sigma\left(\frac{\eta}{\beta+\eta}\right)^{1+\eta} - \left(\frac{\eta}{\beta+\eta}\right)^{\eta} + \frac{\rho}{\delta}$, and $f(1) = \sigma - 1 + \frac{\rho}{\delta}$. So we make the following assumption.

Assumption 1 $0 < \eta \leq 1$, $0 < \frac{\eta}{\sigma(1+\eta)} < 1$, $\frac{\eta}{\sigma(1+\eta)} \neq \frac{\eta}{\beta+\eta}$, $\sigma\left(\frac{\eta}{\beta+\eta}\right)^{1+\eta} - \left(\frac{\eta}{\beta+\eta}\right)^{\eta} + \frac{\rho}{\delta} \neq 0$, *and* $\sigma - 1 + \frac{\rho}{\delta} > 0$.

We now establish that there is a critical value $\bar{\rho}$ such that for all ρ less than $\bar{\rho}$ there are two balanced growth paths. Furthermore, our earlier analysis implies that on each of them the transversality conditions are satisfied. Let $1 - \bar{u} = \frac{\eta}{\sigma(1+\eta)}$, and we have

$$\begin{aligned} f(1-\bar{u}) &= \sigma\left(\frac{\eta}{\sigma(1+\eta)}\right)^{1+\eta} - \left(\frac{\eta}{\sigma(1+\eta)}\right)^{\eta} + \frac{\rho}{\delta} \\ &= \left(\frac{\eta}{\sigma(1+\eta)}\right)^{\eta}\left(\frac{\eta}{1+\eta} - 1\right) + \frac{\rho}{\delta} \\ &= \frac{\rho}{\delta} - \left(\frac{\eta}{\sigma}\right)^{\eta}\left(\frac{1}{1+\eta}\right)^{1+\eta}. \end{aligned}$$

[5] See (18.10), and the second Equation of (18.26) in Appendix I.

If $\left(\frac{\eta}{\sigma}\right)^{\eta}\left(\frac{1}{1+\eta}\right)^{1+\eta} = \frac{\rho}{\delta}$, then $f'(1-\bar{u}) = 0$ and $f(1-\bar{u}) = 0$. Let $\bar{\rho} = \bar{\rho}(\sigma, \delta, \eta)$ be defined as

$$\bar{\rho}(\sigma, \delta, \eta) := \delta\left(\frac{\eta}{\sigma}\right)^{\eta}\left(\frac{1}{1+\eta}\right)^{1+\eta}. \tag{18.12}$$

We can now state the following proposition.

Proposition 1. *Under Assumption 1, we have:*

(i) If $\bar{\rho}(\sigma, \delta, \eta) > \rho$, then there exist two balanced growth paths. On each BGP, the transversality conditions are satisfied.
(ii) If $\bar{\rho}(\sigma, \delta, \eta) = \rho$, then there exists a unique BGP. On the BGP, the transversality conditions are satisfied.
(iii) If $\bar{\rho}(\sigma, \delta, \eta) < \rho$, then there exists no BGP.

Suppose $\bar{\rho}(\sigma, \delta, \eta) \geq \rho$, and so at least one steady state exists for (18.4). Let **J** be a Jacobian of the right-hand side of (18.4) evaluated at a steady state $(X, u, Q) = (X^*, u^*, Q^*)$. **J** is given by the following:

$$\mathbf{J} = \begin{bmatrix} -(1-\beta)AX^{*\beta-1}u^{*1-\beta} & j[1,2] & -X^* \\ 0 & j[2,2] & -\frac{\beta u^*(1-u^*)}{\beta(1-u^*)-\eta u^*} \\ -(1-\beta)A\frac{\beta-\sigma}{\sigma}X^{*\beta-2}u^{*1-\beta}Q^* & j[3,2] & Q^* \end{bmatrix}, \tag{18.13}$$

where $j[1,2] = (1-\beta)AX^{*\beta}u^{*-\beta} + (1+\eta)\delta X^*(1-u^*)^{\eta}$, $j[2,2] = \frac{\delta u^*(1-u^*)^{\eta}}{\beta(1-u^*)-\eta u^*}((1+\gamma)\beta(1-u^*)-\gamma)$, and $j[3,2] = (1-\beta)A\frac{\beta-\sigma}{\sigma}X^{*\beta-1}u^{*-\beta}Q^*$.

We shall provide two parametric examples that satisfy $\bar{\rho}(\sigma, \delta, \eta) > \rho$ so that there exist two balanced growth paths. In the first example, $\sigma < 1$, and we set $(\delta, \beta) = (0.05, 0.25)$ according to Lucas (1988) and Benhabib and Perli (1994). However, in the second example, $\sigma > 1$, and we set $(\delta, \beta) = (0.1, 0.25)$, because a balanced growth rate μ_K of K tends to take smaller value in our model than in the Lucas model for the following two reasons.

1. In the Lucas model, $\mu_K = \frac{1-\beta+\gamma}{1-\beta}\mu_H > \mu_H$, because $\gamma > 0$. In contrast, $\mu_K = \mu_H$ in our model. Therefore for a given balanced growth rate μ_H of H, a balanced growth rate μ_K of K is smaller in our model than in the Lucas model.
2. In the Lucas model, $\mu_H = \delta(1-u^*)$, whereas $\mu_H = \delta(1-u^*)^{1+\eta}$ in our model. Since $0 < 1-u^* < 1$ and $\eta > 0$, $\delta(1-u^*)^{1+\eta} < \delta(1-u^*)$. Thus for a given u^*, a balanced growth rate μ_H of H is smaller in our model than in the Lucas model.

The upper bound of μ_K corresponding to the lower steady state is given by the expression $\delta(1-\bar{u})^{1+\eta} = \delta\left(\frac{\eta}{\sigma(1+\eta)}\right)^{1+\eta}$ which is increasing in δ and η, and decreasing in σ. So we take larger values of δ and η in the second example. In the second example, we choose values of δ and η in such a way that the balanced growth rate μ_K corresponding to the lower steady state is greater than 0.01. In the next

Fig. 18.1

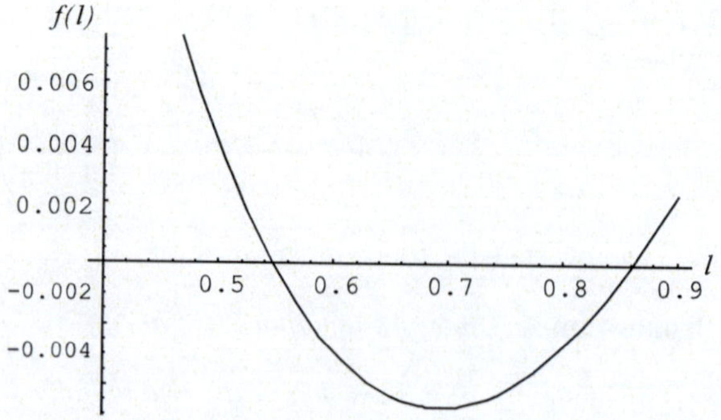

section also, we set $\delta = 0.05$ for the case $\sigma < 1$ and set $\delta = 0.1$ for the case $\sigma > 1$, by the same reason. See Mulligan and Sala-i-Martin (1993, p. 761) and Benhabib and Perli (1994, p. 123). Let l^* be a solution of the equation $f(l^*) = 0$, and let $u^* = 1 - l^*$.

Example 1. First, we consider the case $\sigma < 1$. Set $(A, \beta, \delta, \eta, \rho, \sigma) = (1,\ 0.25,\ 0.05,\ 0.2,\ \frac{77}{2{,}000},\ 0.24)$. This parametric example satisfies Assumption 1. See Fig. 18.1 for the graph of $f = f(l)$. This example has two steady states. In the lower steady state, $l^* \approx 0.547346802426937$, $u^* \approx 0.452653197573063$, and $\mu_K = \mu_H \approx 0.0242596821639116$8. The Jacobian **J** of the lower steady state has two stable roots and one unstable root, and so equilibrium is locally indeterminate near the corresponding BGP. In the higher steady state, $l^* \approx 0.860349851259535$, $u^* \approx 0.139650148740465$, and $\mu_K = \mu_H \approx 0.0417426589802533$3. The Jacobian **J** of the higher steady state has one stable root and two unstable roots, and so equilibrium is locally determinate near the corresponding BGP on which System (18.4) is saddle point stable.

Example 2. Next, we consider the case $\sigma > 1$. Set $(A, \beta, \delta, \eta, \rho, \sigma) = (1, 0.25, 0.1, 0.5, \frac{35}{1{,}000}, 1.1827)$. This parametric example satisfies Assumption 1. See Fig. 18.4 for the graph of $f = f(l)$. This example has two steady states. In the lower steady state, $l^* \approx 0.234800732899719$, $u^* \approx 0.765199267100281$, and $\mu_K = \mu_H \approx 0.0113775609903658$78. The Jacobian **J** of the lower steady state has one stable root and two unstable roots, and so equilibrium is locally determinate near the corresponding BGP on which System (18.4) is saddle point stable. In the higher steady state, $l^* \approx 0.331662160688024$, $u^* \approx 0.668337839311976$, $\mu_K = \mu_H \approx 0.0191004627272037$5. The Jacobian **J** of the higher steady state has three unstable roots, and so equilibrium is locally determinate near the corresponding BGP on which System (18.4) is totally unstable (Fig. 18.2).

Fig. 18.2

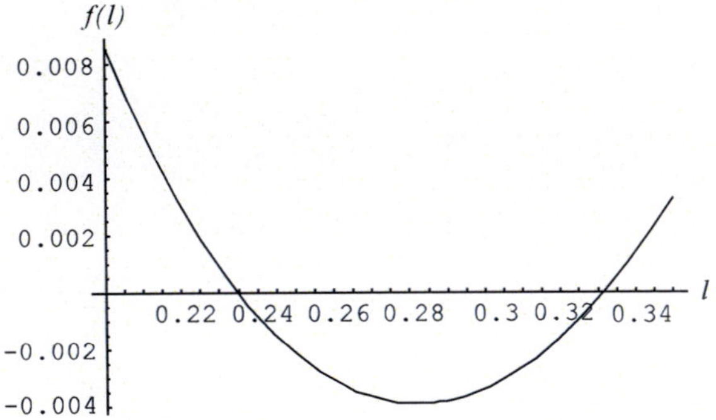

18.4 Homoclinic Orbits

In this section, we appeal to the result of Kopell and Howard (1975) to prove the existence of homoclinic orbits in a number of parametrized examples.

Let $\mathscr{L} = \mathscr{L}(R^3)$ be the set of all 3×3 real square matrices. Let **A** and **I** be elements of $\mathscr{L}(R^3)$ given by

$$\mathbf{A} = \begin{bmatrix} a_1 & a_2 & a_3 \\ b_1 & b_2 & b_3 \\ c_1 & c_2 & c_3 \end{bmatrix} \in \mathscr{L}(R^3), \quad \text{and } \mathbf{I} = \begin{bmatrix} 1 & 0 & 0 \\ 0 & 1 & 0 \\ 0 & 0 & 1 \end{bmatrix}.$$

Let det : $\mathscr{L}(R^3) \to R$, trace: $\mathscr{L}(R^3) \to R$, and $\mathscr{B}$: $\mathscr{L}(R^3) \to R$ be defined as follows:

$$\det(\mathbf{A}) = \begin{vmatrix} a_1 & a_2 & a_3 \\ b_1 & b_2 & b_3 \\ c_1 & c_2 & c_3 \end{vmatrix}, \quad \text{and } \operatorname{trace}(\mathbf{A}) = a_1 + b_2 + c_3, \tag{18.14}$$

$$\mathscr{B}(\mathbf{A}) = \begin{vmatrix} a_1 & a_2 \\ b_1 & b_2 \end{vmatrix} + \begin{vmatrix} b_2 & b_3 \\ c_2 & c_3 \end{vmatrix} + \begin{vmatrix} a_1 & a_3 \\ c_1 & c_3 \end{vmatrix}. \tag{18.15}$$

Suppose $\bar{\rho}(\sigma, \delta, \eta) \geq \rho$, and let **J** be the Jacobian matrix given by (18.13). Then we have

$$\det(\lambda \mathbf{I} - \mathbf{J}) = \lambda^3 - \operatorname{trace}(\mathbf{J})\lambda^2 + \mathscr{B}(\mathbf{J})\lambda - \det(\mathbf{J}). \tag{18.16}$$

From (18.13), we have the following:

$$\det(\mathbf{J}) = -\frac{(1-\beta)AX^{*\beta-1}u^{*1-\beta}Q^*}{\beta - (\beta+\eta)u^*}\frac{\delta u^*(1-u^*)^\eta}{\sigma} \times \beta(\sigma(1+\eta)(1-u^*) - \eta), \tag{18.17}$$

Fig. 18.3

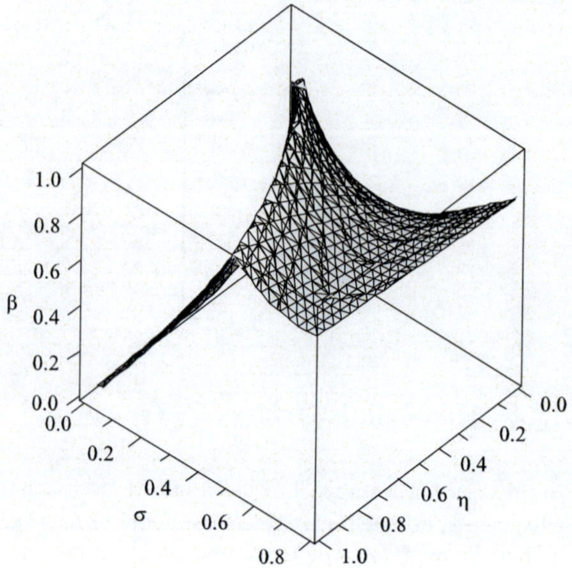

$$\mathscr{B}(\mathbf{J}) = \frac{\delta^2 \eta (1-u^*)^{2\eta}}{\sigma(\beta - (\beta+\eta)u^*)} \times \left[(\beta - \sigma)\left(1 - \frac{\eta}{\sigma(1+\eta)}\right)^2 \right.$$
$$\left. - (1-\beta)\left(\frac{1}{\beta} - \frac{\eta}{\sigma(1+\eta)}\right)\frac{\eta}{1+\eta}\left(\frac{1}{\sigma} - 1\right)\right]. \quad (18.18)$$

If $\rho = \bar{\rho}(\sigma, \delta, \eta)$, and $u^* = 1 - \frac{\eta}{\sigma(1+\eta)}$, $\mathbf{J}$ has a zero eigenvalue, since $\det(\mathbf{J}) = 0$. Consider the following function $g = g(\sigma, \eta, \beta)$.

$$g(\sigma, \eta, \beta) = (\beta - \sigma)\left(1 - \frac{\eta}{\sigma(1+\eta)}\right)^2 - (1-\beta)\left(\frac{1}{\beta} - \frac{\eta}{\sigma(1+\eta)}\right)\frac{\eta}{1+\eta}\left(\frac{1}{\sigma} - 1\right). \quad (18.19)$$

See Fig. 18.3 for the implicit plot of $g(\sigma, \eta, \beta) = 0$ for the case $\sigma < 1$. See Fig. 18.4 for the implicit plot of $g(\sigma, \eta, \beta) = 0$ for the case $\sigma > 1$. Let $\bar{\sigma}$ be a solution of $g(\bar{\sigma}, \eta, \beta) = 0$, where σ is unknown, and η and β are exogenously given control parameters. If $g(\bar{\sigma}, \eta, \beta) = 0$, $\bar{\rho} = \bar{\rho}(\bar{\sigma}, \delta, \eta)$, and $\text{trace}(\mathbf{J}) \neq 0$, then $\mathbf{J}$ has a zero eigenvalue of multiplicity 2.

Consider the following four examples. As mentioned in the previous section, we set $\delta = 0.05$ for the case $\sigma < 1$ and set $\delta = 0.1$ for the case $\sigma > 1$. We cannot freely set $\beta = 0.25$ any more. We choose $(\bar{\sigma}, \eta, \beta)$ that satisfies $g(\bar{\sigma}, \eta, \beta) = 0$. In the next section, we embed System (18.4) into a five-dimensional system in such a way that the resulting system has a zero eigenvalue of multiplicity 4. In this system,

Fig. 18.4

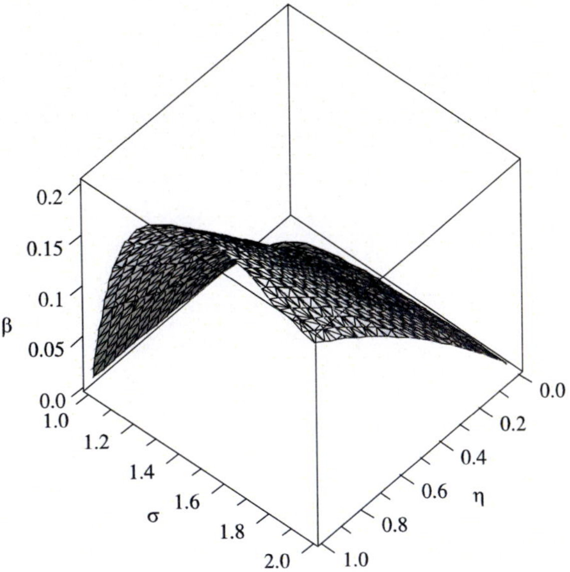

Fig. 18.5

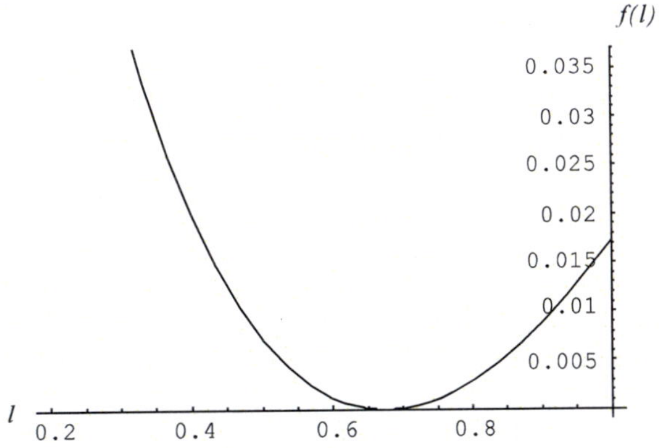

generalized eigenvectors corresponding to the zero eigenvalue of multiplicity 4 do not necessarily behave well numerically. However, we are able to choose the value of the scale parameter A in such a way that numerical pathologies are eliminated.

Example 3. Set $(A, \beta, \delta, \eta) = \left(\frac{1}{20}, 0.8, 0.05, 0.2\right)$. Set $(\bar{\sigma}, \bar{\rho}) \approx (0.246955808450705390, 0.03851542186815400793970015 02)$. This parametric example satisfies Assumption 1, and has a unique steady state. See Fig. 18.5 for the graph of $f = f(l)$. $l^* \approx 0.674884578387759601402795569$, $u^* \approx 0.3251154216122240398597204431$, and $\mu_K = \mu_H \approx 0.031192157098699724$. The Jacobian $\mathbf{J}$ of the unique steady state has a zero eigenvalue of multiplicity 2 and one unstable root. The rank of $\mathbf{J}$ is equal to 2.

Fig. 18.6

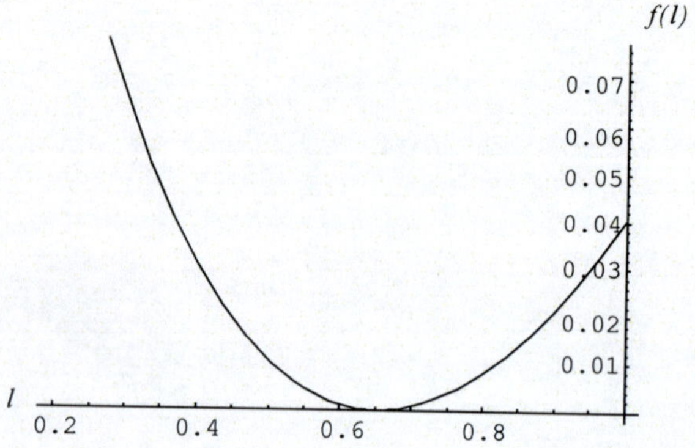

Fig. 18.7

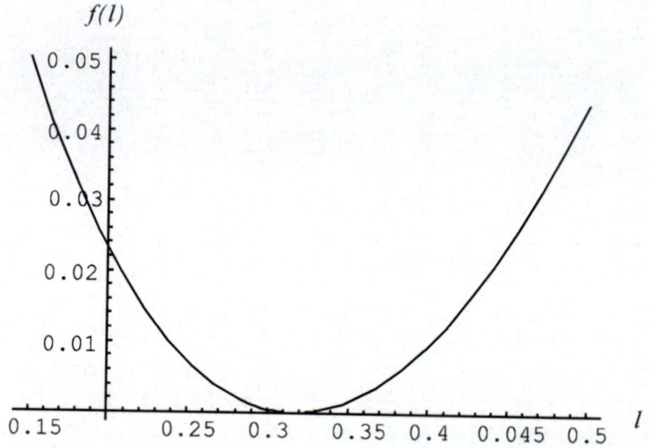

Example 4. Set $(A, \beta, \delta, \eta) = \left(\frac{1}{20}, 0.8, 0.05, 0.4\right)$. Set $(\bar{\sigma}, \bar{\rho}) \approx (0.43819289839923805, 0.0300987608974473603304485574)$. This parametric example satisfies Assumption 1, and has a unique steady state. See Fig. 18.6 for the graph of $f = f(l)$. $l^* \approx 0.6520285626673280824877892552$, $u^* \approx 0.3479714373326719175122107448$, and $\mu_K = \mu_H \approx 0.027475352528442232$. The Jacobian $\mathbf{J}$ of the unique steady state has a zero eigenvalue of multiplicity 2 and one unstable root. The rank of $\mathbf{J}$ is equal to 2.

Example 5. Set $(A, \beta, \delta, \eta) = (0.5, 0.1, 0.1, 0.6)$. Set $(\bar{\sigma}, \bar{\rho}) \approx (1.1827896110067725, 0.0313728719731436989011615924)$. This parametric example satisfies Assumption 1, and has a unique steady state. See Fig. 18.7 for the graph of $f = f(l)$. $l^* \approx 0.3170470864051686353603937729$, $u^* \approx 0.6829529135948313646396062271$, and $\mu_K = \mu_H \approx 0.01591468424199612$. The Jacobian $\mathbf{J}$ of the unique steady state has a zero eigenvalue of multiplicity 2 and one unstable root. The rank of $\mathbf{J}$ is equal to 2.

Fig. 18.8

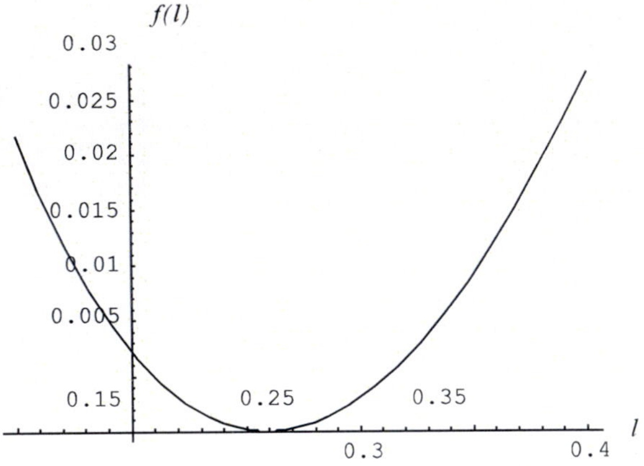

Example 6. Set $(A, \beta, \delta, \eta) = (3, 0.1, 0.1, 0.5)$. Set $(\bar{\sigma}, \bar{\rho}) \approx (1.2871007188910 5315, 0.03392671786031575677302745 8)$. This parametric example satisfies Assumption 1, and has a unique steady state. See Fig. 18.8 for the graph of $f = f(l)$. $l^* \approx 0.25897999157403033633792141 5$, $u^* \approx 0.74102000842596966366 2078585$, and $\mu_K = \mu_H \approx 0.013179511658398618$. The Jacobian $\mathbf{J}$ of the unique steady state has a zero eigenvalue of multiplicity 2 and one unstable root. The rank of $\mathbf{J}$ is equal to 2.

By Examples 3–6, we have the following lemma.

Lemma 1. *Let* $\Theta := R_{++} \times (0, 1) \times R^4_{++}$. *Let* Θ_1 *be defined as* $\Theta_1 := \{(A, \beta, \delta, \eta, \bar{\rho}, \bar{\sigma}) \in \Theta : (A, \beta, \delta, \eta, \bar{\rho}, \bar{\sigma})$ *satisfies Assumption 1,* $g(\bar{\sigma}, \eta, \beta) = 0, \overline{\rho} = \bar{\rho}(\bar{\sigma}, \delta, \eta)$, *trace*$(\mathbf{J}) \neq 0$, $\mathbf{J}$ *has rank 2.*$\}$. *Then* $\Theta_1 \neq \emptyset$.

In other words, there exists a non-empty subset Θ_1 of Θ such that if $(A', \beta', \delta', \eta', \bar{\rho}', \bar{\sigma}')$ belongs to Θ_1 and if we set $(A, \beta, \delta, \eta, \rho, \sigma) = (A', \beta', \delta', \eta', \bar{\rho}', \bar{\sigma}')$, then Assumption 1 is satisfied, System (18.4) has a unique steady state, and Jacobian (18.13) corresponding to this steady state has rank 2 and a zero eigenvalue of multiplicity 2.

In the rest of the present section and the next section, we assume $(A, \beta, \delta, \eta, \bar{\rho}, \bar{\sigma}) \in \Theta_1$. Therefore, $g(\bar{\sigma}, \eta, \beta) = 0$, $\bar{\rho} = \bar{\rho}(\bar{\sigma}, \delta, \eta)$, and trace$(\mathbf{J}) \neq 0$ at $(A, \beta, \delta, \eta, \bar{\rho}, \bar{\sigma})$. Let $\mu = \sigma - \bar{\sigma}$, and $\nu = \rho - \bar{\rho}$. Consider the following system.

$$f_1(X, u, Q, \mu, \nu) = AX^{\beta}u^{1-\beta} - QX - \delta X(1-u)^{1+\eta}$$

$$f_2(X, u, Q, \mu, \nu) = \frac{u(1-u)}{\beta(1-u) - \eta u}(\delta(1-u)^{\eta} - \beta\delta(1-u)^{1+\eta} - \beta Q)$$

$$f_3(X, u, Q, \mu, \nu) = \frac{\beta - (\bar{\sigma} + \mu)}{\bar{\sigma} + \mu} AX^{\beta-1}u^{1-\beta}Q - \frac{\bar{\rho} + \nu}{\bar{\sigma} + \mu}Q + Q^2.$$

Let $F = F(X, u, Q, \mu, \nu)$ be defined as

$$F(X, u, Q, \mu, \nu) = \begin{bmatrix} f_1(X, u, Q, \mu, \nu) \\ f_2(X, u, Q, \mu, \nu) \\ f_3(X, u, Q, \mu, \nu) \end{bmatrix}.$$

Then by (18.4),

$$(\dot{X}, \dot{u}, \dot{Q})^{\mathrm{T}} = F(X, u, Q, \mu, \nu). \tag{18.20}$$

System (18.20) undergoes homoclinic bifurcation as shown by the following proposition.

Proposition 2. *Let $U \subset R^2$ be a small open neighborhood of $(0, 0)$. Let $U_H :=$ $\{(\mu, \nu) \in U : (\dot{X}, \dot{u}, \dot{Q})^T = F(X, u, Q, \mu, \nu)$ has a homoclinic orbit.$\}$, and $U_P :=$ $\{(\mu, \nu) \in U : (\dot{X}, \dot{u}, \dot{Q})^T = F(X, u, Q, \mu, \nu)$ has a periodic orbit.$\}$. Let Θ_1 be the set specified in Lemma 1. Let $\Theta_2 \subset \Theta_1$ be defined as $\Theta_2 := \{(A, \beta, \delta, \eta, \bar{\rho}, \bar{\sigma}) \in \Theta_1 : U_H \neq \emptyset$ and $U_P \neq \emptyset.\}$. Then $\Theta_2 \neq \emptyset$. We can choose U_H and U_P in such a way that U_H is a one-parameter family of equations and that U_P is a two-parameter family of equations.*

Proof. See Appendix II. ■

As shown in Sect. 18.3, the transversality conditions are satisfied on the balanced growth paths, and so for each $(A, \beta, \delta, \eta, \bar{\rho}, \bar{\sigma}) \in \Theta_2$, if we choose U sufficiently small, each path starting from a point on either a homoclinic orbit or a periodic orbit satisfies the transversality conditions. Here we state that U_H is a one-parameter family of equations and that U_P is a two-parameter family of equations. An equation means System (18.20). And the first statement means that if $(\mu_0, \nu_0) \in U_H$, and if ν varies slightly away from ν_0, a homoclinic orbit persists, when μ moves away from μ_0 in an appropriate way. The second statement means that if $(\mu_1, \nu_1) \in U_P$, and if μ and ν vary slightly away from μ_1 and ν_1 independently, a periodic orbit persists. In Appendix II, we have used Example 3. However, we obtain the same result, if we use Example 4, 5, or 6.

18.5 Global Indeterminacy of Equilibrium

In the previous section, we have shown that the dynamics of (18.20) has a homoclinic orbit for some parameter values $(\mu, \nu) \in U_H$ and a periodic orbit for other parameter values $(\mu, \nu) \in U_P$. The mere existence of either a homoclinic orbit or a periodic orbit does not directly imply global indeterminacy of equilibrium, because the dynamics of (18.20) is a three-dimensional system whereas the Jordan curve theorem holds in a two-dimensional system. Suppose that there exists a two-dimensional manifold that is well-located in a three-dimensional ambient space and that is enclosed by either a homoclinic orbit or a periodic orbit.[6] Suppose further that a path of System (18.20) starting from a point on this manifold continues to stay on the manifold. System (18.20) includes one predetermined variable and two non-predetermined variables. The dimension of the manifold is equal to a number that is greater than the number of predetermined variables. Therefore, under these assumptions, equilibrium is globally indeterminate, if the transversality conditions are satisfied. In the present section, we show that there exists a two-dimensional manifold with the properties mentioned above. We shall use the method of the center manifold reduction in order to reduce System (18.20) to a two-dimensional system. See Sect. 3 of Kopell and Howard (1975), and Chap. 6 of Guckenheimer and Holmes (1983) for the center manifold theorem. See Guckenheimer and Holmes (1983) for the center manifold reduction and the bifurcation diagram.[7] Consider the following map, and the dynamics generated by it:

$$L(\mu,\nu,X,u,Q) = \begin{bmatrix} 0 \\ 0 \\ f_1(X,u,Q,\mu,\nu) \\ f_2(X,u,Q,\mu,\nu) \\ f_3(X,u,Q,\mu,\nu) \end{bmatrix}$$

$$(\dot{\mu},\dot{\nu},\dot{X},\dot{u},\dot{Q})^{\mathrm{T}} = L(\mu,\nu,X,u,Q). \tag{18.21}$$

Let $(X,u,Q) = (\bar{X}^*,\bar{u}^*,\bar{Q}^*)$ be a steady state of the dynamics $(\dot{X},\dot{u},\dot{Q})^{\mathrm{T}} = F(X,u,Q,0,0)$. If the Jacobian of $L = L(\mu,\nu,X,u,Q)$ is evaluated at a steady state $(\mu,\nu,X,u,Q) = (0,0,\bar{X}^*,\bar{u}^*,\bar{Q}^*)$, then System (18.21) has a zero eigenvalue of multiplicity 4. Consider the following proposition.

Proposition A *There exist an open set* $W \subset R^4$ *with* $(\bar{X}^*,\bar{u}^*,0,0) \in W$ *and a smooth mapping* $\varphi: W \to R_{++}$ *with* $\bar{Q}^* = \varphi(\bar{X}^*,\bar{u}^*,0,0)$ *such that* $\{(\mu,\nu,X,u,Q) \in R^5: (X,u,\mu,\nu) \in W,\ Q = \varphi(X,u,\mu,\nu).\}$ *constitutes a center manifold of the dynamics (18.21).*

Lemma 2. *Let* Θ_2 *be the set specified in Proposition 2. Let* Θ_3 *be defined as* $\Theta_3 := \{(A,\beta,\delta,\eta,\bar{\rho},\bar{\sigma}) \in \Theta_2$*: Proposition A holds.*$\}$. *Then* $\Theta_3 \neq \emptyset$.

Proof. See Appendix III. ∎

In other words, there exists a non-empty subset Θ_3 of Θ_2 such that if $(A',\beta',\delta',\eta',\bar{\rho}',\bar{\sigma}')$ belongs to Θ_3 and if we set $(A,\beta,\delta,\eta,\bar{\rho},\bar{\sigma}) = (A',\beta',\delta',\eta',\bar{\rho}',\bar{\sigma}')$, then System (18.21) has a four-dimensional center manifold that is characterized by

---

[6] See a two-dimensional manifold $E_{\mu,\nu}$ defined below.

[7] Especially see (3.2.29) and (3.2.32) and Figs. 3.2.6 and 3.2.7 of Guckenheimer and Holmes (1983). The best introduction to these topics is Sects. B5 and C5 of Grandmont (2008), albeit he treats not differential but difference equations.

Proposition A. In Appendix III, we have used Example 3. However, we obtain the same result, if we use Example 4, 5, or 6.

Let $W \subset R^4$ and $\varphi : W \to R_{++}$ be the set and the mapping specified in Proposition A. Then, the following relation holds on $(X, u, \mu, \nu) \in W$:

$$\begin{aligned} & f_3(X, u, \varphi(X, u, \mu, \nu), \mu, \nu) \\ & \quad = \frac{\partial \varphi(X, u, \mu, \nu)}{\partial X} f_1(X, u, \varphi(X, u, \mu, \nu), \mu, \nu) \\ & \quad + \frac{\partial \varphi(X, u, \mu, \nu)}{\partial u} f_2(X, u, \varphi(X, u, \mu, \nu), \mu, \nu). \end{aligned} \tag{18.22}$$

Thus, we obtain the following four-dimensional system on $(X, u, \mu, \nu) \in W$.

$$\begin{bmatrix} \dot{X} \\ \dot{u} \\ \dot{\mu} \\ \dot{\nu} \end{bmatrix} = \begin{bmatrix} f_1(X, u, \varphi(X, u, \mu, \nu), \mu, \nu) \\ f_2(X, u, \varphi(X, u, \mu, \nu), \mu, \nu) \\ 0 \\ 0 \end{bmatrix}, \tag{18.23}$$

with $Q = \varphi(X, u, \mu, \nu)$. System (18.23) constitutes a bifurcation diagram that reflects *the dynamics on the four-dimensional center manifold* of System (18.21). Every recurrent behavior, such as a homoclinic solution and a periodic solution is included in the bifurcation diagram (18.23) by the center manifold theorem.[8] Consider the following two-dimensional dynamics:

$$\begin{bmatrix} \dot{X} \\ \dot{u} \end{bmatrix} = \begin{bmatrix} f_1(X, u, \varphi(X, u, \mu, \nu), \mu, \nu) \\ f_2(X, u, \varphi(X, u, \mu, \nu), \mu, \nu) \end{bmatrix}, \tag{18.24}$$

with $Q = \varphi(X, u, \mu, \nu)$. If $(\mu, \nu) \in U_H$, the dynamics of (18.24) has a homoclinic orbit, and if $(\mu, \nu) \in U_P$, the dynamics of (18.24) has a periodic orbit. If either a homoclinic orbit or a periodic orbit exists, then System (18.24) has a bounded region enclosed by either a homoclinic orbit or a periodic orbit. Suppose $(\mu, \nu) \in U_H \cup U_P \subset U$, and let $D_{\mu,\nu} \subset R^2$ be the set of all points of the bounded region enclosed by either a homoclinic orbit or a periodic orbit of System (18.24) where the boundary of the region is also included in $D_{\mu,\nu}$. Then $D_{\mu,\nu}$ is a compact set. Let Int $D_{\mu,\nu}$ and $\partial D_{\mu,\nu}$ be the set of all interior points of $D_{\mu,\nu}$ and the boundary of $D_{\mu,\nu}$, respectively. If $(\mu, \nu) \in U_H$, then $\partial D_{\mu,\nu}$ is a homoclinic orbit of System (18.24), and if $(\mu, \nu) \in U_P$, then $\partial D_{\mu,\nu}$ is a periodic orbit of System (18.24).[9] Let $E_{\mu,\nu} \subset R^3$ be a two-dimensional manifold defined as

---

[8]Behaviors on either a stable manifold or an unstable manifold could not be recurrent. They are either convergent or divergent.

[9]For a given $(\mu', \nu') \in U_H \cup U_P$, if a homoclinic orbit and periodic orbits coexist, we take the homoclinic orbit as $\partial D_{\mu',\nu'}$, and if there exist multiple periodic orbits without a homoclinic orbit, we take the most outer periodic orbit as $\partial D_{\mu',\nu'}$.

$$E_{\mu,\nu} := \{(X, u, Q) \in R^3 : (X, u) \in \text{Int}D_{\mu,\nu}, Q = \varphi(X, u, \mu, \nu).\},$$

where $(\mu, \nu) \in U_H \cup U_P \subset U$. Then, each path starting from a point on $E_{\mu,\nu}$ continues to stay on $E_{\mu,\nu}$ by (18.22). As shown in Sect. 18.3, the transversality conditions are satisfied on the balanced growth paths, and so if we choose U sufficiently small, each path starting from a point on $E_{\mu,\nu}$ satisfies the transversality conditions. Let $X_{\mu,\nu}^{\min}$ and $X_{\mu,\nu}^{\max}$ be defined as $X_{\mu,\nu}^{\min} :=$ arg min X subject to $(X, u) \in D_{\mu,\nu}$ and $X_{\mu,\nu}^{\max} :=$ arg max X subject to $(X, u) \in D_{\mu,\nu}$, respectively. Because $D_{\mu,\nu}$ is compact, $X_{\mu,\nu}^{\min}$ and $X_{\mu,\nu}^{\max}$ do exist. Let X' be a point such that $X_{\mu,\nu}^{\min} < X' < X_{\mu,\nu}^{\max}$. Let $J(X') := \{(X, u, Q) \in R^3 : X = X'\}$. $E_{\mu,\nu} \cap J(X')$ constitutes a one-dimensional manifold. For a given initial condition $X = X'$, there exists a continuum of equilibria that is composed of the one-dimensional manifold $E_{\mu,\nu} \cap J(X')$. By Definition 1 in Sect. 18.2, the system composed of (18.20) and (18.5) exhibits global indeterminacy of equilibrium for $(\mu, \nu) \in U_H \cup U_P \subset U$. Therefore, we have the following proposition.

Proposition 3. *Let* $U \subset R^2$ *be a small open neighborhood of* $(0, 0)$. *Let* $U_G := \{(\mu, \nu) \in U$: *The system composed of* $(\dot{X}, \dot{u}, \dot{Q})^T = F(X, u, Q, \mu, \nu)$ *and Equation (18.5) exhibits global indeterminacy of equilibrium.*$\}$. *Let* Θ_3 *be the set specified in Lemma 2. Let* $\Theta_4 \subset \Theta_3$ *be defined as* $\Theta_4 := \{(A, \beta, \delta, \eta, \bar{\rho}, \bar{\sigma}) \in \Theta_3 : U_G \neq \emptyset\}$. *Then* $\Theta_4 \neq \emptyset$.

Proof. Since $\Theta_3 \subset \Theta_2$, $U_H \neq \emptyset$, and $U_P \neq \emptyset$ for each $(A, \beta, \delta, \eta, \bar{\rho}, \bar{\sigma}) \in \Theta_3$. Therefore, by Proposition A, $\Theta_4 = \Theta_3$, because for each $(A, \beta, \delta, \eta, \bar{\rho}, \bar{\sigma}) \in \Theta_3$, if we choose U sufficiently small, we can construct a two-dimensional manifold $E_{\mu,\nu}$ for each $(\mu, \nu) \in U_H \cup U_P$ such that each path starting on a point of $E_{\mu,\nu}$ continues to stay on $E_{\mu,\nu}$ and satisfies the transversality conditions and that $E_{\mu,\nu} \cap J(X')$ is a one-dimensional manifold, where $X_{\mu,\nu}^{\min} < X' < X_{\mu,\nu}^{\max}$. By Lemma 2, $\Theta_3 \neq \emptyset$. Thus, $\Theta_4 \neq \emptyset$. ∎

18.6 Conclusion

This paper has studied an endogenous growth model that belongs to the same family as the Lucas model. In the Lucas model an external effect appears in the physical-goods sector, and if a BGP exists, it is unique. In our model, an external effect appears in the educational sector, and this external effect yields multiple balanced growth paths. Based on this multiplicity of balanced growth paths, we have shown that our model undergoes a homoclinic bifurcation and that the model exhibits global indeterminacy of equilibrium. As shown by Chamley (1993) and Benhabib and Perli (1994), there is a family of endogenous growth models that have multiple balanced growth paths due to an external effect in an educational sector. The family might be large. At any rate, our model is the simplest one in this family. By extending our method in a way amenable to a more complex model, one might be able to derive conclusions similar to ours from this model.

18.7 Appendix I

In this appendix, we shall derive (18.7) from (18.4). Let $(X, u, Q) \in R_{++} \times (0, 1) \times R_{++}$ with $u \neq \frac{\beta}{\beta+\eta}$. Let $(X, u, Q) = (X^*, u^*, Q^*)$ be a steady state of (18.4) such that $(X, u, Q) = (X^*, u^*, Q^*) \in R_{++} \times (0, 1) \times R_{++}$ with $u^* \neq \frac{\beta}{\beta+\eta}$. From (18.4), we have the following system of equations.

$$\begin{aligned} &AX^{*\beta-1}u^{*1-\beta} - Q^* - \delta(1-u^*)^{1+\eta} = 0 \\ &\delta(1-u^*)^{\eta} - \beta\delta(1-u^*)^{1+\eta} - \beta Q^* = 0 \\ &\frac{\beta-\sigma}{\sigma}AX^{*\beta-1}u^{*1-\beta} - \frac{\rho}{\sigma} + Q^* = 0. \end{aligned} \tag{18.25}$$

Let $Z^* = AX^{*\beta-1}u^{*1-\beta}$, and we have the followings.

$$\begin{aligned} Q^* &= Z^* - \delta(1-u^*)^{1+\eta} \\ Q^* &= \frac{\delta}{\beta}(1-u^*)^{\eta} - \delta(1-u^*)^{1+\eta} \\ Q^* &= \frac{\rho}{\sigma} - \frac{\beta-\sigma}{\sigma}Z^*. \end{aligned} \tag{18.26}$$

From (18.26), we have the followings:

$$Z^* = \frac{\delta}{\beta}(1-u^*)^{\eta}, \tag{18.27}$$

and

$$\frac{\delta}{\beta}(1-u^*)^{\eta} - \delta(1-u^*)^{1+\eta} = \frac{\rho}{\sigma} - \frac{\beta-\sigma}{\sigma}\frac{\delta}{\beta}(1-u^*)^{\eta},$$

and so

$$\delta(1-u^*)^{1+\eta} - \frac{\delta}{\sigma}(1-u^*)^{\eta} + \frac{\rho}{\sigma} = 0;$$

therefore

$$\sigma(1-u^*)^{1+\eta} - (1-u^*)^{\eta} + \frac{\rho}{\delta} = 0. \tag{18.28}$$

Equation (18.28) is precisely (18.7).

18.8 Appendix II

In this appendix, we shall prove Proposition 2 by means of the homoclinic bifurcation theorem of Kopell and Howard (1975). Let $\mathbf{y}$ and $\mathbf{0}$ be $\mathbf{y} = (y_1, y_2, y_3)$ or $\mathbf{y} = (y_1, y_2, y_3)^{\mathbf{T}}$, and $\mathbf{0} = (0, 0, 0)$ or $\mathbf{0} = (0, 0, 0)^{\mathbf{T}}$, depending on the context. Let $(\mathbf{y}, \mu, \nu) = (y_1, y_2, y_3, \mu, \nu)$. Let V be a sufficiently small open subset of R^5 with $(\mathbf{0}, 0, 0) \in V$. Let $G: V \to R^3$ be $G(\mathbf{y}, \mu, \nu) = G_{\mu,\nu}(\mathbf{y}) = (G_1(\mathbf{y}, \mu, \nu), G_2(\mathbf{y}, \mu, \nu), G_3(\mathbf{y}, \mu, \nu))^{\mathbf{T}}$. Consider the parametrized dynamics given by $\dot{\mathbf{y}} = G_{\mu,\nu}(\mathbf{y})$. We assume $G_{0,0}(\mathbf{0}) = \mathbf{0}$. Let $dG_{\mu,\nu}(\mathbf{0})$, $\bar{g}_1$, $\bar{g}_2$, m_1 and m_2 be defined as $dG_{\mu,\nu}(\mathbf{0}) := \frac{\partial G}{\partial \mathbf{y}}(\mathbf{0}, \mu, \nu)$, $\bar{g}_1 := \frac{\partial G}{\partial \mu}(\mathbf{0}, \mu, \nu)|_{(\mu,\nu)=(0,0)}$, $\bar{g}_2 := \frac{\partial G}{\partial \nu}(\mathbf{0}, \mu, \nu)|_{(\mu,\nu)=(0,0)}$, $m_1 := \frac{\partial}{\partial \mu}\mathscr{B}[dG(\mathbf{0}, \mu, \nu)]|_{(\mu,\nu)=(0,0)}$, $m_2 := \frac{\partial}{\partial \nu}\mathscr{B}[dG(\mathbf{0}, \mu, \nu)]|_{(\mu,\nu)=(0,0)}$. Let $\mathbf{P}$: $R^3 \times R^3 \to R^3$ be defined as

$$\mathbf{P}(\mathbf{x}, \mathbf{z}) := \begin{bmatrix} \frac{1}{2}\mathbf{x}^{\mathrm{T}} \frac{\partial^2 G_1}{\partial \mathbf{y}^2}(\mathbf{0}, 0, 0)\mathbf{z} \\ \frac{1}{2}\mathbf{x}^{\mathrm{T}} \frac{\partial^2 G_2}{\partial \mathbf{y}^2}(\mathbf{0}, 0, 0)\mathbf{z} \\ \frac{1}{2}\mathbf{x}^{\mathrm{T}} \frac{\partial^2 G_3}{\partial \mathbf{y}^2}(\mathbf{0}, 0, 0)\mathbf{z} \end{bmatrix}.$$

Let $R_2: V \to R^3$ be defined as $R_2(\mathbf{y}, \mu, \nu) := G(\mathbf{y}, \mu, \nu) - [dG_{0,0}(\mathbf{0})\mathbf{y} + \mu\bar{g}_1 + \nu\bar{g}_2 + \mathbf{P}(\mathbf{y}, \mathbf{y})] = o(\mu, \nu, y_i y_j)$, and so

$$\dot{\mathbf{y}} = dG_{0,0}(\mathbf{0})\mathbf{y} + \mu\bar{g}_1 + \nu\bar{g}_2 + \mathbf{P}(\mathbf{y}, \mathbf{y}) + R_2(\mathbf{y}, \mu, \nu).$$

We now state the following theorem due to Theorem 7.2 in Kopell and Howard (1975).

Theorem 1. *Let* $\dot{\mathbf{y}} = G(\mathbf{y}, \mu, \nu)$ *be a two-parameter family of ordinary differential equations on* V *with* G *being* C^3*-smooth in all its arguments and* $G_{0,0}(\mathbf{0}) = \mathbf{0}$. *Also assume:*

(i) $dG_{0,0}(\mathbf{0})$ *has rank 2 and a zero eigenvalue of multiplicity 2. Let* $\mathbf{e}$ *and* $\mathbf{l}$ *be the right eigenvector of the zero eigenvalue and the left eigenvector of the zero eigenvalue, respectively.*

(ii) $\det\left(\begin{bmatrix} \mathbf{l}\cdot\bar{\mathbf{g}}_1 & m_1 \\ \mathbf{l}\cdot\bar{\mathbf{g}}_2 & m_2 \end{bmatrix}\right) \neq 0.$

(iii) $[dG_{0,0}(\mathbf{0}), \mathbf{P}(\mathbf{e}, \mathbf{e})]$ *has rank 3.*

Let $U \subset R^2$ be a sufficiently small open set with $(0, 0) \in U$. Let $U_H = \{(\mu, \nu) \in U: \dot{\mathbf{y}} = G(\mathbf{y}, \mu, \nu)$ has a homoclinic orbit.$\}$, and let $U_P = \{(\mu, \nu) \in U: \dot{\mathbf{y}} = G(\mathbf{y}, \mu, \nu)$ has a periodic orbit.$\}$. Then $U_H \neq \emptyset$ and $U_P \neq \emptyset$. The subset $U_H \subset U$ is a one-parameter family of equations, whereas the subset $U_P \subset U$ is a two-parameter family of equations.

Remark 1. Theorem 7.2 in Kopell and Howard (1975) does not explicitly state the degree of smoothness in $G = G(\mathbf{y}, \mu, \nu)$ that one needs. Our system is three-dimensional. For the case where the dimension of a given system is more than two,

Kopell and Howard (1975, p. 342) apply the center manifold reduction to the given system in such a way that the resulting two-dimensional system is C^2-smooth in all its arguments. If one applies the center manifold reduction to a given C^r-smooth system, then the resulting system is C^{r-1}-smooth, where $r \geq 2$. Therefore, we have assumed in Theorem 1 that G is C^3-smooth in all its arguments.

We introduce the following series of definitions in order to apply Theorem 1 to our model. Let $\bar{\sigma}$ be a solution of $g(\bar{\sigma}, \eta, \beta) = 0$, where $g = g(\sigma, \eta, \beta)$ is the function specified in (18.19) and where σ is unknown, and η and β are exogenously given control parameters. Let $\bar{\rho} := \delta\left(\frac{1}{1+\eta}\right)^{1+\eta}\left(\frac{\eta}{\bar{\sigma}}\right)^{\eta}$, $\mu := \sigma - \bar{\sigma}$, and $\nu := \rho - \bar{\rho}$. Let $\bar{u}^* := 1 - \frac{\eta}{\bar{\sigma}(1+\eta)}$, $\bar{X}^* := A^{\frac{1}{1-\beta}}\left(1 - \frac{\eta}{\bar{\sigma}(1+\eta)}\right)\left[\frac{\beta}{\delta}\left(\frac{\bar{\sigma}(1+\eta)}{\eta}\right)^{\eta}\right]^{\frac{1}{1-\beta}}$, and $\bar{Q}^* := \frac{\delta}{\beta}\left(\frac{\eta}{\bar{\sigma}(1+\eta)}\right)^{\eta} - \delta\left(\frac{\eta}{\bar{\sigma}(1+\eta)}\right)^{1+\eta}$. Let $\tilde{X} := X - \bar{X}^*$, $\tilde{u} := u - \bar{u}^*$, and $\tilde{Q} := Q - \bar{Q}^*$, and let

$$\begin{aligned}
G_1(\tilde{X}, \tilde{u}, \tilde{Q}, \mu, \nu) &:= f_1(\bar{X}^* + \tilde{X}, \bar{u}^* + \tilde{u}, \bar{Q}^* + \tilde{Q}, \mu, \nu), \\
G_2(\tilde{X}, \tilde{u}, \tilde{Q}, \mu, \nu) &:= f_2(\bar{X}^* + \tilde{X}, \bar{u}^* + \tilde{u}, \bar{Q}^* + \tilde{Q}, \mu, \nu), \\
G_3(\tilde{X}, \tilde{u}, \tilde{Q}, \mu, \nu) &:= f_3(\bar{X}^* + \tilde{X}, \bar{u}^* + \tilde{u}, \bar{Q}^* + \tilde{Q}, \mu, \nu).
\end{aligned}$$

If we choose V sufficiently small, then $G = G(\tilde{X}, \tilde{u}, \tilde{Q}, \mu, \nu)$ is C^3-smooth in all its arguments. Let $\bar{\mathbf{J}}$ be the Jacobian of $G = G(\tilde{X}, \tilde{u}, \tilde{Q}, \mu, \nu)$ with respect to $(\tilde{X}, \tilde{u}, \tilde{Q})$ evaluated at $(\tilde{X}, \tilde{u}, \tilde{Q}, \mu, \nu) = (\mathbf{0}, 0, 0)$. Let $\bar{g}_1 := \left(0, 0, -\frac{\beta}{\bar{\sigma}^2} A\bar{X}^{*\beta-1}\bar{u}^{*1-\beta}\bar{Q}^* + \frac{\bar{\rho}}{\bar{\sigma}^2}\bar{Q}^*\right)^{\mathbf{T}}$ and $\bar{g}_2 := \left(0, 0, -\frac{1}{\bar{\sigma}}\bar{Q}^*\right)^{\mathbf{T}}$. Let

$$\begin{aligned}
m_1 := &-\frac{\beta}{\bar{\sigma}^2} A\bar{X}^{*\beta-1}\bar{u}^{*1-\beta}\left[\frac{\delta\bar{u}^*(1-\bar{u}^*)^{\eta}}{\beta - (\beta+\eta)\bar{u}^*}((1+\eta)\beta(1-\bar{u}^*) - \eta)\right. \\
&\left. + (1-\beta)\bar{Q}^*\frac{\beta(1-\bar{u}^*)}{\beta - (\beta+\eta)\bar{u}^*} - (1-\beta)A\bar{X}^{*\beta-1}\bar{u}^{*1-\beta} - (1-\beta)\bar{Q}^*\right] \\
&+ \frac{\bar{\rho}}{\bar{\sigma}^2}\left[\frac{\delta\bar{u}^*(1-\bar{u}^*)^{\eta}}{\beta - (\beta+\eta)\bar{u}^*}((1+\eta)\beta(1-\bar{u}^*) - \eta) - (1-\beta)A\bar{X}^{*\beta-1}\bar{u}^{*1-\beta}\right], \\
m_2 := &\frac{1}{\bar{\sigma}}\left[(1-\beta)A\bar{X}^{*\beta-1}\bar{u}^{*1-\beta} - \frac{\delta\bar{u}^*(1-\bar{u}^*)^{\eta}}{\beta - (\beta+\eta)\bar{u}^*}((1+\eta)\beta(1-\bar{u}^*) - \eta)\right].
\end{aligned}$$

Let $\mathbf{P}_i$, $i = 1, 2, 3$, be defined as

$$\mathbf{P}_1 := \begin{bmatrix} p1[1,1] & p1[1,2] & p1[1,3] \\ p1[2,1] & p1[2,2] & p1[2,3] \\ p1[3,1] & p1[3,2] & p1[3,3] \end{bmatrix},$$

$$\mathbf{P}_2 := \begin{bmatrix} p2[1,1] & p2[1,2] & p2[1,3] \\ p2[2,1] & p2[2,2] & p2[2,3] \\ p2[3,1] & p2[3,2] & p2[3,3] \end{bmatrix},$$

$$\mathbf{P}_3 := \begin{bmatrix} p3[1,1] & p3[1,2] & p3[1,3] \\ p3[2,1] & p3[2,2] & p3[2,3] \\ p3[3,1] & p3[3,2] & p3[3,3] \end{bmatrix},$$

where $p1[1,1] = -(1-\beta)\beta A\bar{X}^{*\beta-2}\bar{u}^{*1-\beta}$, $p1[1,2] = p1[2,1] = (1-\beta)\beta A\bar{X}^{*\beta-1}\bar{u}^{*-\beta} + \delta(1+\eta)(1-\bar{u}^*)^\eta$, $p1[1,3] = p1[3,1] = -1$, $P1[2,2] = -(1-\beta)\beta A\bar{X}^{*\beta}\bar{u}^{*-\beta-1} - \delta\eta(1+\eta)\bar{X}^*(1-\bar{u}^*)^{\eta-1}$, $p1[2,3] = p1[3,2] = 0$, $p1[3,3] = 0$, and $p2[1,1] = 0$, $p2[1,2] = p2[2,1] = 0$, $p2[1,3] = p2[3,1] = 0$, $p2[2,2] = \frac{\delta(1-\bar{u}^*)^{\eta-1}(\beta(2-(3+\eta)\bar{u}^*)+(1+\eta)(\beta+\eta)\bar{u}^{*2})}{(\beta-(\beta+\eta)\bar{u}^*)^2}((1+\eta)\beta(1-\bar{u}^*)-\eta) - \frac{\delta\bar{u}^*(1-\bar{u}^*)^\eta(1+\eta)\beta}{\beta-(\beta+\eta)\bar{u}^*}$, $p2[2,3] = p2[3,2] = -\frac{\beta(\beta(1-2\bar{u}^*)+(\beta+\eta)\bar{u}^{*2})}{(\beta-(\beta+\eta)\bar{u}^*)^2}$, $p2[3,3] = 0$, and $p3[1,1] = (\beta-1)(\beta-2)(\frac{\beta-\bar{\sigma}}{\bar{\sigma}})A\bar{X}^{*\beta-3}\bar{u}^{*1-\beta}\bar{Q}^*$, $p3[1,2] = p3[2,1] = -(1-\beta)^2(\frac{\beta-\bar{\sigma}}{\bar{\sigma}})A\bar{X}^{*\beta-2}\bar{u}^{*-\beta}\bar{Q}^*$, $p3[1,3] = p3[3,1] = (\beta-1)(\frac{\beta-\bar{\sigma}}{\bar{\sigma}})A\bar{X}^{*\beta-2}\bar{u}^{*1-\beta}$, $p3[2,2] = -(1-\beta)\beta(\frac{\beta-\bar{\sigma}}{\bar{\sigma}})A\bar{X}^{*\beta-1}\bar{u}^{*-1-\beta}\bar{Q}^*$, $p3[2,3] = p3[3,2] = (1-\beta)(\frac{\beta-\bar{\sigma}}{\bar{\sigma}})A\bar{X}^{*\beta-1}\bar{u}^{*-\beta}$, $p3[3,3] = 2$. Let $\mathbf{P}$: $R^3 \times R^3 \to R^3$ be defined as

$$\mathbf{P}(\mathbf{x},\mathbf{z}) := \begin{bmatrix} \frac{1}{2}\mathbf{x}^{\mathrm{T}}\mathbf{P}_1\mathbf{z} \\ \frac{1}{2}\mathbf{x}^{\mathrm{T}}\mathbf{P}_2\mathbf{z} \\ \frac{1}{2}\mathbf{x}^{\mathrm{T}}\mathbf{P}_3\mathbf{z} \end{bmatrix}.$$

We have sufficient preparations to apply Theorem 1 to our model. For the sake of concreteness, we use the numeric values of Example 3. Set $(A,\beta,\delta,\eta) = \left(\frac{1}{20}, 0.8, 0.05, 0.2\right)$, and $(\bar{\sigma},\bar{\rho}) \approx (0.246955808450705390, 0.0385154218681540079397001502)$. Then, $\bar{\mathbf{J}}$ has rank 2 and a zero eigenvalue of multiplicity 2. Thus the first condition of Theorem 1 is satisfied. $\mathbf{e}$ and $\mathbf{l}$ are given by $\mathbf{e} \approx (18.7639352951194837549267850712, 26.0815261470615170466350348437, 1)^{\mathrm{T}}$, and $\mathbf{l} \approx (-0.377097373476707559552505016 4, 0.232955337107870210821302210 6, 1)$, respectively. We have $\det\left(\begin{bmatrix} \mathbf{l}\cdot\bar{g}_1 & m_1 \\ \mathbf{l}\cdot\bar{g}_2 & m_2 \end{bmatrix}\right) \approx -0.0000593758037579160 \neq 0$. Therefore, the second condition of Theorem 1 is also satisfied. We have

$$\det\left(\begin{bmatrix} \bar{j}[1,1] & \bar{j}[1,2] & \frac{1}{2}\mathbf{e}^{\mathrm{T}}\mathbf{P}_1\mathbf{e} \\ \bar{j}[2,1] & \bar{j}[2,2] & \frac{1}{2}\mathbf{e}^{\mathrm{T}}\mathbf{P}_2\mathbf{e} \\ \bar{j}[3,1] & \bar{j}[3,2] & \frac{1}{2}\mathbf{e}^{\mathrm{T}}\mathbf{P}_2\mathbf{e} \end{bmatrix}\right) \approx 0.000121665229401229 1 \neq 0,$$

where $\bar{j}[i,j]$ is an (i,j)-component of $\bar{\mathbf{J}}$. Thus $\left[\bar{\mathbf{J}}, \mathbf{P}(\mathbf{e},\mathbf{e})\right]$ has rank 3. Therefore, the third condition of Theorem 1 is also satisfied. So, Proposition 2 holds by Theorem 1.

18.9 Appendix III

In this appendix, we shall prove Lemma 2. We continue to use the same notations and the same numeric values (i.e., Example 3) as in Appendix II. Let $\tilde{V} \subset R^5$ be a small open subset with $(0, 0, 0, 0, 0) \in \tilde{V}$ such that $\tilde{V} = \{(\mu, \nu, \tilde{X}, \tilde{u}, \tilde{Q}) \in R^5 \colon (\tilde{X}, \tilde{u}, \tilde{Q}, \mu, \nu) \in V\}$, where V is specified as in Appendix II. Consider the following map $M \colon \tilde{V} \to R^5$, and the dynamics generated by it:

$$M(\mu, \nu, \tilde{X}, \tilde{u}, \tilde{Q}) = \begin{bmatrix} 0 \\ 0 \\ G_1(\tilde{X}, \tilde{u}, \tilde{Q}, \mu, \nu) \\ G_2(\tilde{X}, \tilde{u}, \tilde{Q}, \mu, \nu) \\ G_3(\tilde{X}, \tilde{u}, \tilde{Q}, \mu, \nu) \end{bmatrix} \tag{18.29}$$

$$(\dot{\mu}, \dot{\nu}, \dot{\tilde{X}}, \dot{\tilde{u}}, \dot{\tilde{Q}})^{\mathbf{T}} = M(\mu, \nu, \tilde{X}, \tilde{u}, \tilde{Q}). \tag{18.30}$$

Let $\mathbf{DM}$ be the Jacobian of (18.29) evaluated at $(0, 0, 0, 0, 0)$. It is given by

$$\mathbf{DM} = \begin{bmatrix} 0 & 0 & \mathbf{0} \\ 0 & 0 & \mathbf{0} \\ \bar{g}_1 & \bar{g}_2 & \bar{\mathbf{J}} \end{bmatrix}, \tag{18.31}$$

where $\mathbf{0} = (0, 0, 0)$. By construction, $\mathbf{DM}$ has a zero eigenvalue of multiplicity 4 at $(0, 0, \mathbf{0})$. Thus the system (18.30) has a four-dimensional center manifold.

Let $\mathbf{x}_1 \approx (0,\ 0,\ 18.763935295119483754926785071 2,\ 26.0815261470615170466350348437,\ 1)^{\mathbf{T}}$, $\mathbf{x}_2 \approx (-32.0593409694543283185919930446,\ 1,\ 0,\ 0,\ 0)^{\mathbf{T}}$, $\mathbf{x}_3 \approx (-5.914716521087486 \times 10^{-16},\ 0,\ 9.353014148203561,\ 0,\ 1)^{\mathbf{T}}$, and $\mathbf{x}_4 \approx (63.56177680269106, 0, -67.9963391718935, 1, 0)^{\mathbf{T}}$. Let $\mathbf{0} = (0, 0, 0, 0, 0)^{\mathbf{T}}$. And we have $\mathbf{DMx}_1 = \mathbf{0}$, $\mathbf{DMx}_2 = \mathbf{0}$, $\mathbf{DM}^2\mathbf{x}_3 = \mathbf{0}$, and $\mathbf{DM}^3\mathbf{x}_4 = \mathbf{0}$. Let $\mathbf{e}_5 = (0,\ 0,\ 0,\ 0,\ 1)^{\mathbf{T}}$, and $\mathbf{T} = [\mathbf{x}_1,\ \mathbf{x}_2,\ \mathbf{x}_3,\ \mathbf{x}_4,\ \mathbf{e}_5]$. Then we have $\det(\mathbf{T}) \approx -15505.3 \neq 0$. This implies the following two consequences. (1) $\mathbf{x}_1$, $\mathbf{x}_2$, $\mathbf{x}_3$ and $\mathbf{x}_4$ constitute the basis of the four-dimensional generalized center eigenspace. (2) The generalized center eigenspace intersects $\tilde{Q}$-axis transversely. Let $\tilde{W}$ be a small open subset of R^4 with $(0, 0, 0, 0) \in \tilde{W}$ that is included in the projection of $\tilde{V}$ into the first four coordinates. Recall that $G = G(\tilde{X}, \tilde{u}, \tilde{Q}, \mu, \nu)$ is C^3-smooth in all its arguments. Thus if we choose $\tilde{W}$ with $(0, 0, 0, 0) \in \tilde{W}$ sufficiently small, then there exists a C^2-smooth function $\tilde{\varphi} \colon \tilde{W} \to R$ with $\tilde{\varphi}(0, 0, 0, 0) = 0$ such that $\{(\mu, \nu, \tilde{X}, \tilde{u}, \tilde{Q}) \in \tilde{V} \colon (\mu, \nu, \tilde{X}, \tilde{u}) \in \tilde{W},\ \tilde{Q} = \tilde{\varphi}(\mu, \nu, \tilde{X}, \tilde{u})\}$ constitutes the four-dimensional center manifold of the dynamics (18.30) by the center manifold theorem. Let W be defined as a small open subset of R^4 with $(\bar{X}^*, \bar{u}^*, 0, 0) \in W$ such that $W = \{(X, u, \mu, \nu) \in R^4 \colon (\mu, \nu, X - \bar{X}^*, u - \bar{u}^*) \in \tilde{W}\}$. And let $\varphi \colon W \to R_{++}$ be defined as $\varphi(X, u, \mu, \nu) = \bar{Q}^* + \tilde{\varphi}(\mu, \nu, X - \bar{X}^*, u - \bar{u}^*)$. Note that $Q = \bar{Q}^* + \tilde{Q}$, and so $\varphi(X, u, \mu, \nu) = \bar{Q}^* + \tilde{\varphi}(\mu, \nu, \tilde{X}, \tilde{u}) = Q$. Therefore, there exist an open set $W \subset R^4$ with $(\bar{X}^*, \bar{u}^*, 0, 0) \in W$ and a smooth mapping $\varphi: W \to R_{++}$ with $\bar{Q}^* = \varphi(\bar{X}^*, \bar{u}^*, 0, 0)$ such that $\{(\mu, \nu, X, u, Q) \in R^5: (X, u, \mu, \nu) \in W, Q = \varphi(X, u, \mu, \nu)\}$ constitutes a center manifold of the dynamics (18.21).

References

Benhabib, J., K. Nishimura, and T. Shigoka (2008), "Bifurcation and Sunspots in the Continuous Time Equilibrium Model with Capacity Utilization," International Journal of Economic Theory 4, 337–355.

Benhabib, J. and R. Perli (1994), "Uniqueness and Indeterminacy: On the Dynamics of Endogenous Growth," *Journal of Economic Theory*, 63, 113–142.

Benhabib, J., Schmitt-Grohé, S. and M. Uribe (2001), "The Perils of Taylor Rules," *Journal of Economic Theory*, 96, 40–69.

Caballé, J. and M. Santos (1993), "On Endogenous Growth with Physical and Human Capital," *Journal of Political Economy*, 101, 1042–1067.

Chamley, C. (1993), "Externalities and Dynamics in Models of 'learning or doing,'" *International Economic Review*, 34, 583–609.

Grandmont, J.-M. (2008), "Nonlinear Difference Equations, Bifurcations and Chaos: An Introduction," *Research in Economics*, 62, 122–177.

Guckenheimer, J. and P. Holmes (1983), Nonlinear Oscillations, Dynamical Systems, and Bifurcations of Vector Fields, New York: Springer-Verlag.

Kamien, M. I. and N. L. Schwartz (1991), Dynamic Optimization, 2nd ed., New York, North-Holland.

Kopell, N. and L. N. Howard (1975), "Bifurcations and Trajectories Joining Critical Points," *Advances in Mathematics*, 18, 306–358.

Lucas, R. E. (1988), "On the Mechanics of Economic Development," *Journal of Monetary Economics*, 22, 3–42.

Mattana, P. (2004), The Uzawa-Lucas endogenous growth model, Aldershot, England: Ashgate Publishing Limited Gower House.

Mattana, P. and B. Venturi (1999), "Existence and Stability of Periodic Solutions in the Dynamics of Endogenous Growth," *International Review of Economics and Business*, 46, 259–284.

Mulligan, C. B. and X. Sala-i-Martin (1993), "Transitional Dynamics in Two-Sector Models of Endogenous Growth," *Quarterly Journal of Economics*, 108, 739–773.

Uzawa, H. (1965), "Optimal Technical Change in an Aggregate Model of Economic Growth," *International Economic Review*, 6, 18–31.

Xie, D. (1994), "Divergence in Economic Performance: Transitional Dynamics with Multiple Equilibria," *Journal of Economic Theory*, 63, 97–112.

Bibliography

Azariadis, C. (1981): "Self Fulfilling Prophecies," *Journal of Economic Theory* 25, 380–396.

Basu, S. and J. Fernald (1997): "Returns to Scale in US Production: Estimates and Implications," *Journal of Political Economy* 105, 249–283.

Becker, R. (1980): "On the Long-Run Steady-State in a Simple Dynamic Model of Equilibrium with Heterogeneous Households," *Quarterly Journal of Economics*, 95, 375–382.

Benhabib, J. and R. Farmer (1994): "Indeterminacy and Increasing Returns," *Journal of Economic Theory* 63, 19–41.

Benhabib, J. and R. Farmer (1996): "Indeterminacy and Sector Specific Externalities," *Journal of Monetary Economics* 37, 397–419.

Bewley, T. (1982): "An Integration of Equilibrium Theory and Turnpike Theory," *Journal of Mathematical Economics*, 10, 233–267.

Boldrin, M., and L. Montrucchio (1986): "On the Indeterminacy of Capital Accumulation Paths," *Journal of Economic Theory*, 40, 26–39.

Brock, W. A. and L. Mirman (1972): "Optimal Economic Growth and Uncertainty: The Discounted Case," *Journal of Economic Theory*, 4, 479–513.

Brock, W. and J. Scheinkman (1976): "Global Asymptotic Stability of Optimal Control Systems with Applications to the Theory of Economic Growth," *Journal of Economic Theory* 12, 164–190.

Cass, D. and K. Shell (1976): "The Structure and Stability of Competitive Dynamical Systems," *Journal of Economic Theory* 12, 31–70.

Cass, D. and K. Shell (1983): "Do Sunspots Matter?" *Journal of Political Economy* 91, 193–227.

Epstein, L. (1987): "A Simple Dynamic General Equilibrium Model," *Journal of Economic Theory*, 41, 68–95.

Grandmont, J.-M., P. Pintus and R. De Vilder (1998): "Capital-Labor Substitution and Competitive Nonlinear Endogenous Business Cycles," *Journal of Economic Theory* 80, 14–59.

Li, T.-Y. and J. Yorke (1975): "Period Three Implies Chaos," *American Mathematical Monthly* 82, 985–992.

Lucas, R. (1988): "On the Mechanics of Economic Development," *Journal of Monetary Economics* 22, 3–42.

McKenzie, L. (1976): "Turnpike Theory," *Econometrica* 44, 841–865.

Mulligan, C. (2002): "Capital Interest and Aggregate Intertemporal Substitution," *NBER Working Paper* 9373.

Romer, P. (1986): "Increasing Returns and Long-Run Growth," *Journal of Political Economy* 94, 1002–1037.

Samuelson, P. (1971): "Turnpike Theorems even though Tastes Are Intertemporally Dependent, *Western Economic Journal* 9, 21–26.

J. Stachurski et al. (eds.), *Nonlinear Dynamics in Equilibrium Models*, DOI 10.1007/978-3-642-22397-6,

Sarkovskii, A. (1964): "Coexistence of Cycles of a Continuous Map of the Line Into Itself," *Ukrainian Mathematical Journal* 16, 61–71.

Scheinkman, J. (1976): "On Optimal Steady States of n-Sector Growth Models when Utility is Discounted," *Journal of Economic Theory* 12, 11–30.

Shell, K. (1977): "Monnaie et Allocation Intertemporelle," mimeo, Séminaire d'Econométrie Roy-Malinvaud, Centre National de la Recherche Scientifique, Paris.

Vissing-Jorgensen, A. (2002): "Limited Asset Market Participation and the Elasticity of Intertemporal Substitution," *Journal of Political Economy* 110, 825–853.

Woodford, M. (1986): "Stationary Sunspot Equilibria: the Case of Small Fluctuations Around a Deterministic Steady State," Mimeo, University of Chicago.

Yano, M. (1983): "Competitive Equilibria on Turnpikes in a McKenzie Economy, I: A Neighborhood Turnpike Theorem", *International Economic Review*, 25, 695–717.

Yano, M. (1984): "The Turnpike of Dynamic General Equilibrium Paths and its Insensitivity to Initial Conditions," *Journal of Mathematical Economics*, 13, 235–254.